ANNUAL SURVEY OF PHOTOCHEMISTRY

Volume 3: Survey of 1969 Literature

Annual Survey of Photochemistry

Volume 3: Survey of 1969 Literature

NICHOLAS J. TURRO
GEORGE S. HAMMOND
JOHN F. ENDICOTT
J. CHRISTOPHER DALTON
TIMM L. KELLY
JACK E. LEONARD
DOUGLAS R. MORTON
DAVID M. POND

WILEY–INTERSCIENCE
a division of John Wiley & Sons, Inc., New York – London – Sydney – Toronto

Copyright © 1971, by John Wiley & Sons, Inc.

All rights reserved. Published simultaneously in Canada.

No part of this book may be reproduced by any means, nor transmitted, nor translated into a machine language without the written permission of the publisher.

Library of Congress Catalog Card Number: 70—80463

ISBN 0 471 89327 7

Printed in the United States of America

10 9 8 7 6 5 4 3 2 1

PREFACE

Chemists, like all scientists these days, are faced with an essentially insoluble dilemma: the proliferation of research literature demands that more specialized reviews appear, and in addition to keeping abreast of primary work there is the struggle to keep up with them.

Among the unfortunate characteristics of reviews are lack of comprehensive coverage, failure to define adequately the scope of the material covered, and noncritical compilation of material. One must be sympathetic to the plight of the reviewer who finds himself torn between complete and critical coverage and the continuing deluge of the literature he wishes to cover.

We feel that one possible way to eliminate some of these problems is to define carefully a time period for coverage. This clearly establishes the material to be surveyed and, as a result, a comprehensive (although still not exhaustive) coverage is possible. Thus our aim is to present a summary of progress in certain areas of photochemistry published during the year 1969. Although appropriate references to earlier reviews are given, an absolute minimum discussion of pre- 1969 material is included to satisfy our major goals. The reader is referred to Vol. I and Vol. II of this series for reviews of the photochemical literature of 1967 and 1968, respectively.

We hope that the usefulness of this and future volumes will be increased by the modifications that we will make as a result of the response of our readers. Therefore suggestions are warmly invited to assist us in our attempt to provide those interested in photochemistry with an annual comprehensive summary of the chemical literature.

N. J. T.
G. S. H.

January 1971

CONTENTS

PART 1. ORGANIC PHOTOCHEMISTRY
By Nicholas J. Turro, David M. Pond, Douglas R. Morton, and J. Christopher Dalton

Introduction	3
Chapter One. Saturated Compounds, Ethylenes, and Nonaromatic Compounds	7
I. Saturated Compounds	7
II. Alkenes	12
A. Acyclic Alkenes	12
B. Alicyclic Alkenes	15
1. Small-ring Alkenes	15
2. Medium- and Large-ring Alkenes	19
3. Heteroatom-containing Alkenes	27
III. Conjugated Polyenes	29
A. Acyclic Conjugated Dienes	29
1. Butadiene	29
2. Substituted Butadienes	29
B. Alicyclic Conjugated Dienes	33
1. Cyclopentadiene, Cyclohexadiene, and Related Compounds	33
2. Cyclic *trans*-fused Dienes and Trienes	36
C. Acyclic Conjugated Trienes	38
D. Alicyclic Conjugated Trienes	38
E. Heterocyclic Dienes and Related Compounds	41
Chapter Two. Aromatic Hydrocarbons and Derivatives	44
I. Photochemistry of Benzene and Related Aromatic Compounds	44
A. Benzene and Its Simple Derivatives	44
1. Benzene and Alkyl Benzenes	44

		2. Aromatic Halides	46
		3. Phenols, Anilines, and Related Compounds	48
		4. Biphenyl Derivatives	51
	B.	Conjugated Aromatics	51
		1. Styrene and Its Derivatives	51
		2. Stilbene Derivatives	57
II.	Polynuclear Aromatic Compounds		59
	A.	Naphthalene Derivatives	59
	B.	Polycondensed Aromatic Compounds	64

Chapter Three. Carbonyl Compounds ... 67

I.	Alkyl Carbonyl Compounds		67
	A.	Aldehydes and Ketones	67
		1. Alkyl Aldehydes	67
		2. Alkyl Ketones	70
		3. Alkyl Ketones with α-Substituents	79
		4. Cyclic Ketones	80
II.	Aryl Aldehydes and Ketones		95
	A.	Aryl Aldehydes	95
	B.	Aryl Alkyl Ketones	96
	C.	Diaryl Ketones	106
III.	Carboxylic Acids and Derivatives		112
	A.	Esters and Acids	112
	B.	Cyclic Derivatives of Carboxylic Acids	117
	C.	Amides and Related Compounds	119
	D.	Sulfur Compounds	123
	E.	Ketenes and Isocyanates	125

Chapter Four. Conjugated Enones, Diones, and Related Compounds 126

I.	α, β-Unsaturated Carbonyl Compounds		126
	A.	Acyclic Enones, Esters, and Amides	126
	B.	Conjugated Cyclic Enones	131
		1. Three-, Four-, and Five-ring Enones	131
		2. Cyclohexenones and Larger Ring-conjugated Enones	134
		3. Cyclohexadienones and Trienones	145
		4. Enediones, Enediacids, and Related Compounds	151
		5. Benzoquinone and Related Compounds	156
		6. α-Dicarbonyl Compounds	158
		7. Purines, Pyrimidines, and Related Compounds	162

Chapter Five. Nitrogen-containing Chromophores 164

    I. Unsaturated Nitrogen Compounds 164
        A. Azines, Oximes, and Related Compounds 164
        B. Azo Compounds 168
        C. Azides, Triazolines, and Related Compounds 172
        D. Diazo and Diazonium Compounds 175
   II. Organic Compounds Having NO, NO_2, and ONO Functions 177
        A. Nitrocompounds and Organic Nitrites 177
        B. N-Oxides and Related Compounds 180
  III. N-Heteroaromatics 183
        A. Pyridine Derivatives 183
        B. Purines, Quinolines, Acridines, and Related Compounds . 184
    References 185

PART 2. PHOTOPHYSICAL PROCESSES OF ORGANIC COMPOUNDS: FORMATION AND DECAY OF ELECTRONICALLY EXCITED STATES
By Jack E. Leonard and George S. Hammond

Introduction 215

    I. Formation of Electronically Excited States: Light Absorption 215
        A. Absorption Spectroscopy of Molecules: Ground State to Spin-allowed Excited States 216
        B. Absorption Spectroscopy of Spin-forbidden Transitions .. 232
        C. Absorption Spectroscopy of Electronically Excited States 235
        D. Polymolecular Effects in Absorption Spectroscopy 236
   II. Properties of Excited States 243
  III. Decay of Electronically Excited States 247
        A. Radiative Decay of Excited States 247
            1. Molecular Luminescence 247
            2. Environmental Effects on Radiative Decay of Excited Molecules 263
        B. Pseudounimolecular Radiationless Deactivation 267
        C. Deactivation of Excited Molecules by Ground-state Molecules 283
        D. Deactivation of Excited Molecules by Processes Involving Two Excited Molecules: Annihilation Processes 289
    References 290

PART 3. PROGRESS IN THE EMISSION SPECTROSCOPY AND PHOTOCHEMISTRY OF COORDINATION COMPLEXES
By John F. Endicott and Timm L. Kelly

Introduction 325

Chapter One 327

    I. Emission Spectroscopy, Energy Transfer, and Sensitized Reactions 327
    II. Photochemistry of Transition-Metal Complexes 331
        A. d^2 Systems 331
        B. Chromium (III) 332
        C. d^4 Systems 333
        D. d^5 Systems 333
        E. Cobalt (III) and Other d^6 Systems 333
        F. d^8, d^9, and d^{10} Systems 336
        References 337

Author Index 341

ANNUAL SURVEY OF PHOTOCHEMISTRY

Volume 3: Survey of 1969 Literature

Part I

Organic Photochemistry

NICHOLAS J. TURRO, DAVID M. POND,
DOUGLAS R. MORTON, and
J. CHRISTOPHER DALTON

*Department of Chemistry,
Columbia University, New York*

INTRODUCTION

Part 1 of this volume is an attempt to survey the organic photochemical literature for the year 1969. Emphasis has been placed on the synthetic utility and mechanistic aspects of the reactions reported. Yields and critical evaluation are given when appropriate. References to earlier literature are generally not given except references to significant review articles (see below) and past volumes in this series. *Annual Survey of Photochemistry 1967* and *Annual Survey of Photochemistry 1968* should be consulted for comprehensive reviews of the organic photochemical literature reported during 1967 and 1968.

As in previous volumes, photooxidations, which probably involve the attack on a ground-state organic molecule by some form of singlet oxygen, are generally not included in this section. Photoreactions are classified here by the concept of an excited chromophore. The simplest systems are discussed first and then those of increasing complexity—monoolefins, then dienes, then higher polyenes, for example. Although the survey is comprehensive, articles were omitted if, in the authors' opinion, the structure of the products or starting materials was not properly defined.

All the major Western journals were surveyed by a page-to-page search. Russian and Japanese journals were surveyed by way of abstract surveys.

A list of recent literature reviews, books, and journals of interest to organic photochemists follows:

(i) N. J. Turro, "Molecular Photochemistry," W. A. Benjamin, New York, 1965.
(ii) R. O. Kan, "Organic Photochemistry," McGraw-Hill, New York, 1966.
(iii) J. G. Calvert and J. N. Pitts, Jr., "Photochemistry," Wiley, New York, 1966.
(iv) D. C. Neckers, "Mechanistic Organic Photochemistry," Reinhold, New York, 1967.
(v) A. Schonberg, "Preparative Organic Photochemistry," Verlag Chemie, New York, 1968.
(vi) "Advances in Photochemistry," ed. W. A. Noyes, Jr., G. S. Hammond, and J. N. Pitts, Jr., vol. I, 1963; vol. II, 1964; vol. III, 1964; vol. IV; 1966; vol. V, 1968; vol. VI, 1968; vol. VII, 1969; Wiley, New York.

(vii) Cycloadditions: R. Steinmetz, *Fortschritte Chem. Forschung*, **7**, 445 (1967); O. L. Chapman, *Org. Photochem.*, **1**, 283 (1968); W. Dilling, *Chem. Rev.*, **69**, 845 (1969).

(viii) Photoalkylations: D. Elad, *Fortschritte Chem. Forschung*, **7**, 428 (1967); "The Chemistry of the Ether Linkage," ed. S. Patai, Wiley, New York, 1967, p. 353.

(ix) Photochromism and reversible photoisomerizations. E. Fischer, *Fortschritte Chem. Forschung*, **7**, (1967).

(x) Photooximation: M. Pape, *ibid.*, **7**, 559 (1967).

(xi) "Organic Photochemistry," ed. O. L. Chapman, vol. I, Marcel Dekker, New York, 1968; vol. II, 1969.

(xii) Quinone photochemistry: J. M. Bruce, *Quart. Rev.*, **21**, 405 (1967).

(xiii) PhotoFries: D. Bellus and P. Hrdlovic, *Chem. Rev.*, **67**, 599 (1967); N. P. Shusheriva and P. Y. Levina, *Russian Chem. Rev.*, **37**, 198 (1968).

(xiv) "Reactivity of the Photoexcited Organic Molecule," Solvay Conference, Wiley-Interscience, New York, 1967.

(xv) Photochemical reaction mechanisms: B. Capon, M. J. Perkins, and L. W. Rees, "Organic Reaction Mechanisms," 1965, 1966, 1967, 1968, 1969, Wiley-Interscience, New York.

(xvi) Photochemistry of ethers: D. Elad in "Chemistry of the Ether Linkage," ed. S. Patai, Wiley, New York, 1967, p. 353.

(xvii) Tropolones: K. F. Koch, "Advances in Alicyclic Chemistry," ed. H. Hart and G. J. Karabatsos, Academic, New York, 1966, p. 258; D. J. Pasto, *Org. Photochem.*, **1**, 155 (1968).

(xviii) Photochemistry of carbonyl compounds: J. N. Pitts, Jr., and J. K. S. Wan, "The Chemistry of the Carbonyl Group," ed. S. Patai, Wiley-Interscience, New York, 1966, p. 823; A. J. Davis and R. B. Cundall, *Prog. Reaction Kinetics*, **4**, 149 (1967).

(xix) Photochemistry of unsaturated systems: R. N. Warrener and J. B. Bremmer, *Rev. Pure Appl. Chem.*, **16**, 117 (1966).

(xx) Nucleic acid derivatives: J. G. Burr, *Advan. Photochem.*, **6**, 193 (1968); H. E. Johns, *Methods in Enzymology*, **16**, 253 (1969); E. Fahr, *Ann. Chim.*, **8**, 578 (1969); T. Goto, *Pure Appl. Chem.*, **17**, 421 (1968).

(xxi) Oxetane formation: D. R. Arnold, *Advan. Photochem.*, **6**, 301 (1968); L. L. Muller and J. Hamer, "1,2-Cycloaddition Reactions," Wiley-Interscience, New York, 1967, p. 111.

(xxii) Organic molecules in the triplet state: P. J. Wagner and G. S. Hammond, *Adv. Photochem.*, **5**, 21 (1968); N. J. Turro, *J. Chem. Ed.*, **46**, 3 (1969).

(xxiii) Conjugated cyclic ketones: K. Schaffner, *Advan. Photochem.*, **4**, 81 (1966); P. E. Eaton, *Acc. Chem. Research*, **1**, 65 (1968).

(xxiv) Dienes and polyenes: R. Srinivasan, *Adv. Photochem.*, **4**, 113 (1966); M. Mousseron *ibid.*, **4**, 195 (1966); G. J. Fonken, *Org. Photochem.*, **1**, 197 (1968).

(xxv) Small rings conjugated to carbonyls: A. Padwa, *ibid.*, **1**, 91 (1968).

(xxvi) Cyclohexadienones: P. J. Kropp, *ibid.*, **1**, 1 (1968).

(xxvii) Stilbenes: F. R. Stermitz, *ibid.*, **1**, 248 (1968); M. Scholz, F. Dietz, and M. Muhlstadt, *Z. Chem.*, **7**, 329 (1967); E. V. Blackburn and C. J. Timmons, *Quart. Rev.*, **23**, 482 (1969).

(xxviii) Intramolecular ring closure: W. L. Dilling, *Chem. Rev.*, **66**, 373 (1966).

(xxix) Aromatic halides: R. K. Sharma and N. Kharasch, *Angew. Chem. Int. Ed. Eng.*, **7**, 36 (1968); N. Kharasch and J. L. Day, *Quart. Reports Sulfur Chem.*, **3**, 177 (1968).

(xxx) General review of photochemistry: A. C. Day, *Ann. Rept. Chem. Soc. for* 1967,

64B, 161 (1968); A. C. Day and E. J. Forbes, *Ann. Rept. Chem. Soc. for* 1968, 65B, 187 (1969).
(xxxi) Photosensitization by energy transfer: N. J. Turro, J. C. Dalton, and D. S. Weiss, *Org. Photochem.*, 2, 1 (1969).
(xxxii) Photodimerizations: D. J. Trecker, *ibid.*, 2, 63 (1969).
(xxxiii) Photochemistry of heteroaromatic nitrogen compounds: P. Beak and W. R. Messer, *ibid.*, 2, 117 (1969).
(xxxiv) Photoadditions to multiple bonds: D. Elad, *ibid.*, 2, 168 (1969).
(xxxv) Photochemistry of olefins: D. H. Scharf, *Fortscritte Chem. Forschung*, 11, 216 (1969); J. S. Swenton, *J. Chem. Ed.*, 46, 7 (1969); N. J. Turro, *Photochem. Photobio.*, 9, 555 (1969); J. A. Marshall, *Acc. Chem. Research*, 2, 33 (1969).
(xxxvi) Cycloheptadienes: L. B. Jones and V. K. Jones, *Fortschritte Chem. Forschung*, 13, 307 (1969).
(xxxvii) Nucleophilic photosubstitution: F. Pietra, *Quart. Rev.*, 23, 504 (1969).
(xxxviii) α-Diketones: M. B. Rubin, *Fortscritte Chem. Forschung*, 13, 251 (1969).
(xxxix) Organic azides: G. L'Abbe, *Chem. Rev.*, 69, 345 (1969).
(xL) Barton reaction: R. H. Hesse, *Adv. Free Radical Chem.*, 3, 83 (1969).

In addition to these, a new journal, *Molecular Photochemistry*, edited by A. A. Lamola, of Bell Telephone Laboratories, helps to fill the need for a place in which discourse between spectroscopists and photochemists can occur.

A new book, "Energy Transfer and Organic Photochemistry," by A. A. Lamola and N. J. Turro has been published.

A number of useful books on various aspects of luminescence have appeared: "Luminescence in Chemistry," ed. E. J. Bowen, Van Nostrand, London, 1968; "Fluorescence Theory, Instrumentation and Practice," ed. G. G. Guilbault, Arnold, London, 1967; C. A. Parker, "Photoluminescence of Solutions," Elsevier, Amsterdam, 1968; "Molecular Luminescence," ed. E. C. Lim, W. A. Benjamin, New York, 1969; R. S. Becker, "Theory and Interpretation of Fluorescence and Phosphorescence," Wiley-Interscience, New York, 1969; N. Zander, "Phosphorimetry," Academic, New York, 1968; "Fluorescence and Phosphorescence," ed. D. M. Hercules, Wiley-Interscience, New York, 1966.

The Chemical Society (London) has initiated a series of comprehensive reports on the progress of photochemistry. The first volume, "Photochemistry," ed. D. Bryce-Smith, has appeared.

The papers presented at a Conference on Primary Photochemical Processes at Frankfurt in May 1969 are given in *Ber. Bunsenges. Phys. Chem.*, 73, 732–911 (1969).

CHAPTER ONE: SATURATED COMPOUNDS, ETHYLENES, AND NONAROMATIC COMPOUNDS

I. SATURATED COMPOUNDS

The photolyses of liquid pentane and 2,2,4-trimethylpentane [1] and the photochemical carbonylation and oxidation of aqueous methanol [2] at 1470 Å have been reported.

Irradiation of the iodogalactopyranose **1** in methanolic sodium hydroxide results in a quantitative formation of **2**:

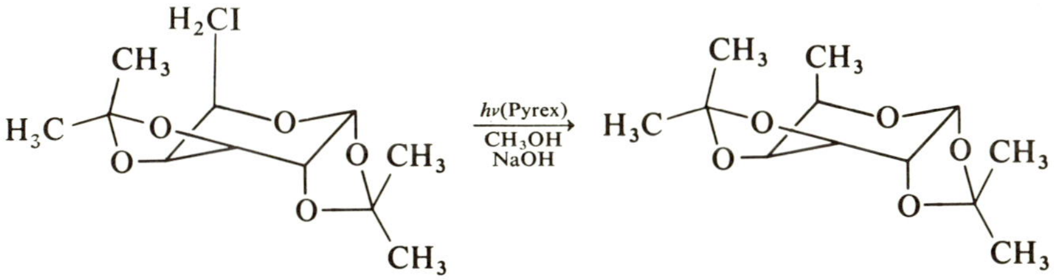

1 → hv(Pyrex), CH₃OH, NaOH → **2**

and irradiation in poorly hydrogen-donating solvents affords appreciable amounts of **3** in addition to **2** [3]:

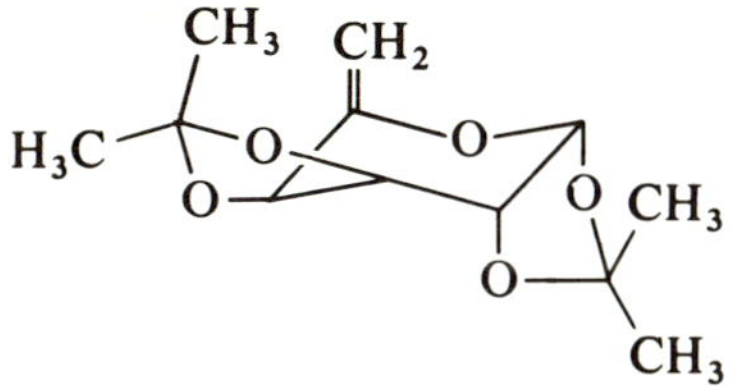

3

The photolyses of ozonides continue to be of interest (Volume 1, page 5; Volume 2, page 7). For example, ozonide **4**, when photolyzed at low temperature in a cooled tube [4], affords the benzo-Dewar benzene **5**:

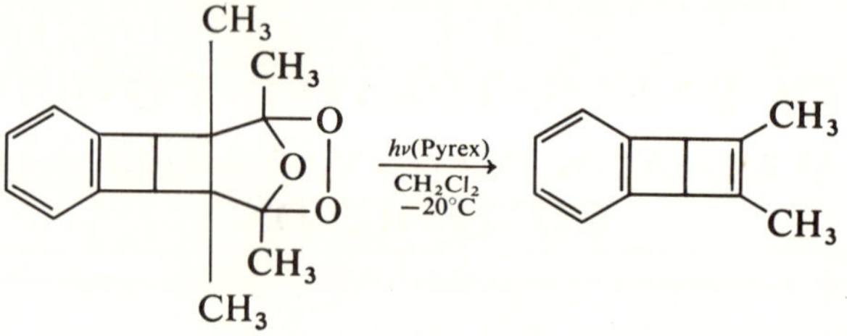

and a similar irradiation of **6** [5] affords the Dewar benzene **7** and **8**:

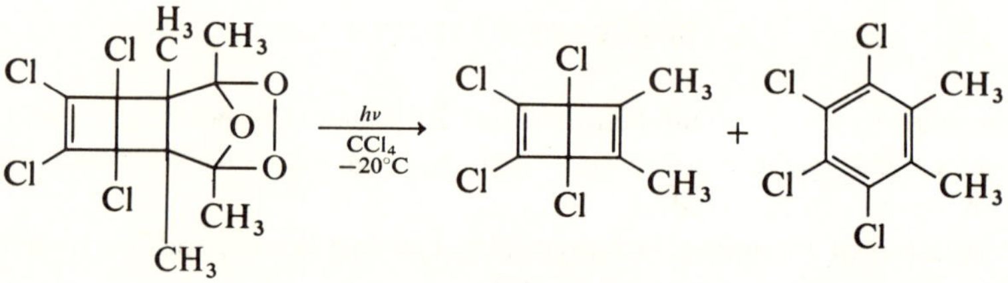

Finally, irradiation of ozonide **9** in pentane solution [6] yields tricyclooctadiene **10** in addition to ketone **11**, acetic acid, and acetic anhydride:

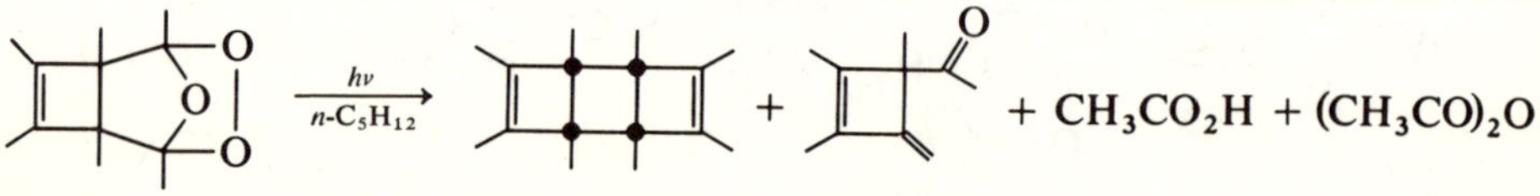

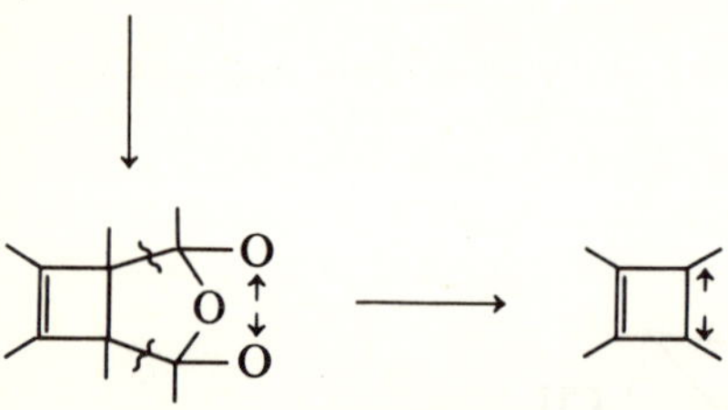

The symmetrical octadiene **10** is presumably formed from the dimerization of singlet tetramethylcyclobutadiene.

The photolysis of peroxides has been reported to result in fragmentation and rearrangement products. For example, photolysis of **12** affords naphthalene **13** and acetylene in addition to bis epoxide **14** [7]:

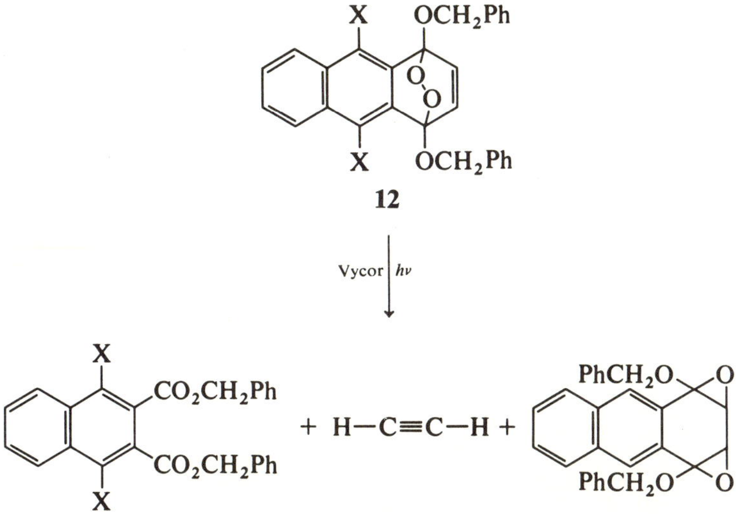

12

$\xrightarrow{h\nu}$ Vycor

13 + H—C≡C—H + **14**

X = H or Ph

Similarly, irradiation of peroxide **15** affords bis epoxide **16** in good yield [8]:

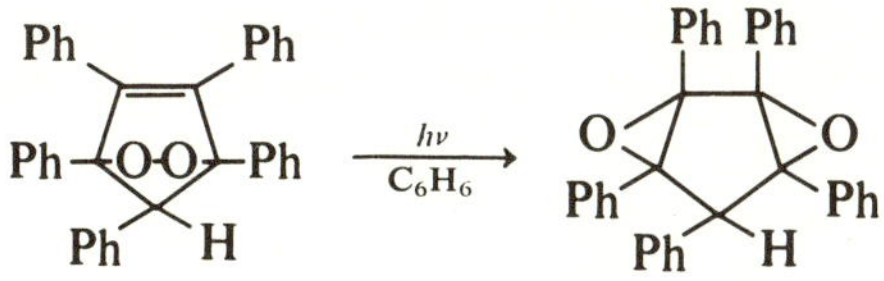

15 $\xrightarrow[C_6H_6]{h\nu}$ **16**, 70%

On the other hand, photolysis of *endo*-peroxide **17** yields anthraquinone **18** and 10,10′-bianthronyl **19** [9]:

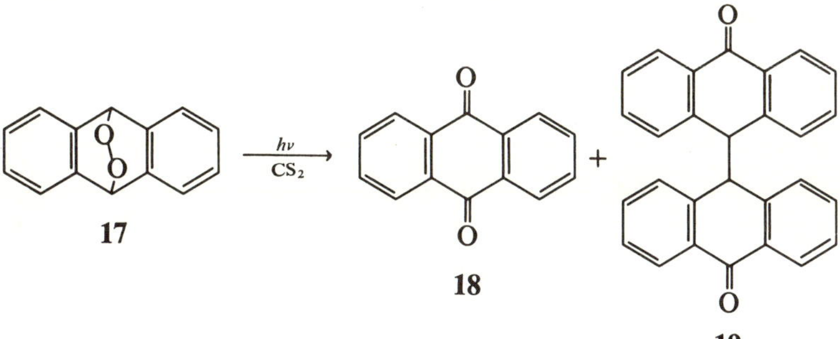

17 $\xrightarrow{h\nu}_{CS_2}$ **18** + **19**

Photolysis of polypropylene hydroperoxides at 3650 Å under high vacuum [10] results in the initial cleavage of the oxygen-oxygen bond followed by β-cleavage to yield hydrocarbon and carbonyl fragments.

Photochemically induced radical and ionic heterocyclizations of olefinic [11] and acetylenic [12] thiols have been reported. Photolysis of phenacylsulfonium salts in aqueous solution [13] results in initial PhCOCH$_2$—S$^+$(CH$_3$)$_2$X$^-$ cleavage to afford a phenacyl radical and (CH$_3$)$_2$S$^+$. Photolysis of methylpenicillin (S)-sulfoxide methyl ester in acetone [14] results in a clean inversion to the (R)-sulfoxide. Irradition of α-lipoic acid in various solvents has been reported to cause initial homolysis of the disulfide bond followed by hydrogen atom chain transfer [15]. The nature and number of products obtained seem to depend on the solvent. Photolysis of bicyclic sulfide **20** in isooctyl phosphite [16] yields cis-fused hydrocarbon **21** in addition to cyclooctene and cyclooctane:

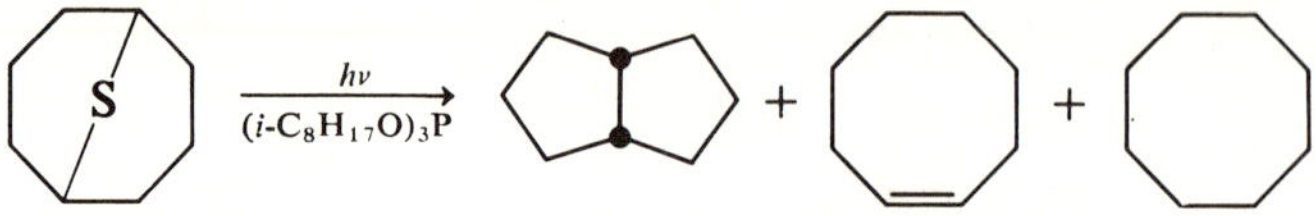

Similarly, irradiation of 7-thiabicyclo[2.2.1]heptane [16] affords cyclohexene as the major product in addition to cyclohexane and 1,5-hexadiene.

Photolyses of steroidal tertiary amines result in the homolysis of a C—N bond followed by disproportionation, hydrogen atom abstraction from solvent (Volume 1, page 6), and incorporation of dissolved oxygen [17]. In similar fashion, irradiation of steroidal N-chloroamine **22** causes initial N—Cl cleavage (Volume 2, page 7) followed by radical displacement and hydrogen atom abstraction [18]:

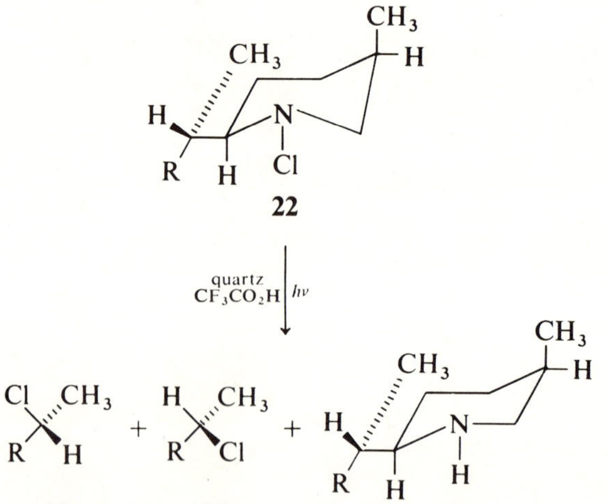

Irradiation of chloroamide **25a** yields **25b** [19]:

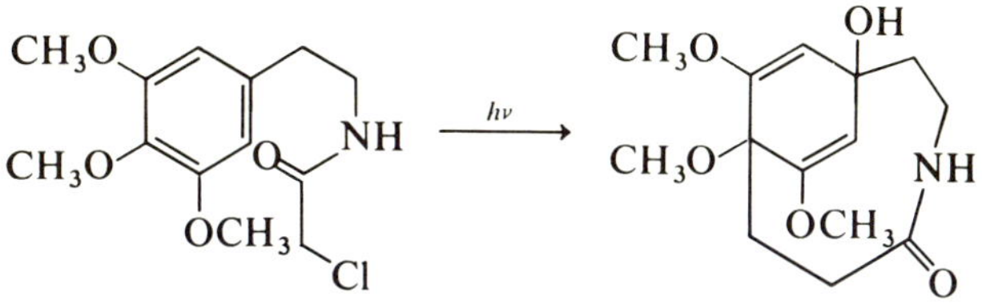

25a → **25b, 35%**

Photolyses of tetrahydroisoquinoline *N*-tosylates in neutral or basic media [20] afford 3,4-dihydroisoquinolines in moderate yields. When photolyses are carried out in basic media in the presence of sodium borohydride, good yields of the corresponding amines are obtained [21]. Photolysis of desmethoxypyrodelphonine **25c** in the presence of sodium borohydride yields olefin **25d** [22]:

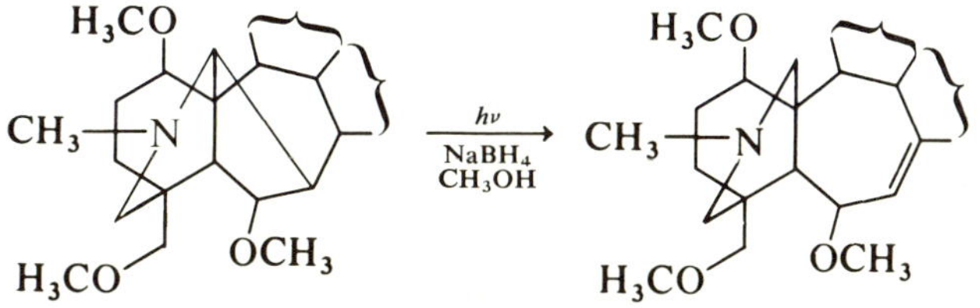

25c → **25d**

Irradiation of alcohol **25e** affords **25f** and olefin **25g** [23]:

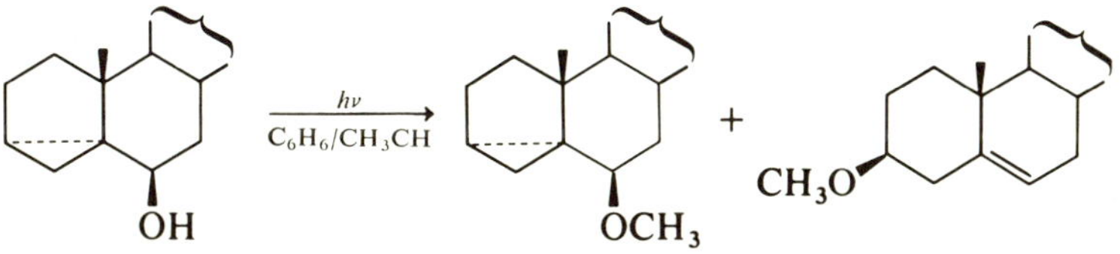

25e → **25f** + **25g**

An ionic photorearrangement is postulated.

Irradiation of α-amyrin acetate **26** in aqueous dioxane in the presence of *N*-bromosuccinimide and $CaCO_3$ [24] yields enone **27** in high yield:

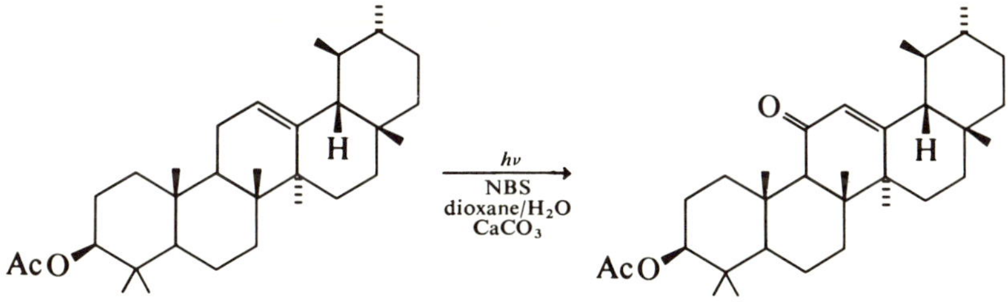

26 → **27, 98%**

Similar oxidations of β-amyrin acetate and teraxeryl acetate are also reported.

Photolysis of allyl chloride in the liquid state at 27°C and in the solid state at 77°K results in initial C—Cl bond cleavage [25].

The photochemical additions of $BrCCl_3$ [26], $BrCH(CN)_2$ [27], $C_2H_5SO_2NCl_2$ [28], and $ClNSF_2$ [29] to olefins have been reported. The direct and sensitized irradiations of 1,4-, 1,3-, 1,2-, and 2,3-dichlorobutanes [30] yield HCl as the principal product, and irradiation of poly(vinyl chloride) results in a stable concentration of polyenyl free radicals and an increase in the concentration of conjugated double bonds [31, 32].

Irradiation of *n*-butane solutions of CNCl yields 1-cyanobutane and HCl [33].

The photoisomerization of triethyl phosphite to diethylphosphite and diethyl ethylphosphonate has been reported [34]. The reaction of *t*-butyl-hydroperoxide with triethylphosphite leads to chemiluminescence [35].

Irradiation of methanol (1850 Å) gives a number of products resulting from the initial cleavage of the O—H bond [36].

II. ALKENES

A. Acyclic Alkenes

A review of the ultraviolet spectra and excited states of ethylene and its alkyl derivatives has appeared [37]. In addition, reviews discussing olefin photochemistry (38, 39, 40) and the Woodward-Hoffmann rules [41] have also appeared.

The photosensitized isomerizations of alkenes [42], for example, 2-butene [43], [44, 45], 2-pentene [46], and 1,2-dichloroethylene [47] have been reported.

Irradiation of **27a** results in isomerization to **27b** [48]:

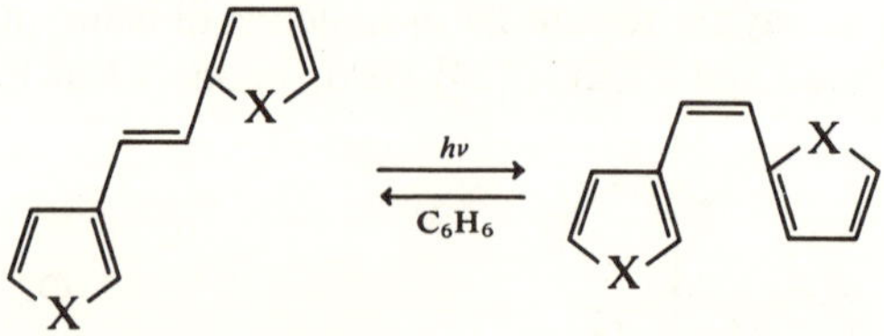

27a, X = O, S   **27b**

Irradiation of 1,2-polybutadiene **28** yields cyclized products **29** to **31** [49] (Volume 1, page 8):

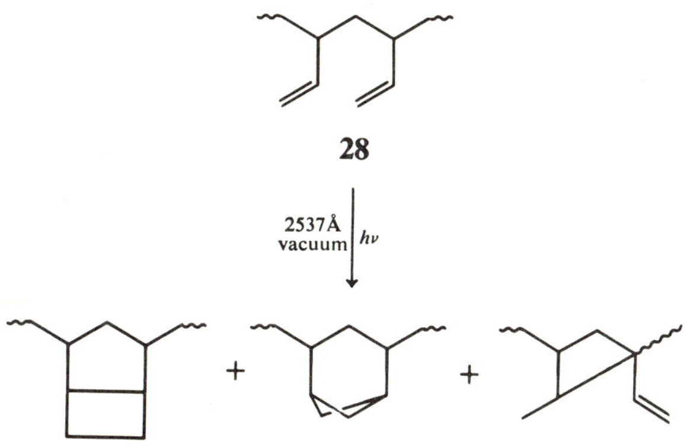

ORGANIC PHOTOCHEMISTRY

28

$$\xrightarrow[\text{vacuum}]{2537\text{Å} \;\; h\nu}$$

29   +   **30**   +   **31**

Similar results are described for the irradiation of 3,4-polyisoprene. Irradiation of *cis*- and *trans*-2-butene with 2288 Å light yields the cyclobutane dimers stereospecifically [50]. Close contact is required for these dimerizations, because dimerization is suppressed on dilution of the 2-butene in hydrocarbon solvents. Other dimerizations have appeared [51].

Intramolecular photosensitization in ketone **31a** results in cycloaddition to give **31b** [52]:

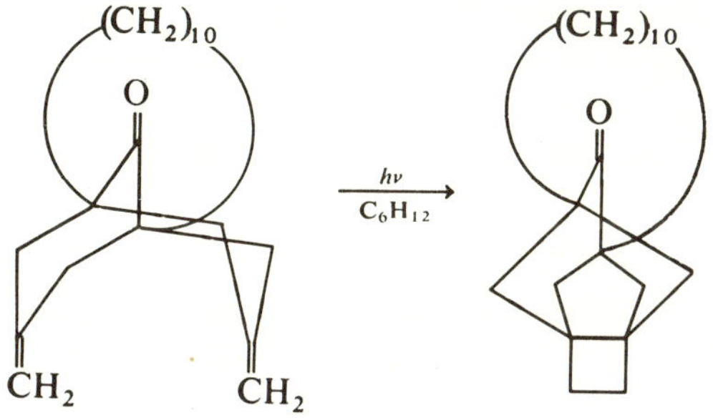

31a $\xrightarrow[C_6H_{12}]{h\nu}$ **31b, 73%**

Sensitized photolysis of *trans*-crotyl chloride **32** yields **33** and the isomeric cyclopropanes **34**:

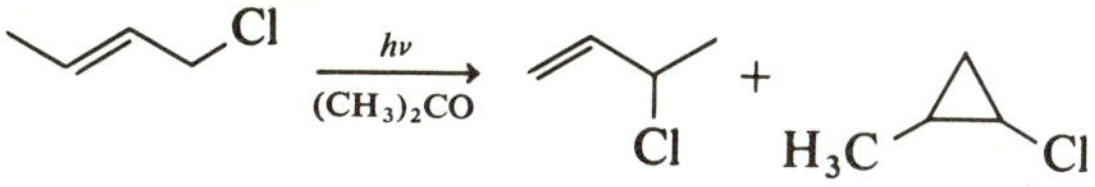

32 $\xrightarrow[(CH_3)_2CO]{h\nu}$ **33, 13%**   +   **34**, *cis*, 13%
                                                            trans, 15%

The mechanism is believed to involve the vibrationally excited carbonium ion–halide ion ion pair **35** or a triplet biradical pair **36**:

35 **36**

Similarly, the sensitized photolyses of allyl chloride (see also Section I) and allyl bromide afford the corresponding halocyclopropanes in 19 and 11% yields, respectively [53].

Photolysis of alkene **37** results in the formation of cyclopropane **38** under conditions of both sensitized and direct excitation [54]:

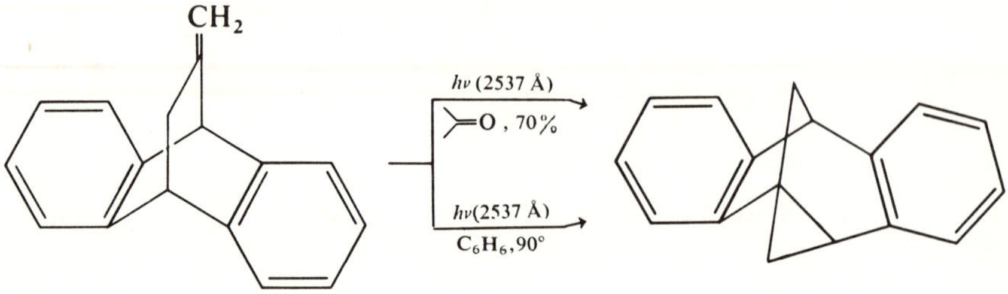

37 **38**

The mechanism of the formation of **38** is most conveniently rationalized in terms of a di-π-methane rearrangement (Volume 1, page 18; Volume 2, pages 13 to 14).

Photolysis of epoxyalkene **39** affords cumulene **40**, allene **41**, and oxetane **42** [55]. Oxetane **42** was shown to be a secondary photoproduct from cumulene **40** and excited acetone, and the formation of **40** and **41** is postulated to occur by initial C—O bond cleavage to biradical **43**. Elimination of acetone would result in carbene formation and the subsequent formation of **40**, and rearrangement to biradical **44** would explain the formation of both **40** and **41**:

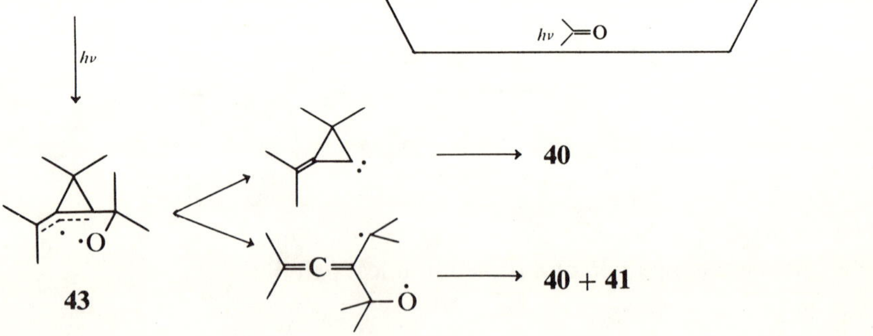

39 **40**, 19% **41**, 13% **42** **43** **44**

Attempts to trap the hypothetical carbene were not successful.

The synthesis of bicyclo[2.1.1]hexane derivatives has been reported by way of the irradiation of activated 1,5-hexadienes [56].

Irradiation of trimethylenemethane iron tricarbonyl in pentane solution yields 1,4-dimethylenecyclohexane [57]. No evidence for methylene cyclopropane was obtained (see also Volume 1, page 78).

Photolysis of β-pinene in the presence of acetic anhydride and oxygen results in numerous oxidation products [58]. The primary process is postulated to be hydrogen atom abstraction from the secondary carbon alpha to the carbon-carbon double bond.

Photopolymerization of acetonitrile in the presence of isobutyl vinyl ether is reported to be accelerated by a contact charge-transfer complex [59].

B. Alicyclic Alkenes

1. SMALL-RING ALKENES

The mercury-sensitized photolysis of cyclobutene **45** yields the coupled diene **46** [60]:

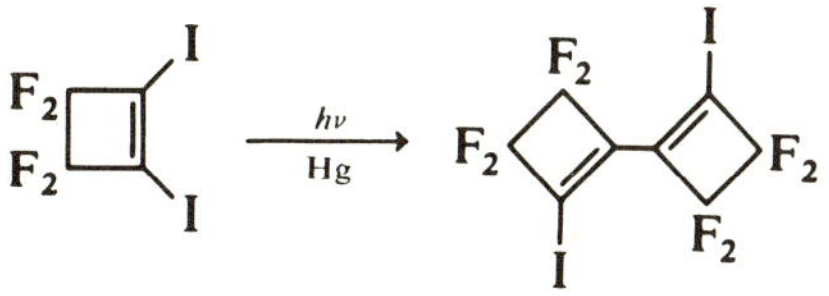

Irradiation of cyclobutene **47** in the presence of triphenylene affords *cis*-olefin **48**:

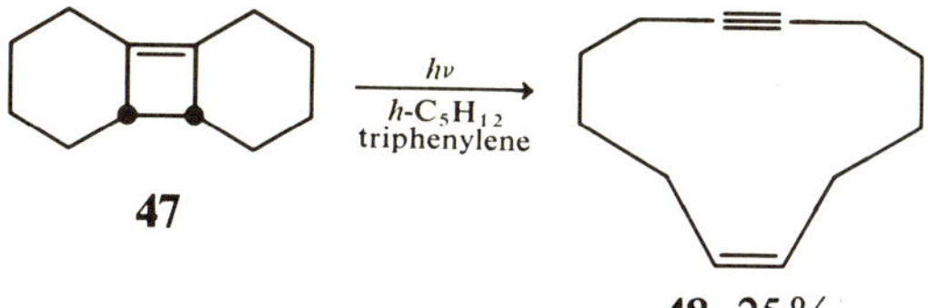

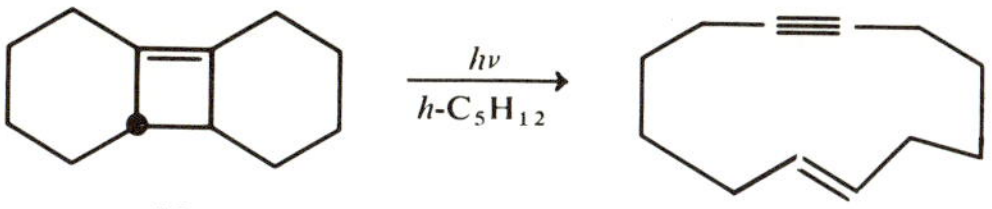

and irradiation of **49** affords *trans*-olefin **50** [61]. These results agree with the Woodward-Hoffmann rules for a concerted photochemical, four-electron, $2\sigma + 2\pi$ process.

The sensitized photolysis of bicyclic alkene **51** (Volume 2, page 10) results in an intramolecular addition to yield **52** [62]:

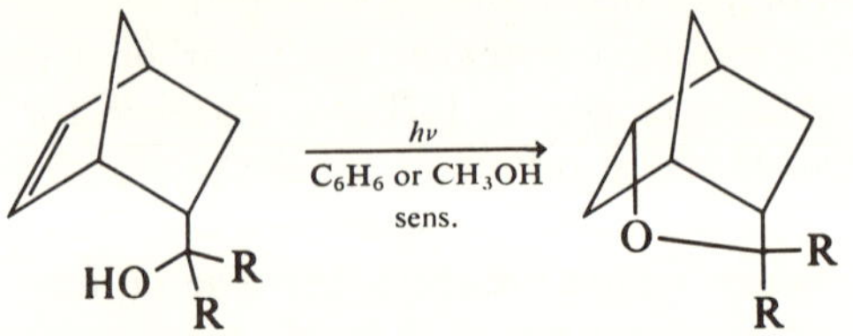

$$\text{51} \xrightarrow[\text{sens.}]{h\nu, \, C_6H_6 \text{ or } CH_3OH} \text{52}$$

R = H, D or CH$_3$

An ionic mechanism is postulated.

Irradiation of diene **53** affords norbornadiene **54**, which is subsequently converted to the corresponding quadricyclene **55** [63]:

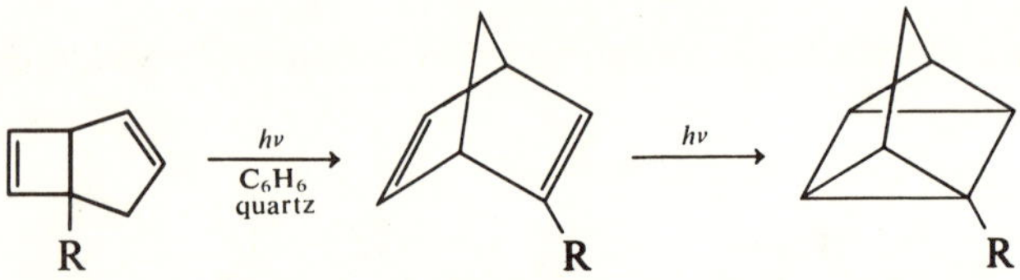

$$\text{53} \xrightarrow[\text{quartz}]{h\nu, \, C_6H_6} \text{54} \xrightarrow{h\nu} \text{55}$$

R = CO$_2$CH$_3$

The photoisomerizations of norbornadienes to quadricyclenes have been reported (Volume 1, pages, 14, 15, 143; Volume 2, pages 11 to 14). For example, irradiation of diene **55a** yields **55b** [64]:

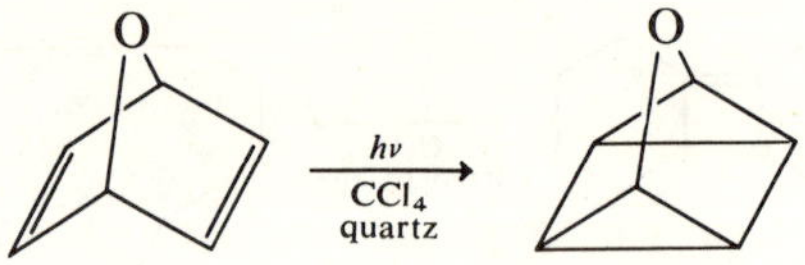

$$\text{55a} \xrightarrow[\text{quartz}]{h\nu, \, CCl_4} \text{55b}$$

and irradiation of the iron tricarbonyl **56** in ether-isopentane-alcohol at 77°K affords **57** in low yield [65]:

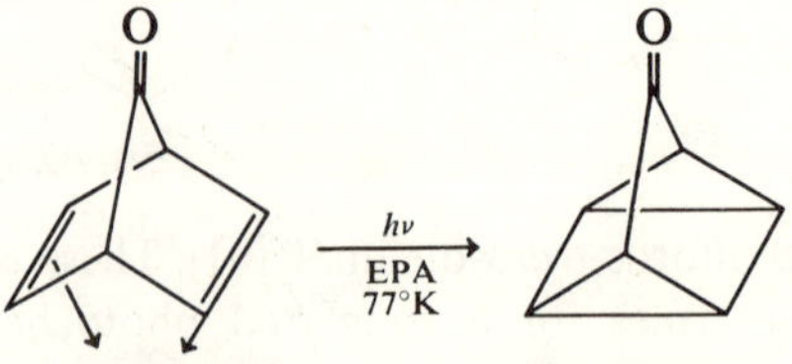

$$\text{56 (Fe(CO)}_3\text{)} \xrightarrow[\text{77°K}]{h\nu, \, EPA} \text{57, 0.5-5\%}$$

The probable intermediacy of norbornadiene-7-one in the latter reaction is also supported by the presence of benzene (photodecarbonylation product) and diphenylanthracene **58** when photolyses are conducted in the presence of 1,3-diphenylisobenzofuran (trapping agent):

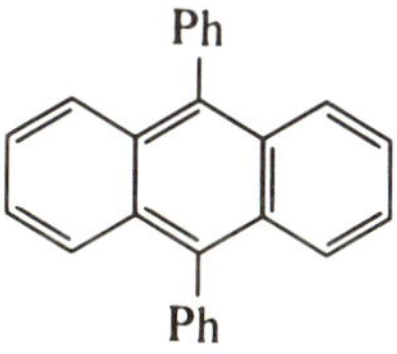

Direct irradiation of 7-oxabenzonorbornadiene **58a** yields oxepin **58b**, and sensitized irradiation results in dimerization [66]:

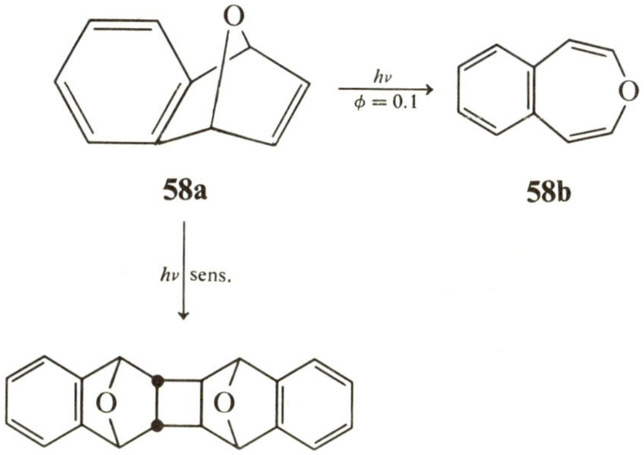

Similar examples have also appeared [67].

The sensitized di-π-methane rearrangement of benzonorbornadiene **59** affords vinylcyclopropane **60** [68]:

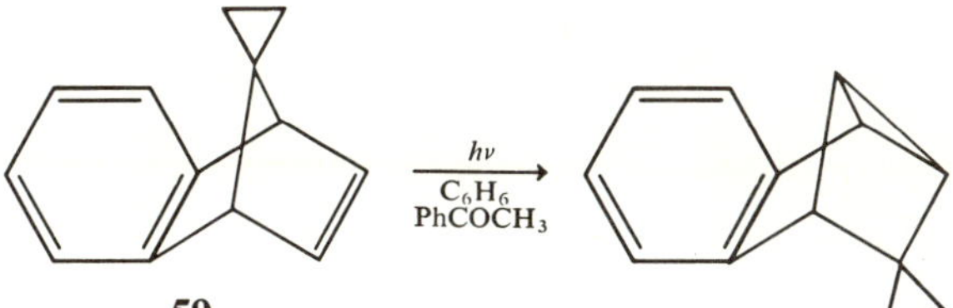

The presence of the cyclopropane ring serves to accelerate the rearrangement relative to benzonorbornadiene itself [69].

Further examples of photochemical "cage" compound formation (Volume 1, pages 11 to 14; Volume 2, page 10) have been reported. Sensitized photolysis of ketone **61** quantitatively affords a mixture of isomers **62** and **63** [70]:

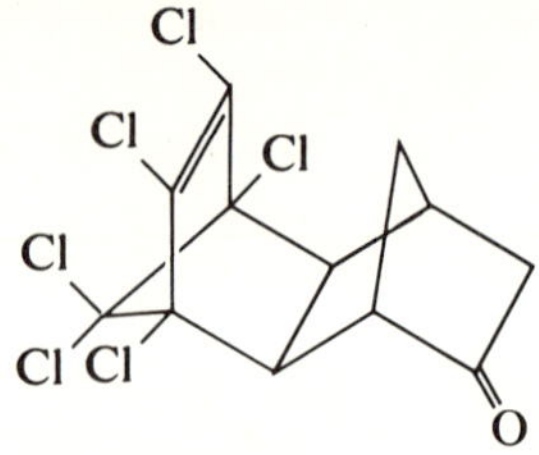

61

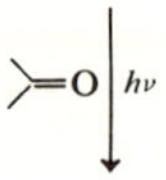

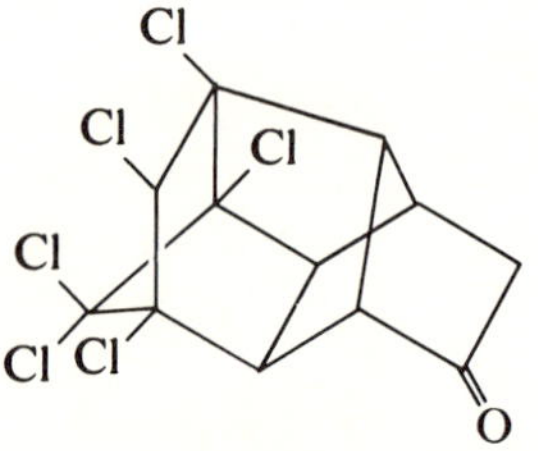

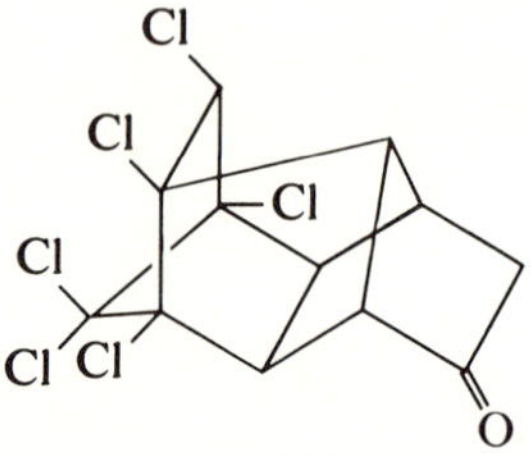

62, 50% + **63**, 50%

The direct [71] and sensitized [72] photolyses of the α-, β-, and δ-chlordans, heptachlor, β-dihydroheptachlor, nonachlordan, chlorden, and isodrin [73] have been reported. For example, β-chlordan **64** yields **65**:

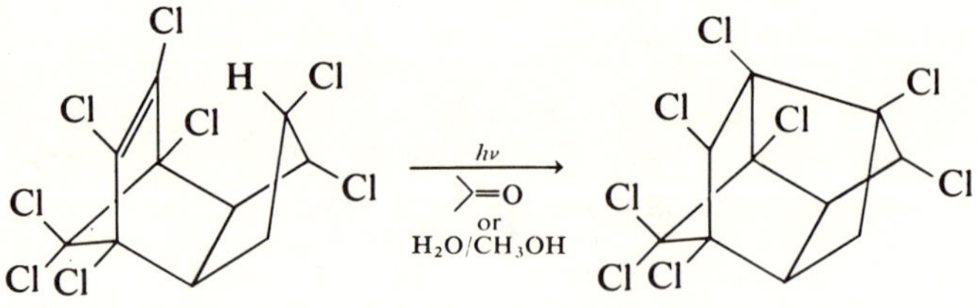

64 → **65**

and irradiation (sensitized or direct) of heptachlor **66** yields **67**:

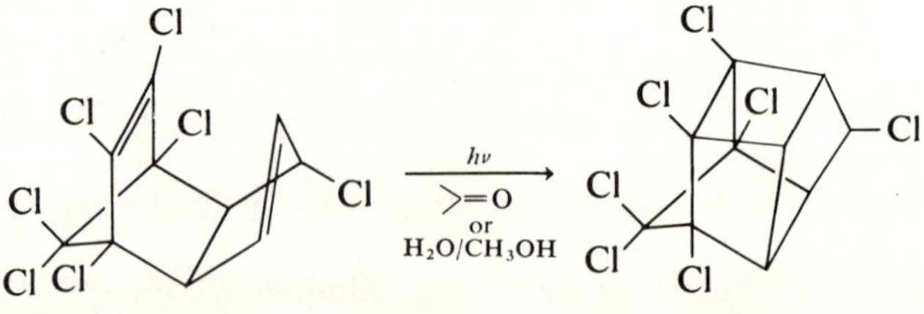

66 → **67**

Interestingly, photolysis of γ-chlordan **68** does not result in "cage" compound formation:

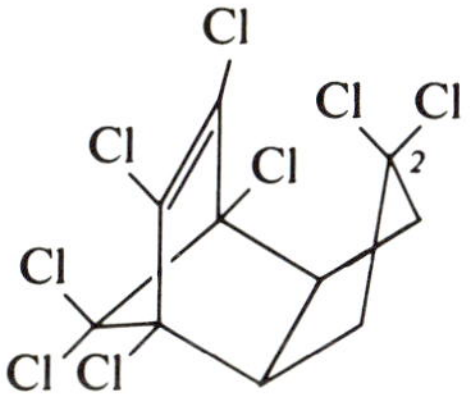

68

Similar results for α-chlordan and nonachlordan [72] suggest that the presence of a *syn*-hydrogen atom on carbon-2 is essential for photocyclization to occur [see also 74].

Photolysis of similar polychloro compounds in oxygenated hexane results in the reduction of the C—Cl bonds [75].

Photocyclization of the Diels-Alder dimer of tetrachlorocyclopentadiene [76] has been used to eliminate certain possible structures for this dimer. The sensitized photocyclizations of dicyclopentadienes **69**, **71**, and **73** to **70** and **72** have been used in the syntheses of *syn*- and *anti*-pentacyclo[5.3.0.0^{2,5}.0^{3,9} 0^{4,8}]dec-6-yl *p*-toluenesulfonates [77]:

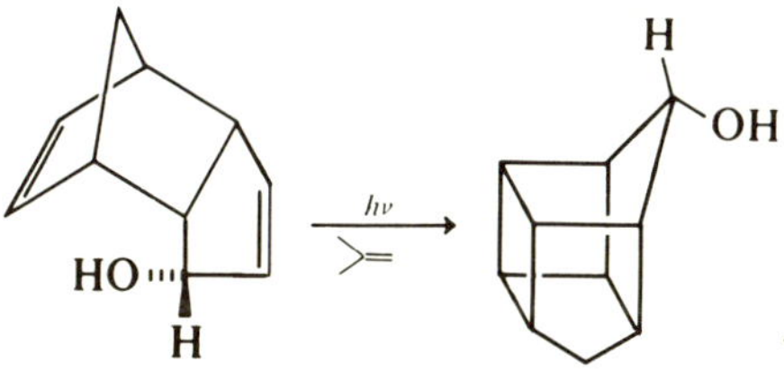

69 → **70, 20%**

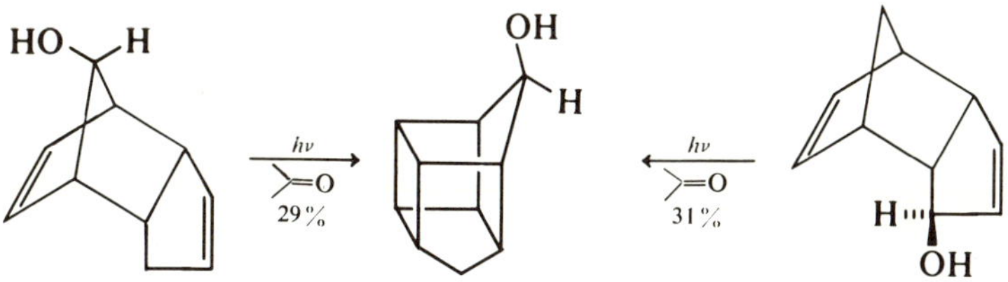

73 → **72** ← **71**
    29%   31%

1. MEDIUM- AND LARGE-RING ALKENES

The photosensitized ionic additions to cyclohexenes have recently been reviewed [78] (Volume 1, pages 14 to 20; Volume 2, page 12). A study of the effects of ring size and substitution on the photosensitized addition of alcohols to cycloalkenes has appeared [79] (see also Chapter 2). Interestingly, irradiation of cyclohexene **74** results in an intramolecular addition to give epoxide **75** [80]:

74 **75**, 10%

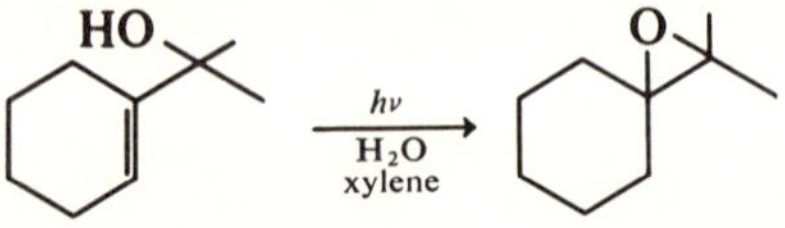

and a similar irradiation of **76** affords methylenecyclohexane **77** and acetaldehyde as primary photoproducts [62]:

76 **77**

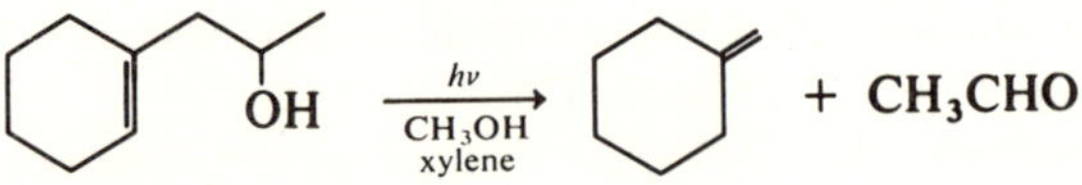

The intermediacy of oxetane **78** in this reaction is not well supported but cannot be rigorously excluded:

78

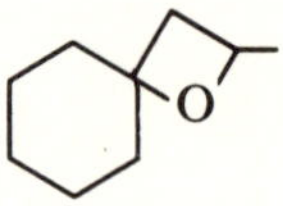

The sensitized photolysis of cholesterol **79** in the presence of water yields diol **80** and oxetane **81** [81]:

79 **80**, 54% **81**, 21%

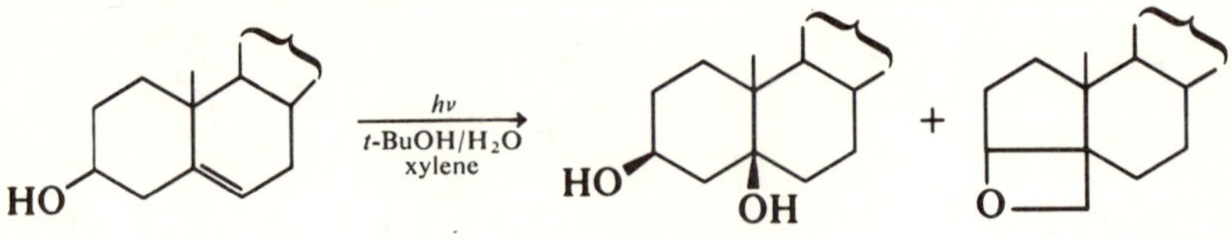

Identical products were obtained from the sensitized photolysis of 4-cholesten-3β-ol. Deuterium studies suggest that **81** is formed from the initial tertiary carbonium ion by a Meerwein-Wagner rearrangement followed by an attack of the alcohol oxygen on the newly formed primary carbonium ion.

Sensitized photolyses of 10-methyl-1(9)-octalins results in isomerization [82]. For example, xylene-sensitized irradiation of octalin **82** in the presence of acetic acid affords **83** and **84**:

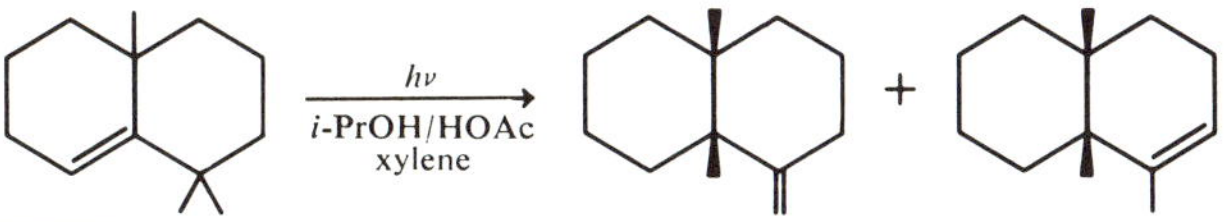

82 $\xrightarrow[\text{xylene}]{\substack{h\nu \\ i\text{-PrOH/HOAc}}}$ **83**, 50% + **84**, 3%

A mechanism that involves the initial protonation of the octalin to produce a C-9 cation is postulated. Stereospecific migration of an adjacent methyl group then affords another tertiary cation that subsequently deprotonates to yield the observed products.

Irradiation of 4- and 5-cholestene in hydroxylic solvents reveals an interesting dependence of product stereochemistry on the presence or absence of an aromatic sensitizer [83]. Irradiation of **85** in methanol-cyclohexane, for example, yields hydrocarbon **86**, and methanol adducts **87** and **88**:

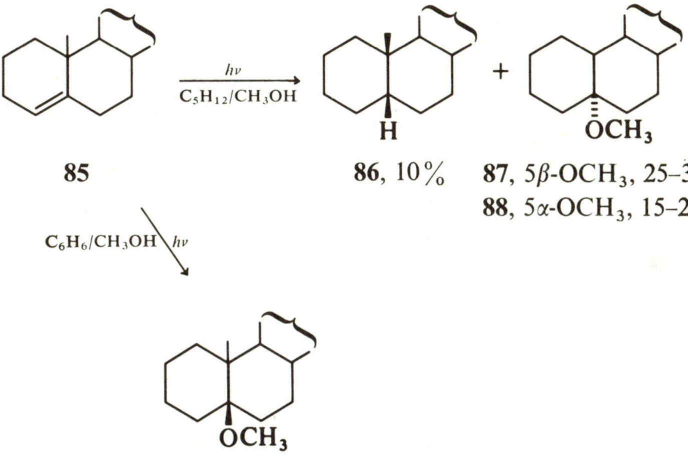

85 $\xrightarrow[C_5H_{12}/CH_3OH]{h\nu}$ **86**, 10%   **87**, 5β-OCH$_3$, 25–30%
                                                    88, 5α-OCH$_3$, 15–20%

$C_6H_6/CH_3OH \searrow h\nu$

87 (OCH$_3$)

Sensitized irradiation of **85** in methanol-benzene, however, yields only **87** [see also 84].

Photochemical "cage" compound formation from derivatives of pentacyclo[4.4.0.0^{2,5}.0^{3,8}.0^{4,7}]deca-9-ene has been reported. Some examples (**89** → **90** [85, 86], **91** → **92** [85], **93** → **94** [85, 86]), are given (see also this chapter, Section II.B.1):

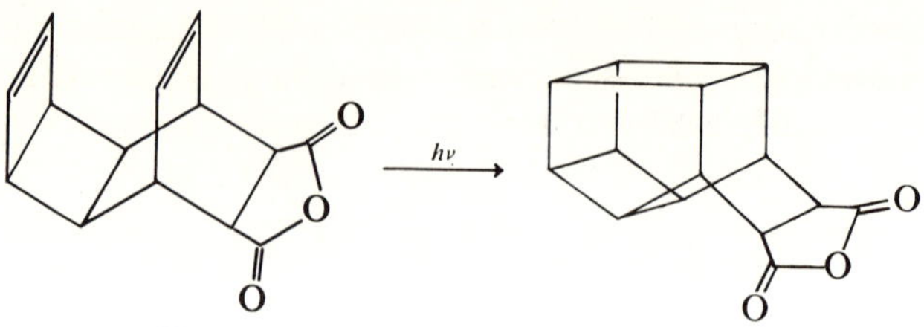

89 → **90**

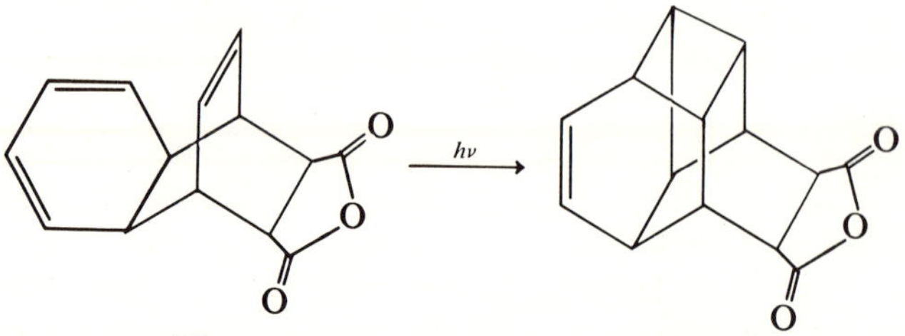

91 → **92**, 63%

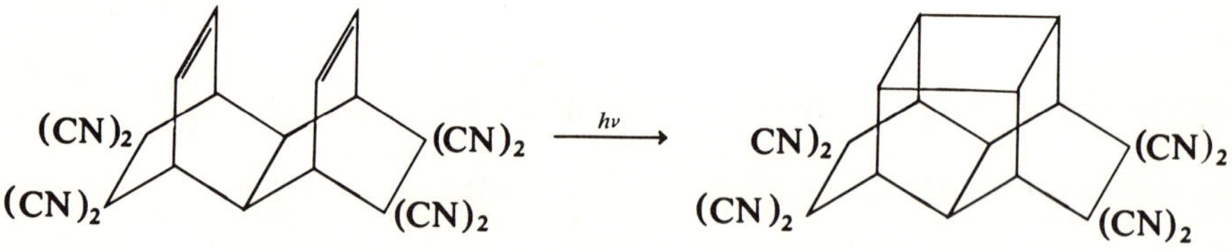

93 → **94**

Quadricyclene formation from trifluoromethylated norbornadienes has been reported (for example, **95 → 96**) [87]:

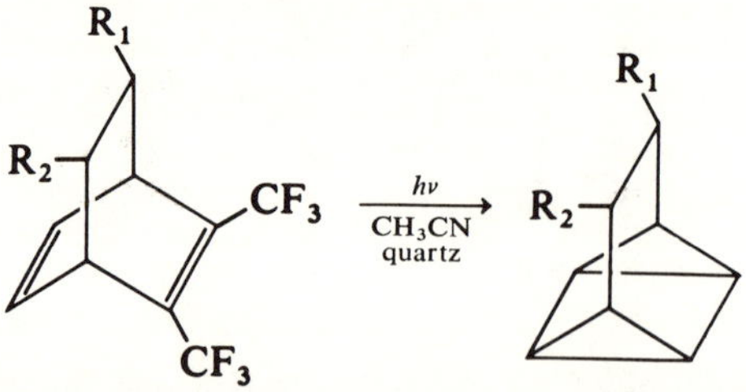

95a, $R_1 = R_2 = H$
95b, $R_1 R_2 = -CH_2-$
95c, $R_1 R_2 = -CH=CH-$

96

Interestingly, acetone does not sensitize these cyclizations.

ORGANIC PHOTOCHEMISTRY

Further examples of barrelene [88-91] and benzobarrelene [89, 90, 92-95] photorearrangements have been reported (Volume 1, pages 18, 19; Volume 2, pages 13, 14). Barrelene **97** yields cyclooctatetraene **98** by way of a singlet excited state and semibullvalene **99** by way of a triplet state [88]:

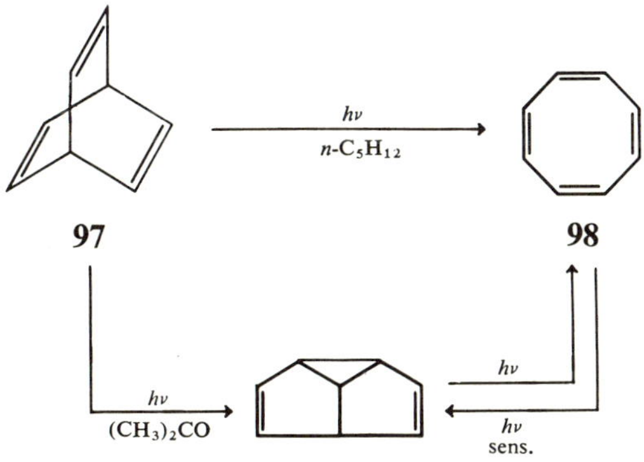

Although barrelene has three equivalent double bonds, deuterium studies demonstrate that only two are structurally involved in the proposed mechanism (for example, biradical **100** and not **101**):

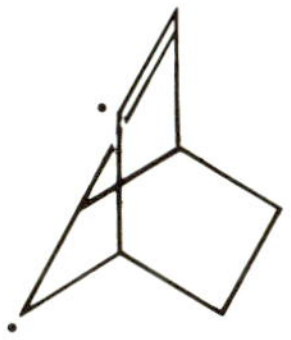

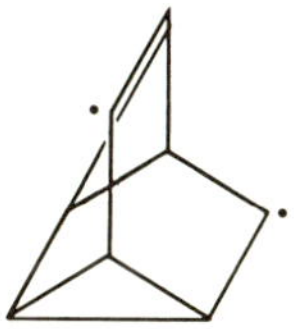

Benzobarrelene **102**, on sensitized excitation, affords benzosemibullvalenes **103** and **104**:

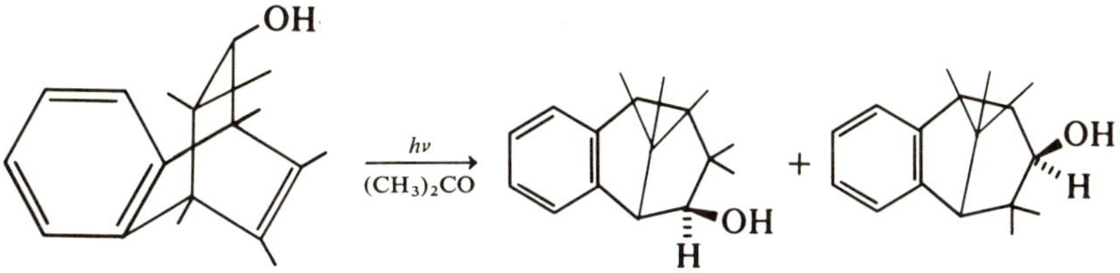

syn alcohol **105**, however, affords only **106** [94, 93]:

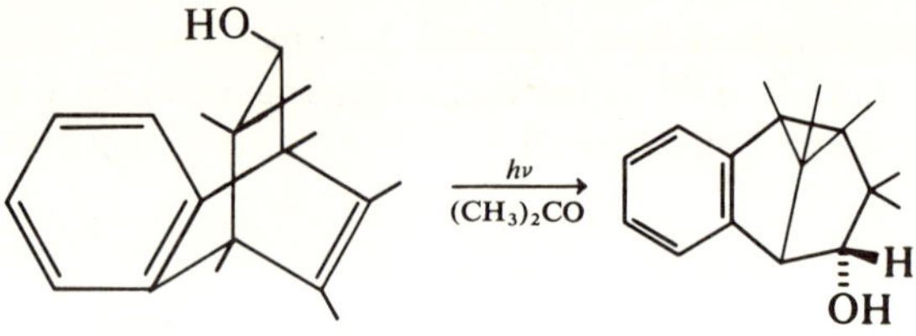

105 → (hv, (CH₃)₂CO) → **106, 68%**

Hydrogen bonding or charge-transfer interaction are discussed as possible reasons for the observed regiospecificity in the latter example.

Sensitized irradiation of per-deuterio-1,2-naphthobarrelene **107** affords semibullvalenes **108** and **109**, both products resulting from initial α-naphthyl-vinyl bridging:

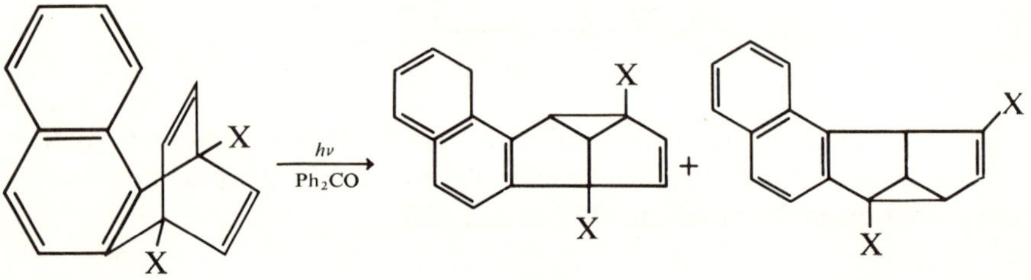

107, X = H → (hv, Ph₂CO) → **108** + **109**

On the other hand, per-deuterio-2,3-naphthobarrelene **110** affords only semibullvalene **111**, a product that is attributable to initial vinyl-vinyl bridging [96]:

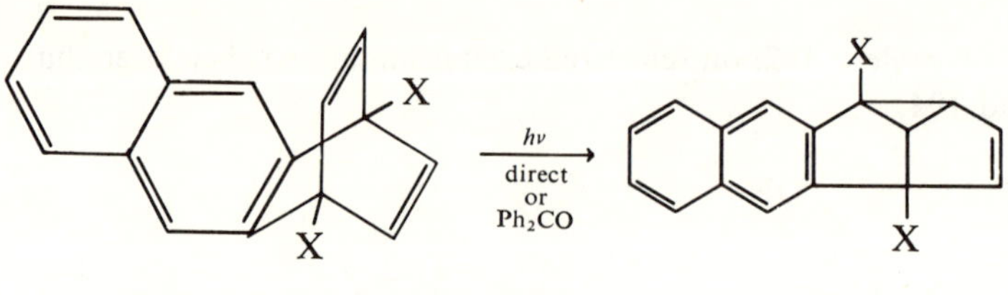

110, X = H → (hv, direct or Ph₂CO) → **111**

Since it would be expected that the energy would be localized in the naphthalene system ($E_T = 61$ kcal/mole), it is apparent that the β position of naphthalene displays a reluctance to participate in the bridging process. This reluctance has been attributed to a residual aromaticity still present in the triplet excited state.

The direct or sensitized irradiation of benzobicyclooctadiene **112** yields **113**, a result that contrasts with other di-π-methane rearrangements that do not proceed by way of S_1 (direct excitation) [97, 98]:

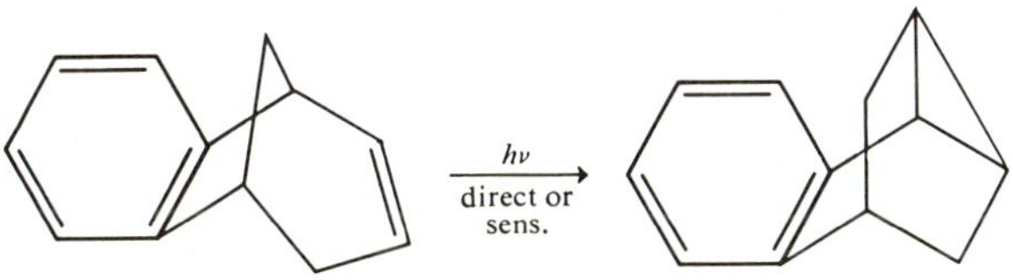

112 → (hv, direct or sens.) **113**, 94%

Bicyclo[3.2.2]nonatriene **114** undergoes photosensitized rearrangement to **115** and **116**:

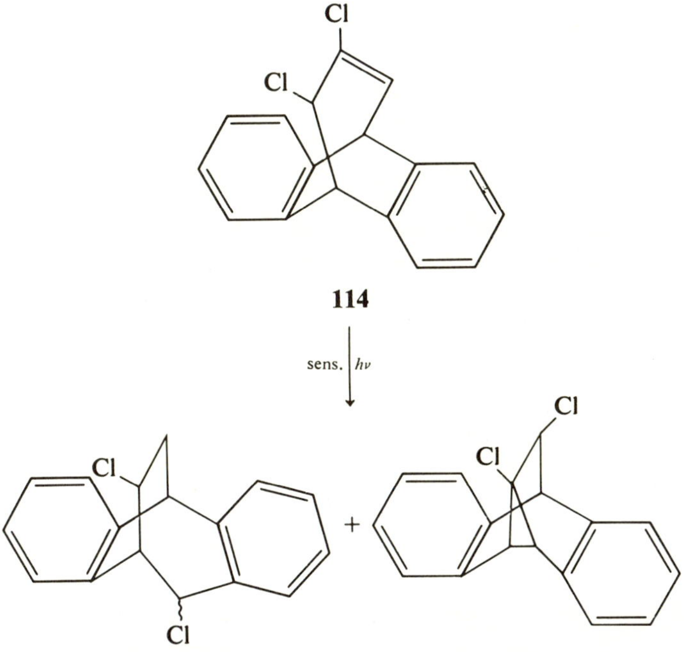

114

sens. | hv

115b, *endo*
115c, *exo*

116

A carbonium ion–chloride ion ion pair has been postulated as a possible intermediate [99].

A method using the sensitized photoisomerization of *cis*-cyclooctene is reported to yield 97% *trans*-cyclooctene [100].

Photosensitized irradiation of *cis,trans*-1,5-cyclooctadiene affords the *cis,cis* isomer [101].

Irradiation of copper complex **117** affords tricyclo[3.3.0.0^{2,6}]octane **118** in addition to copper (I) complexes of *cis,trans*- and *trans,trans*-cyclooctadienes (COD) [102] (see also Volume 1, page 21):

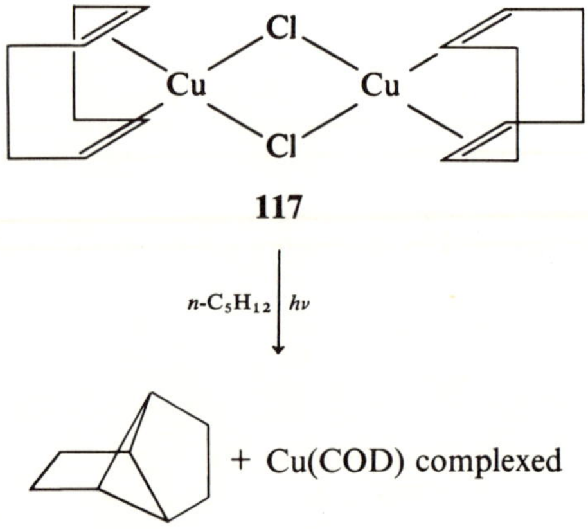

117

n-C$_5$H$_{12}$ | $h\nu$

118 + Cu(COD) complexed

The structures of photoisomers of caryophyllene and isocaryophyllene have been studied by x-ray techniques [103] (see also Volume 2, page 13).

Irradiation of methylenecycloheptene **119** affords **120** on direct excitation:

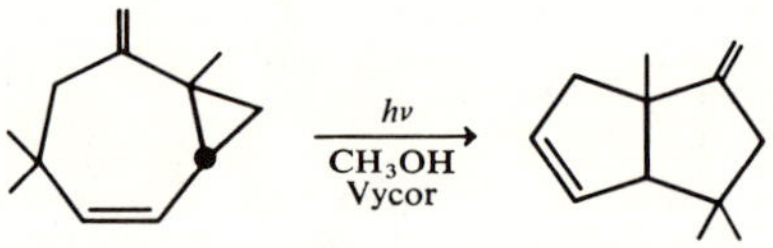

119 $\xrightarrow[\text{Vycor}]{h\nu \atop \text{CH}_3\text{OH}}$ **120, 56%**

and **121** undergoes photosensitized rearrangement to **122** [104]:

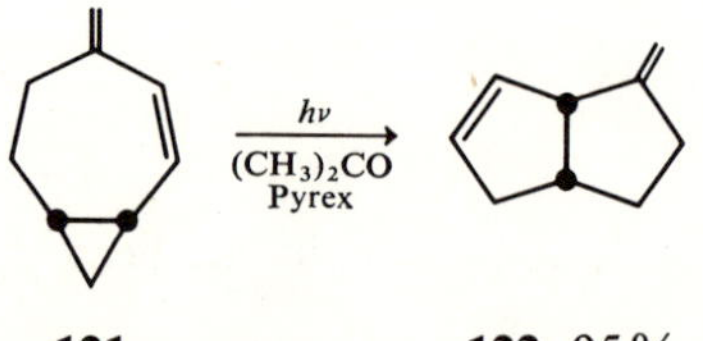

121 $\xrightarrow[\text{Pyrex}]{h\nu \atop (\text{CH}_3)_2\text{CO}}$ **122, 95%**

Cyclic and alicyclic allenes have been reported to undergo benzene-sensitized photorearrangement in the vapor phase [105, 106]. For example, allene **123** yields **124** ($\Phi = 0.17$):

123 **124**

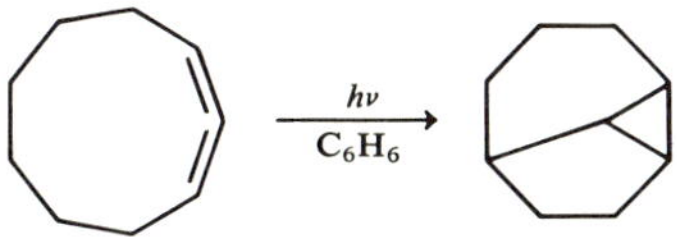

and alkene-allene **125** undergoes a photosensitized Cope rearrangement to **126** ($\Phi = 0.02$) [106]:

125 **126**

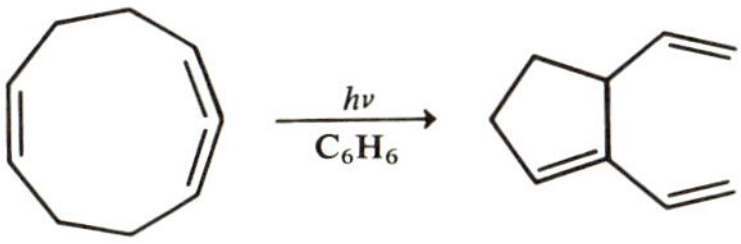

Cyclohexa-1,4-diene is believed to quench excited triplet states of arakyl ketones vibrationally [107].

3. HETEROATOM-CONTAINING ALKENES

Photolysis of dihydrofuran **127** yields cyclopropyl aldehydes **128** and **129**:

127 **128** **129**

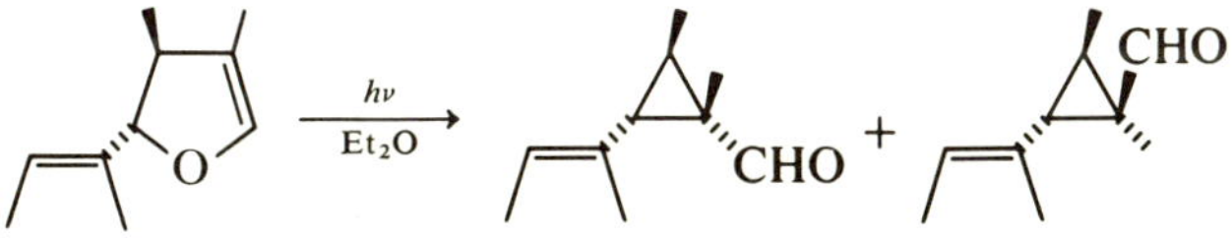

and **130** affords **131** and **132** [108]:

130 **131** **132**

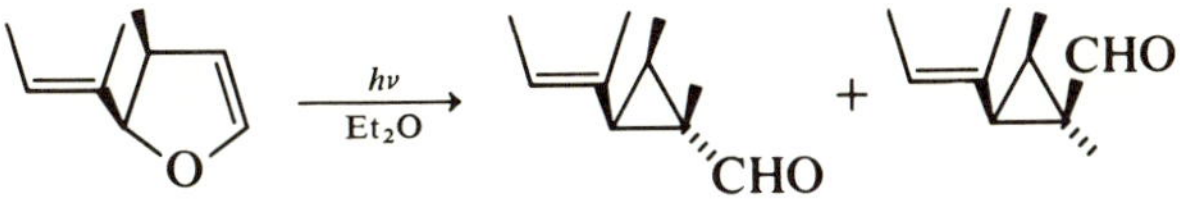

On extended irradiation, all four aldehydes are observed from either starting material. It is proposed that two mechanisms are involved: (1) the concerted formation of cyclopropyl aldehydes from the corresponding dihydrofuran and (2) the photoisomerization of the cyclopropyl aldehydes by way of biradical intermediates.

Irradiation of methanolic solutions of dihydrofuran **133** yields adducts **134** to **136** [109]:

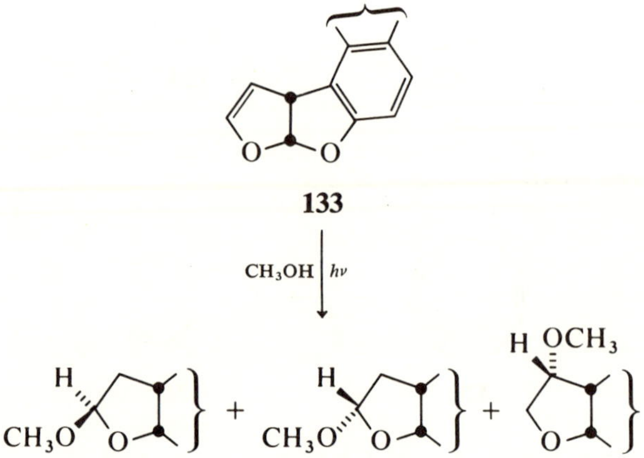

133

↓ CH₃OH | hν

134, 24% + **135, 30%** + **136, 38%**

Cis-fused dihydrofuran **137** photorearranges to aldehydes **138** and **139** in addition to cycloheptatriene, phenylacetaldehyde, and benzene [110]:

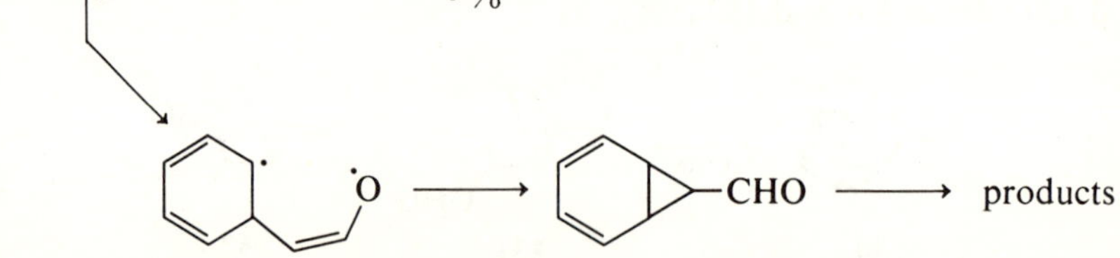

137 →(hν/Et₂O) **138, 12%** + **139, 24%** + 24%
+ PhCH₂CHO   + C₆H₆·
9%

| hν
↓

→ →-CHO → products

The mechanism presumably involves initial C—O cleavage and subsequent rearrangement to a cyclopropyl aldehyde.

Photolysis of 1,1-diphenyl-1-stannacyclohepta-2,6-diene yields a cyclic and a linear dimer [111].

III. CONJUGATED POLYENES

A. Acyclic Conjugated Dienes

1. BUTADIENE

Photocyclization of 1,3-butadiene to cyclobutene has been reviewed [112, 113, 40, 39, 41], and problems associated with the Woodward-Hoffmann disrotatory excited-state rotation have been discussed.

Irradiation of 1,3-butadiene in methanol yields a 1:1 mixture of cyclobutyl methyl ether and cyclopropylmethyl methyl ether [114]. If base-washed glassware is used, some bicyclobutane is formed (see also Volume 2, page 16).

2. SUBSTITUTED BUTADIENES

Photolysis of 2-cyano-1,3-butadiene yields 1-cyanobicyclo[1.1.0]butane and 1-cyanocyclobutene [115].

The benzophenone-sensitized irradiation of the 2,4-hexadienes results in photoisomerization [116, 117]. The results are interpreted in terms of a common triplet mechanism with a 1,4-biradical geometry, or rapidly equilibrating allylmethylene biradicals:

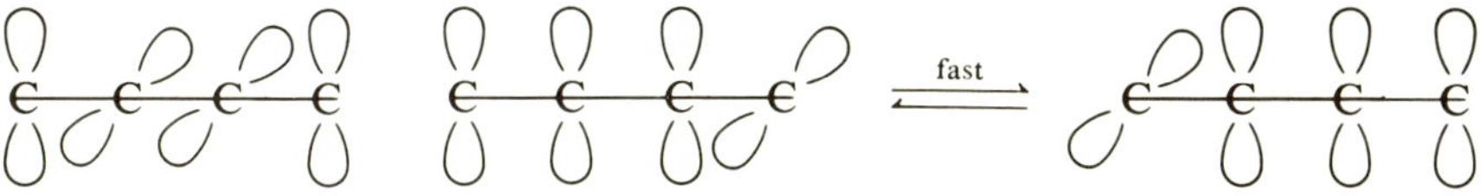

140   **141**

Diene **142** undergoes intramolecular, photosensitized *cis-trans* isomerization to diene **143**:

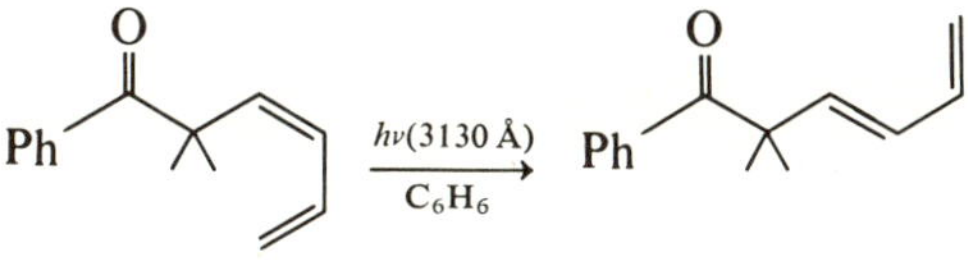

142   **143**

Interestingly, the quantum yield for isomerization and the composition of the photostationary state are unaffected by the presence of several quenchers [118].

Irradiation of myrcene **144** yields cyclobutene **145** and bicyclic compounds **146** and **147**:

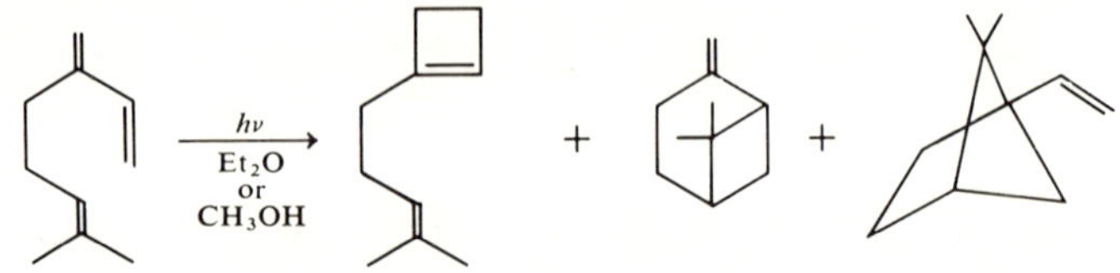

144   **145**, 81–83%   **146**, 14–16%   **147**, 3%

In the case in which methanol was used as a solvent, no enhancement of the triplet product **147** was observed, which demonstrates that methanol does not increase the rate of intersystem crossing to T_1 [119].

Sensitized irradiation of either *cis*- or *trans*-heptatriene **148** yields the same ratio of **149** and **150** (Volume 1, page 25):

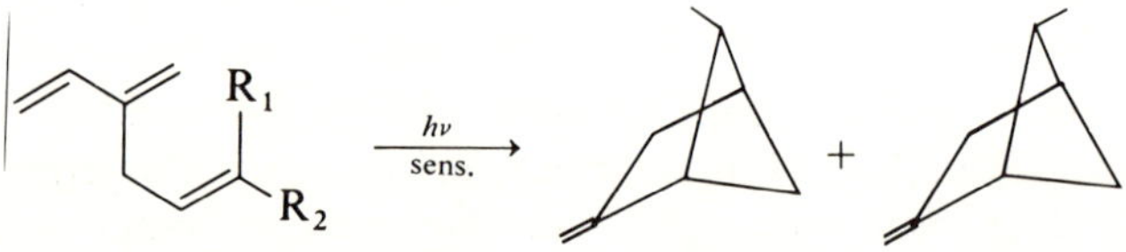

148, $R_1 = H$; $R_2 = CH_3$
or $R_1 = CH_3$; $R_2 = H$ **149**   **150**

Calculations concerning the electronic state of intermediate **151** suggest that it cyclizes from S_0. Elimination of intermediates such as **152** is based on the following experimental rule [120]: "In the cyclization of an excited triplet state, intermediates which contain two unpaired electrons in one ring structure cannot exist, while in the cyclization of an excited singlet state the restriction mentioned above does not exist":

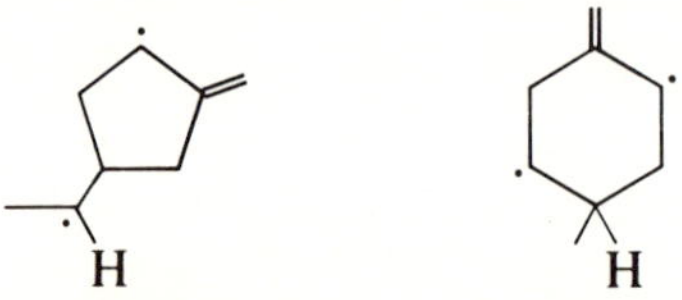

151   **152**

Sensitized irradiation of β-farnesene **153** yields **154** and **155**, and direct irradiation affords smaller yields of **154** and **155** in addition to cyclobutene **156** and a bicyclic compound A:

153 $\xrightarrow[\text{sens.}]{h\nu}$ **154**, 45–47% + **155**, 53–55%

153 $\xrightarrow[\text{quartz}]{h\nu}$ **154** (15–25%) + **155** (29–30%) + **156**, 30–56% + A 5–10%

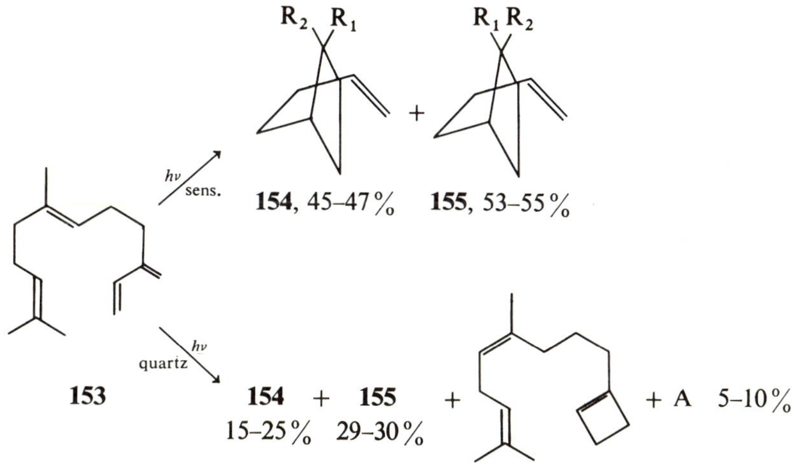

Compound A has been tentatively identified both as bicyclo[3.2.0]heptane **159** [121] and as the epimeric mixture **160** + **161** [122]:

A = **159** or **160** + **161**

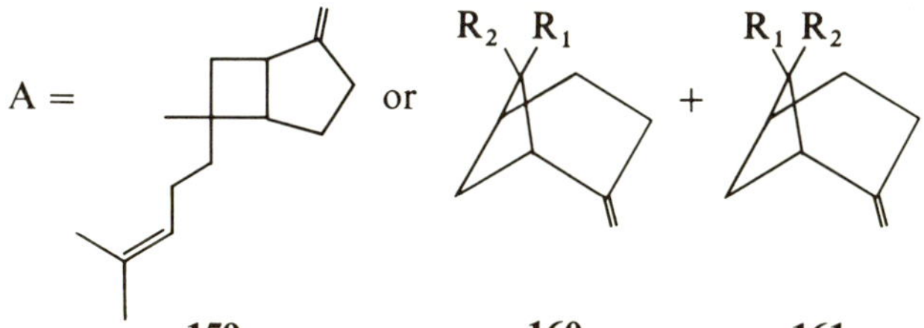

$R_1 = CH_3$; $R_2 = CH_2CH_2CH = C(CH_3)_2$

Interestingly, sensitized or direct irradiation of α-farnesene **162a** results in isomerization only about the 3,4 bond:

162a $\xrightarrow[C_6H_6\text{ direct or sens.}]{h\nu}$ **162a** + **162b**

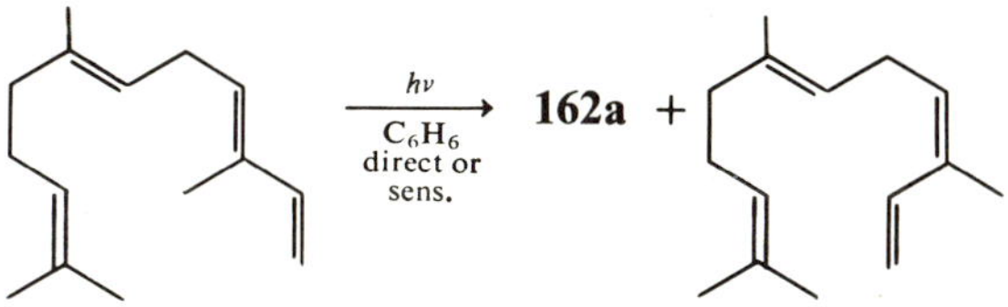

Direct irradiation of **163a** gives **163b**, and sensitization leads only to *cis-trans* isomerization [123]:

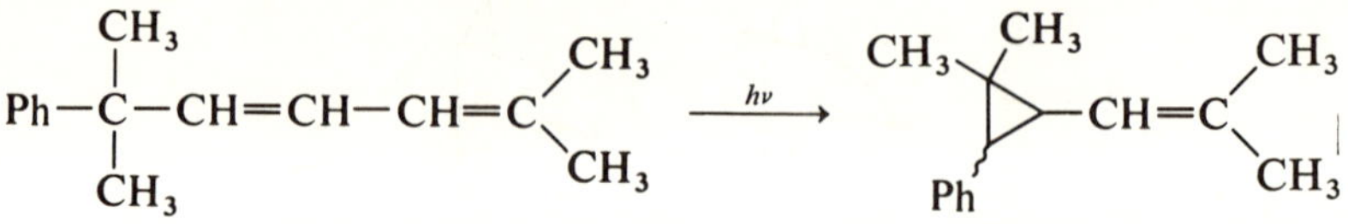

$$Ph-\underset{CH_3}{\underset{|}{\overset{CH_3}{\overset{|}{C}}}}-CH=CH-CH=C\overset{CH_3}{\underset{CH_3}{\diagdown}} \xrightarrow{h\nu} \underset{Ph}{\overset{CH_3\diagup CH_3}{\triangle}}-CH=C\overset{CH_3}{\underset{CH_3}{\diagdown}}$$

163a   **163b**

The triene **164** is photocyclized to cyclobutene **165** [124]:

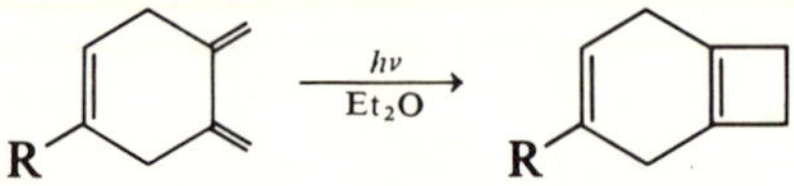

164, R = H or CH$_3$ **165**, 60%

Photoisomerizations of 1,2-dimethylenecyclobutanes have been reported (for example, **165a** → **166**, **167** → **168**, and **169** → **170**) [125, 126]:

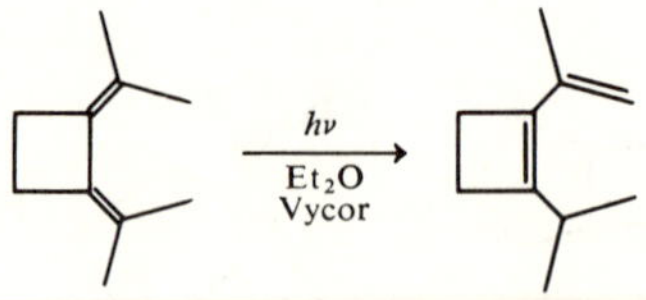

165a   **166**

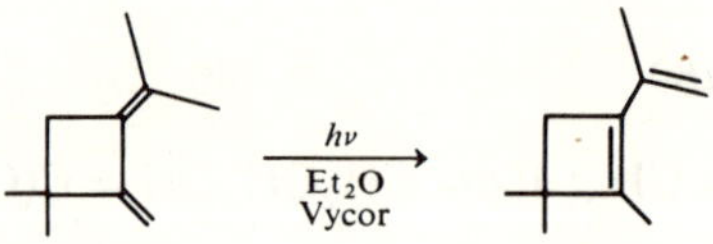

167   **168**

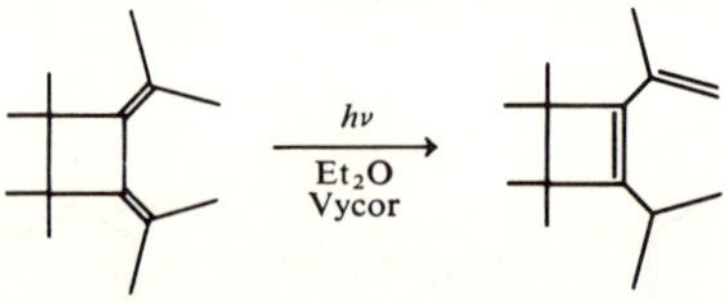

169   **170**

Examination of the relative rates of isomerization (**169** > **165** > **167**) confirms what would be expected on the basis of steric effects if 1,5-hydrogen migration were occurring antarafacially in the exited state.

Photoinitiated polymerization of perfluorobutadiene yields low-molecular-weight polymers [127].

B. Alicyclic Conjugated Dienes

1. CYCLOPENTADIENE, CYCLOHEXADIENE, AND RELATED COMPOUNDS

Direct excitation of spirocyclopentadiene **171** affords fulvene **172**, apparently by way of biradical intermediate **173**:

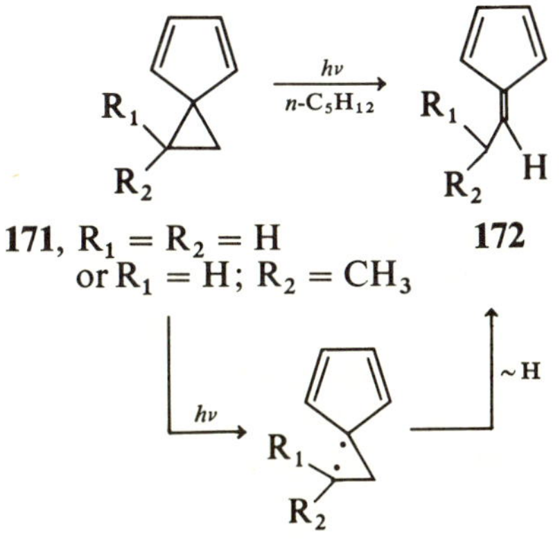

171, $R_1 = R_2 = H$ **172**
or $R_1 = H$; $R_2 = CH_3$

173

When $R_1 = R_2 = CH_3$, the expected fulvene is not observed [128].

Irradiation of α-phellandrene **174** yields *cis,trans*- and *cis,cis*-3,7-dimethylocta-1,3,5-trienes **175** and **176** [129] as primary photoproducts:

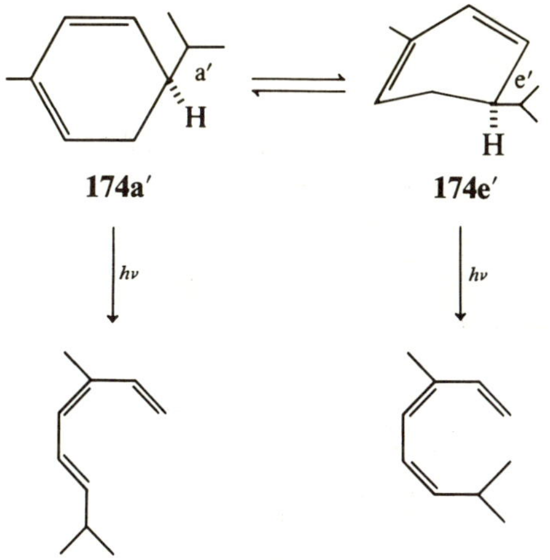

174a′   **174e′**

175   **176**

The temperature dependence of the ratio of *cis,trans-*:*cis,cis-*, in addition to CD and ORD studies, suggests that ground-state conformations determine which isomeric octatriene is formed.

Irradiation of 1,3-cyclohexadiene in the presence of benzaldehyde leads to a mixture of dimers [130].

Irradiation of *trans*-fused cyclohexadiene **177** yields the *cis*-fused isomer **179** presumably by way of triene **178** [131]:

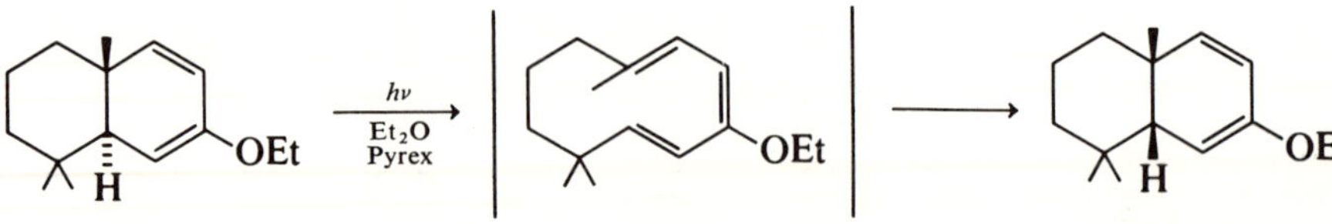

Sensitized irradiation of *trans*-cyclohexadiene **180** yields benzene, diacetoxyethylenes **181** and **182**, and a dimer [132] (Volume 2, page, 18). It is postulated that isomerization of the olefin moiety can occur by either (1) the decomposition of triplet **180** to ground-state benzene and triplet olefin **183** followed by the decay of **183** to *cis*- and *trans*-olefins or (2) a C—C bond cleavage to biradical **184** followed by decomposition to benzene, **181**, and **182**:

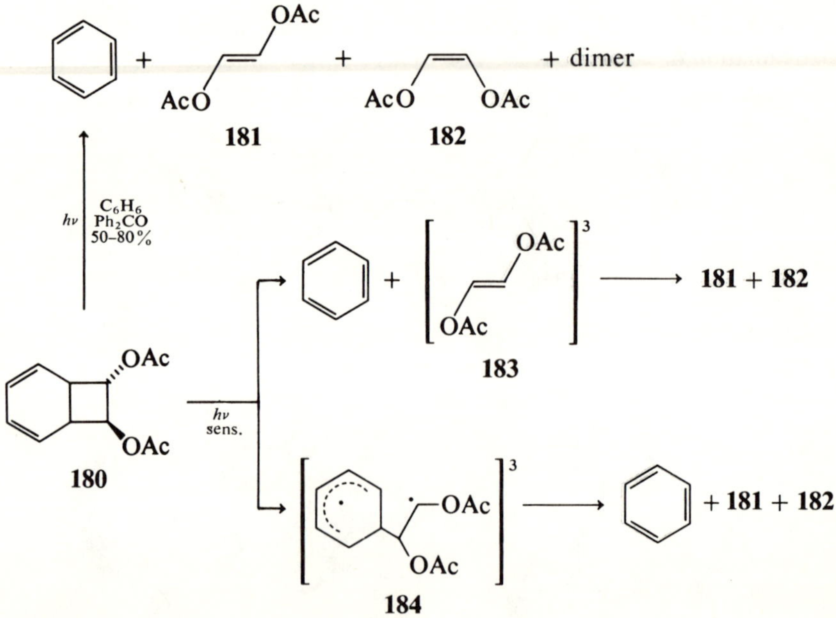

Photochemical cyclization of 9,10-dihydronaphthalene systems has been reported (Volume 1, pages 30, 31; Volume 2, page 18). For example, irradiations of **185a** to **e** afford hexahydro compounds **186a** to **e** [133, 134, 135]:

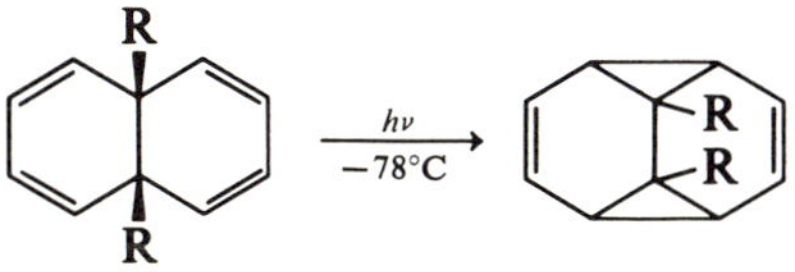

185a, RR = —CO—O—CO— **186**
185b, RR = —CO—NH—CO—
185c, RR = —CO—N(CH_3)—CO—
185d, RR = —CO—O—CH_2—
185e, RR = —CH_2—O—CH_2—

Only for **185b** and **185c** were the corresponding hexahydrocompounds isolable.

Irradiation of *cis*-9,10-dihydronaphthalene **187** yields pentaenes **188** and **189** in addition to **190** to **194** [136]:

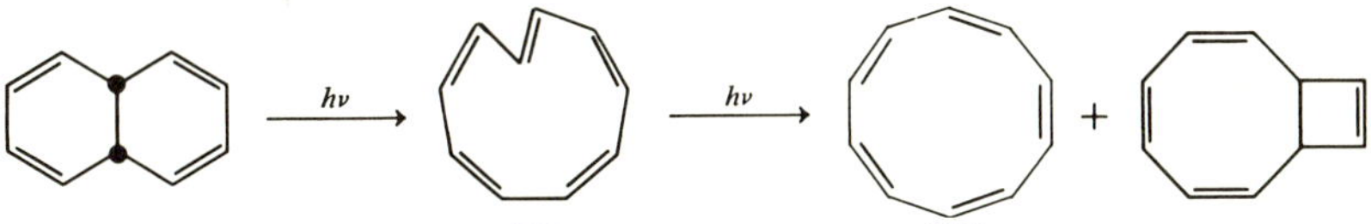

187 **188** **189** **190**

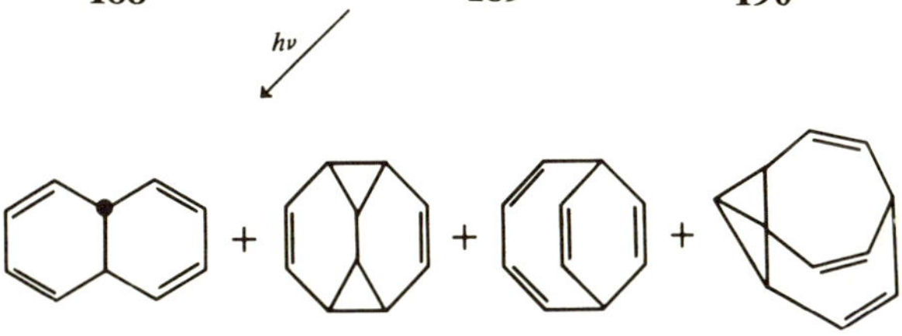

191 **192** **193** **194**

Irradiation of dihydronaphthalene **195** yields naphthalene and acetylene [137]:

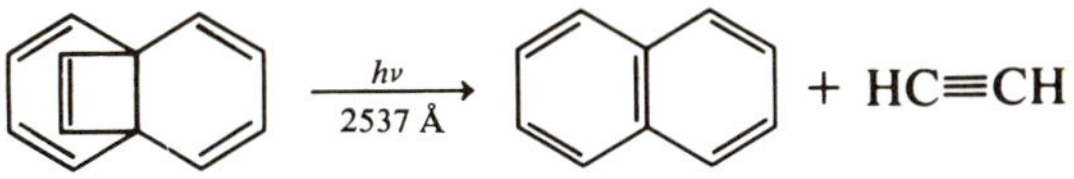

195

Direct excitation of tetraene **196** results in electrocyclic closure to **197** and **198** [138]:

196 **197** **198**

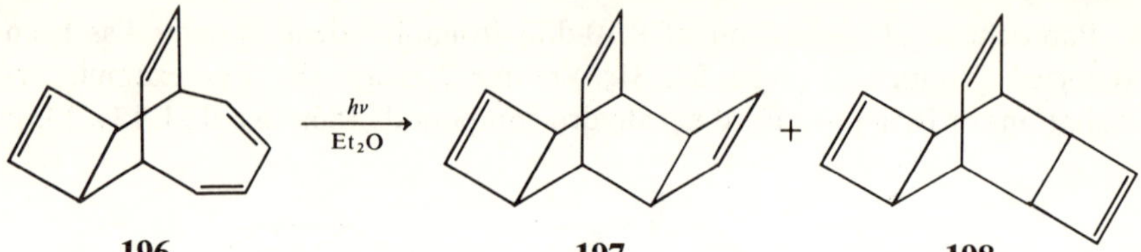

and irradiation of prebullvalene **199** affords isomeric bullvalenes **200** and **201** [134]:

199 **200** **201**

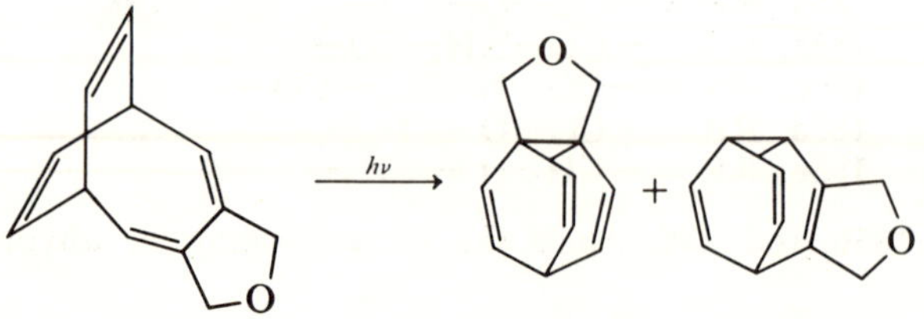

Irradiation of *cis,cis*-1,3-cyclooctadiene yields *cis,trans*-1,3-cyclooctadiene in addition to *cis,cis*-1,4-cyclooctadiene and *cis*-fused bicyclo[4.2.0]oct-2-ene [139].

2. CYCLIC *TRANS*-FUSED DIENES AND TRIENES

Irradiation of methanolic solutions of diene **202** yields adduct **203** [140] (see also Volume 1, pages 33,34; Volume 2, p. 19):

202 **203**, 39%

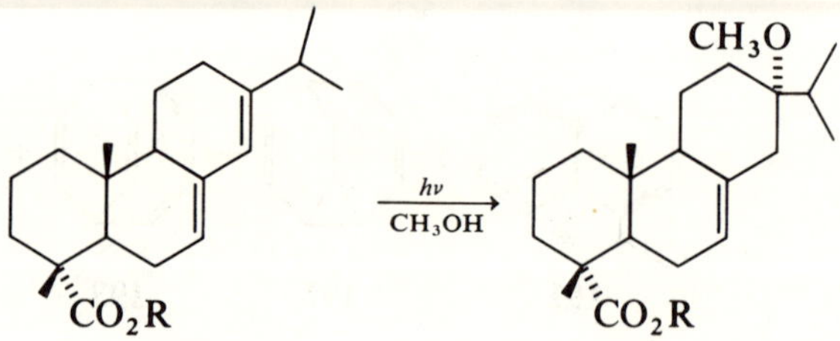

In close analogy to the photolysis of 4,4-diphenyl-2-cyclohexenone, irradiation of diene **203a** yields olefins **203b** and **203c** as primary photoproducts and diene **203d** as a secondary product [141]:

203a **203b** **203c** **203d**

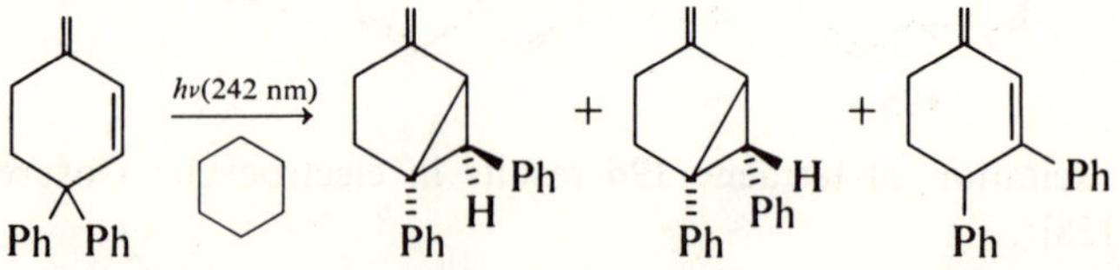

Unlike the enone case, however, the diene triplet is unreactive. The photolysis of a number of bicyclo[6.1.0]polyenes has been reported [142].

Irradiation of germacrene D **204** yields β-bourbonene **205** as the major product in addition to lesser amounts of α-bourbonene and β-copaene [143]:

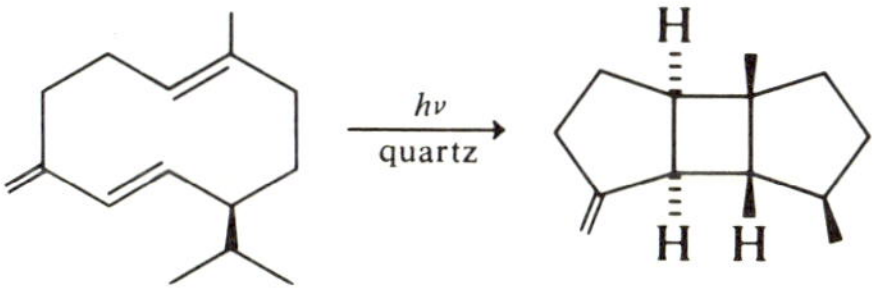

204 → (hv, quartz) → **205**

The facility of cyclization coupled with the observed transannular interaction (UV) suggests that germacrene D predominantly assumes conformation **204a** [see also 104]:

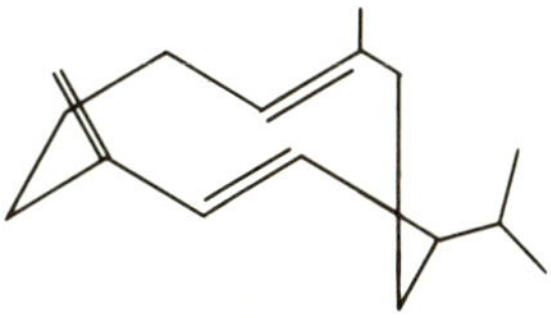

204a

Irradiation of triene **205a** yields bicyclic compounds **205b** and **c** as primary photoproducts. Subsequent irradiation leads to the formation of **205d** to **f** [144, 145]:

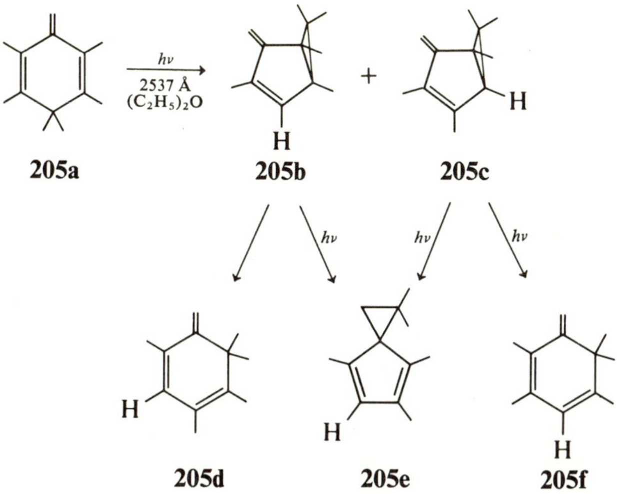

205a →(hv, 2537 Å, (C$_2$H$_5$)$_2$O)→ **205b** + **205c**

205d   **205e**   **205f**

C. Acyclic Conjugated Trienes

Photosensitized isomerization of allocimenes is reported to occur in the presence and absence of oxygen [146].

Irradiation of 3,30-diacetoxyoleana-9(11),12-diene **205** yields triene **206**, which is subsequently photolyzed to cyclobutene **207** [147]:

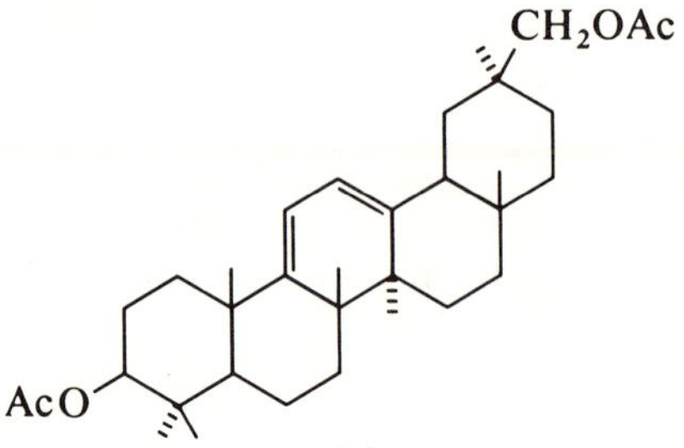

205

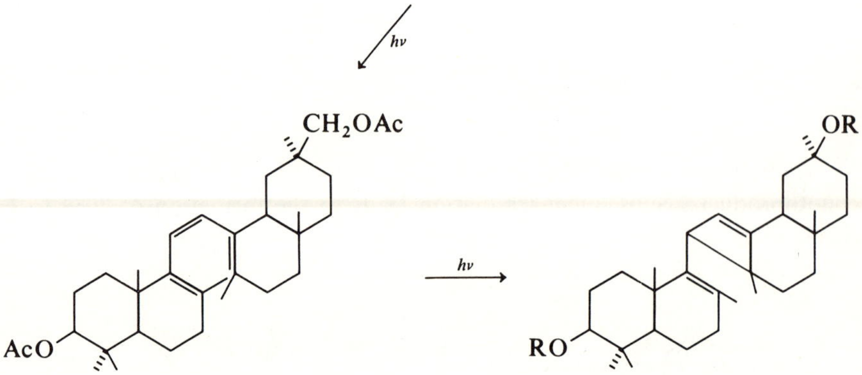

206 **207**, R = H, OAc

The photolysis of a series of *o*-divinylbenzenes has also appeared [148].

D. Alicyclic Conjugated Trienes

A review of the photochemical reactions of cycloheptatrienes and related compounds has appeared [149].

Irradiation of cycloheptatriene **208** yields **209** and **210** as primary photoproducts [150]. Cycloheptatriene **210** is formed by a selective [1,7] sigmatropic methyl migration and is subsequently cyclized to diene **211** (see also Volume 1, pages 37, 38; Volume 2, page 19):

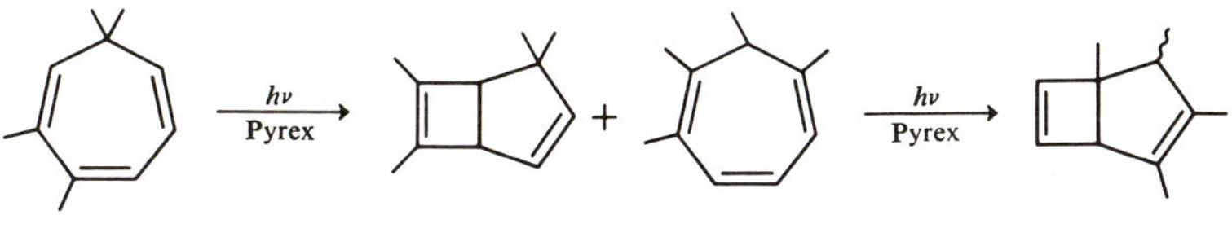

208 → **209** + **210** → **211**

Irradiation of 6-methylcyclohepta-2,4,6-triene **212** yields predominantly the 1-methyl isomer **213** and smaller amounts of the 2-methyl isomer **214** [151]. Similarly, **215** affords predominantly **214** and lesser amounts of **213**:

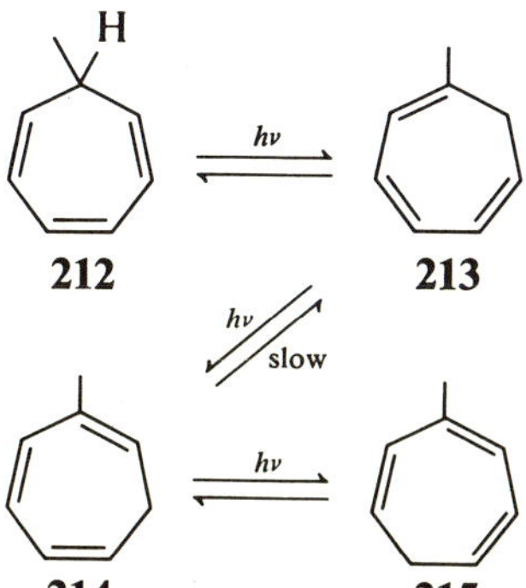

212 ⇌ **213**

hν / slow

214 ⇌ **215**

This specificity is discussed in terms of transition state **216** and electron densities at carbons $1 \leftrightarrow 7$:

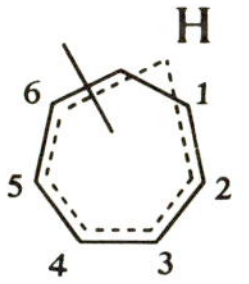

216

In contrast to **208** and **210** no bicyclo[3.2.0]heptadienes were detected for **212** to **215**.

Irradiation of cycloheptatriene-^{14}C **217** results in a partial [1,7] sigmatropic phenyl migration to isomers **218** and **219** [152]:

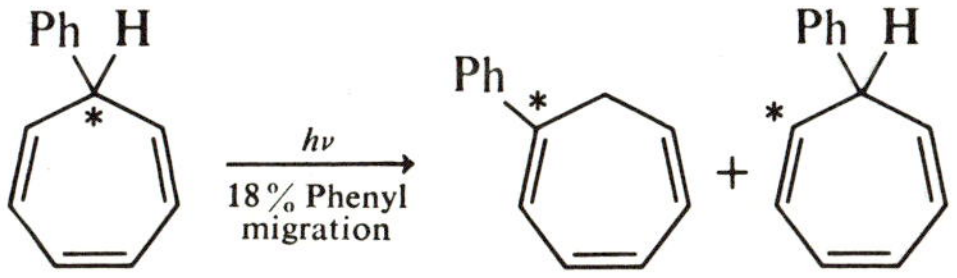

217 →(hν, 18% Phenyl migration) **218** + **219**

This result demonstrates that there is less discrimination between phenyl and hydrogen migration in the excited state.

Biscycloheptatriene **220** undergoes sigmatropic rearrangement by way of its singlet state to **221**. Further irradiation affords the doubly rearranged product **222** [153]:

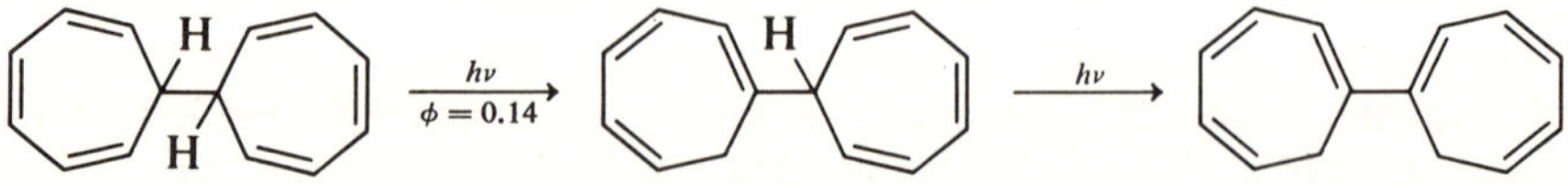

$$\mathbf{220} \xrightarrow[\phi=0.14]{h\nu} \mathbf{221} \xrightarrow{h\nu} \mathbf{222}$$

Photolysis of all-*cis*-cyclooctatriene in an inert gas matrix at 20°K yields benzene, ethylene, *cis,cis*-1,3,5,7-octatetraene, and a strained isomer as primary photoproducts. Bicyclo[4.2.0]octa-2,4-diene affords the same products [154].

Irradiation of triene **223** at low temperature affords predominantly isomers **224** to **226**, which are in photoequilibrium with one another and with **223** [155, 156]:

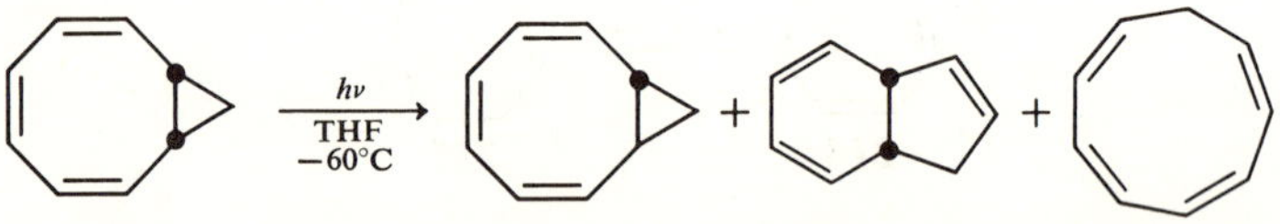

$$\mathbf{223} \xrightarrow[-60°C]{h\nu \atop THF} \mathbf{224}, 5\% + \mathbf{225}, 9-12\% + \mathbf{226}, 5-6\%$$

Irradiation of cyclobutane dimers from thermal dimerization of cyclooctatetraenes results in the formation of [16]annulenes [157].

Irradiation of triene **227** affords the potential 10 π electron system **228** [158–161] (see also Volume 2, page 20):

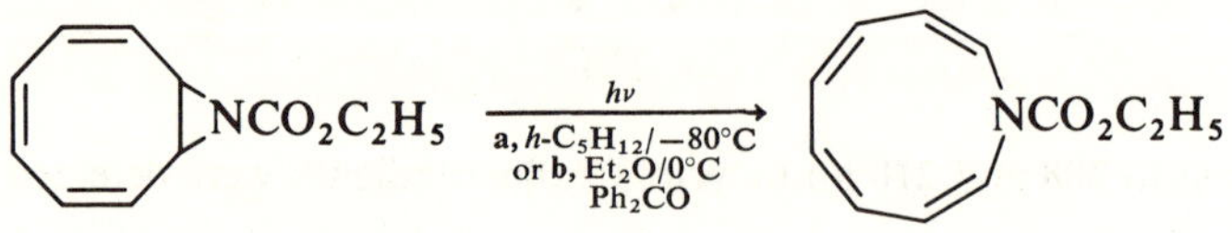

$$\mathbf{227} \xrightarrow[\text{a, }n\text{-}C_5H_{12}/-80°C \atop \text{or b, Et}_2O/0°C \atop Ph_2CO]{h\nu} \mathbf{228}$$

In similar fashion, **229** affords **230** [162]:

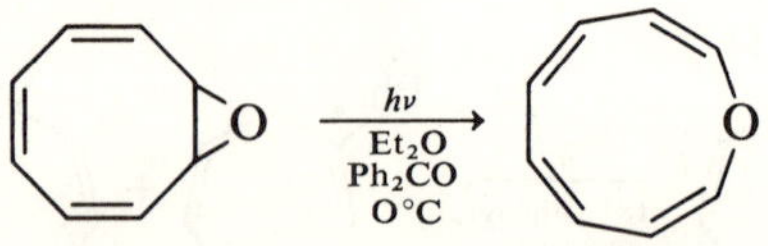

$$\mathbf{229} \xrightarrow[\substack{Et_2O \\ Ph_2CO \\ 0°C}]{h\nu} \mathbf{230}$$

Low-temperature photolysis of **231** yields naphthalenes **233** in addition to prebullvalenes **232**. Subsequent photolysis of **232** yields bullvalenes **234** [163]:

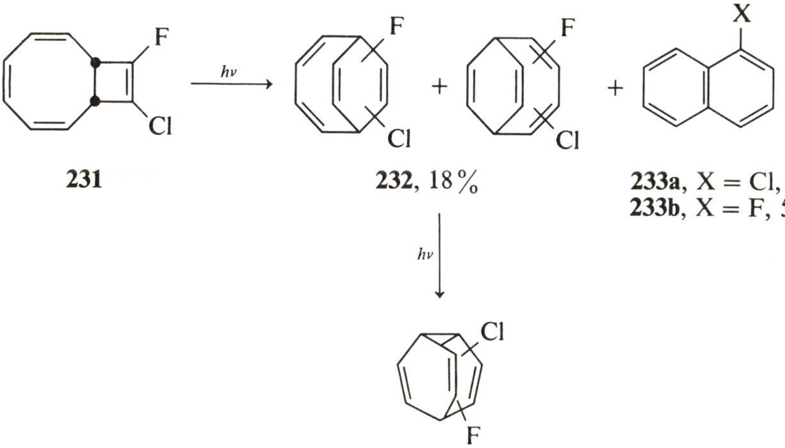

231   **232**, 18%   **233a**, X = Cl, 52%
                          233b, X = F, 5%

234

Low-temperature photolysis of **235** affords **236** and **237** as primary photoproducts [164, 165]:

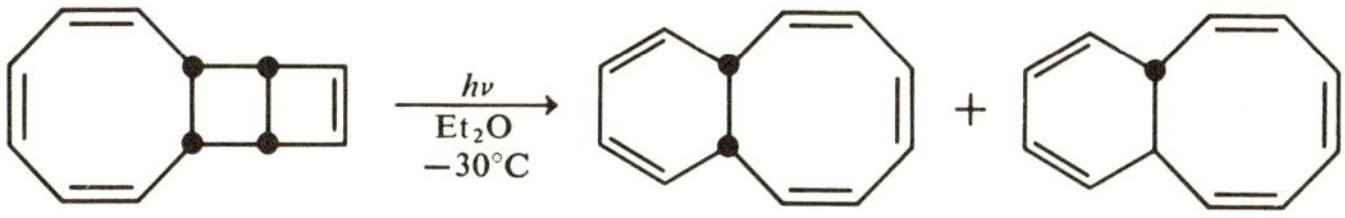

235   **236**   **237**

Sensitized and direct irradiation of all-*cis*-cyclononatetraene affords *cis*-fused bicyclo[6.1.0]nona-2,4,6-triene [166]. Cyclononatetraenide ion demonstrates enhanced basicity on irradiation [156] and affords photoproducts identical to those obtained from *cis*-fused **225** or triene **223**.

E. Heterocyclic Dienes and Related Compounds

Photoinduced rearrangement of *N*-benzylpyrroles is found to be consistent with a direct 1,3-migration and not through a 2H-pyrrole (volume 1, page 41). Thus, photolysis of optically active pyrrole **238** yields **239** and **240** with 54% optical activity:

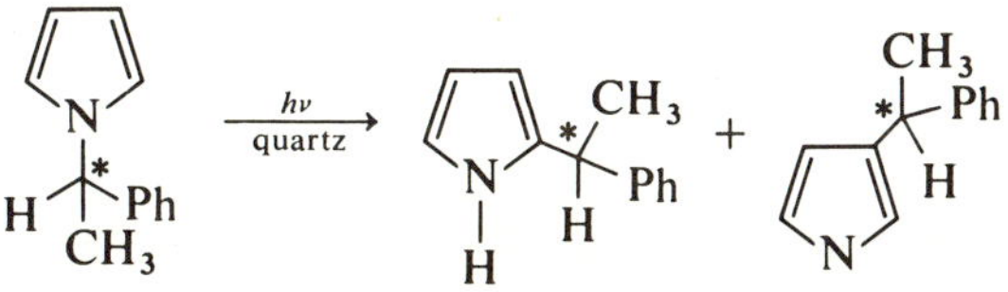

238   **239**, 12%   **240**, 3%

In addition, **238** was not racemized [167].

Irradiation of thiophene in the presence of butadiene affords a 1:1 adduct [168]. On the other hand, irradiation of thiophenes in the presence of primary alkyl amines yields *N*-alkylpyrroles [169].

Mercury complex **241** undergoes photolysis in benzene to 3-phenylthiophene, perhaps by way of intermediate **242** [169a]:

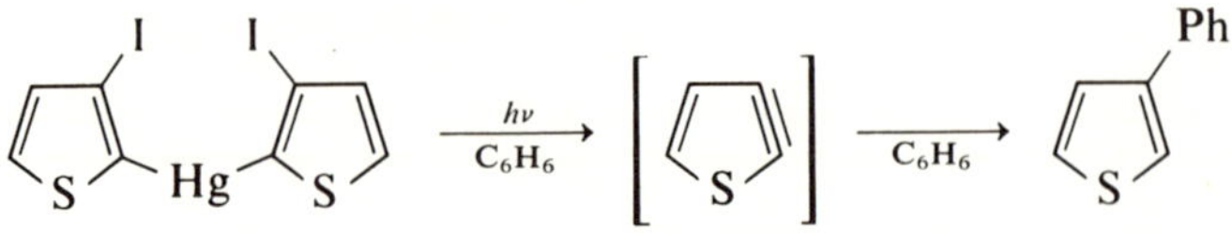

241 **242**

Sensitized photolysis of the furan **243** results in the formation of lactone **244** in addition to several minor products [170]:

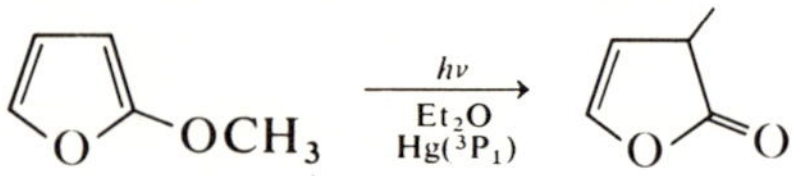

243 **244**, 10–20%

Annulene **245** undergoes photorearrangement to phenanthrene and phenol [171]:

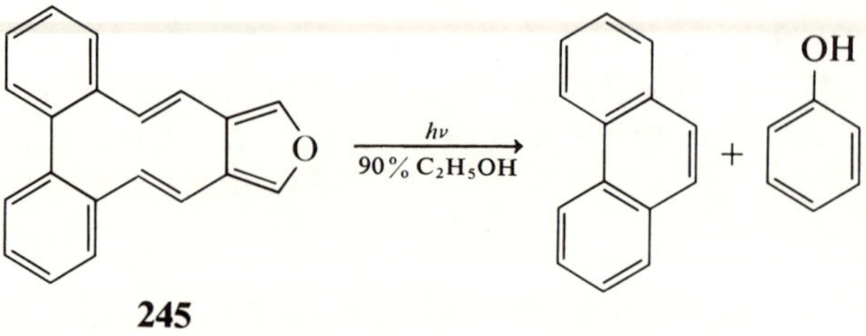

245

Photolysis of dihydrofuran **246** yields cycloheptatriene:

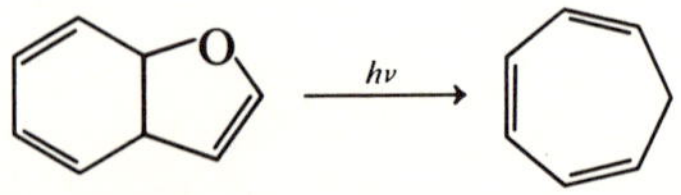

246

and a similar irradiation of oxonin **247** yields exclusively cyclooctatetraene oxide [172, 67]:

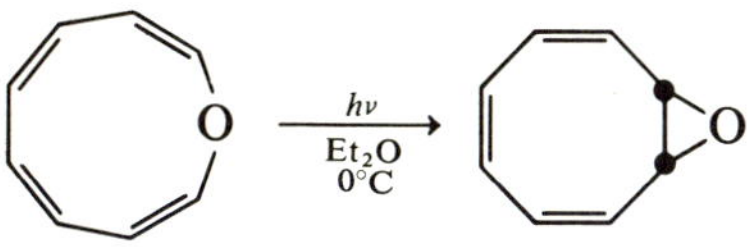

247

A similar result has been reported for N-carbethoxylazonin [161].

Irradiation of methanolic solutions of triene **248** yields valence tautomers **249** and **250** [173]:

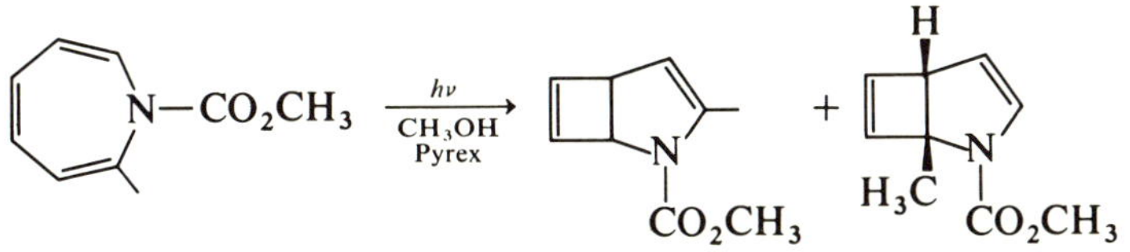

248 **249** **250**

Similar results were obtained for the 3- and 4-methyl derivatives.

CHAPTER TWO: AROMATIC HYDROCARBONS AND DERIVATIVES

I. PHOTOCHEMISTRY OF BENZENE AND RELATED AROMATIC COMPOUNDS

A. Benzene and Its Simple Derivatives

1. BENZENE AND ALKYL BENZENES

A review of the photovalence isomerization reactions of benzenes [174] and a further study of photosensitization by benzene [175] have appeared. It has been suggested that the Woodward-Hoffmann rules do not apply to photochemical-concerted cycloadditions to benzenes [176]. The intramolecular 1,3-cycloaddition of the benzene unit to the C=C group of **1** occurs [177] in a process that seems to proceed with the retention of stereochemistry:

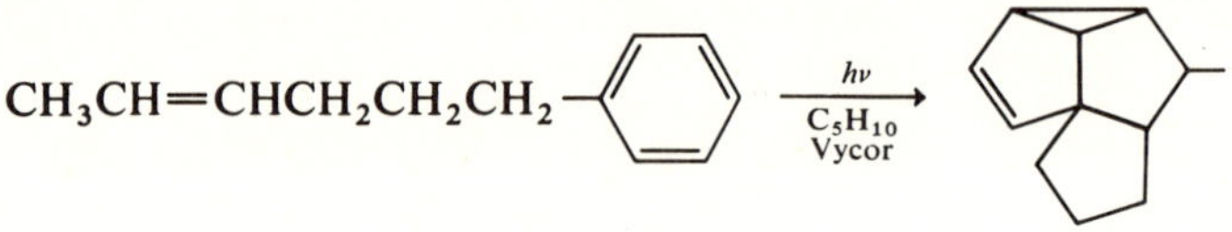

$$CH_3CH=CHCH_2CH_2CH_2-\text{[phenyl]} \xrightarrow[\text{Vycor}]{\substack{h\nu \\ C_5H_{10}}} \text{[product]}$$

1   **2**

Photolyses of degassed benzene solutions lead to low yields of linear polyenes and dimers [178].

Irradiation of **3** results in a reverse Diels-Alder reaction to form anthracene **4** and styrene [179]:

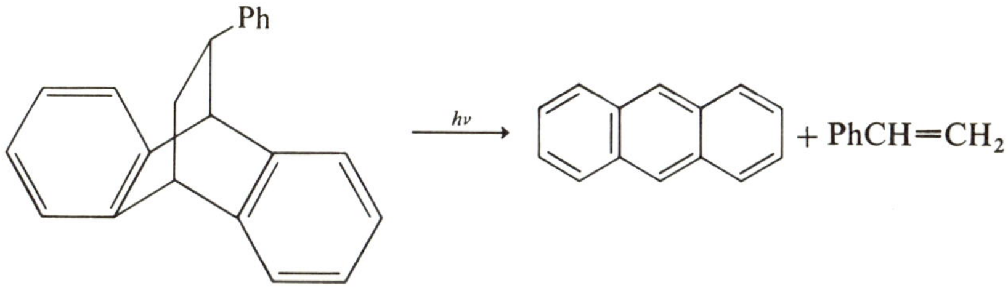

The [2.2]metacyclophane **5** is photocyclized to **6** and dehydrogenated derivatives of the latter [180]:

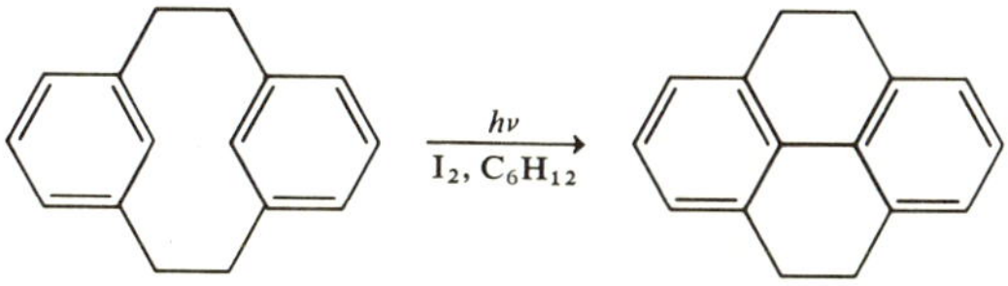

A similar technique has recently been reported in the synthesis of the indole skeleton [181].

A number of interesting rearrangements of benzene-containing, nonconjugated C=C groups have appeared [99, 90, 94, 97, 69] and are discussed in Chapter 1. Irradiation of the optically active paracyclophane **7** results in racemization. A benzvalene intermediate has been suggested to explain this reaction. Irradiation of **7** in acetone results in cleavage to **8** [182]:

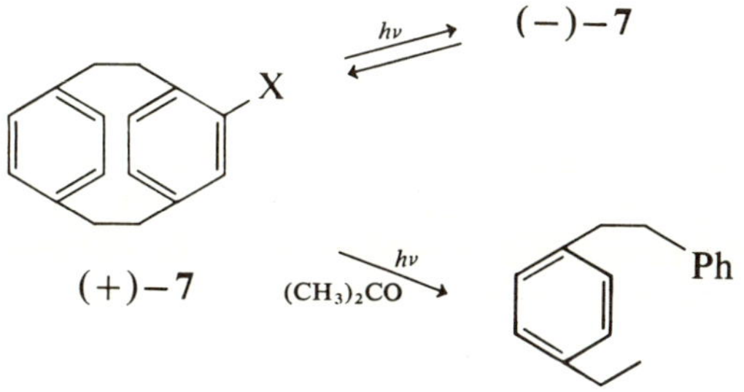

Triptycene **9** was found to undergo a rapid rearrangement (probably triplet) to the norcaradiene **11**, possibly by way of the intermediate **10** [183, 184]:

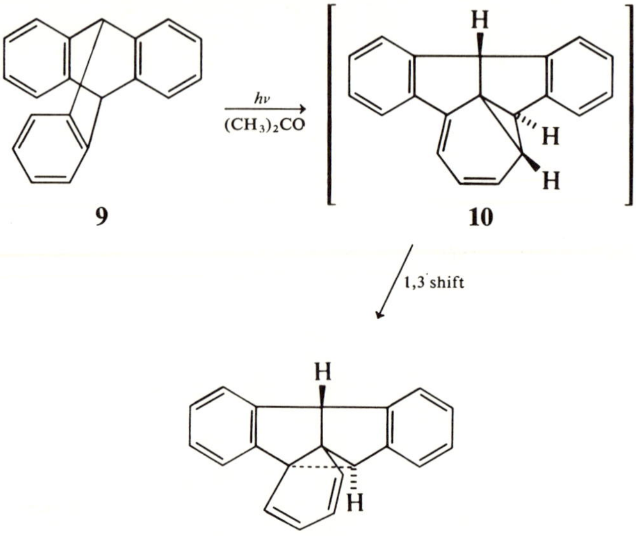

A detailed report of the photoaddition of benzene and dimethyl acetylenedicarboxylate has appeared [185]. The vapor-phase photolysis of benzene and perfluoro-2-butyne results in a 1:1 adduct formation, and in solution, a 2:1 adduct is the predominant product [89]. The photoaddition of benzene to thiomaleic anhydride has been reported (see Chapter 4 for related reactions).

Irradiation of benzene and haloethylenes at room temperature results in the formation of octatetraenes [186]. At low temperature or in viscous solvents, the addition of chloroalkanes to the benzene ring occurs.

Irradiation of alkylbenzenes in the presence of trihaloboranes yields *o*-, *m*- and *p*-substitution products [187]. Several additional reports on the photooxidation of benzenes have appeared [188, 189].

The low-temperature photochemistry of 3,3,4,4-tetraphenyl-1,2-benzocyclobutene has been reported in detail [189a], (Volume 2, page 25).

2. AROMATIC HALIDES

Perfluorobenzene can be photoisomerized to its Dewar isomer (Volume 1, page 49) [190, 191]. It also undergoes photocycloaddition reactions as shown [191]:

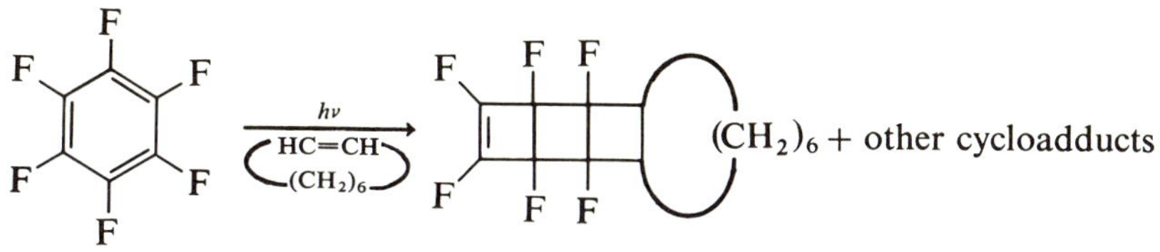

[Scheme showing perfluorobenzene + HC=CH-(CH₂)₆- undergoing hv to give compound **12** (CH₂)₆ + other cycloadducts]

12

The photohalogenation of perfluorostyrene has appeared [192].

Photoarylation by way of aryl halides (Review: Volume 2, page 25, Ref. 113) has continued to receive attention [193, 194, 195, 196]. In addition, a number of examples of the photoreduction of aryl halides have appeared [197, 198, 199, 200, 201]. For example, **13 → 14** [199]:

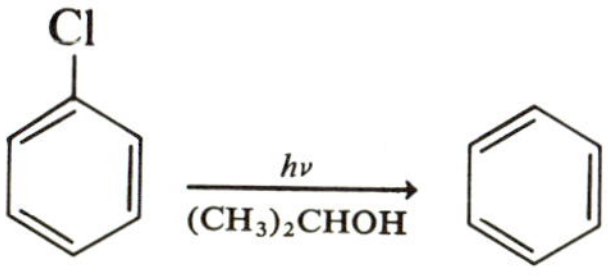

[Scheme: chlorobenzene **13** → benzene **14, 72%** via hv, (CH₃)₂CHOH]

Evidence has been obtained for the intermediacy of an excited charge-transfer complex in the photoreduction of halobenzenes by dimethylaniline [197].

Intramolecular arylation occurs when **15** is irradiated [202]:

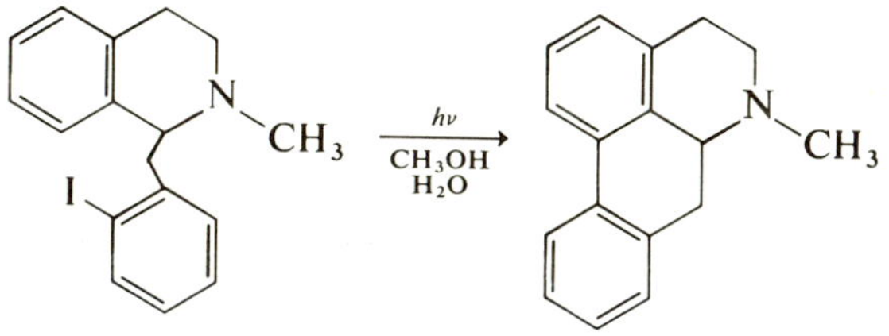

[Scheme: compound **15** → **16, 15%** via hv, CH₃OH, H₂O]

Irradiation of p-chlorophenol in basic aqueous solution (review of photonucleophilic aromatic substitution [203]) yields **18** to **20** [204]:

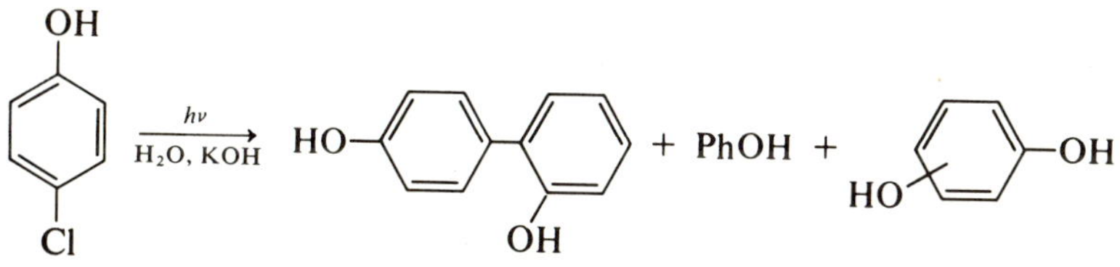

[Scheme: **17** (p-chlorophenol) → **18** + PhOH (**19**) + **20** via hv, H₂O, KOH]

Irradiation of bromobenzene in the presence of borane **21** results in the formation of **22** to **24** [205]:

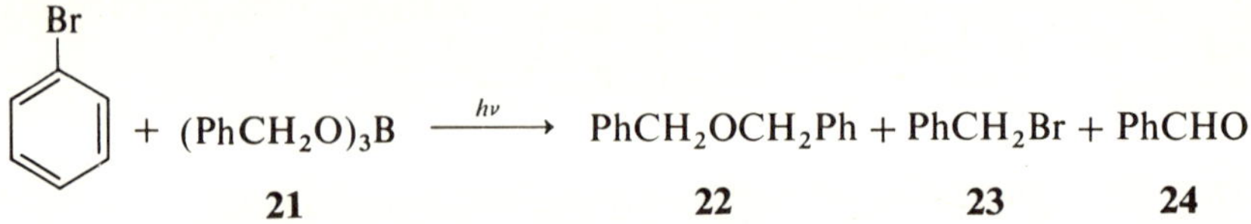

$$\text{PhBr} + (\text{PhCH}_2\text{O})_3\text{B} \xrightarrow{h\nu} \text{PhCH}_2\text{OCH}_2\text{Ph} + \text{PhCH}_2\text{Br} + \text{PhCHO}$$
$$\hspace{2.5cm} \textbf{21} \hspace{3cm} \textbf{22} \hspace{2cm} \textbf{23} \hspace{1.5cm} \textbf{24}$$

Charge-transfer spectra of halogen atoms and the halogenation reactions of a series of aromatics have been studied by a flash spectroscopy technique [206].

3. PHENOLS, ANILINES, AND RELATED COMPOUNDS

Aryl pinacols are cleaved photochemically into ketyl radicals [207]. Several phenyl-substituted oxiranes have been found to add methanol photochemically, for example, **25** → **26** [208]:

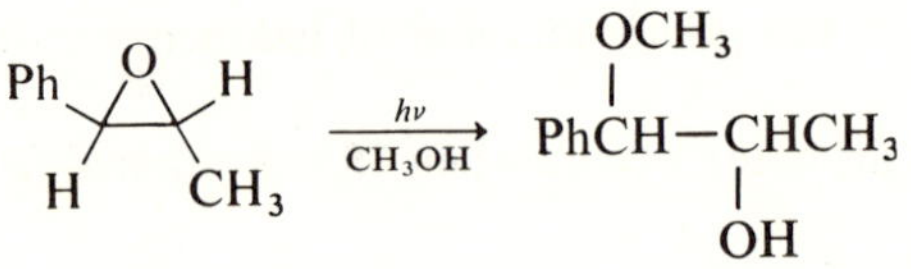

$$\text{Ph-CH(O)CH-CH}_3 \xrightarrow[\text{CH}_3\text{OH}]{h\nu} \text{PhCH(OCH}_3)\text{-CH(OH)CH}_3$$
$$\hspace{2cm} \textbf{25} \hspace{4cm} \textbf{26}$$

On the other hand, aryl carbenes are generated on photolysis of a variety of aryl-substituted oxiranes, carbonates, sulfites, oxalates, phosphoranes, and dioxolanes [209, 210, 211].

Studies on the behavior of acridane [212], diacridane [213], triphenyl amine, diphenylamine, carbazole [214, 215, 216, 217], and anilines [214, 218, 219, 220, 221] have appeared.

Flash photolysis and triplet quenching have been used to study the mechanism of the photocyclization of diphenylamines to carbazoles [219]. Although quenching due to triplet energy transfer was observed when biphenyl was used as the triplet energy acceptor, the diphenylamine (triplet) to carbazole rearrangement was too fast to detect the triplet state of the amine directly.

The fluorescence of methoxybenzene is quenched in aqueous solutions at much higher pH's than required for ground-state molecules, thus confirming the increased basicity of the excited state [222].

Photolysis of aromatic hydrazo compounds has been reported [222a]. Hydrazobenzene, for example, yields aniline and azobenzene on irradiation in *n*-hexane at 246 nm.

Several reports on the photolyses of phenols [196, 223, 224, 224a] and aryl thiols [225, 226, 227] have been published. The structure of photothebainehydroquinone, originally thought to be a dimer [228], has been shown by x-ray study to be **26b** [229]:

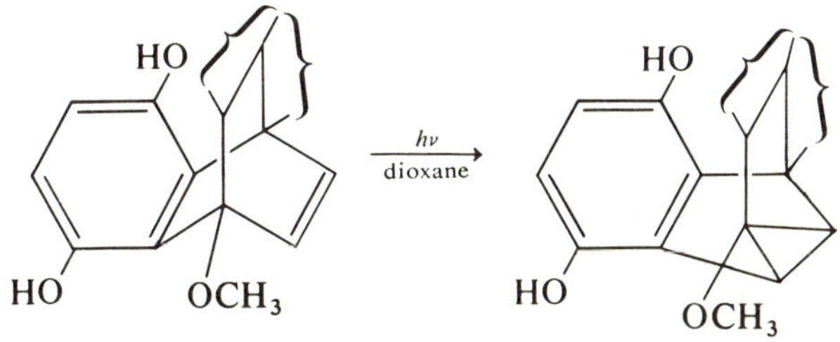

Analogous reactions occur on irradiation of **26c** and **26d**:

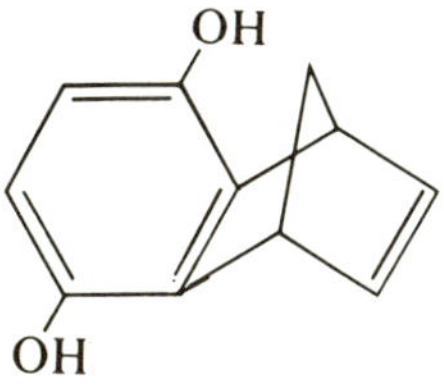

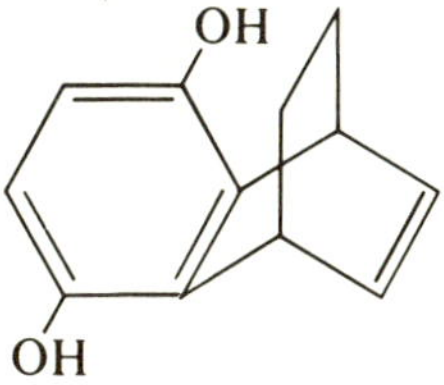

Irradiation of 2-phenoxy-1-phenylethanol in the presence of various carbonyl compounds yields phenol and acetophenone as decomposition products [229a].

Sulfide **27** is reported to yield **28** and **29** when irradiated [230]:

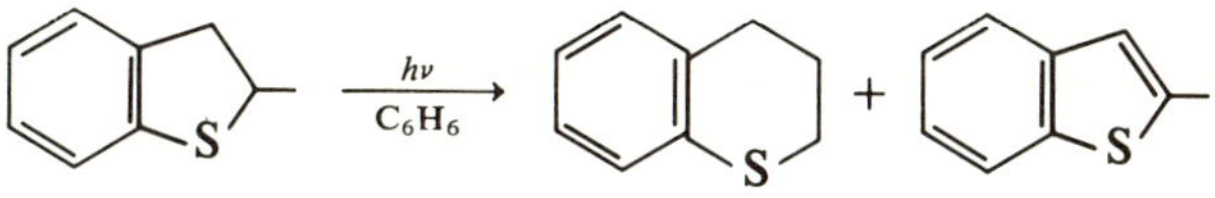

The latter compound is a minor product in hydrocarbon solvents but becomes an important product in chlorinated solvents.

Perfluoroalkyl benzenes undergo smooth photoisomerization, for example, **30 → 31 + 32 + 33** [231, 232]:

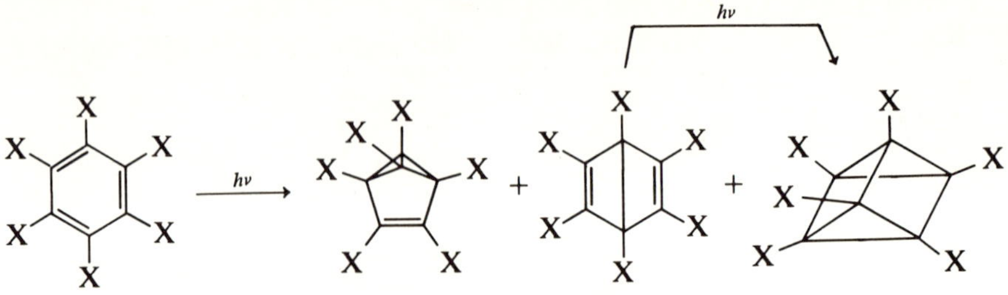

30, X = CF₃   **31**, 43%   **32**, 31%   **33**, 5%

Note, however, **30**(X = C_2F_5) → **33**(X = C_2F_5, 96%) [232].

A review of the nature of electronic excitation and the mechanism of photodecomposition of triphenylmethanes has appeared [233].

Irradiation of benzonitrile and 2,3-dimethylbutadiene is claimed to yield the 1:1 adduct **36** [234]:

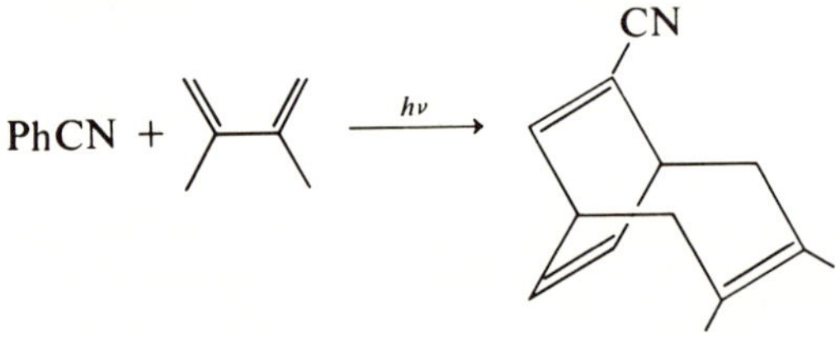

PhCN + (**34**)   (**35**) $\xrightarrow{h\nu}$ (**36**)

4-Cyanophenol is converted to **38** and **39** by irradiation [235]:

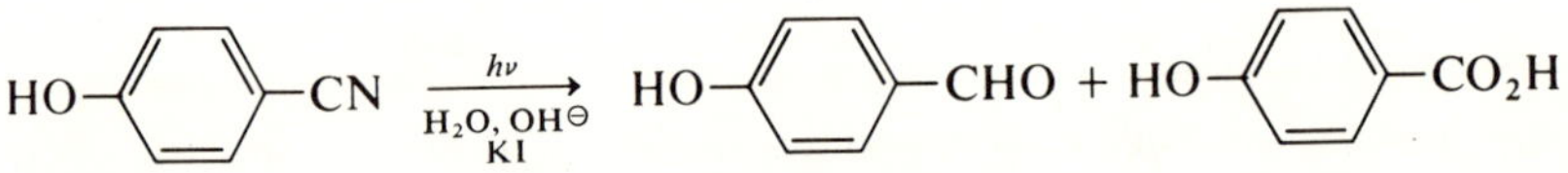

HO–C₆H₄–CN $\xrightarrow[\text{KI}]{h\nu,\; H_2O,\; OH^{\ominus}}$ HO–C₆H₄–CHO + HO–C₆H₄–CO₂H

37   **38**, 80%   **39**

Photolysis of **40** in benzene yields **41** [236] in addition to sulfur-containing products:

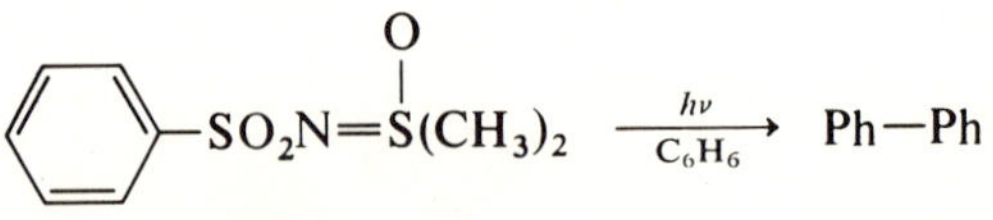

Ph–SO₂N=S(CH₃)₂ $\xrightarrow[C_6H_6]{h\nu}$ Ph–Ph

40   **41**, 74%

Irradiation of triphenylphosphinecarbethoxymethylene in cyclohexene results in the formation of benzene, phenylcyclohexane, bi-2-cyclohexen-1-yl, ethyl cyclohexylacetate, and ethyl 2-cyclohexene-1-acetate [237]. The initial step is believed to involve P—Ph cleavage.

4. BIPHENYL DERIVATIVES

Methylbiphenyls undergo photoisomerization [236]. Further examples of the formation of triphenylenes from halo-o-terphenyls have appeared [238]. Numerous derivatives were found to photocyclize, although a 4-NO_2 group strongly hindered cyclization. With iodocompounds, no oxidant is necessary to effect the cyclodehydrogenation. The photoconductivity spectrum of biphenyl has also been reported [239].

The excited state pK values for fluorene and its 9-phenyl, 9-ethoxycarbonyl, and 9-cyano derivatives in their excited S_1 states appear to be $>10^{22}$ more acidic than the values for S_o. A smaller acidity enhancement has been noted for T_1 relative to S_o [240]:

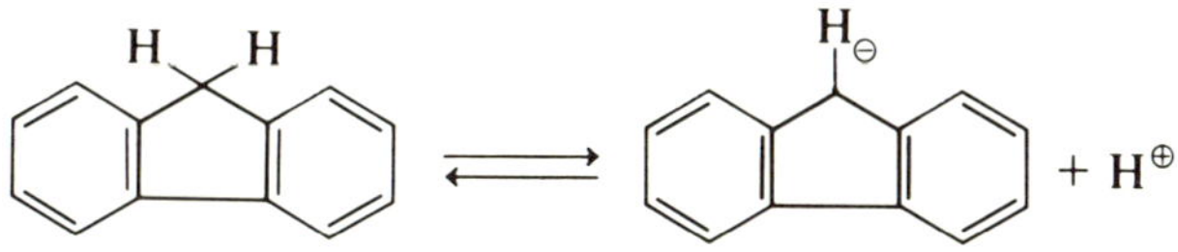

p$K(S_o)$ = 20.5, p$K(S_1)$ = −8.5, p$K(T_1)$ = 5

The spirofluorene to dibenzofulvene rearrangement is reported to be intramolecularly sensitized by a 2-cyclopentenone moiety [241].

B. Conjugated Aromatics

1. STYRENE AND ITS DERIVATIVES

Irradiations of solutions of styrene in trimethylamine result in the formation of adduct **44** [242]:

$$\text{PhCH=CH}_2 + (C_2H_5)_3N \xrightarrow{h\nu} \text{PhCH—CHN}(C_2H_5)_2$$
$$\text{42} \qquad \text{43} \qquad \qquad \overset{|}{\text{CH}_3}\ \overset{|}{\text{CH}_3}$$
$$\qquad\qquad\qquad\qquad\qquad \text{44}$$

The addition of dioxane to substituted styrenes is sensitized by benzophenone [243].

A study of the dimerization of indene [244], photosensitized by different compounds, reveals that the sensitized dimerization is relatively inefficient [245].

Further work on the direct and photosensitized addition of acrylonitrile to indene has also appeared (Volume 1, page 57) [246, 247].

Irradiation of methanolic solutions of cyclic ethylenes (acetophenone as sensitizer) results in addition for ring sizes of 5,6,7, and 8 [79]:

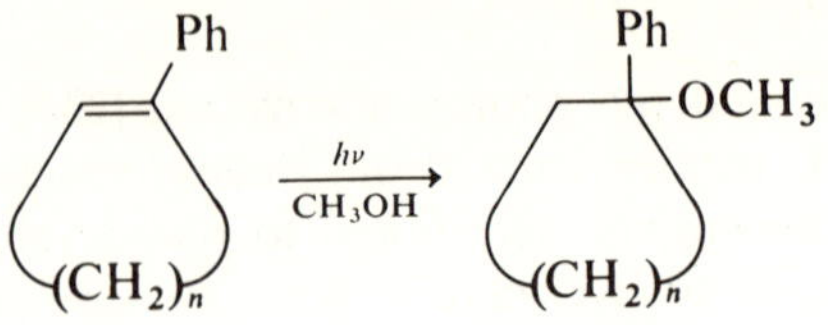

45, $n = 3, 4, 5, 6$ **46**

A similar reaction occurs in acetic acid, and, in addition, a small amount of photoreduction and methylation occurs [248].

Further studies on various aspects of the photochemistry of spiropyrans and chromenes have been published [249, 250, 251, 252, 253, 254, 255, 256]. Excitation of the chromene (for example, **47**) into successively higher vibrational levels decreases the yield of fluorescence from these compounds but increases the extent of photoreaction [250]:

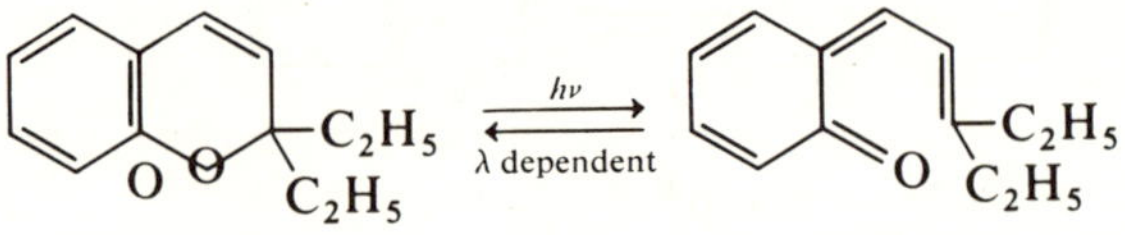

47 **48**

Irradiation of 1,1-diphenylethylene and 2,3-dihydropyran yields the adducts **49** to **51** [257]:

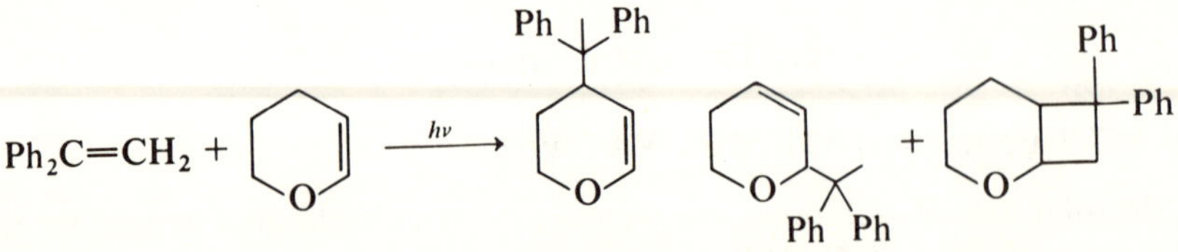

49, 10% **50**, 15% **51**, 30%

1,2-Dihydronaphthalene is photorearranged to **53** [258]:

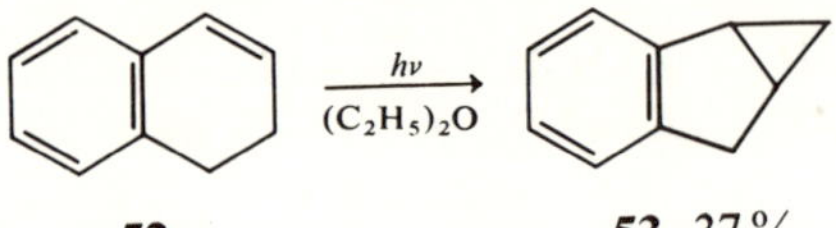

52 **53**, 37%

Lower yields of tetralin and 1,4-dihydronaphthylene are obtained. Separate labelling of the olefinic and of a pair of methylene hydrogen atoms in **52** demonstrated that this photorearrangement proceeds by ring opening and reclosure rather than by hydrogen migration.

On the other hand, irradiation of **54** results in the formation of a number of unexpected products, several of which may derive from an initial rearrangement to **55** [259]:

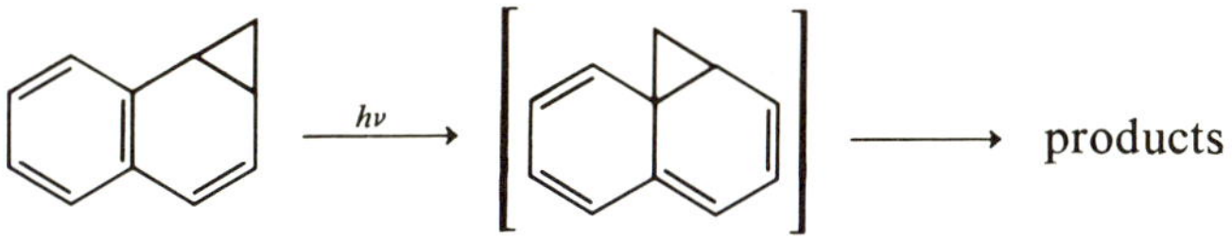

ORGANIC PHOTOCHEMISTRY

$$54 \xrightarrow{h\nu} [55] \longrightarrow \text{products}$$

Derivatives of indoles appear to undergo a self-quenching process that involves excitation energy dissipation through the formation of hydrogen bonds or possibly a proton-transfer mechanism [260, 261].

It has been proposed that indoles undergo phototautomerization [262]. Photoarylation of 3-iodobenzothiophene has been reported [263]. Benzothiophene **56** undergoes photoaddition to acetylenes to generate **58** [264, 265, 266]:

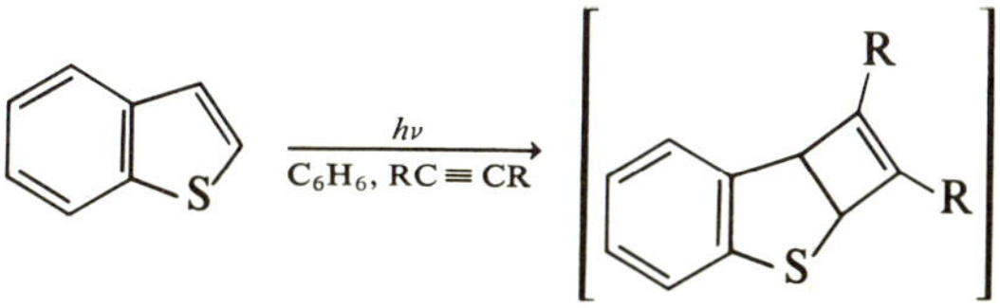

$$56 \xrightarrow[C_6H_6, RC \equiv CR]{h\nu} [57] \xrightarrow{h\nu} 58, 100\%$$

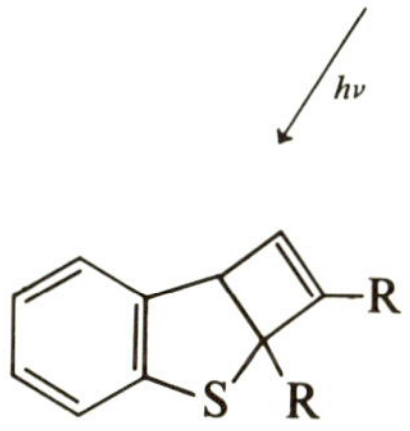

and other heterocyclopentadiene derivatives undergo photodimerization [267] and ring opening in the presence of oxygen [268]. Irradiation of the cyclopentadiene **58a** results in dimerization to **58b** [267, 269]:

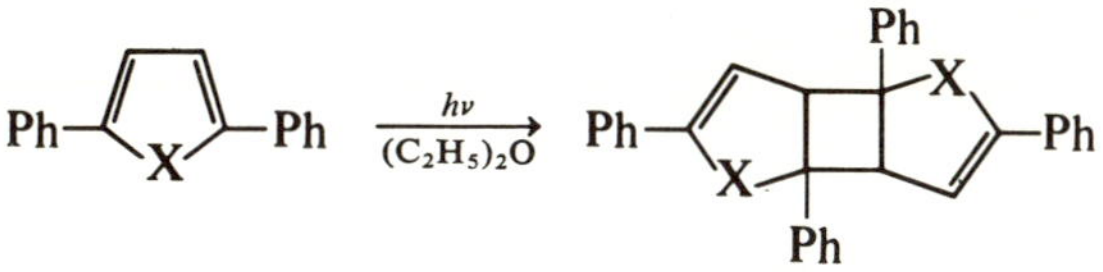

$$\text{58a} \xrightarrow[(C_2H_5)_2O]{h\nu} \text{58b}$$

$$X = (CH_3)_2Si, (CH_2)_2Ge, Ph-P$$

Irradiation of tryptophan **59** in acetic acid yields the cyclized products **60** and **61** [270]:

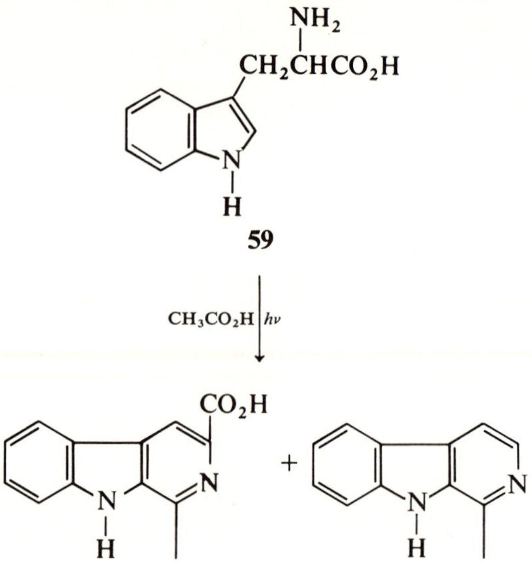

59

CH₃CO₂H | hν

60, 63%

+

61, 16%

Irradiation of DDT **62** in hexane results in the reduction of the C=C bond [271], and in methanol in the presence of oxygen, the C=C bond is oxidized to a carbonyl function with concomitant cyclization to **64** [272]:

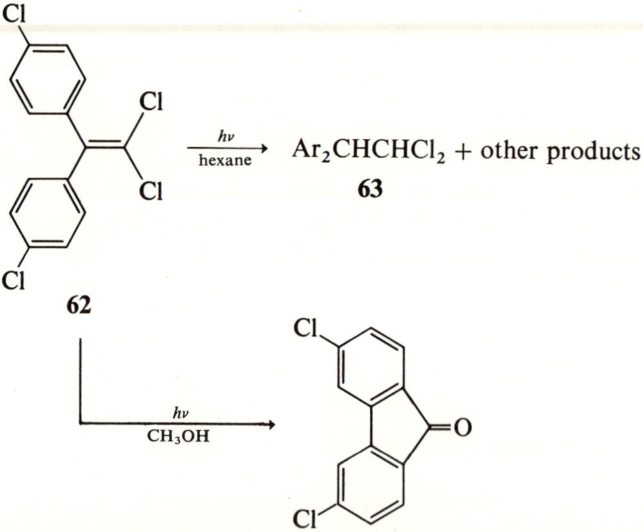

62 $\xrightarrow[\text{hexane}]{h\nu}$ Ar$_2$CHCHCl$_2$ + other products

63

62 $\xrightarrow[\text{CH}_3\text{OH}]{h\nu}$ **64**

The diene **65** undergoes a di-π-methane rearrangement to the cyclopropane **66** by way of a singlet state [273]:

65 **66**

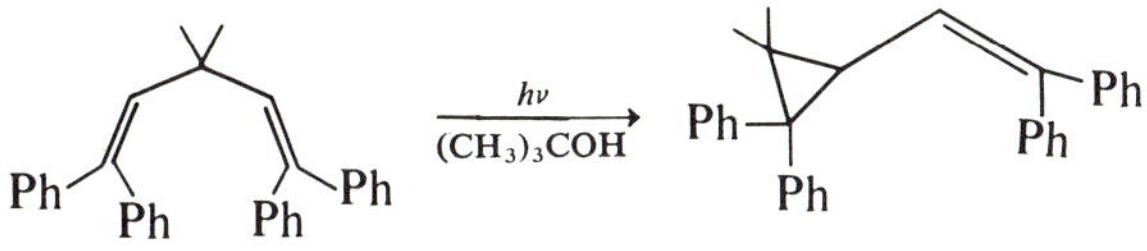

Triplet sensitization does not effect this reaction.

In contrast, the sulfide **67** is photocyclized by direct irradiation to **68** and **69**:

67 **68**, 30% **69**

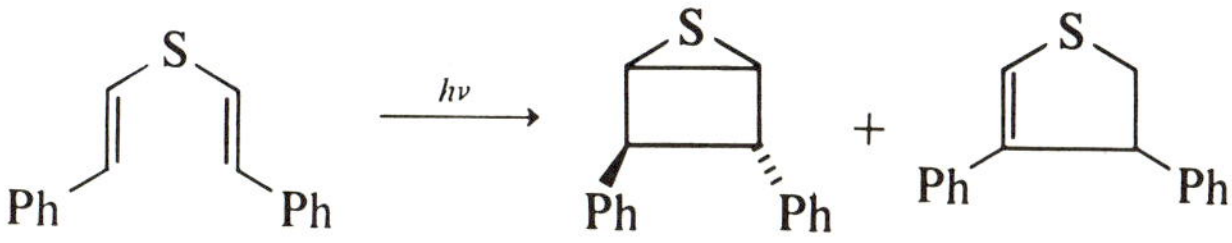

and sensitizers only effect *cis-trans* isomerization [274; see also 274a].

Irradiation of the cyclopentadiene **70** causes decarboxylation and formation of **71** [275]:

70 **71**, 87%

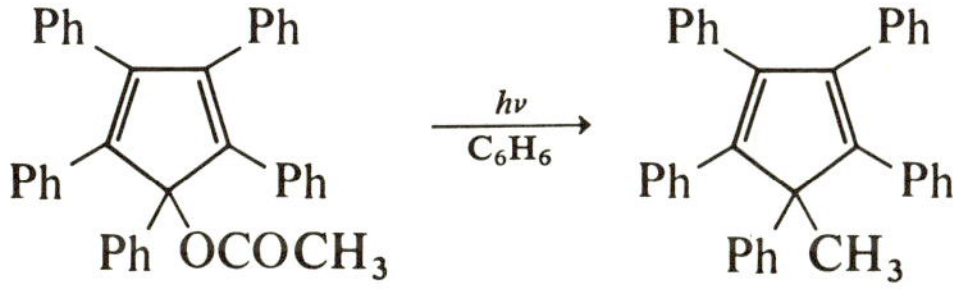

The cyclohexadiene **72** is photoisomerized to **73** [276], as expected from orbital symmetry considerations:

72 **73**

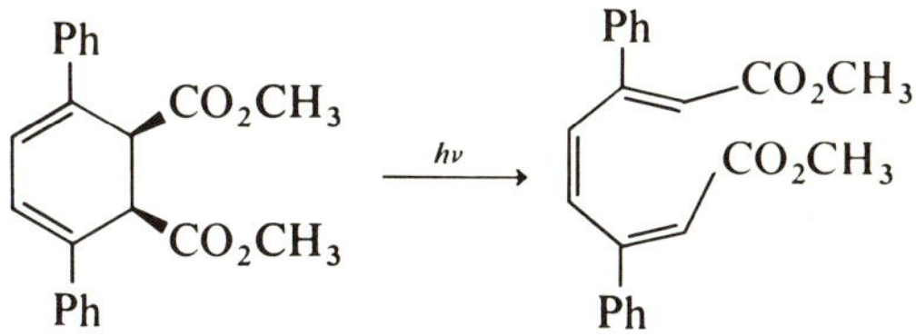

The photo-*cis-trans* isomerization of some 1,4-diaryl-1,3-butadienes has been discussed [277].

The norcaradiene **74** is photolyzed to **75** and photoisomerized to **76** [278]:

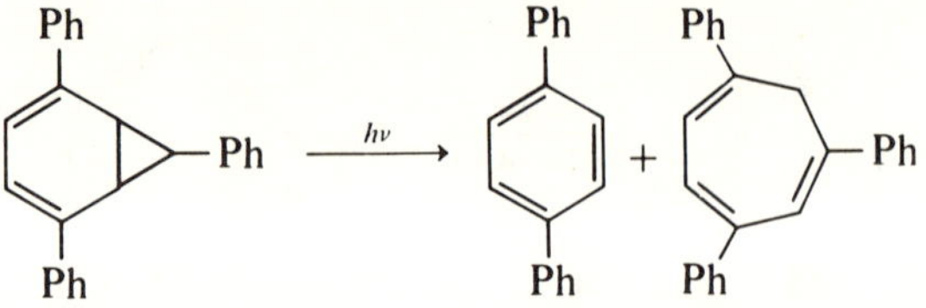

Phenyl iodo acetylene may be photoarylated by benzene [279].

Examples of photorearrangements of 1,2-diyne benzene derivatives have appeared (for example, **77** → **78** + **79**) [280, 281]:

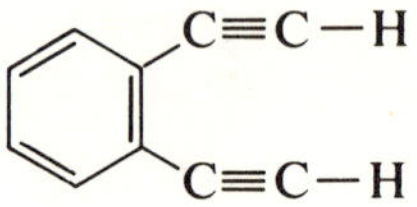

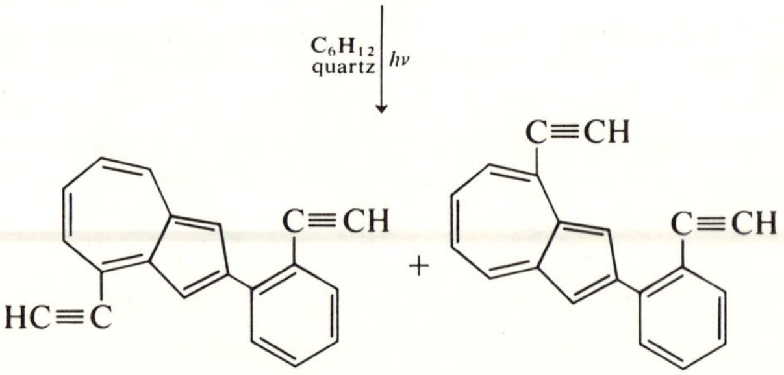

The benzocyclooctatetraene **80** is photoisomerized to **81** [282] (see also Volume 2, page 25):

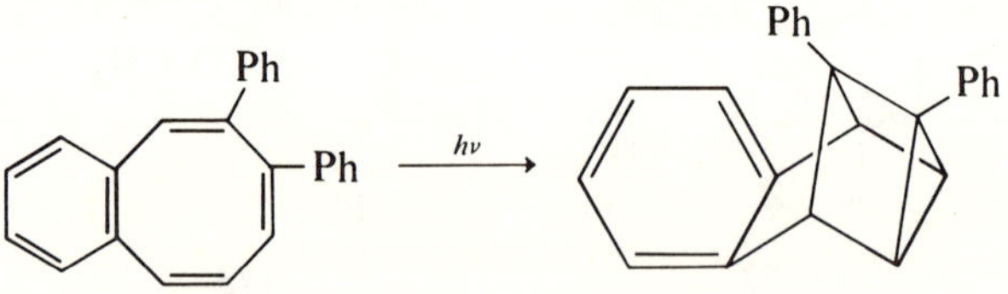

The tetraphenylcyclooctatetraene **82** undergoes photoisomerization to **83** [283]:

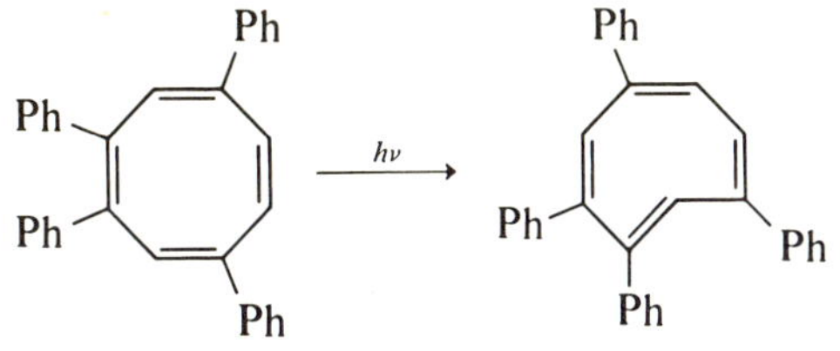

ORGANIC PHOTOCHEMISTRY 57

82 → **83** (hv)

A full paper on the photodimerization of tetrabenzoheptafulvene [284] has appeared.

2. STILBENE DERIVATIVES

Further mechanistic studies [285, 286, 287, 288, 289, 290] on stilbene isomerization have led to divergent conclusions concerning the state(s) involved in the isomerization (see also Chapter 1, Section II.A). One group favors a triplet-state path [291], and another a singlet-state intermediate [292].

The quantum yield of the *cis* to *trans* photoisomerization of stilbenes decreases with increasing heavy-atom substituents [288], but the quantum yield of the reverse *trans* to *cis* reaction is practically not influenced by the heavy-atom effect.

Photoisomerization of stilbene, directly excited by irradiation within the singlet-triplet absorption band, is enhanced by high pressures of oxygen [285] or by methyl iodide [291]. No phenanthrene was observed to form, a result consistent with the proposed singlet mechanism of cyclization (Volume I, page 61). The quantum yields for *cis* to *trans* and *trans* to *cis* isomerizations were 0.22 and 0.42, respectively [285].

The photocyclization [Review: 293] of stilbenes to phenanthrenes [294, 295, 296, 297, 298, 299] continues to find synthetic usefulness [300, 301, 302], including the preparation of a (13)-helicene [303, 302].

Irradiation of stilbene and 2,3-dihydropyran leads to the formation of stilbene dimers and **84** [304] (Volume 2, page 36):

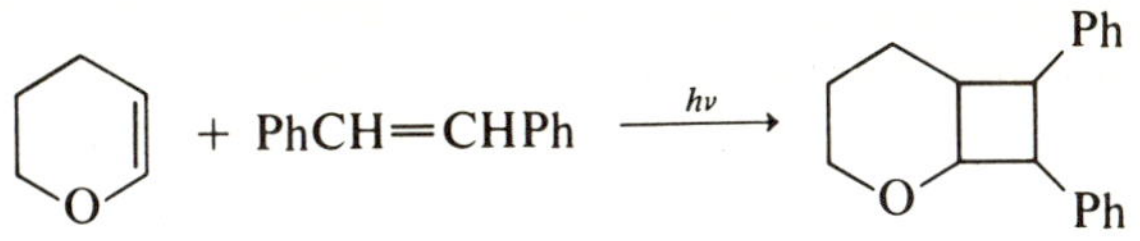

dihydropyran + PhCH=CHPh →(hv) **84**

The photodimerization of 1,2,3-triphenylcyclopropene (Volume 1, page 63) leads to the formation of stereoisomeric tricyclo[3.1.0.0]hexanes [305]. Irradiation of the salt of the tosylhydrazone from 1,2,3-triphenylpropenone yields the same dimers. Dimerization to **87** and rearrangement to **88** may be sensitized by sensitizers having energy greater than 50 kcal/mole:

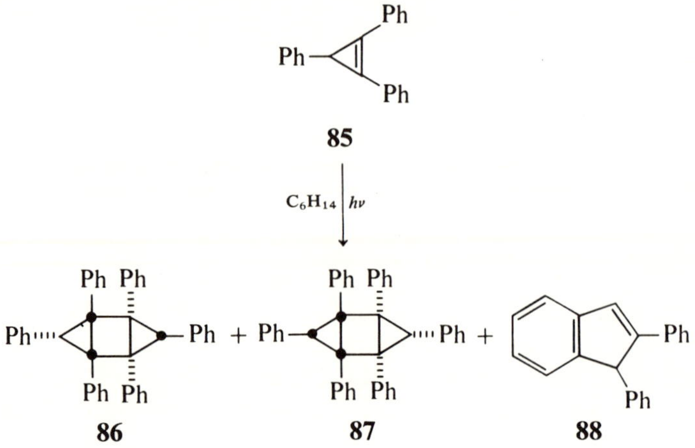

Irradiation of 2,5-distyrylpyrazine and 1,4-bis(β-pyridyl-(2)-vinyl)benzene results in polymerization to linear structures having recurring cyclobutane rings in the main chain [306, 307].

The boron stilbene analog **89** undergoes photocyclization [308]:

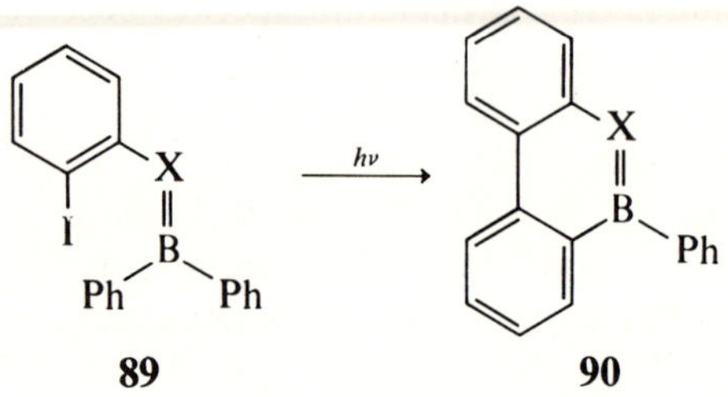

It was found that 1,2-bis(4-pyridyl)ethylene undergoes photoreduction of the C=C bond in cyclohexane [309]. Photohydration of bis methyl iodide salts of 1,2-bis(4-pyridyl)ethylene has also been reported [310]. This reaction is not sensitized, although sensitizer quenching seems to occur. Similarly, bifluorenylidene is photoreduced in hexane/methanol [311, 295, 312]. The sensitized *cis-trans* isomerization of 1-(α-naphthyl)-2-(4-pyridyl)ethylene in metaloporphyrin complexes has been reported [313].

The pK_a values for ground- and excited-state β(2-pyridyl) styrene have also been reported [314].

II. POLYNUCLEAR AROMATIC COMPOUNDS

A. Naphthalene Derivatives

The stereochemistry of the 2-methoxynaphthalene photodimer (Volume 2, page 36) has finally been unambiguously determined by x-ray analysis as the *trans* (centrosymmetric) 1,4-1',4'-adduct **91** [315]:

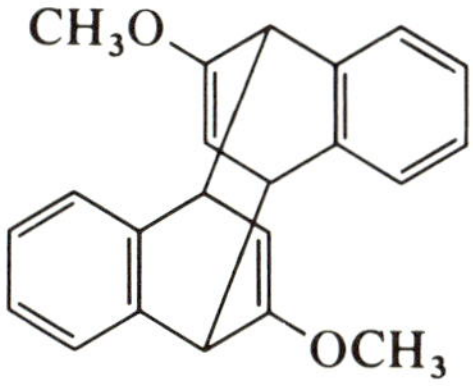

91

Irradiation of a solution of 2-methoxynaphthalene in methanol/water, however, yields, on the basis of spectral analysis and chemical degradation, compound **93** [316]:

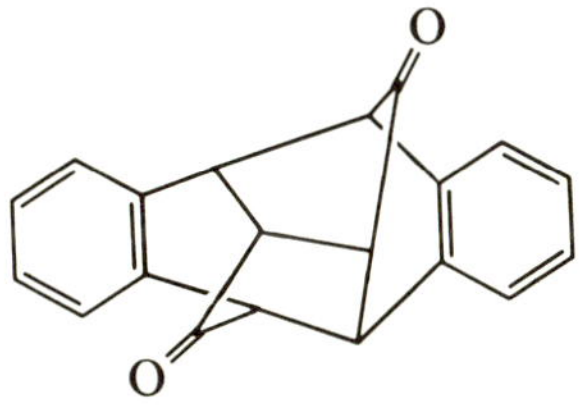

93

which corresponds to a dimerization to **92** followed by hydrolysis and oxidative 3,3 bond formation:

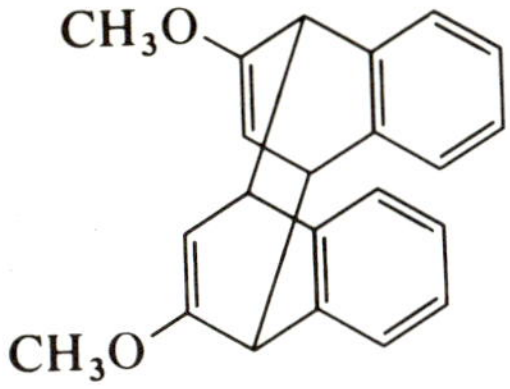

92

Thus, one must conclude that either the dimerization follows a different path in methanol/water from that in benzene or the structure **93** is incorrectly assigned.

The intersystem crossing efficiencies of a number of aromatic hydrocarbons have been determined by the measurement of sensitized biacetyl phosphorescence in cyclohexane, ethanol, and benzene [317].

The intermediacy of exciplexes has been proposed in the dimerization of 2-alkoxynaphthalenes [318] and the photocycloaddition of diphenylacetylene to naphthalene [282] (Volume 2, page 37). The latter reaction has been studied as a function of concentration, wavelength of exciting light, and temperature [282]. The fluorescence quenching of naphthalenes by quadricyclene is proposed to proceed by way of a charge-transfer complex rather than by transfer of vibrational energy [319]. The photoreduction of 1- and 2-nitronaphthalene (isopropanol) in the presence of H^+ and Cl^- has been investigated [320]. The optical spin polarization of the triplet-state naphthalene has recently been reported [321].

Irradiation of the *anti*(2.2)paracyclophane **94** results in conversion to the *syn*-isomer **95** [322]:

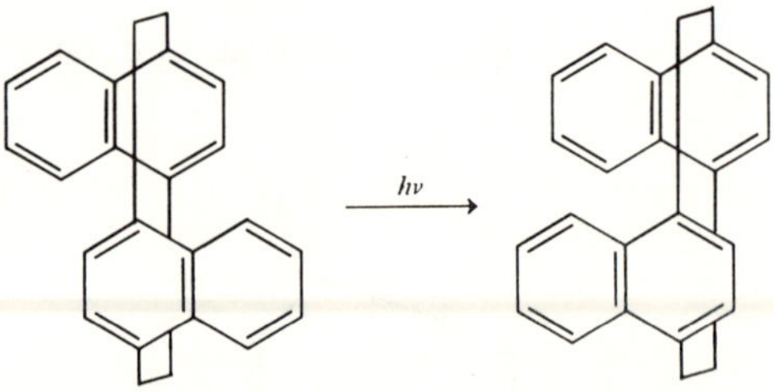

2,3-Naphthobarrelene **99** and 1,2-naphthobarrelene **96** undergo direct and photosensitized rearrangements to semibullvalenes [96]. In the case of **96**, labeling experiments indicate that α-naphthyl-vinyl bridging initiates the rearrangement:

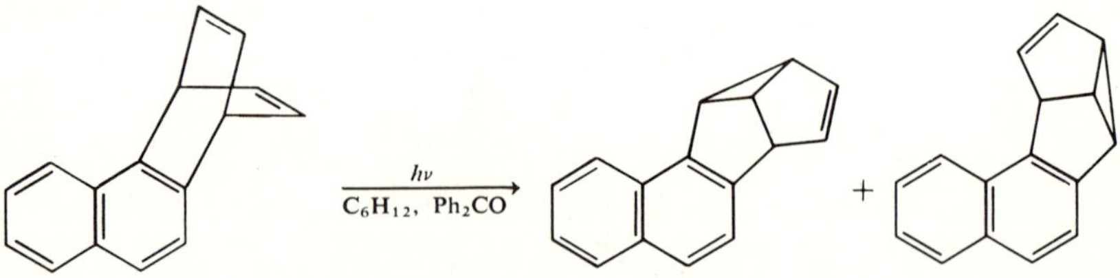

in contrast to **99** which rearranges by way of vinyl-vinyl bonding:

99 **100**

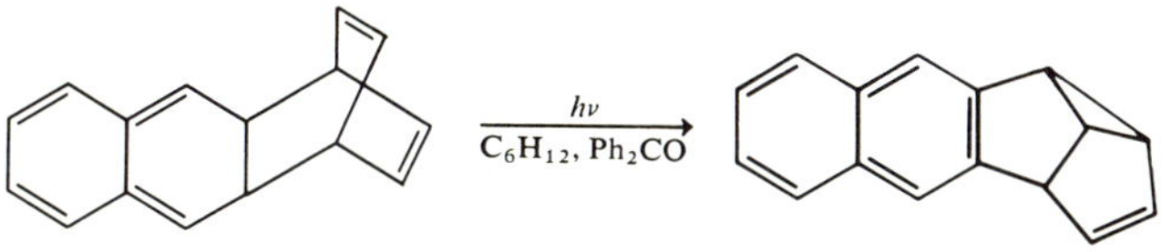

It is suggested that the latter reaction may be guided by the reluctance of the β-naphthyl position to involve itself in bridging.

Irradiation of a solution of **102** in molten naphthalene yields adducts **103** to **106** [185]:

+ $CH_3O_2CC{\equiv}CCO_2CH_3$

101 **102**

90° | hν ↓

103 **104**

105 **106**

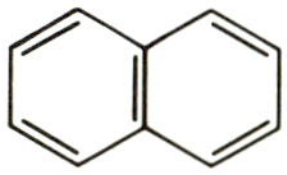

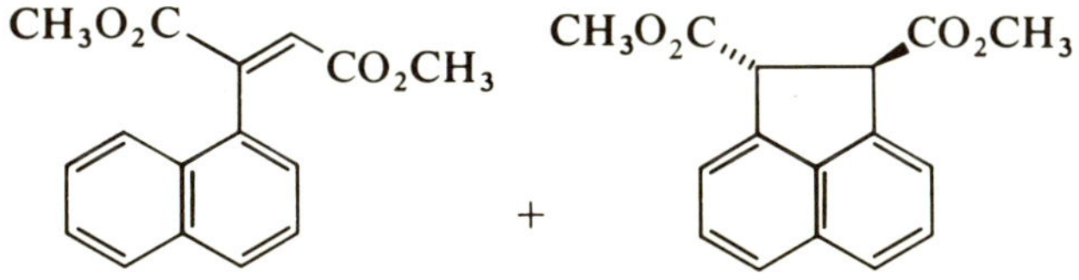

It was shown that the irradiation of **103** leads to **104**.

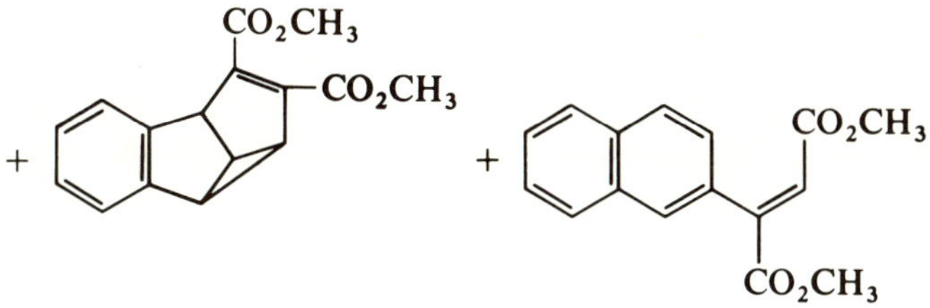

A detailed report of the photoaddition of naphthalene to acrylonitrile (Volume 2, page 37) has appeared [246; see also 323].

Reports have appeared concerning photocyclizations of halostyrylnaphthalenes [324, 303], a mechanistic study of the direct and photosensitized isomerization of β-styrylnaphthalene [325], photolytic rearrangement and cyclization of halophenyl naphthyl ethers (for example, 107 → 108) [326]:

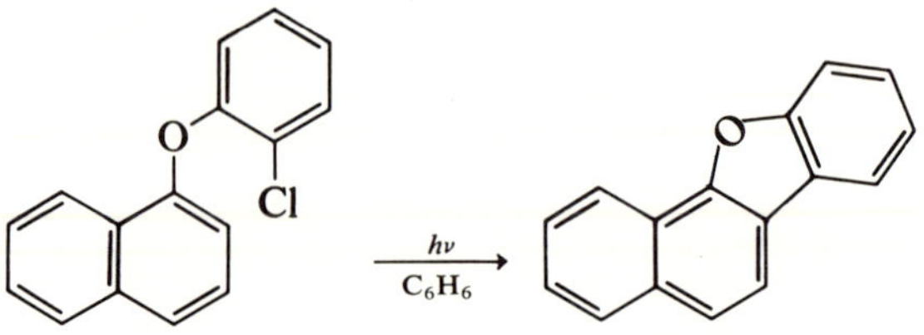

$$107 \xrightarrow[C_6H_6]{h\nu} 108, 79\%$$

and halophenylnaphthalenes (for example, 109 → 110) [327]:

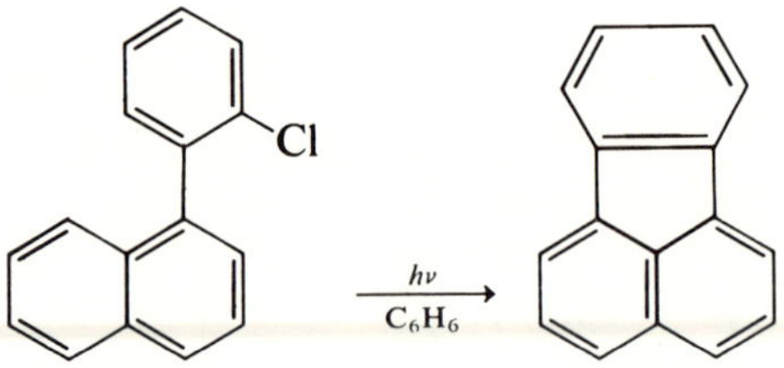

$$109 \xrightarrow[C_6H_6]{h\nu} 110, 59\%$$

and photonucleophilic substitution of a methoxynaphthalene [328].

The photoring closures of a series of *trans*-1,2-dinaphthylethylenes have been reported [329]. Ring closure was specific with respect to possible rotational conformers.

The photodimerization of acenaphthylene [330] and haloacenaphthylenes [331] and the photodecomposition of the *cis* dimer of acenaphthalene [332] have been studied. Irradiation of acetonitrile solutions of acenaphthylene in the presence of tertiary amines yields acenaphthene, in addition to the *cis*-acenaphthalene photodimer [333].

The photoconversion of sulphoxide **111** into the sulphines **112** and **113** proceeds by way of a triplet-state reaction [334, 335]. The sulphines, in turn, are photolyzed, in a singlet-state reaction, to the ketone **114** (Volume 2, page 38):

111 **112** **113** **114**

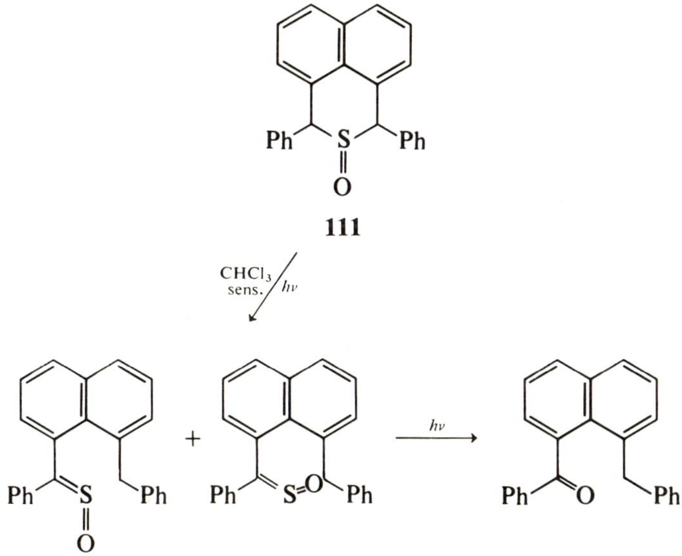

Irradiation of 1,2-divinylnaphthalenes causes rearrangements to cyclized structures [336], as with 1,3-diynyl [337] and 1,3-divinylnaphthalenes [336]:

115 **116** **117**

118 **119** **120**

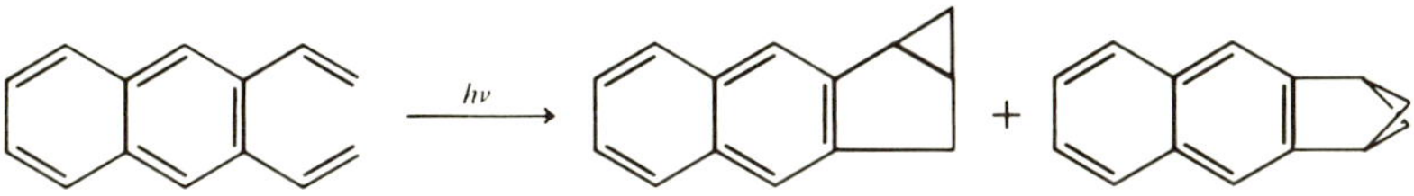

Electronic energy transfer in copolymers of styrene and 1-vinylnaphthalene has been reported [338].

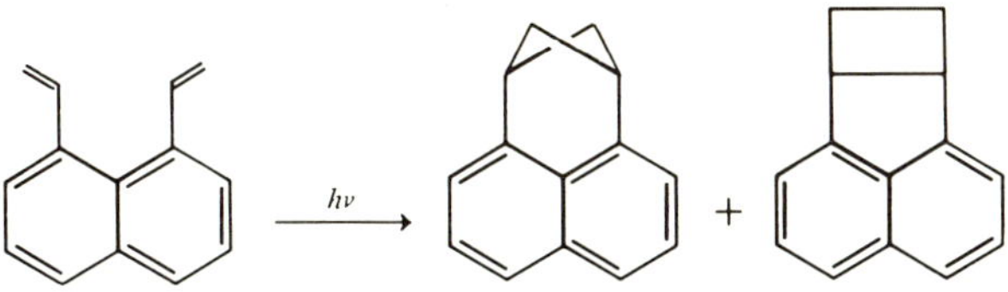

A study of the photochemical substitution of deuterium for hydrogen in 1-methylnaphthalene, 2,6-dimethylnaphthalene, and anthracene has appeared [339].

B. Polycondensed Aromatic Compounds

Further examples of the dimerization of 9-substituted anthracenes show that head-to-tail dimerization is generally favored [340, 341, 342, 343, 344]. Photocross additions of different 9,10-disubstituted anthracenes can be made to compete with dimerization [345, 346]. The photoaddition of anthracenes to benzo(a)anthracene has been observed [347]. Anthracene is photoreduced by dimethyl aniline [333, 348]. The photorearrangements of substituted barrelenes and norbornadienes have been reported to be sensitized by T_2 of various anthracenes [90, 89] (see also Chapter 1, Sections II.B.1 and II.B.2, and Ref. 349). Ultraviolet and fluorescence spectra of a number of styryl substituted anthracenes have appeared [350].

Stilbene-to-phenanthrene photocyclizations have been used to synthesize helicenes [351, 302, 303] (for example, **121** → **122**) [302]:

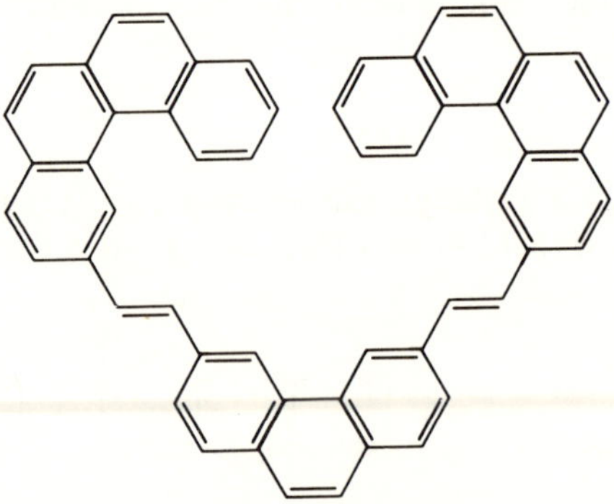

121

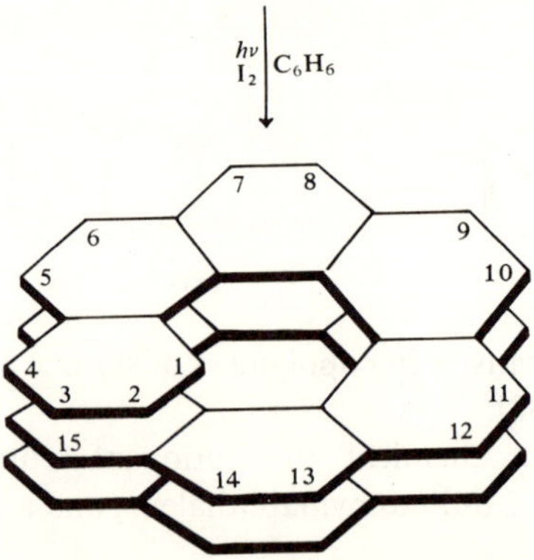

$\xrightarrow{h\nu, I_2}$ C_6H_6

122

and **123** → **124** [303] and benzoannulenes [352]:

123 **124**

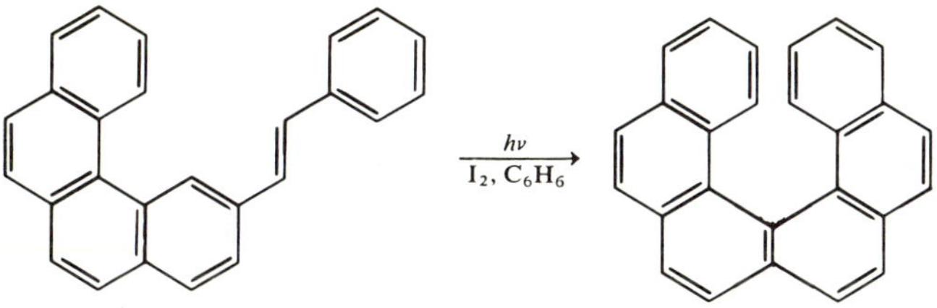

Diphenyl acetylene undergoes cycloaddition to the 9,10 positions of phenanthrenes, for example, **125** → **127** [353]:

125 **126** **127**

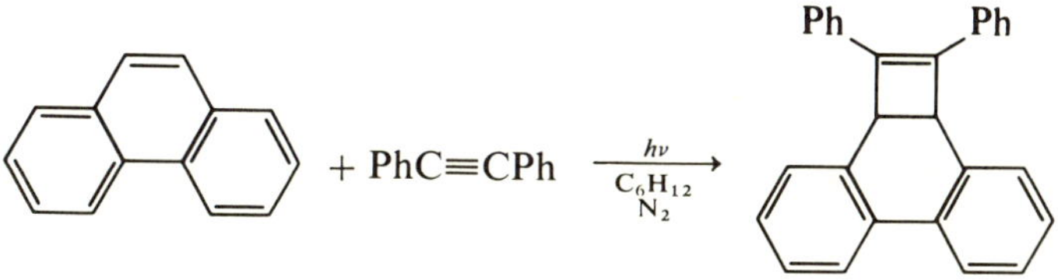

In the case of **128**, a cyclobutane dimer **130** was also isolated:

128

129

130

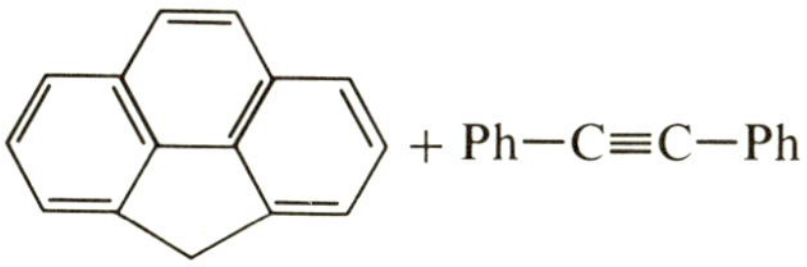

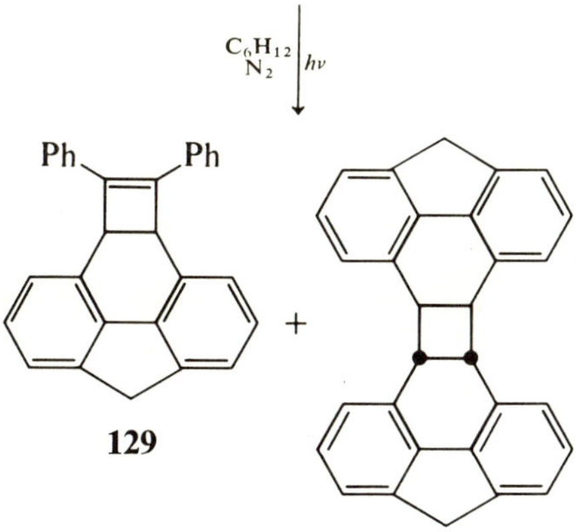

Irradiation of azulene vapor results in a very low yield of naphthalene [354].

The electronic absorption and fluorescence of phenylethynyl-substituted acenes have been reported [355].

Several spectroscopic studies of pyrene and substituted pyrenes have appeared [356, 357, 358, 359, 360]. The fluorescence spectrum of coronene has also been reported [361].

CHAPTER THREE: CARBONYL COMPOUNDS

I. ALKYL CARBONYL COMPOUNDS

A. Aldehydes and Ketones

1. ALKYL ALDEHYDES

Several reviews of various aspects of the photochemistry of carbonyl compounds have appeared [362, 363, 364, 365].

The photoadditions of propionaldehyde to ethyl vinyl ether [399, 401] have been reported [365a]. The predominance of the 3 isomer is consistent with preferred Markownikoff addition to the most electron-rich carbon of the double bond by the electron-poor oxygen atom of the n, π^* excited state (see below, Sections I.2, I.4, II.1, and II.2 for further examples):

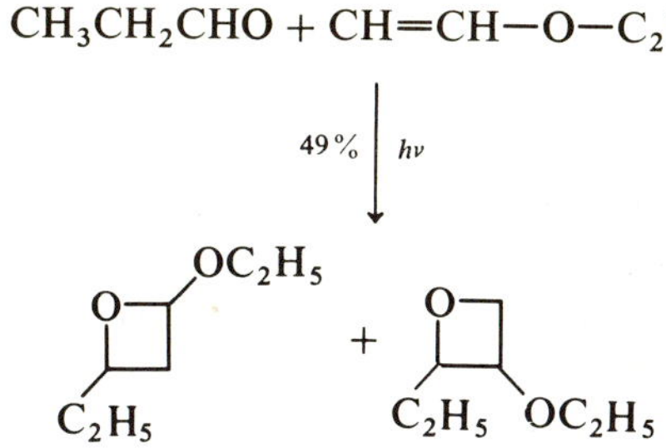

$$CH_3CH_2CHO + CH=CH-O-C_2H_5$$

$$\downarrow 49\% \; h\nu$$

[oxetane with OC$_2$H$_5$ and C$_2$H$_5$] + [oxetane with C$_2$H$_5$ and OC$_2$H$_5$]

1, 16%   **2, 84%**

Irradiation of propionaldehyde in the presence of 1,3-cyclohexadiene gives oxetanes **3** and **4**, in addition to dimers of the diene:

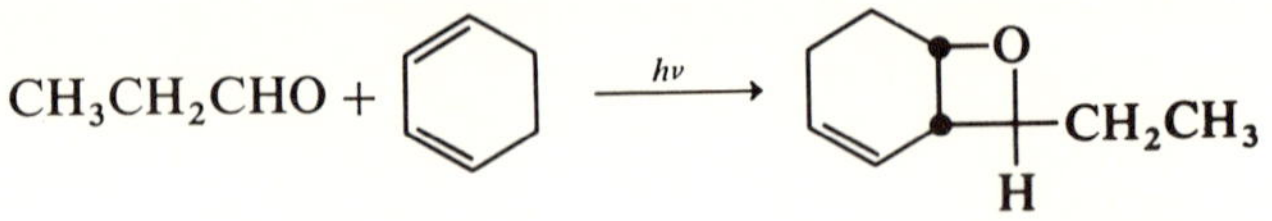

$$CH_3CH_2CHO + \underset{}{\bigcirc} \xrightarrow{h\nu} \underset{H}{\overset{O}{\bigcirc\!\!\!\!\square}}\!\!-CH_2CH_3$$

3, trans, 80%
4, cis, 20%

1,3-Pentadiene, isoprene, 2,3-dimethylbutadiene, and 1,3-cyclopentadiene also gave oxetanes [366]. Stern-Volmer plots of the quantum yields for (1) oxetane formation and (2) dimers as a function of 1,3-cyclohexadiene concentration reveal that the dimer yield decreases as the diene concentration increases; the oxetane yield increases linearly as the diene concentration increases. Thus, different precursors probably exist for the formation of oxetane and dimer. Since diene dimers almost certainly are formed by way of diene triplets, it seems that propionaldehyde singlets add to ground-state diene molecules. This interpretation is also consistent with the observations that phenanthrene does not sensitize oxetane formation and that the fluorescence of propionaldehyde is quenched by 1,3-cyclohexadiene (see below, Section I.A.2). Intramolecular photocycloadditions of aldehydes to 1,3 dienes, yielding oxetanes, have also been observed [367].

It seems that β,γ-unsaturated aldehydes undergo efficient photodecarbonylation [368, 368a]. A particularly striking case is the optically active aldehyde **5**, which is shown to photolyze to **6** and **7** as well as a trace of the geometric isomer **8**:

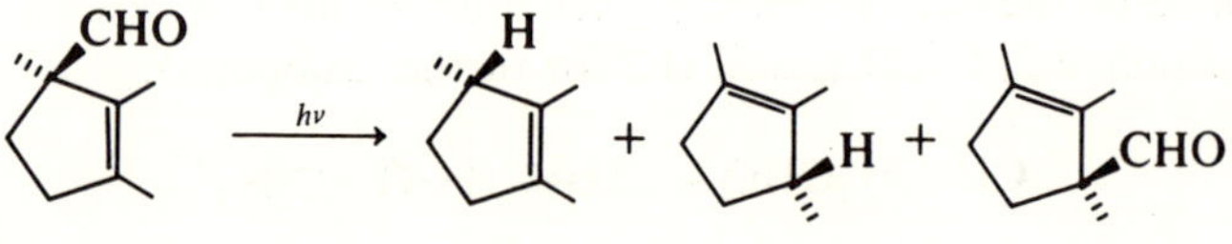

$$\underset{5}{\text{CHO}} \xrightarrow{h\nu} \underset{6}{\text{H}} + \underset{7}{\text{H}} + \underset{8}{\text{CHO}}$$

The decarbonylation was shown to be unimolecular and to occur with partial racemization. This rearrangement appears to be a singlet reaction. It was found that tri-*n*-butylstannane is capable of partially photoreducing **5** to the corresponding alcohol **9**, in competition with photodecarbonylation:

CHOH

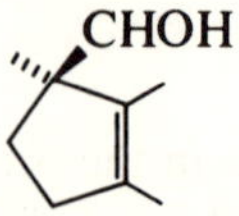

9

Irradiation of a dilute acetone solution of 5-hexenal **10** results in the formation of 2-methylcyclopentanone **11** and cyclohexanone **12**:

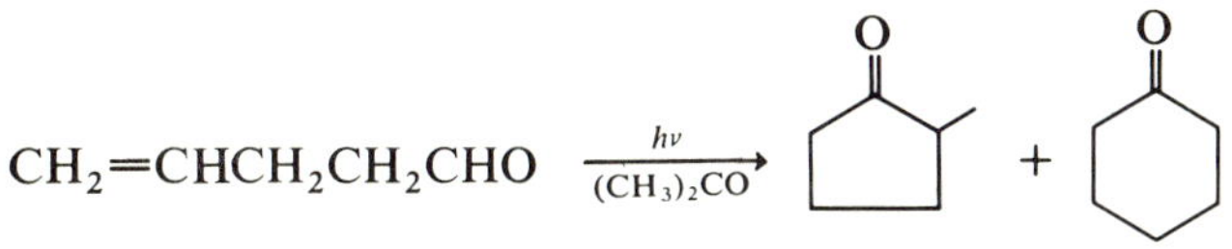

$$CH_2=CHCH_2CH_2CHO \xrightarrow[(CH_3)_2CO]{hv}$$

10   **11, 4%** + **12, 2%**

although neither product was observed on irradiation of **10** in *n*-pentane [369]. Acetone sensitization of this reaction could involve either energy transfer or chemical sensitization. The cyclization of **10** provides an attractive explanation for the unexpected formation of 2-methylcyclopentanone in low yield on irradiation of cyclohexanone. It was originally suggested that this ring contraction reaction represented a new primary photochemical process of cyclic ketones [370].

In contrast to the decarbonylations, which are characteristic of β,γ-unsaturated aldehydes, γ,δ-unsaturated aldehydes have been found [370a] to undergo intramolecular cyclization [371, 372] to yield polycyclic oxetanes (for example, **13 → 14** and **15 → 16**):

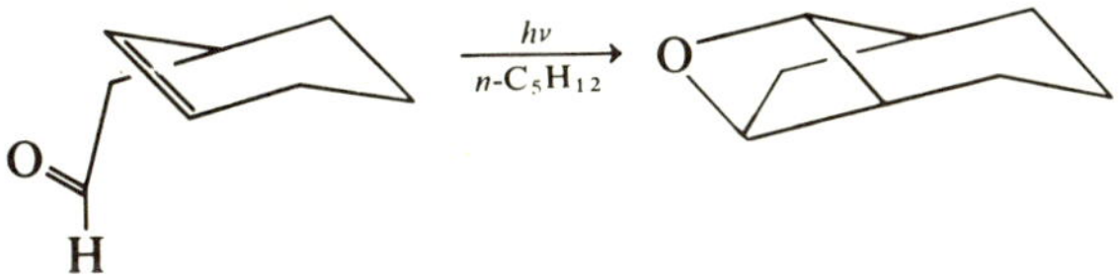

13 $\xrightarrow[n\text{-}C_5H_{12}]{hv}$ **14, 40%**

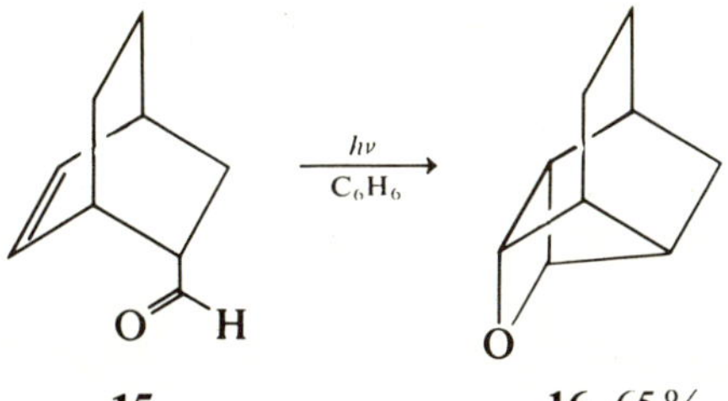

15 $\xrightarrow[C_6H_6]{hv}$ **16, 65%**

The cyclopropylaldehyde **17** (and the analogous methyl ketone) were found to photorearrange to **18** and **19** as well as add to solvent to yield **20** [373]. These reactions contrast with the more common ring cleavages previously observed for conjugated cyclopropylcarbonyl compounds. All these reactions involve conjugative opening of the cyclopropane ring. The formation of a ketene (precursor to **20**) has been shown to occur intramolecularly. The reactions may be envisaged as proceeding by way of the biradical **21**:

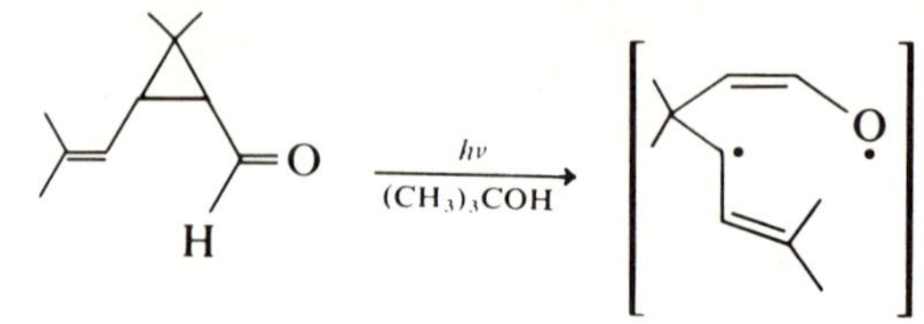

17 → (hv / (CH₃)₃COH) → **21**

↓

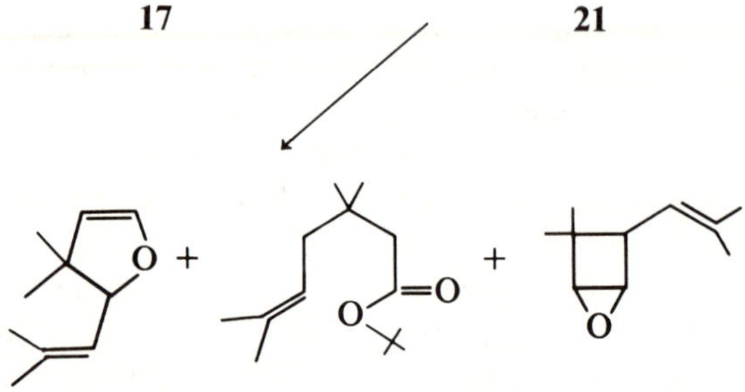

18, 61%   **20**, 14%   **19**, 11%

2. ALKYL KETONES

A precise value for E_3 of acetone remains unknown, although recent work suggests a value close to 80 kcal/mole [374, 220]. It seems that an earlier report that acetone fluorescence is that of an excimer is incorrect [375] (see Volume 2, page 41). Evidence has appeared indicating that the thermal decomposition of 1,2-oxaoxetanes results in an electronically excited ketone or aldehyde [376, 377]. The thermal decomposition of **22** has been used to "photosensitize" the *cis-trans* isomerization of stilbene and a cyclohexadiene rearrangement [377; see also 378]:

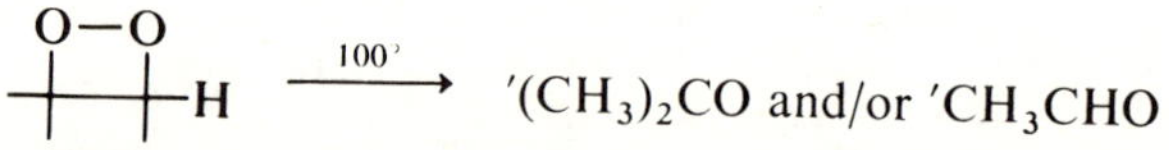

22 —100°→ '(CH₃)₂CO and/or 'CH₃CHO

The formation of CH_3COCH_2, when aqueous solutions of acetone and hydrogen peroxide are irradiated, has been supported by ESR measurements [379].

A preliminary report of the photoreduction of acetone by benzhydrol has been published [380]. Irradiation of the dioxolane **23** in the presence of acetone and *t*-butanol results in hydrogen abstraction followed by hydroperoxide formation **25** (381, 382). In the absence of oxygen, cleavage products are observed **27**:

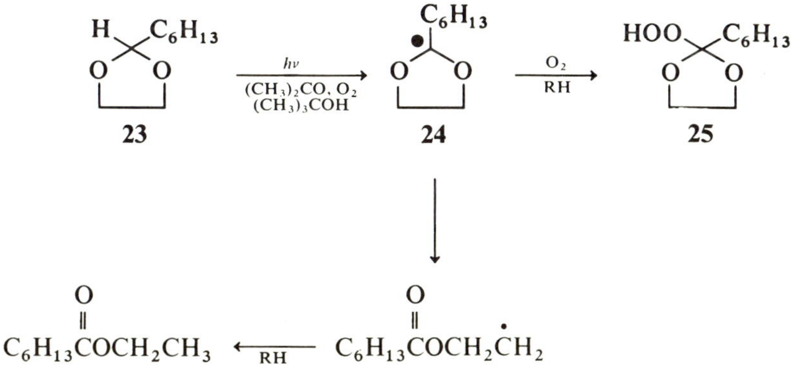

$$\underset{\mathbf{23}}{\overset{H}{\underset{O\_\_\_O}{\bigvee}}\!\!{}^{C_6H_{13}}} \xrightarrow[\substack{(CH_3)_2CO, O_2 \\ (CH_3)_3COH}]{h\nu} \underset{\mathbf{24}}{\overset{C_6H_{13}}{\underset{O\_\_\_O}{\bigvee}}^{\bullet}} \xrightarrow[RH]{O_2} \underset{\mathbf{25}}{\overset{HOO}{\underset{O\_\_\_O}{\bigvee}}\!\!{}^{C_6H_{13}}}$$

$$\underset{\mathbf{27}}{C_6H_{13}\overset{O}{\overset{\|}{C}}OCH_2CH_3} \xleftarrow[RH]{} \underset{\mathbf{26}}{C_6H_{13}\overset{O}{\overset{\|}{C}}OCH_2\overset{\bullet}{C}H_2}$$

Acetone sensitization of the cyclization of **28** to **29** and **30** may proceed by way of initial hydrogen abstraction (by excited acetone) from the carbon alpha to the carboethoxy1 group, although energy transfer to the double bond followed by internal abstraction and cyclization also may occur [383, 369]:

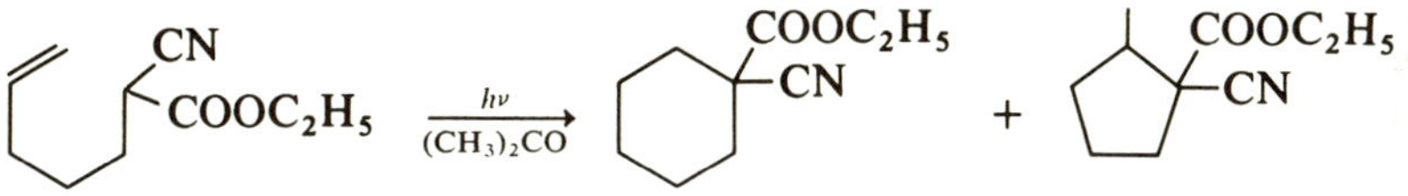

28 $\xrightarrow[(CH_3)_2CO]{h\nu}$ **29** + **30**

Further examples of acetone-sensitized alkylations have appeared [384, 385, 382, 386]. Acetone-initiated photoalkylations of glycine dipeptides with isobutene, 1-butene, and toluene may be realized in yields up to 60% [386, 385]. A small amount of asymmetric induction was noted [385]. 1,3-Dioxolane may be added to unsaturated carbohydrates by way of photosensitization [382].

Irradiation of acetone solutions of 2-alkoxyoxetanes (for example, **31**) yields β-lactones **32** presumably by way of hydrogen abstraction-cleavage sequence [387]. It is somewhat surprising that the radical **33** does not ring open to **34** (see, however, Volume 1, page 71, and Refs. 281, 382):

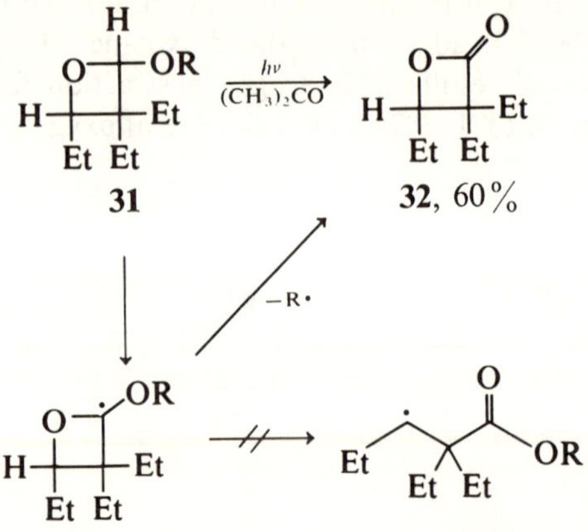

$$\underset{31}{\overset{\text{H}}{\underset{\text{Et Et}}{\text{H}\!\!-\!\!\!\overset{\displaystyle\text{O}\!-\!\!\overset{\text{H}}{\text{C}}\!-\!\text{OR}}{\underset{}{\text{C}}}\!\!-\!\!\text{Et}}}} \xrightarrow[(CH_3)_2CO]{h\nu} \underset{32,\ 60\%}{\overset{\text{O}\!-\!\!\overset{\text{O}}{\text{C}}}{\underset{\text{Et Et}}{\text{H}\!\!-\!\!\text{C}\!-\!\text{Et}}}}$$

$$\underset{33}{\overset{}{\underset{\text{Et Et}}{\text{H}\!\!-\!\!\text{C(OR)}\!-\!\text{Et}}}} \xrightarrow{-\!\!\!/\!\!\!\!/\!\!\!\rightarrow} \underset{34}{\text{Et}\!-\!\!\overset{\cdot}{\text{C}}(\text{Et})\!-\!\text{C(O)OR}}$$

Irradiation of acetone solutions of 1,4-dichlorobutane results in the formation of 1,3-dichlorobutane and HCl [388]. The yield of HCl depends on the concentration of acetone. Although a competition between excited excimer and monomer excitation transfer has been proposed to explain these results, it seems to us that a more probable explanation is that at high concentrations acetone reacts preferentially with itself rather than 1,4-dichlorobutane. The previously reported results on acetone lifetimes are probably related [389, 390]. Acetone also sensitizes the conversion of trimethylphenylammonium halides to benzene [330].

Two interesting new products **35** and **36** are formed from the direct irradiation of acetone [391], in addition to the previously unreported ketoalcohol **37**:

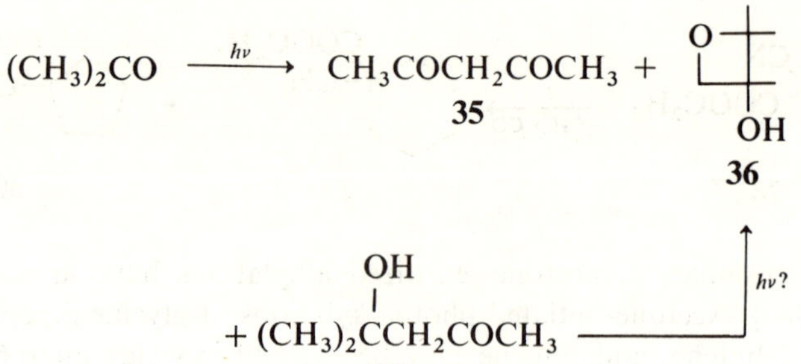

$$(CH_3)_2CO \xrightarrow{h\nu} \underset{35}{CH_3COCH_2COCH_3} + \underset{36}{\text{oxetane}}$$

$$+ \underset{37}{(CH_3)_2CCH_2COCH_3\ (\text{OH})} \xrightarrow{h\nu?} 36$$

The genesis of the oxetane **36** is not known for certain, but it may result from the addition of excited acetone to acetone enol or (more likely) from a secondary photorearrangement of **37** (see Volume 2, page 61):

$$(CH_3)_2CO + CH_2\!\!=\!\!C(OH)CH_3 \xrightarrow{h\nu} 36$$

Several significant papers dealing with the type II photorearrangement of alkanones have appeared [392, 393, 394]. The slope of Stern-Volmer plots for quenching of the type II photoelimination of 2-pentanone by 1,3-pentadiene depend on the percentage of conversion. Thus, at 3% conversion, $k_q\tau = 380$, and at 20% conversion, $k_q\tau = 60$ [392, 392a]. The effect is less pronounced for 2-hexanone. The origin of this effect is not certain, but it may result from the generation of quenching impurities, which would effectively shorten τ. The revised value for k_{II} (6×10^7 sec^{-1}) for 2-pentanone triplets now brings the reactivity of 2-pentanone closer to that for butyrophenone (8×10^6 sec^{-1}, Volume 2, page 60). The dependence of $k_q\tau$ on the percentage of conversion may also result from decreases in 1,3-pentadiene concentration (see above, Section I.A.1), since reaction of the ketone with the diene probably occurs [393].

In a pair of significant papers [393, 394], it has been found that (1) racemization of **38** occurs from the triplet but not from the singlet, (2) a large amount of the inefficiency toward the type II reaction for alkanones is the result of S_1 deactivation, (3) the fraction of S_1 deactivation increases as the γ-hydrogen goes from 1° to 2° to 3°. Less racemization of **38** occurs in *t*-butanol than in hexane, along with a concomitant increase in type II products:

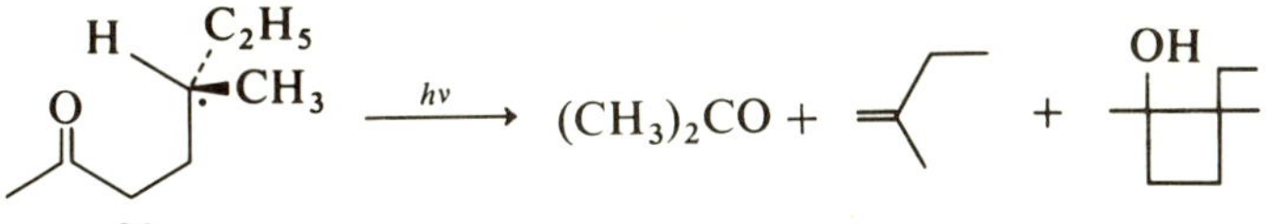

$$\underset{\mathbf{38}}{\text{[structure]}} \xrightarrow{h\nu} (CH_3)_2CO + \text{[alkene]} + \text{[cyclobutanol-OH]}$$

Table I summarizes data [394] that strongly suggests that the rate of nonradiative decay from excited-singlet states of alkanones with γ-hydrogens increases as such hydrogens change from primary to secondary to tertiary. A reasonable interpretation of these data is that this singlet nonradiative process involves chemical reversal from an initially formed singlet biradical intermediate. It should also be noted that product formation is more efficient from $^3(n,\pi^*)$ than $^1(n,\pi^*)$ state:

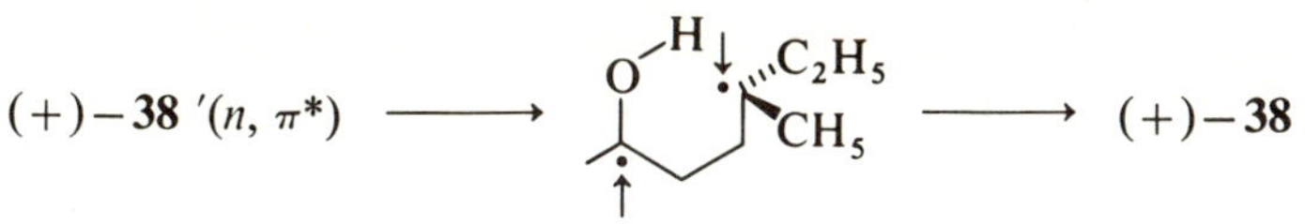

$$(+)-\mathbf{38}\ '(n, \pi^*) \longrightarrow \text{[biradical intermediate]} \longrightarrow (+)-\mathbf{38}$$

A theoretical analysis of photoreactivity [395] also suggests that photochemically the most reactive atoms in a molecule are those involved in high-frequency (for example, C—H) vibrations and that the rate of nonradiative decay might well depend on C—H bond strength.

TABLE I
Photoreactivity of Alkyl Ketone Excited Singlet States toward Intramolecular γ-Hydrogen Abstraction [a]

	2-Pentanone	2-Hexanone	5-Methyl-2-hexanone
$k_q\tau_S$, M^{-1} [b]	20.2	7.3	4.1
τ_S, nsec [c]	2.02	0.73	0.41
ϕ_{ST}	0.63	0.27	0.11
k_{ST}, sec^{-1}	3.1×10^8	3.7×10^8	2.7×10^8
ϕ_r^s [d]	0.025	0.10	0.10
k_r^s, sec^{-1} [e]	1.2×10^7	1.4×10^8	2.4×10^8
ϕ_S [f]	0.35	0.63	0.79
k_S [g]	1.7×10^8	8.5×10^8	1.9×10^9
ϕ_r^t [h]	0.36	0.23	0.09

[a] Taken from Ref. 394.
[b] Slope of Stern-Volmer plot for quenching of singlet reaction by biacetyl.
[c] Singlet state lifetime assuming k_q equals 1×10^{10} M^{-1} sec^{-1} in n-hexane.
[d] Quantum yield for the formation of both acetone and cyclobutanol from singlet state reaction.
[e] Rate constant for singlet reaction.
[f] Quantum yield for nonradiative decay from the singlet state.
[g] Rate constant for nonradiative decay from the singlet state.
[h] Quantum yield for the formation of both acetone and cyclobutanol from triplet state reaction.

Irradiation of 6-hepten-2-one **39**, in both the absence and presence of 1,3-pentadiene, yields **40** to **44** [394]:

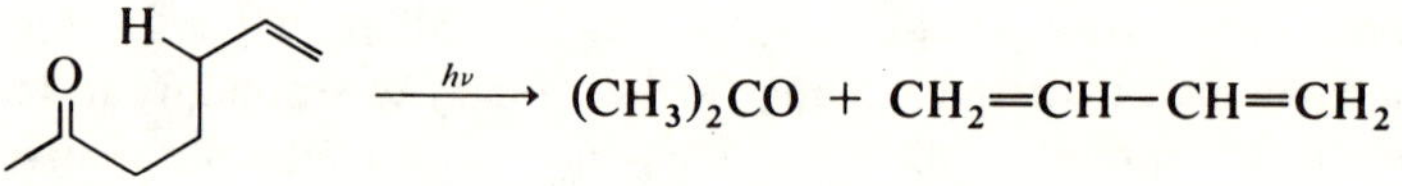

$$\mathbf{39} \xrightarrow{h\nu} (CH_3)_2CO + CH_2=CH-CH=CH_2$$

39   **40**   **41**

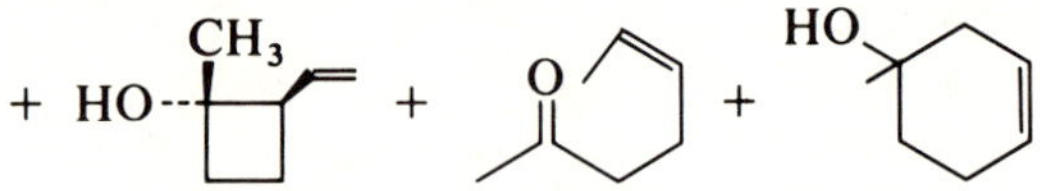

+ HO—cyclobutane(CH₃)   +   **43** (cyclohexenone)   +   **44** (HO-cyclohexene)

42   **43**   **44**

Thus, the formation of **44** seems best consistent with the intermediacy of a singlet biradical intermediate.

The discussion above may be applied to account for the observation that (1) the quantum efficiency of the photolysis of n-butyl t-butyl ketone is relatively low compared with other t-butyl ketones; that is, the reversal of type II from $^1(n,\pi^*)$ serves as an energy sink, and the quantum efficiency of the type II

ORGANIC PHOTOCHEMISTRY

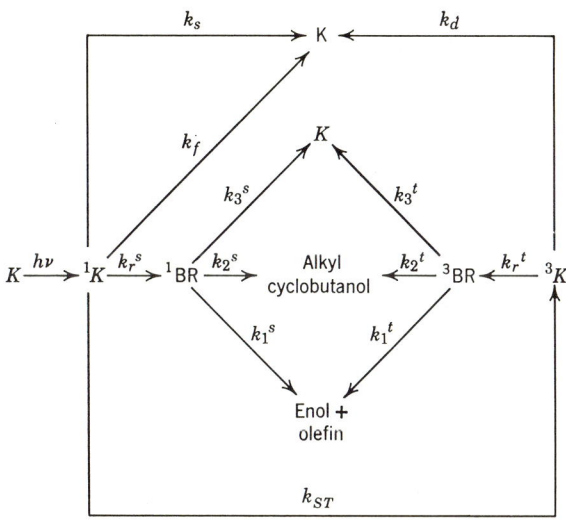

Scheme 1 Summary of processes involved in the type II reactions of branched alkanones.

^1K S_1 state of alkanone
^3K T_1 state of alkanone
^{1}BR Biradical formed by intramolecular γ-hydrogen abstraction from S_1
^{3}BR Biradical formed by intramolecular γ-hydrogen abstraction from T_1

Rate Constant	Process
k_f	Fluorescence from S_1
k_s	Nonradiative decay from S_1 to S_o
k_{ST}	Intersystem crossing from S_1 to T_1
k_r^s	Intramolecular γ-hydrogen abstraction from S_1
k_d	Nonradiative decay from T_1 to S_o
k_r^t	Intramolecular γ-hydrogen abstraction from T_1
k_1^s	β cleavage from ^{1}BR
k_2^s	Closure from ^{1}BR
k_3^s	Return of hydrogen to γ-carbon from ^{1}BR
k_1^t	β cleavage from ^{3}BR
k_2^t	Closure from ^{3}BR
k_3^t	Return of hydrogen to γ-carbon from ^{3}BR

photoreaction of 2-hexanone-5,5-d$_2$ is enhanced over that for 2-hexanone; that is, a γ-deuterium is less reactive in $^1(n,\pi^*)$ than a γ-hydrogen, allowing Φ_{ST} to increase and resulting in a higher percentage of the more efficient $^3(n,\pi^*)$ reaction.

Scheme I summarizes the processes now believed to be involved in the type II reactions of branched alkanones [see also 396].

Oxidation of polypropylene polymers is believed to generate polymeric ketones of structures **45** and **48**. Irradiation of **45** and **48** results in both type I and type II scission products [397]:

In contrast to the photochemistry of laurolenal **5** [368], the methyl ketone homolog **49** is photoracemized, probably by way of a reversible sigmatropic 1,3-acetyl migration [398], in the excited singlet state [see also 398a]:

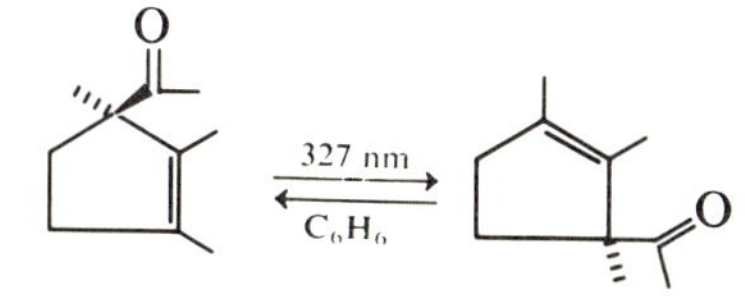

$$(+)-49 \underset{C_6H_6}{\overset{327 \text{ nm}}{\rightleftarrows}} (-)-49$$

In the presence of tri-*n*-butyl stannane some photoreduction occurs. Sensitization by acetone gives rise to a new photoisomer **50**:

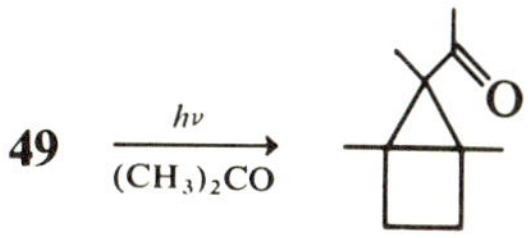

$$49 \xrightarrow[(CH_3)_2CO]{hv} \mathbf{50}$$

It seems that (1) $S_1 \to T_1$ crossing in **49** is negligible, (2) the S_1 state of **49** undergoes 1,3-acetyl shifts and photoreduction, and (3) the T_1 state of **49** undergoes predominantly a 1,2-acetyl shift and cyclopropane ring closure (see also Chapter 3, Section I.A.4).

The mechanisms of interaction of alkyl ketones and ethylenes and the synthetic potential of these reactions [399, 400, 401] continue to attract attention [402, 403, 46; Review: 404].

Study of the *cis-trans* isomerization of 2-pentene sensitized by acetone suggests that a Schenck (biradical) intermediate is involved somewhat but that the triplet excitation transfer represents the major path of the isomerization [46] (see also Section II.B).

A discussion of the synthetic aspects of the photocycloadditions of acetone and other alkyl ketones to 1,2-dicyanoethylene, maleic anhydride, and 1-methoxy-1-butene has appeared [400], as have examples of the intramolecular photocyclizations of γ,δ-unsaturated ketones [371, 372].

Irradiation of acetone in the presence of **51** at $-75°$ has been shown to form **53** by way of the oxetane intermediate **52** [405]:

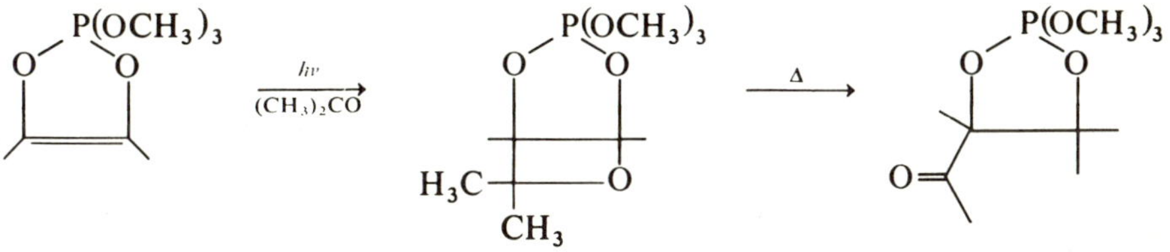

$$\mathbf{51} \xrightarrow[(CH_3)_2CO]{hv} \mathbf{52} \xrightarrow{\Delta} \mathbf{53}$$

The allene ketone **54** yields **55** and **56** on irradiation [406]:

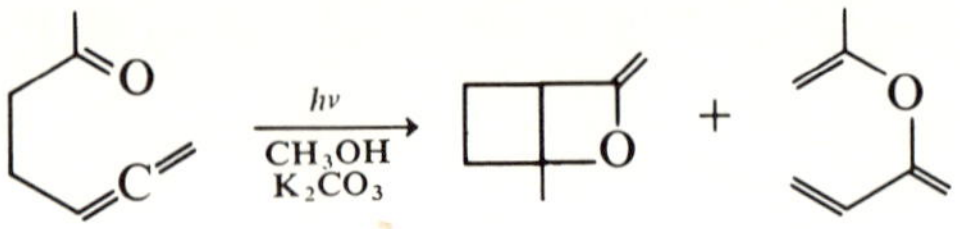

$$\mathbf{54} \xrightarrow[\text{K}_2\text{CO}_3]{\text{CH}_3\text{OH}, h\nu} \mathbf{55}, 90\% + \mathbf{56}, 6\%$$

The 5-acylnorbornene **57** (R = CH$_3$) is photocyclized to **58**, as are the R = CH$_2$CH$_3$, CH$_2$C$_6$H$_5$, H, and aryl derivatives [372]:

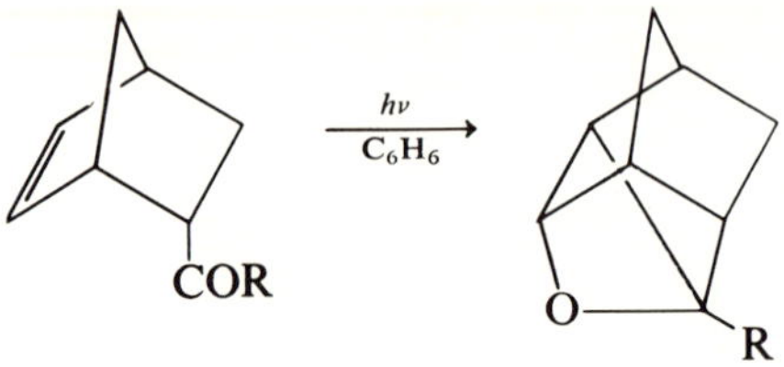

57 $\xrightarrow[\text{C}_6\text{H}_6]{h\nu}$ **58**

5-Acylbicyclo[2.2.2]octenes undergo analogous cyclizations [371].

Further examples of the photoring cleavage of conjugated cyclopropylalkanones have appeared [407]. In pentane, the *cis*-methyl cyclopropyl ketone **59** rearranges to 1-hexen-5-one **60** presumably through initial intramolecular γ-hydrogen abstraction:

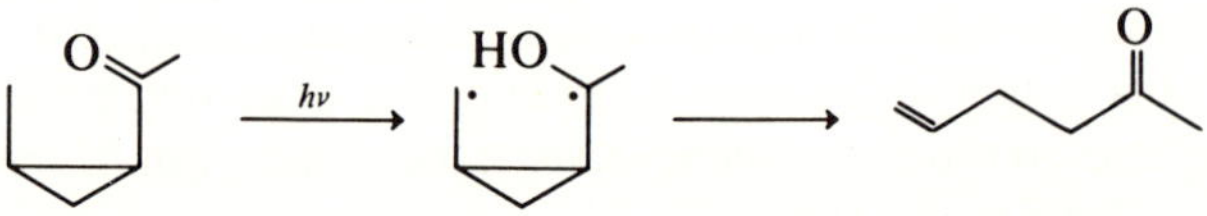

59 $\xrightarrow{h\nu}$ → **60**

but the *trans*-cyclopropyl ketone **61** yields **59**, **62**, and **63**:

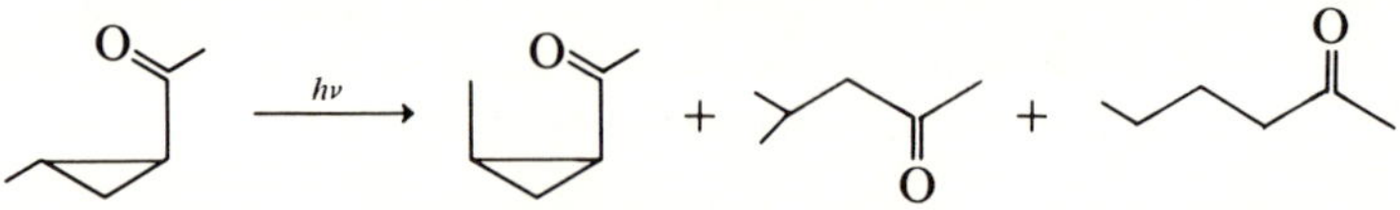

61 $\xrightarrow{h\nu}$ **59** + **62** + **63**

The latter two compounds probably result from initial ring opening followed by hydrogen abstraction from the solvent [408].

Direct and sensitized photolyses of sulfoxides have been reported [409, 396].

3. ALKYL KETONES WITH α-SUBSTITUENTS

Irradiation of 3-hydroxy-2-butanone yields acetaldehyde and ethanol as major products, the former product possibly arising by way of a five-membered transition state [410]:

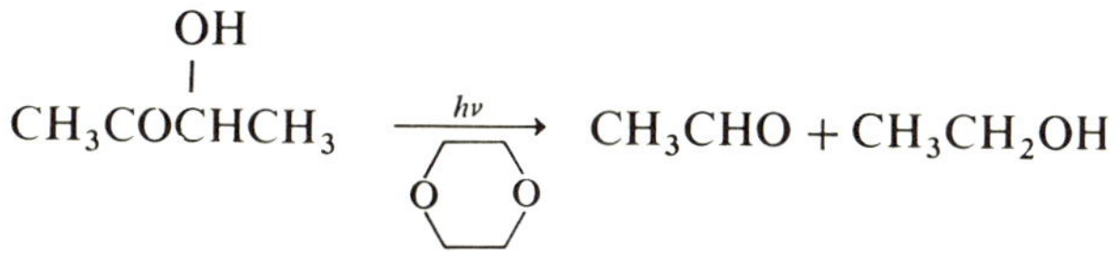

$$\text{CH}_3\text{COCHCH}_3 \text{ (OH)} \xrightarrow{h\nu} \text{CH}_3\text{CHO} + \text{CH}_3\text{CH}_2\text{OH}$$

The ketone **64** rearranges to **66** on irradiation, possibly by way of the cyclic intermediate **65** [411]; **64** also undergoes photoreduction in cyclohexane or cyclohexene:

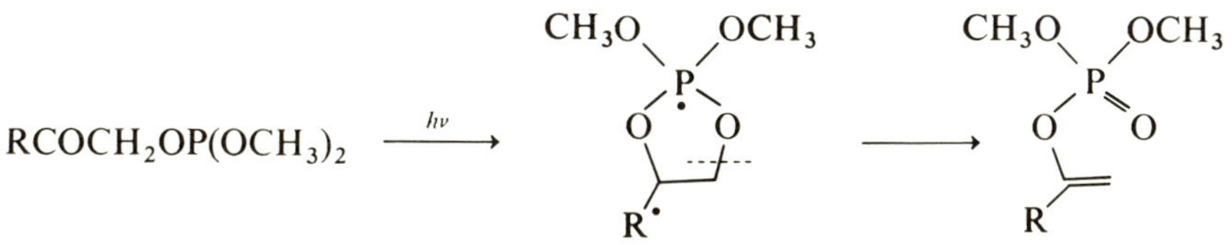

$$\text{RCOCH}_2\text{OP(OCH}_3)_2 \xrightarrow{h\nu} \mathbf{65} \longrightarrow \mathbf{66}$$

64   **65**   **66**

β-Ketophosphonates (for example, **67**) are photoreduced to β-hydroxyphosphonates **68** efficiently in ether [412]:

$$\text{CH}_3\text{COCH}_2\text{P(OC}_2\text{H}_5)_2 \xrightarrow[\text{(C}_2\text{H}_5)_2\text{O}]{h\nu} \text{CH}_3\text{CHOHCH}_2\text{P(OC}_2\text{H}_5)_2$$

67   **68**, 85%

Irradiation of ethyl acetoacetate **69** in ether yields mainly the corresponding β-hydroxy ester **70** [413, 412]:

$$\text{CH}_3\text{COCH}_2\text{CO}_2\text{C}_2\text{H}_5 \xrightarrow[\text{(C}_2\text{H}_5)_2\text{O}]{h\nu} \text{CH}_3\text{CHOHCH}_2\text{CO}_2\text{C}_2\text{H}_5$$

69   **70**, 70%

A value of $k_r \sim 10^7\ M^{-1}\ \text{sec}^{-1}$ is estimated for photoreduction of **67** and **69** by ether.

The ketoester **69** also undergoes photoreductive addition to the ketofunction in methanol [413]:

$$\mathbf{69} \xrightarrow[R_2CHOH]{h\nu} CH_3-\underset{\underset{R_2COH}{|}}{\overset{\overset{OH}{|}}{C}}-CH_2CO_2C_2H_5$$

4. CYCLIC KETONES

To date, the only reported photoreaction of cyclopropanones is photodecarbonylation (for example, **71 → 72** [414]):

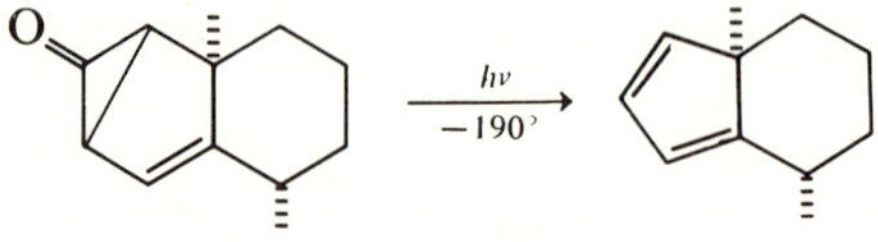

71   **72**

A study of the photoreduction of cyclobutanone by tributyltin hydride in the presence of piperylene has been reported [414a]. It is proposed that in solution most cyclobutanone photochemistry occurs from the singlet manifold.

Irradiation of a methanolic solution of **73** or its isomer **74** yields the same distribution of products, derived from dimethyl ketene and tetramethylcyclopropanone, under conditions such that **73** and **74** are not interconverted. These results were interpreted as evidence that the intermediate **75** is common to both photolyses [415; see also 415a]:

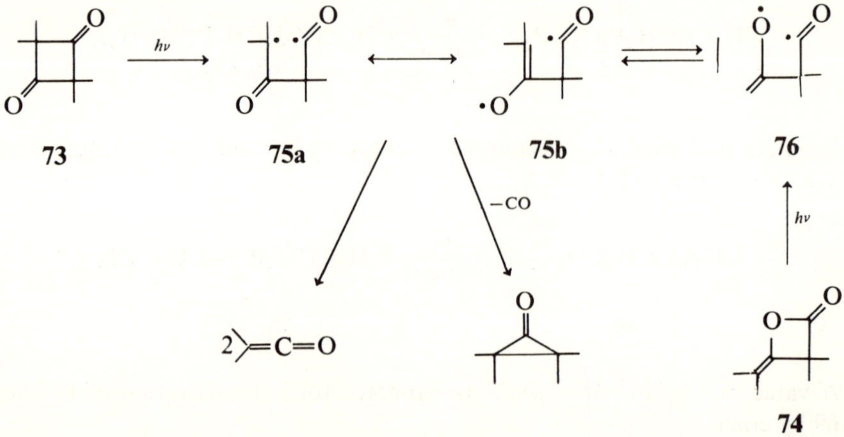

73   **75a**   **75b**   **76**

$2\rangle = C=O$

74

The diazabicycloheptanone **77** is photorearranged to **78** and **79**. In addition, **80** and benzil are formed [416]. Only **80** is straightforwardly derived from a standard photoreaction of cyclobutanones [416a]. However, **78**, **79**, and benzil could come from rearrangement and dimerization of the expected decarbonylated biradical **81**:

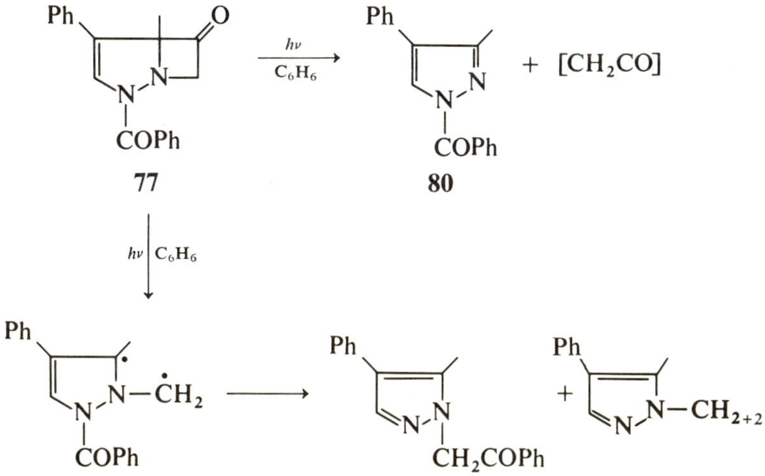

$$77 \xrightarrow[C_6H_6]{h\nu} 80 + [CH_2CO]$$

$$77 \xrightarrow[C_6H_6]{h\nu} 81 \longrightarrow 78 + 79$$

Irradiation ($\lambda > 2800$ Å) of the thiaspiro[2.3]hexan-5-one **82** in methanol leads to **84** and **85** in a wavelength-dependent reaction [417]:

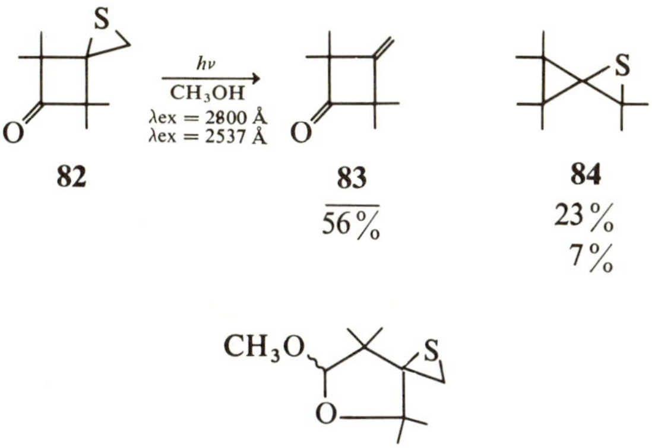

$$82 \xrightarrow[\substack{\lambda_{ex} = 2800 \text{ Å} \\ \lambda_{ex} = 2537 \text{ Å}}]{h\nu \\ CH_3OH} 83 + 84 + 85$$

83: 56%
84: 23%, 7%
85: 37%, 12%

With 2537 Å light, **83** becomes the major reaction product (56%), with smaller yields of **84** and **85**. It appears that sulfur extrusion is preferred at shorter wavelengths, thus indicating that intermolecular and intramolecular energy from the n,σ^* state of the sulfur atom, excited by short wavelength, is inefficiently transferred to the lower-energy n,π^* state of the carbonyl group.

Photolysis of 7,7-dichlorobicyclo[3.2.0]hept-2-en-6-one **86** in methanol results in the formation of **87** to **89** as the major products [418]:

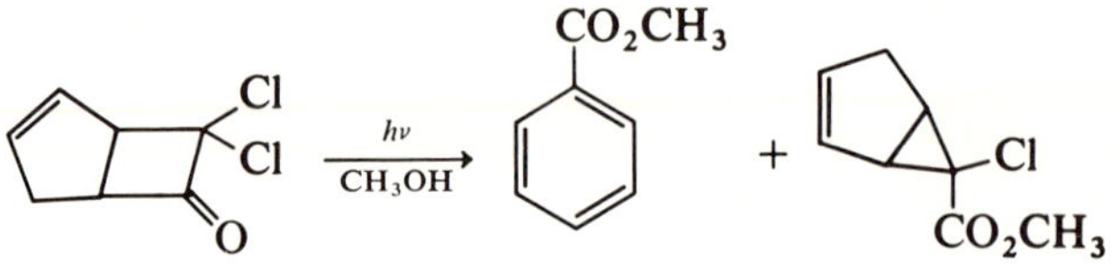

Homoallylic assistance for the loss of chlorine has been suggested to explain formation of **87** and **88**.

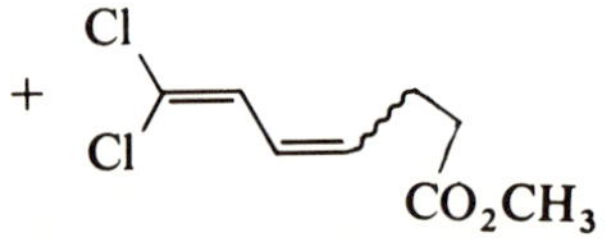

The photorearrangement of a number of saturated cyclic ketones to alkenals is quenched by 1,3-dienes [419]. Stern-Volmer analysis allows estimation of the rates at which ketone triplets undergo type I cleavage. Table II summarizes the rate and quantum-yield data. Since enal formation is completely quenchable by 1,3-dienes, cleavage is assumed to occur exclusively by way of a triplet-state reaction. These results are completely compatible with the hypothesis that α cleavage of the ketone triplet yields an acyl-alkyl biradical that disproportionates to yield alkenal (and ketene that was not analyzed) and recouples to yield ground-state ketone. The ease of α cleavage depends on the stability of the acyl-alkyl radical pair.

Of special interest are (1) the enhancement of enal formation with 2-alkyl substitution, (2) the lowering of enal formation efficiency (but not reactivity toward cleavage) with 3-alkyl substitution, (3) the sharp decrease in enal formation efficiency with 2-phenyl substitution.

Cycloaddition reactions of cyclic alkanones to vinyl ethers [401] and electron deficient ethylenes [400] have been reported.

TABLE II

Kinetic Data for Photoisomerization of Cyclic Ketones to Enals [a]

Ketone	Φ_a [b]	$k_q\tau_T$, M^{-1} [c]	$1/\tau_T$, sec^{-1} [d]
Cyclopentanone	0.24	47	1.1×10^8
Cyclohexanone (C)	0.09	152	3.0×10^7
3-Methyl-C	0.033	209	2.5×10^7
3,5-Dimethyl-C	0.005	206	2.4×10^7
3,3,5-Trimethyl-C	0.002	200	2.5×10^7
2-Methyl-C	0.42	10.6	4.7×10^8
2,6-Dimethyl-C	0.40	5.4	9.3×10^8
2,2-Dimethyl-C	0.41	2.8	1.8×10^9
2-Phenyl-C	0.04	15.2	3.3×10^8

[a] Taken from Ref. 419.
[b] Quantum yield for the formation of unsaturated aldehyde in benzene.
[c] Slope of Stern-Volmer plot for quenching of unsaturated aldehyde formation by 1,3-pentadiene.
[d] Assuming k_q in benzene equals 5×10^9 M^{-1} sec^{-1}.

Irradiation of fenchone in methanol containing a trace of *p*-toluenesulfonic acid leads to the formation of **91** to **94** [420, 421]:

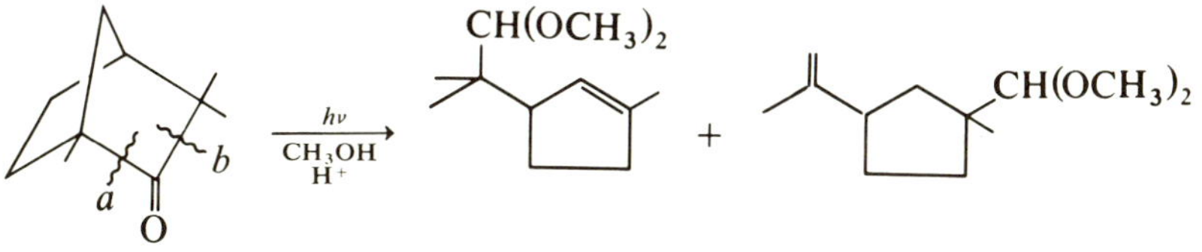

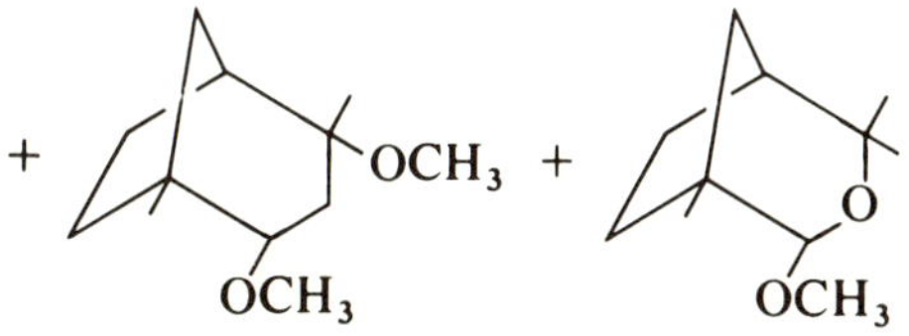

The formation of **91** and **92** is anticipated (Volume 1, page 49). It seems that **93** is derived from **92**, possibly from an acid-catalyzed cyclization. The formation of **94** is relevant to the relatively rare photoconversion of a cyclopentanone to a cyclic acetal. Interestingly, in the case of camphenilone **95** only products derived from cleavage of bond *b* are detected [see also 421a]:

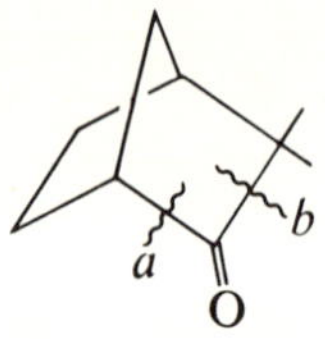

95

In addition to the expected photodecarbonylation, which occurs in C_6H_6, **96** undergoes photoreduction of its double bond in isooctane [422]:

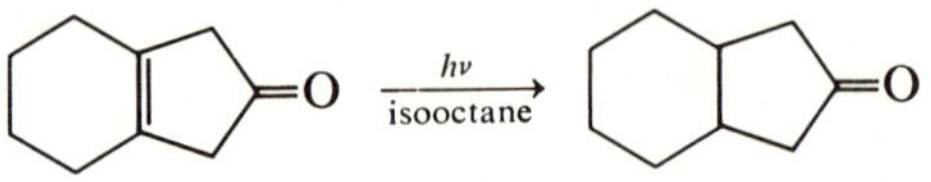

$$\mathbf{96} \xrightarrow[\text{isooctane}]{h\nu} \mathbf{97}, cis, ? \quad \mathbf{98}, trans, ?$$

Although the benzonorcamphor **99** undergoes rearrangement to the aldehyde **100**:

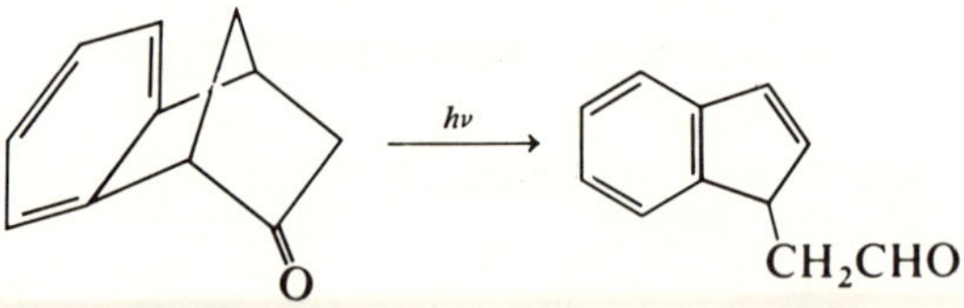

99 $\xrightarrow{h\nu}$ **100**, 80%

irradiation of **101** and **103** in acetone yields the rearranged ketones **102** and **104**, respectively [423]:

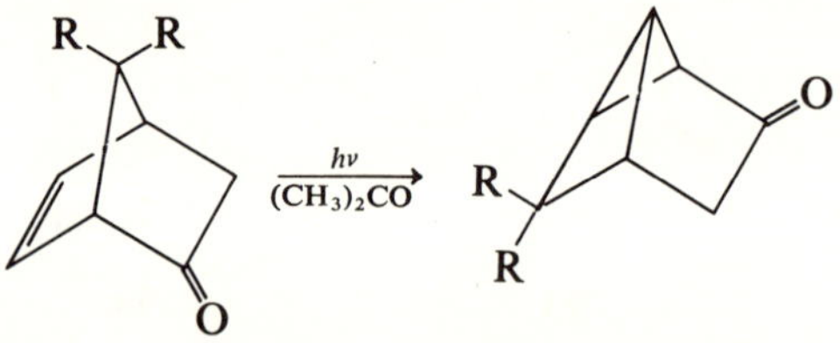

$$\xrightarrow[(CH_3)_2CO]{h\nu}$$

101, R = H
103, R,R = —CH$_2$CH$_2$—

102, R = H, 75%
104, R = —CH$_2$CH$_2$—, 60%

Irradiation of the diene-iron tricarbonyl complex **105** at 77°K results in a low yield of quadricyclanone [65], presumably produced by photoclosure of **106**:

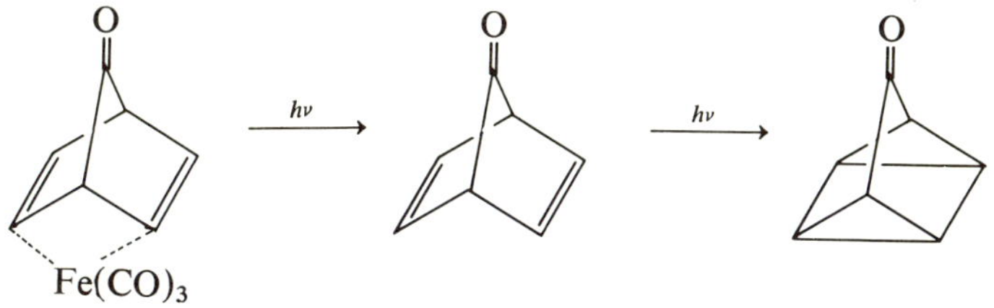

ORGANIC PHOTOCHEMISTRY

105 → hv → **106** → hv →

The tricyclic dienone **107** is photodecarbonylated to **108** and photorearranged to **109** [424]:

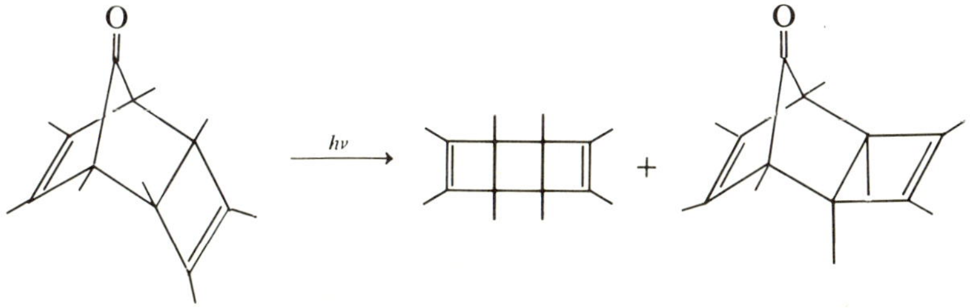

107 → hv → **108** + **109**

A number of reports of photodecarbonylation of bicyclo[2.2.1]-hept-2-en-7-ones have appeared [425, 426, 427]. An extension of this reaction has been used to obtain **111** from **110** [427]:

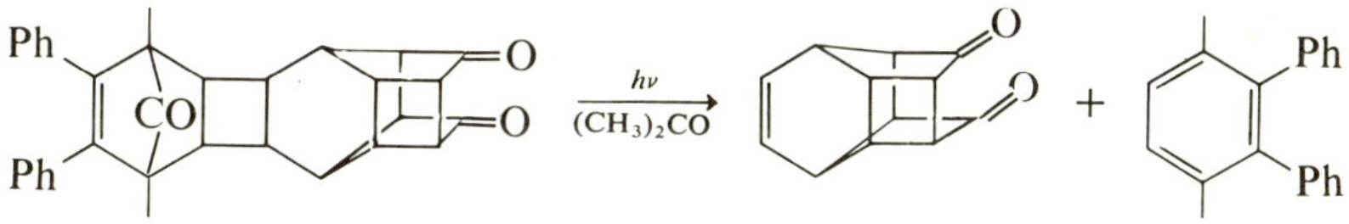

110 → hv, $(CH_3)_2CO$ → **111** + (Ph-substituted product)

The cyclopropyl ketone **112** is photorearranged to **113** and **114** [428]:

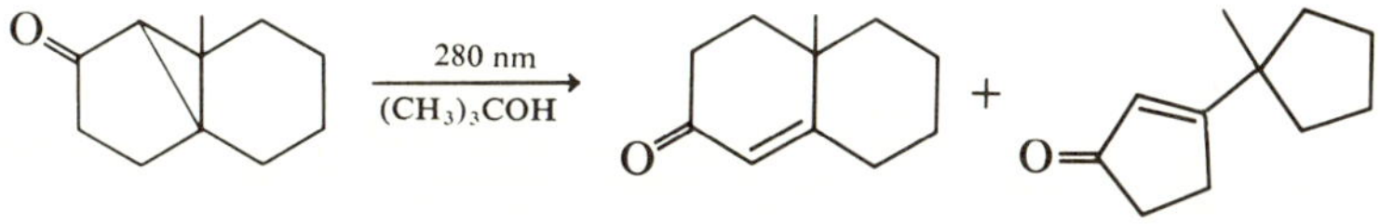

112 → 280 nm, $(CH_3)_3COH$ → **113** + **114**

The photolysis of 2-methylcyclohexanone yields, contrary to an earlier [428a] report, both *trans*- and *cis*-5-heptenal as major products [429]. Since it was shown that **116** and **117** are not readily interconverted under the reaction conditions, it was concluded that the unsaturated aldehydes are generated from a biradical precursor of sufficient lifetime to cause loss of stereochemistry:

115 **116** **117**

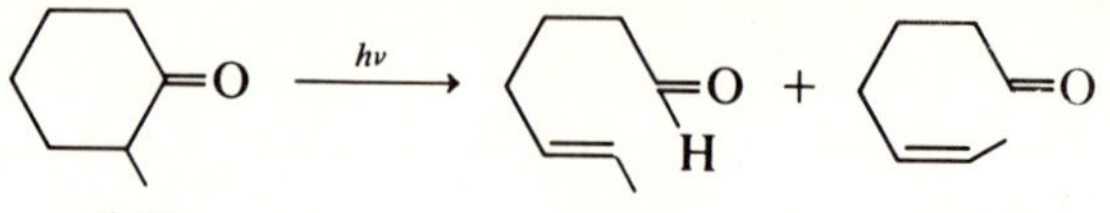

Evidence has also been reported that suggests that the irradiation of *cis*- and *trans*-2,3-dimethylcyclohexanone in methanol-benzene leads to unsaturated aldehydes, esters, and epimers by way of a common biradical intermediate [430]:

118 **120** **121**

120, 121 → **122,** **123,** **119**

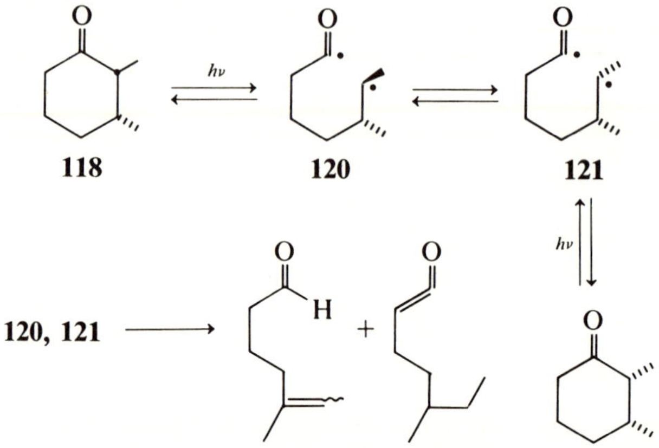

A reformation of the cyclobutanol structure produced by the irradiation of 11-oxalanstarrol has appeared [431].

Irradiation of 2-cyclopropylcyclohexanone [**124**] yields ring-expanded *cis*- and *trans*-4-cyclononenones as major products. In addition, the aldehyde **128** was also formed [432]:

124, **125**

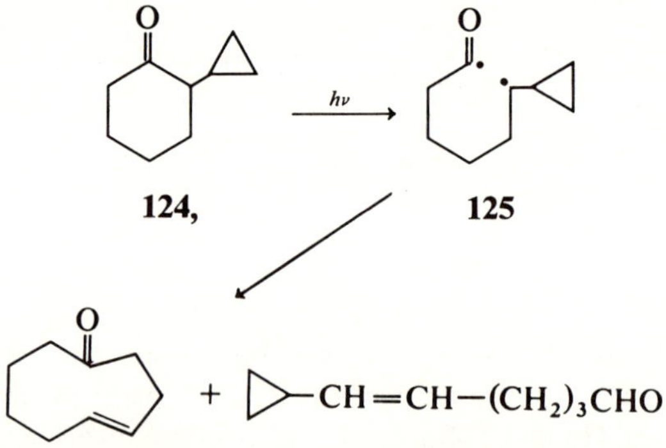

126, *trans*, 44%
127, *cis*, 29%

128, 21%

Irradiation of the spiroketone **129** results in the formation of **130**:

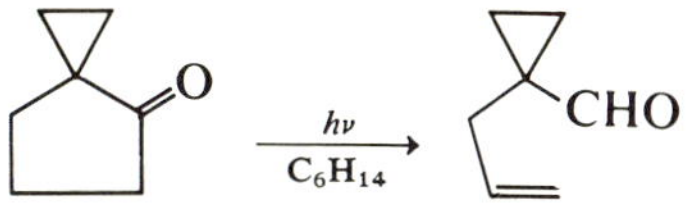

129 →(hν, C₆H₁₄) **130, 34%**

Analogously, **131** yields **132**, but, in addition, **133** is also formed [433]:

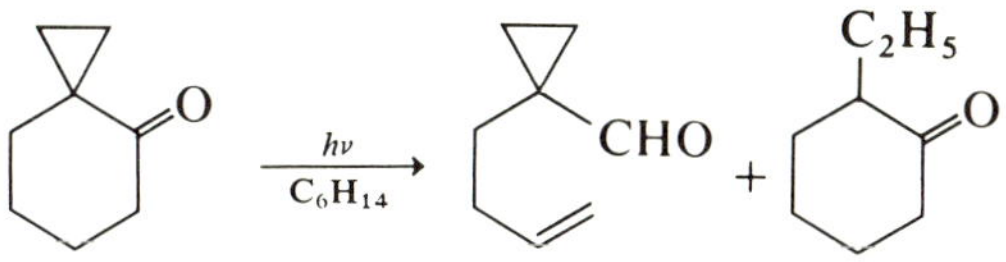

131 →(hν, C₆H₁₄) **132, 37%** + **133, 10%**

The cyclopropyl ketone **134** and its homolog **136** yield the ring-expanded products **135** and **137**, respectively:

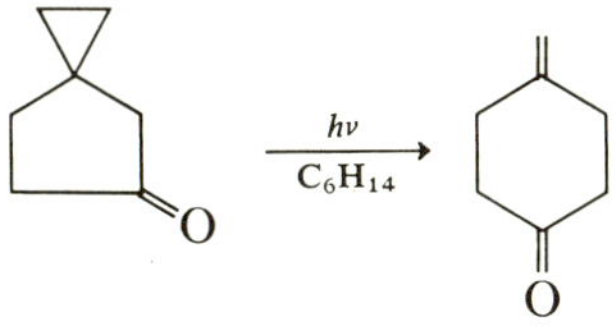

134 →(hν, C₆H₁₄) **135, 33%**

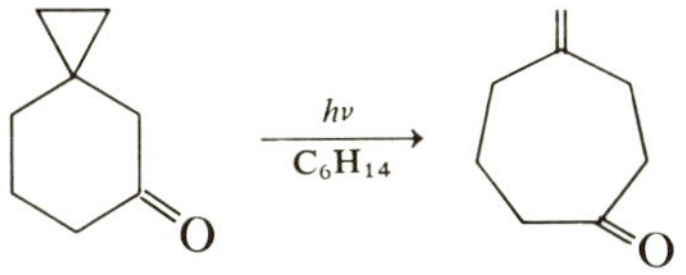

136 →(hν, C₆H₁₄) **137, 22%**

Both ring expansion and unsaturated aldehyde formation may be explained on the basis of initial α-cleavage, followed by rearrangement. In the case of biradical **139**, apparently ring opening is much faster than hydrogen abstraction:

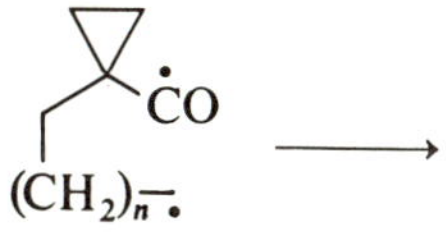

138 ⟶ **130** or **132**

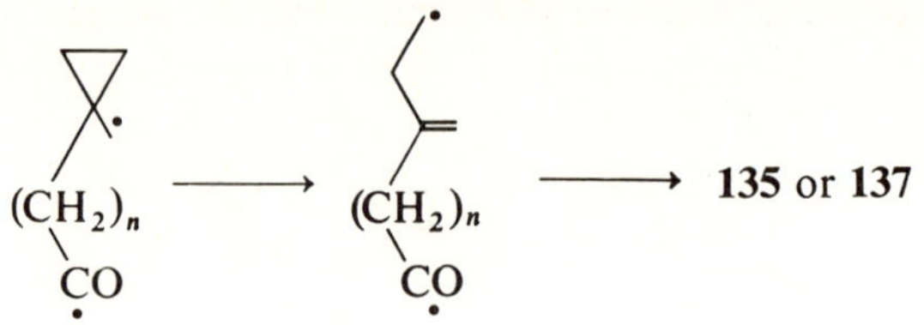

The photoreduction of bicyclo[4.1.0] and bicyclo[3.1.0]alkan-2-ones [408] in 2-propanol results in the reduction of the "outside" cyclopropyl bond. The ketyl radical **145** is presumably involved in this photoreduction:

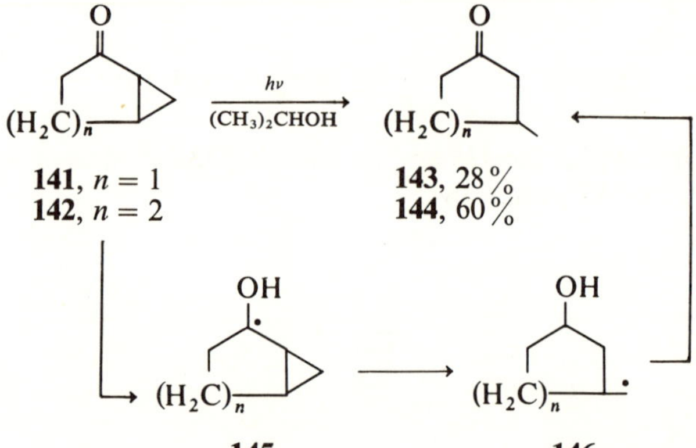

141, $n = 1$
142, $n = 2$

143, 28%
144, 60%

Irradiation of *cis*-caran-5-one **147** results in the formation of the ring-opened products **148**, **149**, and **150**:

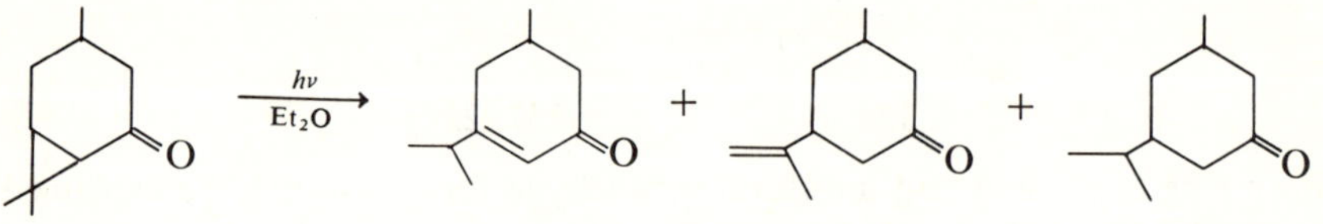

Although the first two products are anticipated, the last is an unexpected reduction product [434].

Further examples of the photolyses of conjugated epoxyketones have appeared [435].

Irradiation of **151**, from the photolysis of the enone with tetramethylethylene (TME), apparently results in α cleavage on the less substituted side of the carbonyl group to form an enal **152** that undergoes subsequent photocycloaddition to TME to yield **153** [436]:

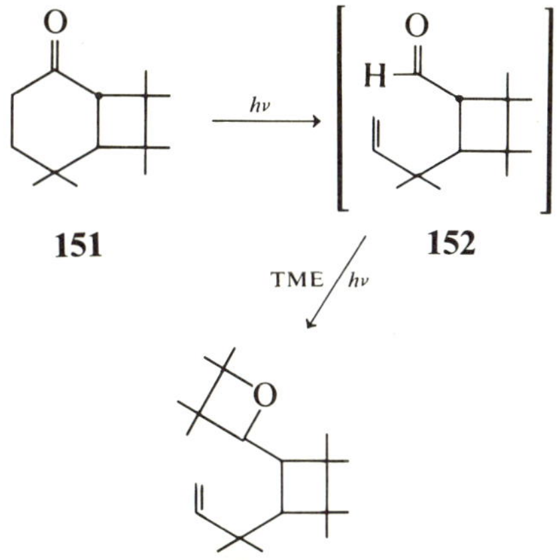

151 → **152**

TME/hν

153

Cleavage on the more substituted side of the carbonyl group may result in a biradical that closes to regenerate **151**.

The β,γ-unsaturated enone **154** is partially isomerized by 256 nm light to the α,β-unsaturated enone **155**:

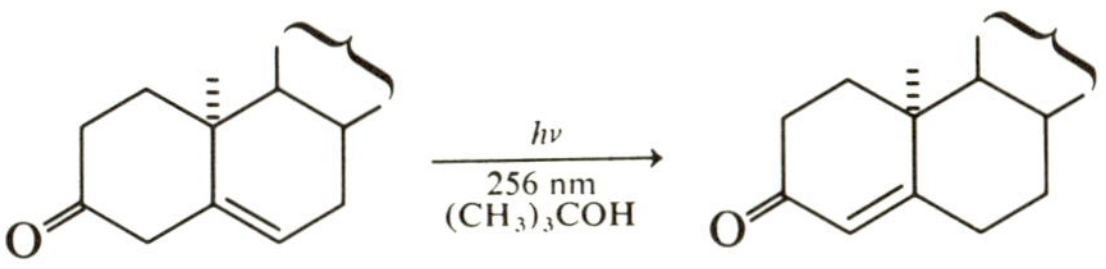

154 $\xrightarrow[(CH_3)_3COH]{hν, 256\ nm}$ **155**

The reaction does not proceed with >300 nm excitation [437].

Irradiation of β,γ-unsaturated ketones, such as **156**, in which the double bond occupies an exocyclic position with respect to the carbonyl group, causes a reversible rearrangement to unsaturated bridged bicyclic ketones, such as **157** [438] (see also Volume 2, page 52):

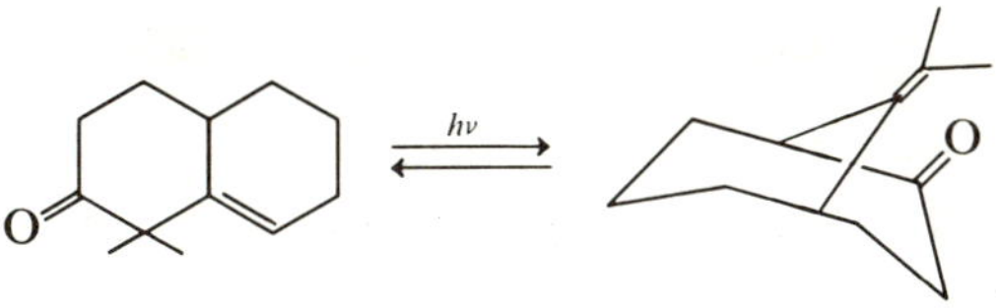

156, 77% ⇌ **157**, 23%

The β,γ-unsaturated ketosteroid **158**, however, undergoes rearrangement to the oxetane **160**, presumably by way of the dienal **159** [439]:

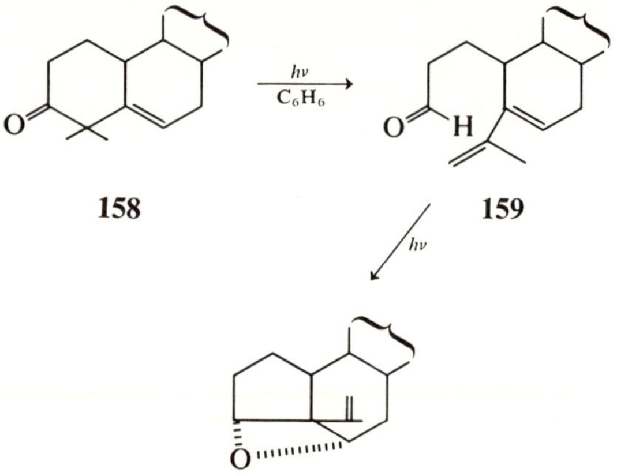

158 $\xrightarrow{h\nu}{C_6H_6}$ **159** $\xrightarrow{h\nu}$ **160**, 30%

Acetone solutions of **158** cause rearrangement to the cyclopropylketone **161** [367]:

158 $\xrightarrow{h\nu}{(CH_3)_2CO}$ **161**, 19%

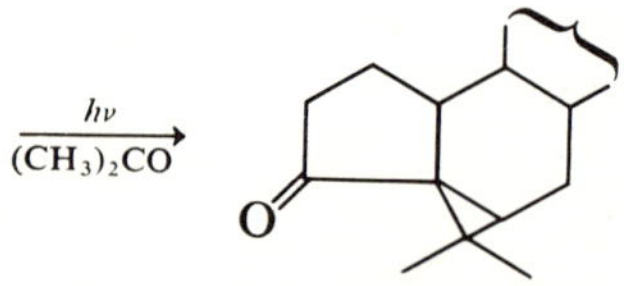

Irradiation of the β,γ-unsaturated ketosteroid **162**, which differs from **158** in that it has an additional angular methyl group, yields the oxetane **163**:

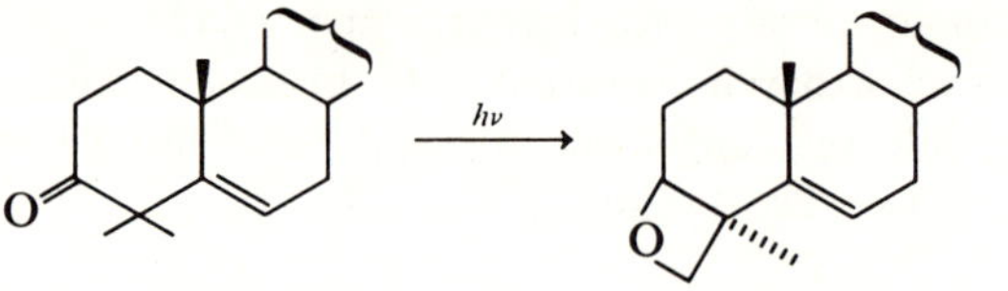

162 $\xrightarrow{h\nu}$ **163**

The difference in behavior of **158** and **162** has been attributed to steric interactions between the 19-methyl group and the diene function in the intermediate dienal [439].

Irradiation of the bicyclic enone **164** results in a stereospecific [2 + 2 + 2 + 2] ring opening to yield the ketene **165**, which is trapped as its methanol adduct [440]:

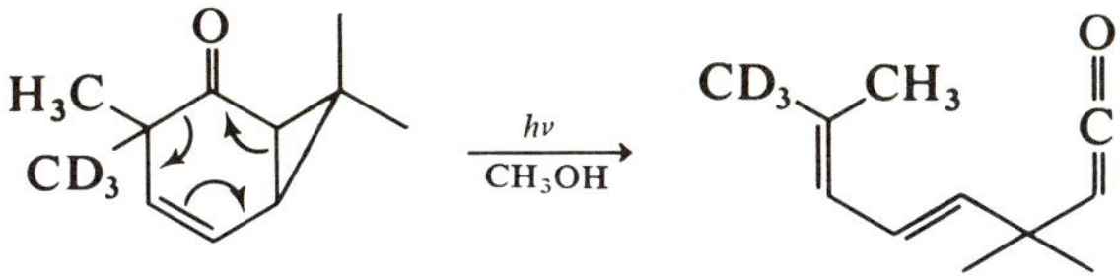

[Structures 164 → 165 shown]

The bicyclic enone **166** undergoes what is formally a di-π-methane rearrangement to **167** and loss of ketene when irradiated in acetone [398, 441; see also 442]:

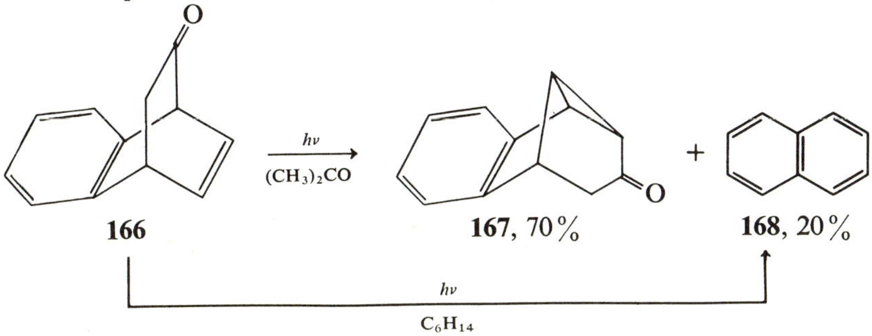

[Structures 166 → 167, 70% + 168, 20%; with secondary hν/C₆H₁₄ pathway]

Direct excitation results mainly in the loss of ketene. Irradiation of the dibenzo analog of **166** also causes a loss of ketene to yield anthracene.

Direct irradiation of **169** yields **170** and **171**. The ratio **170:171** was initially high, but **170** is photolyzed to **171** [442a]. Irradiation of **169** in acetone yields **172**:

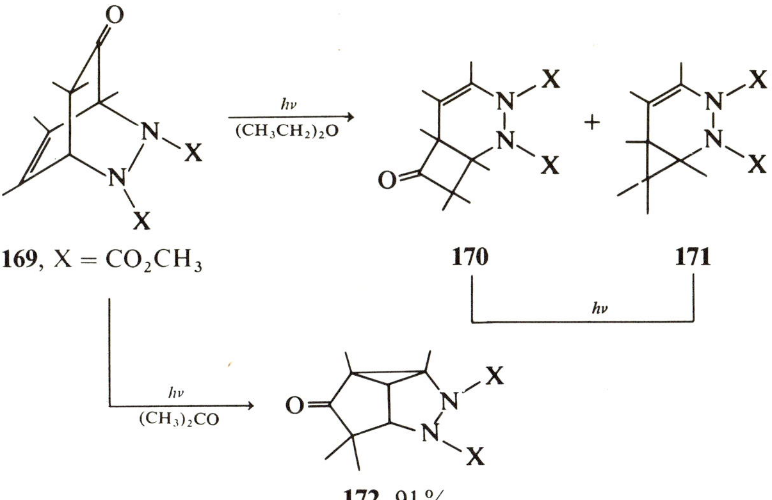

169, X = CO₂CH₃   **170**   **171**

172, 91%

On irradiation, 4-phenyl-3-chromanone **173** rearranges to 4-phenyldihydrocoumarin **176** [443], possibly by way of the cyclopropanone **175**:

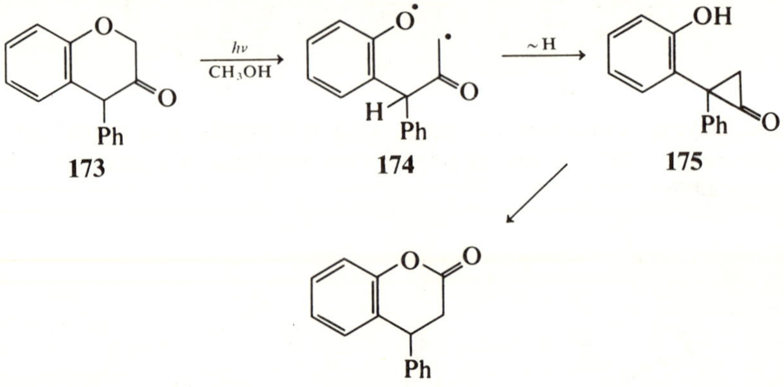

176, 35%

An ESR study of the lowest triplet state of coumarin and related compounds has appeared [444].

Irradiation of the pyranosidulose **177** results in the formation of the enone **179**, possibly by way of the type II enol **178**, which then loses methanol [445; see also 445a]:

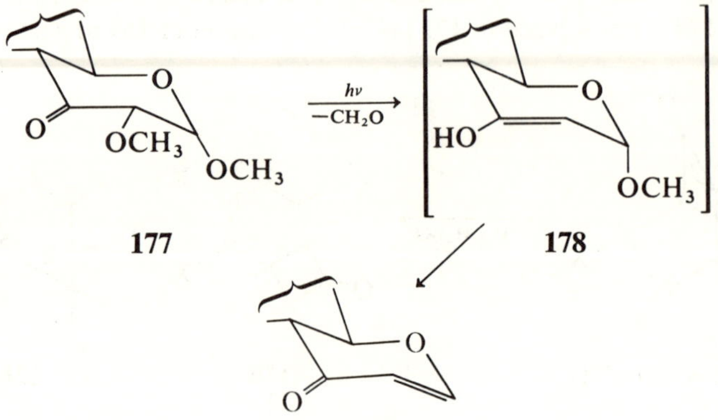

Compounds related to **177** but not having available γ-hydrogens undergo photodecarbonylation [446].

A number of studies of β-thiocycloalkanones (see Volume 2, page 56) have appeared [447, 448, 449, 450, 451, 452]. Several of these are listed, for example, **180** → **181** and **182** [448]; **183** → **184** and **185** [448]; **186** → **187** [452]:

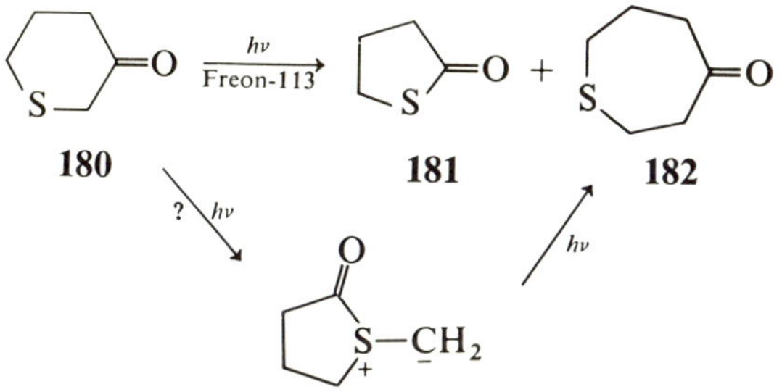

180 → **181** + **182** (hv, Freon-113)

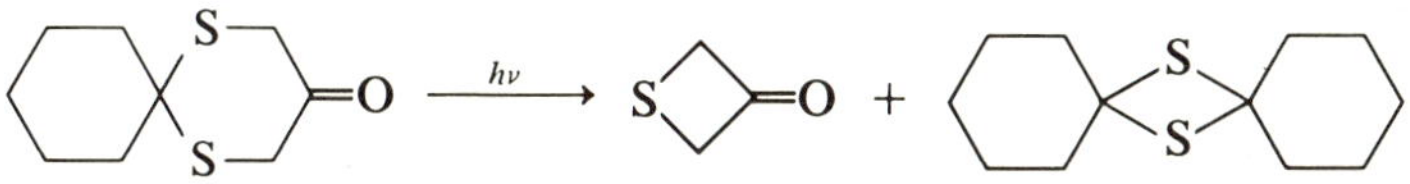

183 → **184** + **185**

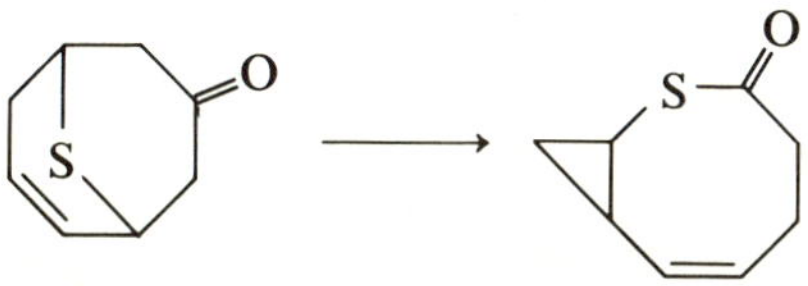

186 → **187**, 35%

Some further interesting photoreactions of unsaturated cycloheptanone derivatives have appeared [453, 454, 455]. The rearrangements of **188** → **189** [453, 456] and **190** → **191** + **192** merit citation:

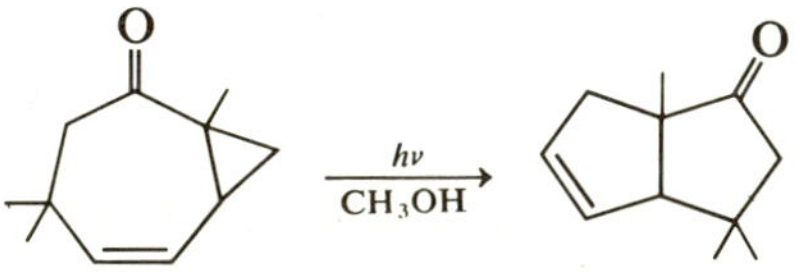

188 → **189**, 20% (hv, CH$_3$OH)

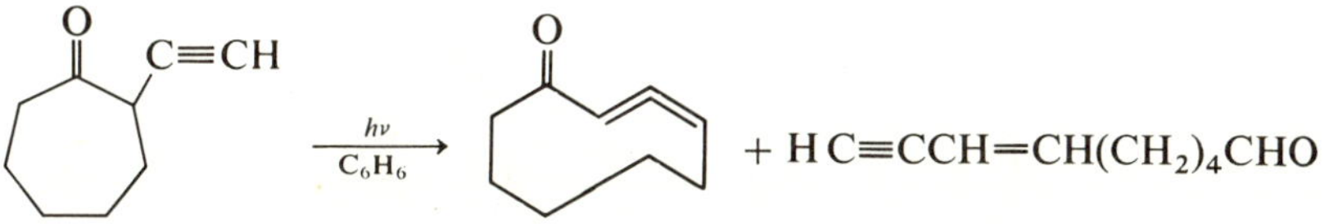

190 → **191** + HC≡CCH=CH(CH$_2$)$_4$CHO **192** (hv, C$_6$H$_6$)

These reactions could not be sensitized or quenched.

Irradiation of *cis*-fused ketone **192a** in ether gives rise to photoreduction products. However, irradiation of **192a** in *t*-butanol produces rearrangement products presumably by way of initial α cleavage:

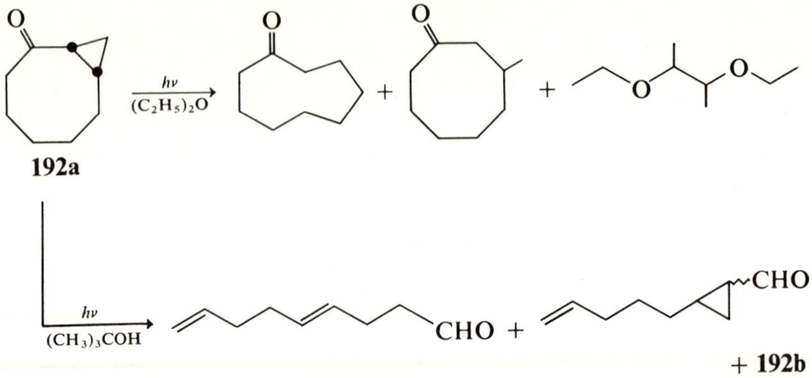

Photolysis of the *trans*-isomer **192b** leads to a cyclopropyl ester. The intermediacy of ketene **192c** in this reaction is supported by the location of the deuterium atom in the product formed in *t*-butanol-*O*-*d* [456a]:

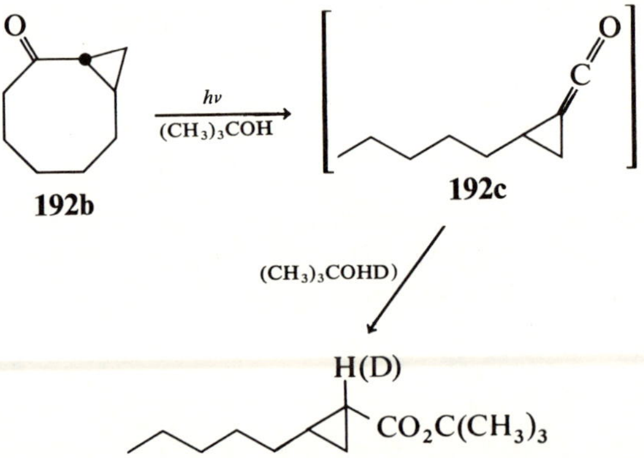

An earlier report of an intramolecular double-bond cycloaddition that may proceed by way of intramolecular energy transfer from an excited carbonyl group [167, 457, 458, 458a] has been followed by the example of **193 → 194** [52]:

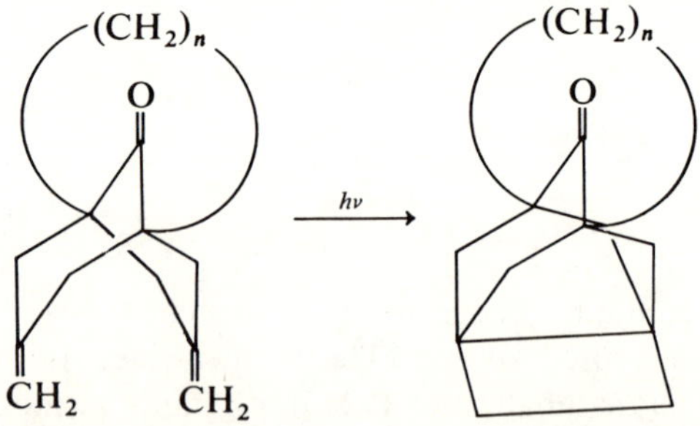

Photolysis of ketone **194a** in methanol results in the formation of ester **194b** presumably by way of a ketene intermediate [458b]:

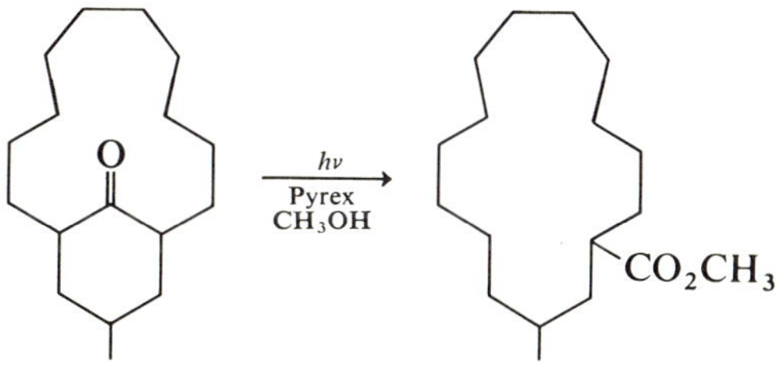

194a → **194b**, 33%
(hν, Pyrex, CH₃OH)

II. ARYL ALDEHYDES AND KETONES

A. Aryl Aldehydes

The pinacol **196** is formed when toluene solutions of 9-anthraldehyde **195** are irradiated [459]:

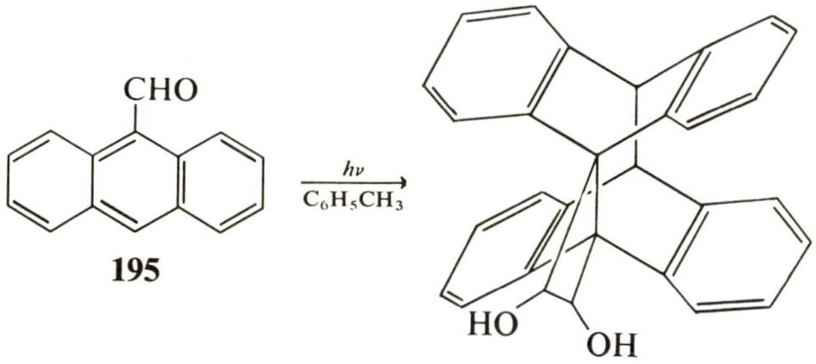

195 → **196**
(hν, $C_6H_5CH_3$)

It has been suggested that **196** is formed by way of the photoreduction of **195** to yield a pinacol followed by intramolecular photocyclization. It was found that a plot of log (Φ_X/Φ_H) for photoreduction of para-substituted toluenes (X = para substituent other than hydrogen) versus σ^+ yields a ρ^+ value of -1.31 [459]. The negative ρ^+ value suggests an electrophilic attack on toluene by the excited anthraldehyde, which may react by way of its n,π^* state.

Irradiation of benzaldehyde in the presence of ethyl vinyl ether yields the oxetanes **197** and **198** [401]:

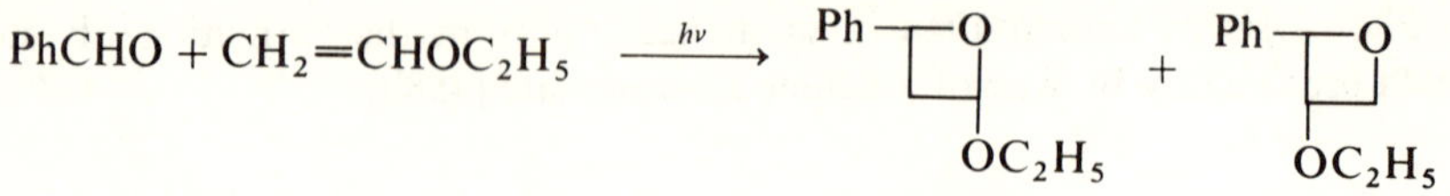

$$\text{PhCHO} + \text{CH}_2\text{=CHOC}_2\text{H}_5 \xrightarrow{h\nu} \underset{\underset{\text{197, 25\%}}{\text{OC}_2\text{H}_5}}{\text{Ph}\!\!-\!\!\square\!\!-\!\!\text{O}} \;+\; \underset{\underset{\text{198, 60\%}}{\text{OC}_2\text{H}_5}}{\text{Ph}\!\!-\!\!\square\!\!-\!\!\text{O}}$$

Irradiation of *o*-nitrobenzaldehyde results in oxidation to *o*-nitrobenzoic acid [344].

The photochemical decompositions of 1-naphthylacetic acid [459a] and its sodium salt [459b] have been reported.

B. Aryl Alkyl Ketones

The observation of phosphorescence from acetophenone in purified, degassed hydrocarbon or fluorinated solutions has opened up a new method of determining kinetics of acetophenone reactions [460, 461, 462; see also 462a].

Although the self-quenching of aryl ketones having n,π^* triplet states is generally negligible, certain aryl ketones having lowest π,π^* triplet states have been found to undergo strong self-quenching, an observation of importance for consideration when such compounds are used as photosensitizers [463].

Comparison of the decay ratio for photosensitized *cis-trans* isomerization of 2-pentene by acetophenone, with the decay ratio achieved by other sensitizers, suggests that the Schenck addition-elimination mechanism (Volume 2, page 56) for isomerization may hold for sensitization by acetophenone [46]. Since oxetane formation is a minor reaction, the biradical intermediate must collapse to alkene and ketone much more efficiently than it cyclizes.

A number of further studies of the photoreduction of acetophenone [464, 465, 466] and substituted acetophenones [467, 468, 469, 470, 471] have been reported. Irradiation of acetophenone in ether-cyclohexane mixtures leads to the formation of **199**, in addition to the expected photoreduction products [464] (see Table VI, this chapter, for kinetic data on photoreduction in isopropanol:

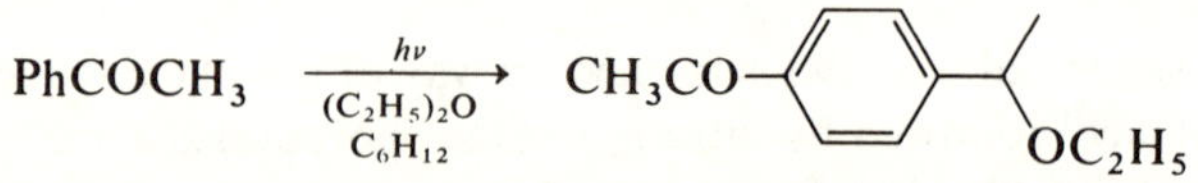

$$\text{PhCOCH}_3 \xrightarrow[\substack{(C_2H_5)_2O \\ C_6H_{12}}]{h\nu} \text{CH}_3\text{CO}\!-\!\!\langle\!\!\bigcirc\!\!\rangle\!\!-\!\!\text{CH(OC}_2\text{H}_5)$$

199, 3%

Benzylacetophenone has been found to undergo slow photoreduction to a pinacol with isopropanol [468], although the rate of pinacolization seems abnormally slow.

The type II reactions of butyrophenone continue to attract considerable attention [472, 473, 474, 475, 476, 477, 478, 479, 470]. A polar substituent effect has been found to operate on the collapse of the 1,4-biradical intermediate in type II cleavage of butyrophenones [473]. As shown in Table III,

TABLE III

Substituent Effects on Quantum Yields for Type II Photoelimination of n-Butyl Phenyl Ketones [a]

Ketone	$\Phi(C_6H_6)$ [b]	$\Phi(alc)$ [c]	β [d]
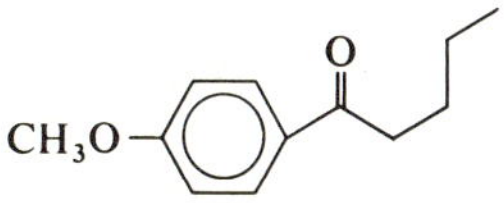 CH$_3$O-C$_6$H$_4$-CO-n-Bu	0.18	0.26	0.69
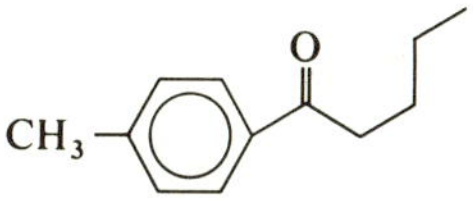 CH$_3$-C$_6$H$_4$-CO-n-Bu	0.39	0.80	0.50
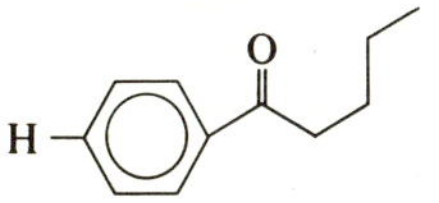 H-C$_6$H$_4$-CO-n-Bu	0.40	1.00	0.40
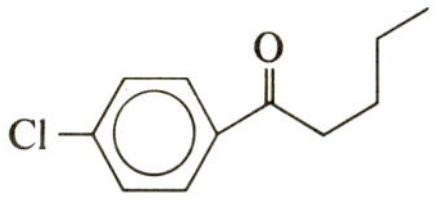 Cl-C$_6$H$_4$-CO-n-Bu	0.30	0.80	0.37
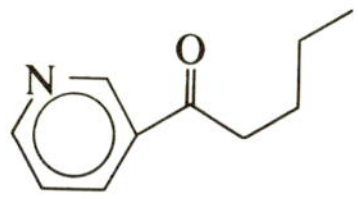 2-pyridyl n-butyl ketone	0.29	1.00	0.29
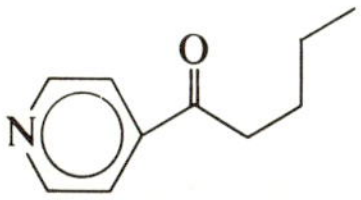 4-pyridyl n-butyl ketone	0.18	1.00	0.18

[a] Taken from Ref. 473.
[b] Total quantum yield for reaction of ketone in benzene.
[c] Maximum total quantum yield for reaction of ketone reached in benzene with added t-butyl alcohol.
[d] The probability that the biradicals formed by intramolecular γ-hydrogen abstraction in benzene proceed on to products rather than revert back to the initial ketone.

electron-withdrawing substituents, which increase the reactivity toward the initial hydrogen abstraction, favor reversion of the biradical to the initial ketone by disproportionation and thus result in a lower overall quantum yield for product formation. This is another indication of the lack of correlation between quantum yields and reactivities.

β,γ-Diphenylbutyrophenone undergoes type II cleavage to yield acetophenone and stilbene [477, 476, 478], the latter being mainly (>98%) trans:

$$\text{PhCOCH}_2\text{CH(Ph)CH}_2\text{Ph} \xrightarrow{h\nu} \text{PhCOCH}_3 + \text{PhCH}=\text{CHPh}$$
$$\qquad\qquad 200 \qquad\qquad\qquad\qquad 201 \qquad\quad 202$$

This result has been interpreted as evidence that the type II triplet biradical does not collapse concertedly to acetophenone enol and triplet stilbene, since the latter would be expected to produce a much lower trans/cis ratio.

TABLE IV

Type II Photoelimination of β- and γ-Phenyl Butyrophenones [a]

Structure	Φ_{II} [b]	τ_T, 10^{-9} sec
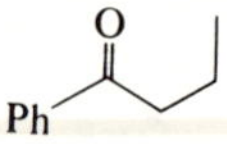 Ph-CO-CH₂-CH₂-CH₃	0.45	130
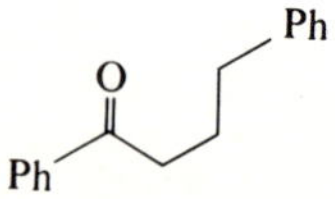 Ph-CO-CH₂-CH(Ph)-CH₃	0.50	2.5
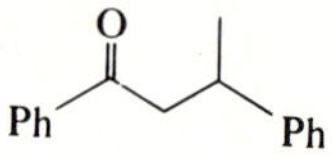 Ph-CO-CH(Ph)-CH₃	0.001	0.5
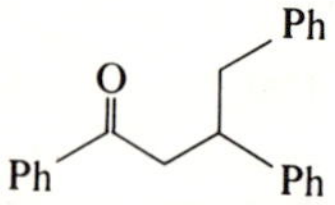 Ph-CO-CH(Ph)-CH₂-Ph	0.09, 0.08,[d] 0.11 [e]	0.5 [a,d,e]

[a] Except where noted, taken from Ref. 478.
[b] Quantum yield for type II reaction.
[c] Triplet state lifetime.
[d] From Ref. 476
[e] From Ref. 477

It also appears that β-phenyl ketones undergo type II cleavage with relatively low efficiency (see Table IV). This low efficiency may result from some very rapid deactivation process involving the nonconjugated 2-phenyl group [478].

TABLE V
Photoreactivity of Pyridyl Ketones toward Type II Photoelimination [a]

Compound	$1/\tau_T$, 10^8 sec^{-1} [b]	$\Phi_{II}(C_6H_6)$ [c]	$\Phi_{II}(t\text{-BuOH})$ [d]
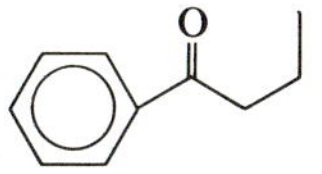 phenyl propyl ketone	0.076	0.35	0.80
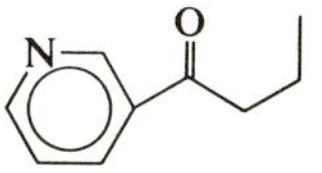 2-pyridyl propyl ketone	0.26	0.28	0.48
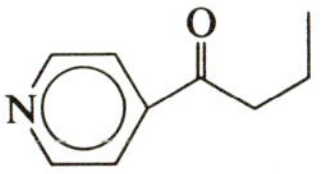 4-pyridyl propyl ketone	0.62	0.23	0.62
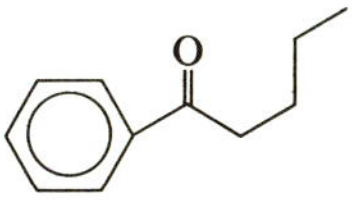 phenyl butyl ketone	1.4	0.33	0.90
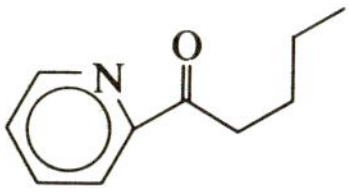 2-pyridyl butyl ketone	1.9	0.16	0.79
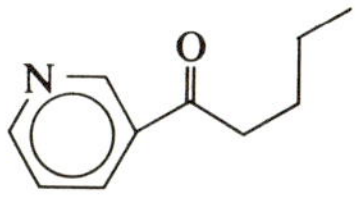 3-pyridyl butyl ketone	3.1	0.23	0.57
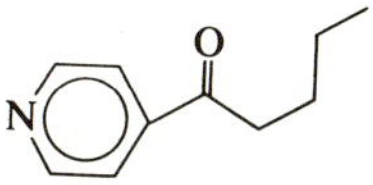 4-pyridyl butyl ketone	6.8	0.25	0.57

[a] From Ref. 475.
[b] Reciprocal of triplet lifetime.
[c] Quantum yield for acetophenone or acetylpyridine formation in benzene.
[d] Quantum yield for acetophenone or acetylpyridine formation in *tert*-butanol.

Both α-aminoacetophenones [470] and γ-aminoacetophenones [479] undergo type II reactions:

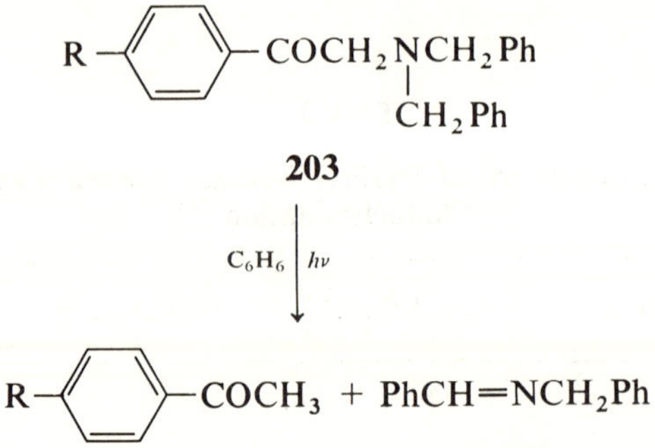

R—C$_6$H$_4$—COCH$_2$NCH$_2$Ph
          |
          CH$_2$Ph
203

↓ C$_6$H$_6$, hv

R—C$_6$H$_4$—COCH$_3$ + PhCH=NCH$_2$Ph
204   **205**

PhCOCH$_2$CH$_2$CH$_2$N(CH$_3$)$_2$
206

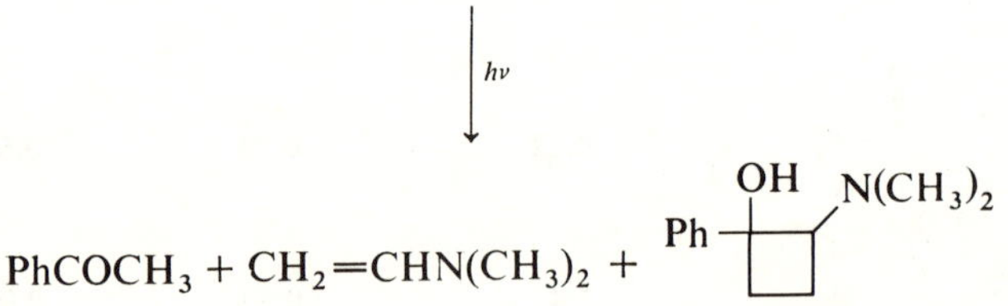

↓ hv

PhCOCH$_3$ + CH$_2$=CHN(CH$_3$)$_2$ + Ph—[cyclobutane with OH, N(CH$_3$)$_2$]
207   **208**

It has been suggested that electron transfer followed by proton transfer best explains the reactivity of **203** (R= Ph), which appears to have a $T_1(\pi,\pi^*)$ state [470, 479].

The results of a quantitative study of the type II reaction of pyridyl ketones are summarized in Table V [475]:

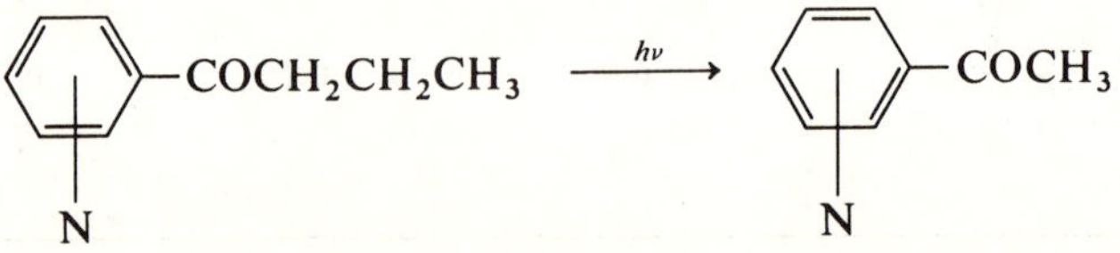

pyridyl—COCH$_2$CH$_2$CH$_3$ —hv→ pyridyl—COCH$_3$

Irradiation of certain 2,4,6-trimethylphenyl ketones **209** results in the formation of benzocyclobutenols [480] when the R group is not too bulky. On the other hand, when the steric bulk of the R group is sufficient to twist the acyl group out of conjugation, fragmentation of the carbonyl-alkyl bond becomes significant. The intermediacy of **211** is indicated by deuterium exchange and oxygen trapping experiments [481]:

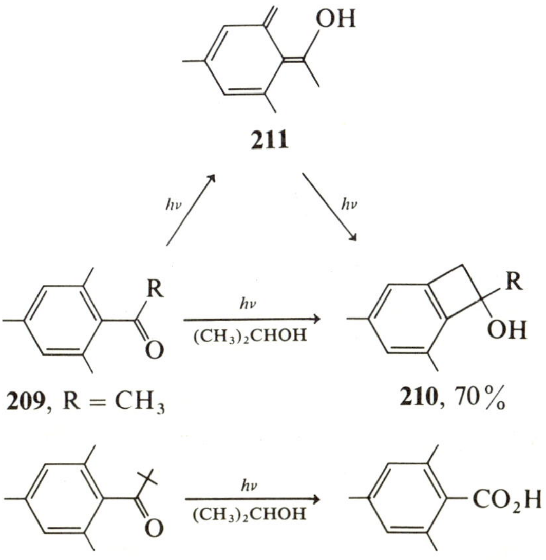

211

209, R = CH$_3$ → (via (CH$_3$)$_2$CHOH, hv) → **210**, 70%

212 → (via (CH$_3$)$_2$CHOH, hv) → **213**, 48% (-CO$_2$H)

These results may be explained as the consequence of the stereoelectronic requirements for intramolecular hydrogen abstraction. In the case in which R is quite bulky (**212**), the expected conformation **212a** is inappropriate for photoenolization, and instead type I cleavage predominates [480]:

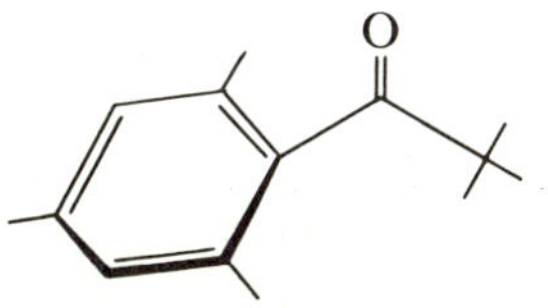

212a

The reactivity toward the type I cleavage also increases as substitution makes R· a more stable radical.

Irradiation of pivalophenone **214** results in the formation of isobutylene and hydrobenzoin **215** [482], presumably by way of type I cleavage and the subsequent photoreduction of intermediary benzaldehyde:

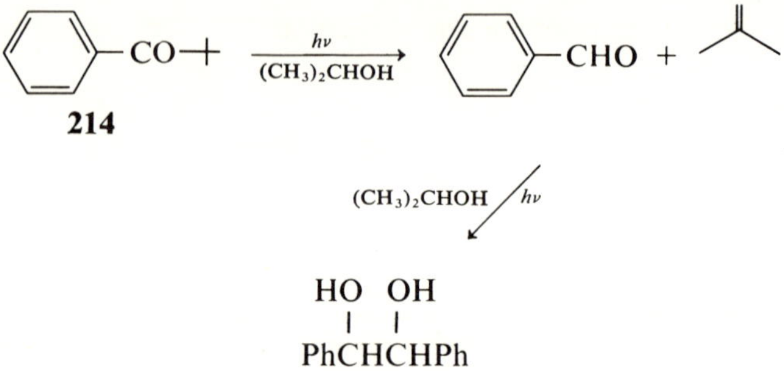

$$\text{Ph-CO-+} \xrightarrow[(CH_3)_2CHOH]{h\nu} \text{Ph-CHO} + (CH_3)_2C=O$$

214

$$\downarrow (CH_3)_2CHOH / h\nu$$

$$\begin{array}{c} \text{HO OH} \\ |\ \ | \\ \text{PhCHCHPh} \end{array}$$

215, 42%

Cyclobutyl phenyl ketone **216** undergoes photoreduction to pinacol **217** in isopropanol ($k_r \sim 10^5\ M^{-1}\ \text{sec}^{-1}$), and in benzene **218**, **219**, and **220** are formed ($k_r \sim 6 \times 10^3\ M^{-1}\ \text{sec}^{-1}$) in addition to **217** [483]:

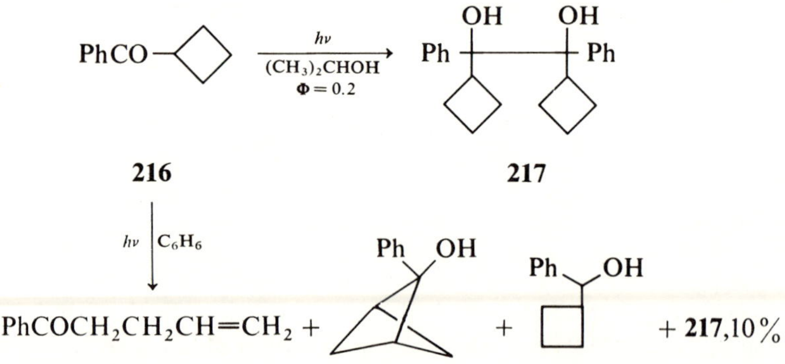

$$\text{PhCO-}\square \xrightarrow[\substack{(CH_3)_2CHOH \\ \Phi = 0.2}]{h\nu} \text{Ph}\underset{\square}{\overset{OH}{\mid}}\!\!\!-\!\!\!\underset{\square}{\overset{OH}{\mid}}\text{Ph}$$

216   **217**

$$\downarrow h\nu\ |\ C_6H_6$$

$$\text{PhCOCH}_2\text{CH}_2\text{CH=CH}_2\ +\ \triangle\!\!\!-\!\!\!\overset{Ph}{\underset{OH}{}}\ +\ \square\!\!\overset{Ph}{\underset{OH}{}}\ +\ \mathbf{217}, 10\%$$

218, 24%   **219, 38%**   **220, 8%**

The inefficiency and low rate of the photoisomerization of **216** have been attributed to the unfavorable geometry for initial hydrogen abstraction available to this ketone.

The phenylcyclobutyl ketone **221** undergoes isomerization to **222**, which in turn undergoes ring opening to **223** [483]:

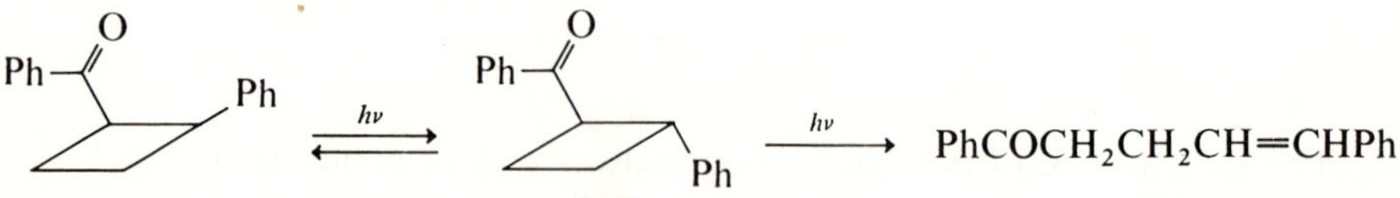

$$\underset{\mathbf{221}}{\text{Ph-CO-}\square\!\!\!-\!\!\!\text{Ph}} \xrightleftharpoons{h\nu} \underset{\mathbf{222}}{\text{Ph-CO-}\square\!\!\!-\!\!\!\text{Ph}} \xrightarrow{h\nu} \underset{\mathbf{223}}{\text{PhCOCH}_2\text{CH}_2\text{CH=CHPh}}$$

Cyclopentyl phenyl ketone **224** undergoes type II cleavage to **225** ($k_r = 7 \times 10^6\ \text{sec}^{-1}$), which in turn is photoisomerized to **226** and **227** ($k_r = 4 \times 10^7\ \text{sec}^{-1}$) [484]:

PhCO–[cyclopentyl] $\xrightarrow{h\nu}$ PhCOCH$_2$CH$_2$CH$_2$CH=CH$_2$

224   **225**

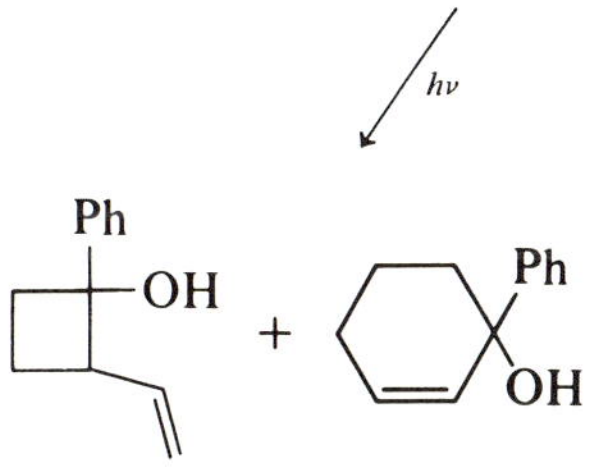

$\downarrow h\nu$

[cyclobutanol with Ph, OH, vinyl] + [cyclohexenol with Ph, OH]

226   **227**

Although the cyclohexyl phenyl ketone **229** does not display type II reactivity, **228** undergoes a "double type II" cleavage [445]:

PhCO–[cyclohexyl]   PhCO–[cyclohexyl]–

228   **229**, no type II

$h\nu \downarrow$

PhCO(CH$_2$)$_4$CH=CH$_2$ $\xrightarrow{h\nu}$ PhCOCH$_3$ + [diene]

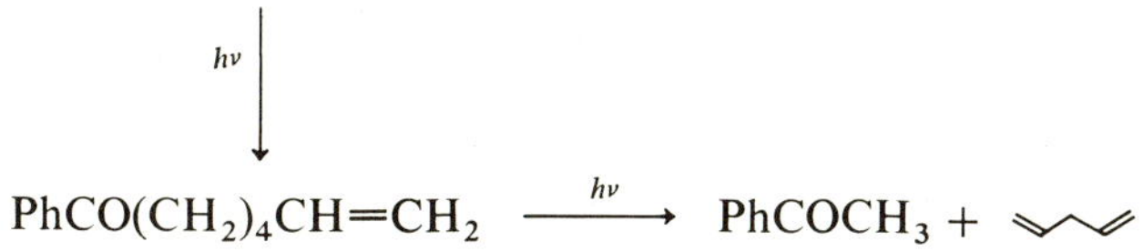

230

The differing emission properties of indanone and its alkylated derivatives have been ascribed to enolization reactions [485].

Intramolecular photocyclizations of the bicyclic ketones **231** and **232** have been reported [371, 372]:

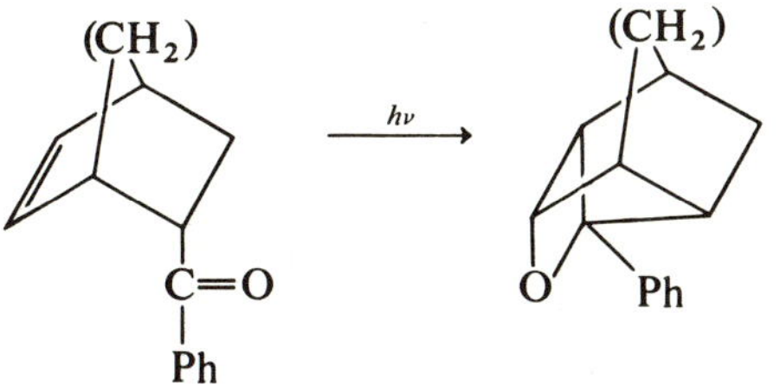

[bicyclic ketone structure with (CH$_2$)$_n$ bridge and C=O, Ph] $\xrightarrow{h\nu}$ [cage structure with (CH$_2$)$_n$, O, Ph]

231, $n = 1$ **233**, $n = 1$, 20%
232, $n = 2$ **234**, $n = 2$

Irradiation of the dione **235** results in isomerization to a compound for which the structure **236** has been proposed [486]:

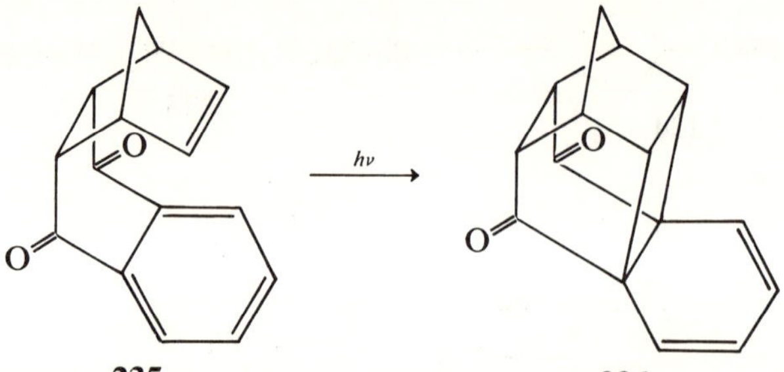

235 → **236**

Study of the series **237**, revealed that for $n = 1$ or $n = 2$, the only important photoreaction was energy transfer from the acetophenone group to the β-methylstyryl group, as evidenced by the *trans-cis* isomerization of the latter [487]. For $n = 3$ or $n = 4$, β cleavage becomes a significant reaction, but energy transfer still predominates:

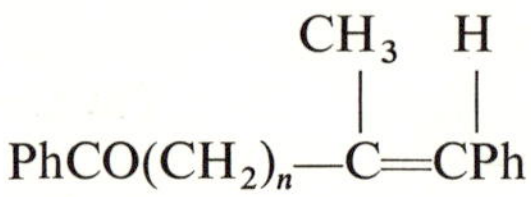

$$\text{PhCO(CH}_2)_n\text{—}\overset{\overset{\displaystyle CH_3}{|}}{C}=\overset{\overset{\displaystyle H}{|}}{C}\text{Ph}$$

237, $n = 1, 2, 3$ or 4

The ketone **238** is photoreduced to the pinacol **239** by ether [412]:

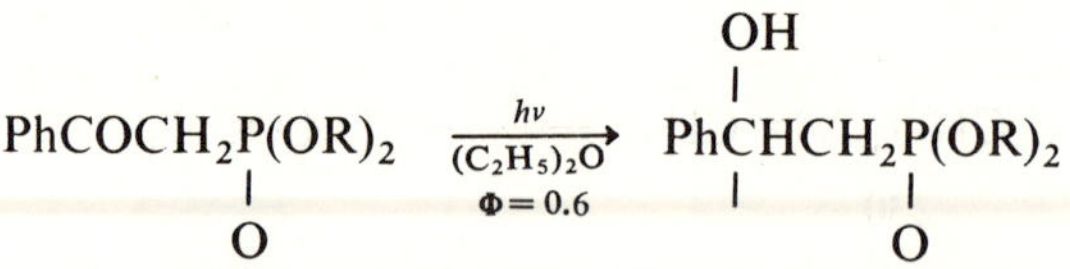

$$\text{PhCOCH}_2\underset{\underset{\displaystyle O}{|}}{P}(OR)_2 \xrightarrow[\Phi = 0.6]{\underset{(C_2H_5)_2O}{hv}} \text{PhCH(OH)CH}_2\underset{\underset{\displaystyle O}{|}}{P}(OR)_2$$

238   **239**

Irradiation of α-thia (and related) acetophenones generally results in β cleavage followed by secondary reactions of the radicals produced [488, 489], for example, **240** → **241** + **242** [see, however, 491]:

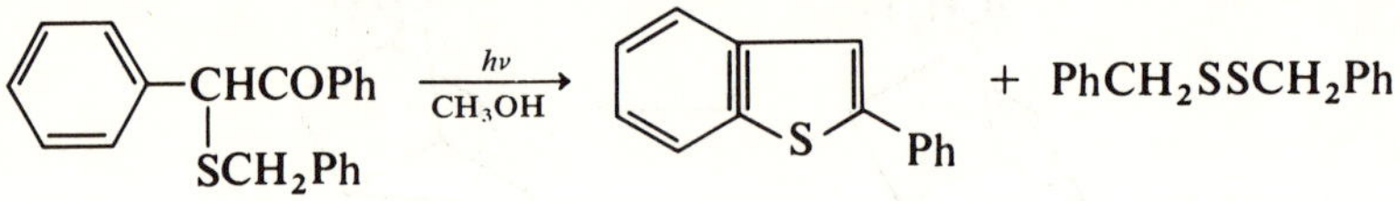

PhCH(SCH$_2$Ph)COPh $\xrightarrow{\underset{CH_3OH}{hv}}$ (benzothiophene-2-Ph) + PhCH$_2$SSCH$_2$Ph

240   **241**   **242**

Irradiation of **243** yields **244**, which undergoes type II cleavage to yield **245** [492]:

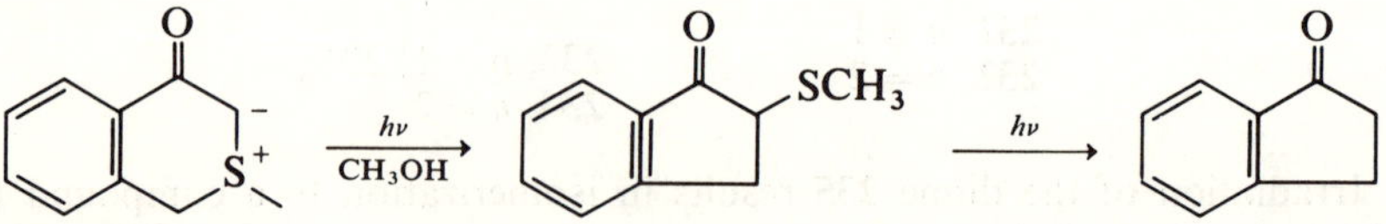

243 $\xrightarrow{\underset{CH_3OH}{hv}}$ **244** $\xrightarrow{hv}$ **245**

In contrast to the reported inertness of naphthyl ketones toward intermolecular oxetane formation, **246** undergoes intramolecular cyclization to yield **247** [372]:

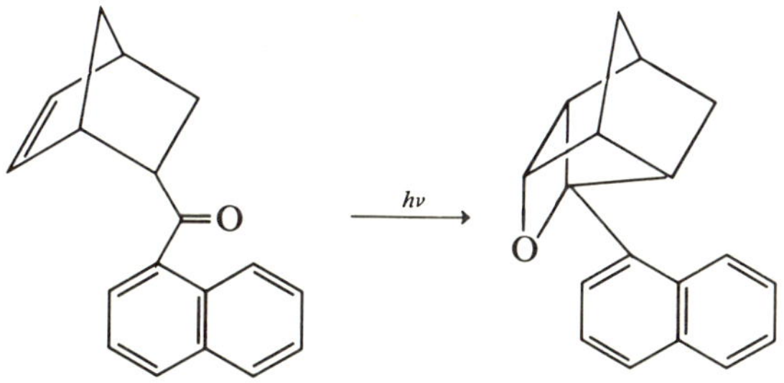

246 $\xrightarrow{h\nu}$ **247**

Irradiation of 2-acetylnaphthalene **248** and *trans*-methyl cinnamate **249** yields *cis*-methylcinnamate **251** and the $2\pi + 4\pi$ photoadduct **250** [493]:

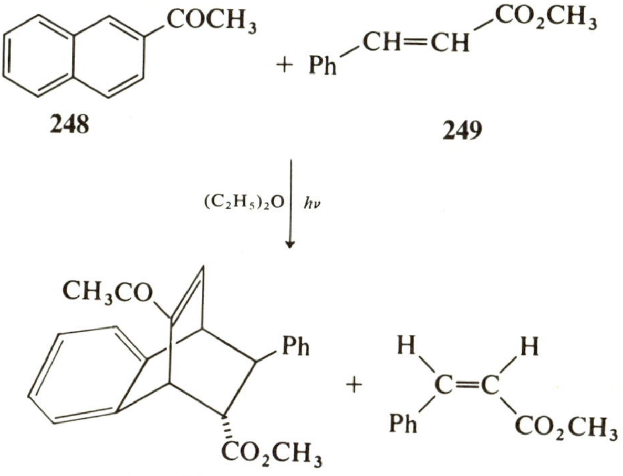

248 + Ph–CH=CH–CO_2CH_3 **249**

$(C_2H_5)_2O \mid h\nu$

250 + **251**

Since the photoisomerization of the cinnamate takes place much more rapidly than the photocycloaddition reaction, **250** could arise from the addition of **248** to either **249** or **251**. This is one of the rare examples of known photochemistry involving a reaction of the naphthalene nucleus. Compound **248** has a lowest $^3(\pi,\pi^*)$ state, which is known to be relatively inert [471] to the usual photoreactions of aryl ketones having a lowest $^3(n,\pi^*)$ state [see also 493a]. The formation of **250** provides additional evidence that excitation energy is not localized in the carbonyl group. ESR studies of the triplet states of naphthyl carbonyl compounds further confirm this thesis [494].

Benzil and phenanthraquinone sensitize the photorearrangements of the spiroketones **252** to dioxenes **253** and fragmentation products [495]:

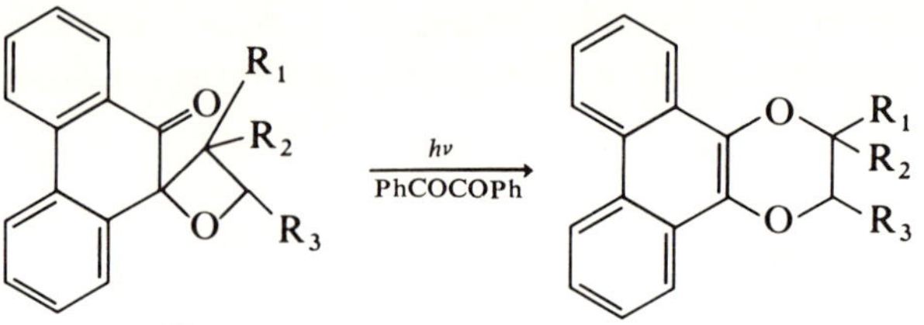

[Structure 252] →(hv, PhCOCOPh)→ [Structure 253]

This sensitization may represent a low-energy triplet energy transfer that results in the rupture of a weak single bond, although it may also be an example of chemical sensitization. Interestingly, the same array of products is achieved by the irradiation of phenanthraquinone and $R_1R_2C{=}CHR_3$ (Volume 2, page 97, Chapter 4).

C. Diaryl Ketones

The phosphorescence emission of benzophenone in isooctane has been reported and may be of use as a tool in mechanistic studies of benzophenone photochemistry [460, 461, 462; see also 317].

The intermolecular photoreduction of benzophenone [496, 497, 498, 499, 500, 501, 502, 503, 504, 505, 506, 507, 508, 509, 471] and substituted benzophenones [510, 511, 512, 513, 514, 471, 467] has continued to attract attention. Table VI summarizes some recent kinetic data on the photoreduction of benzophenones by isopropanol [467].

A rather remarkable result was discovered by labeling benzophenone or benzhydrol [499]. It was found that, contrary to expectations, the photoreaction of benzophenone with benzhydrol to yield benzpinacol does not proceed by way of the direct combination of free ketyl radicals [506]. Thus, **254** leads only to **255** at low conversions:

$$2(C_6D_5)_2CO + (C_6H_5)_2CHOH \xrightarrow[C_6H_6]{hv} (C_6D_5)_2\underset{\underset{255}{|}}{\overset{\overset{OH}{|}}{C}}{-}\underset{\underset{}{|}}{\overset{\overset{OH}{|}}{C}}(C_6D_5)_2 + (C_6H_5)_2CO$$
$$\text{254}$$

Just as striking is the result that irradiation of perdeuterobenzophenone **254** and benzpinacol yields almost exclusively perdeuterobenzpinacol (at low conversions):

$$2(C_6D_5)_2CO + (C_6H_5)_2\overset{\overset{OH}{|}}{C}{-}\overset{\overset{OH}{|}}{C}(C_6H_5) \xrightarrow[\Phi \sim 0.1]{\underset{C_6H_6}{hv}}$$
$$\text{254}$$

$$2(C_6H_5)_2CO + (C_6D_5)_2\overset{\overset{OH}{|}}{C}{-}\overset{\overset{OH}{|}}{C}(C_6D_5)_2$$
$$\text{255}$$

TABLE VI

Kinetic Data on Photoreduction of Benzophenones and Acetophenones by Isopropanol

Ketone	Φ [a]	$k_r \times 2\,M \times 10^{-5}$	$k_t \times 10^{-5}$
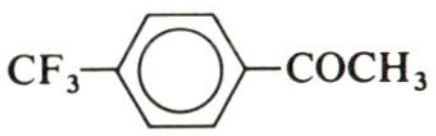 CF_3–C$_6$H$_4$–$COCH_3$	0.72	55.2	21.0
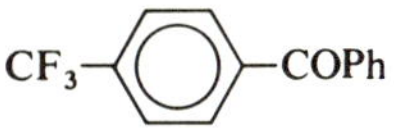 CF_3–C$_6$H$_4$–$COPh$	0.62	43.8	27.7
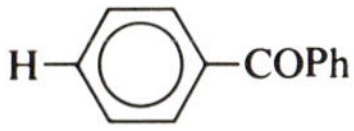 H–C$_6$H$_4$–$COPh$	0.72	25.0	7.28
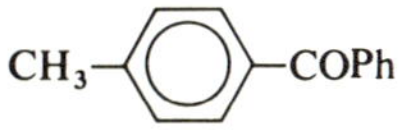 CH_3–C$_6$H$_4$–$COPh$	0.70	17.0	6.15
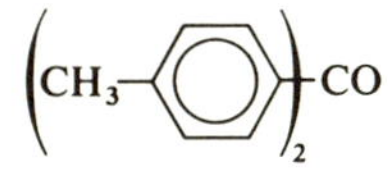 $(CH_3$–C$_6$H$_4)_2$–CO	0.71	13.6	5.13
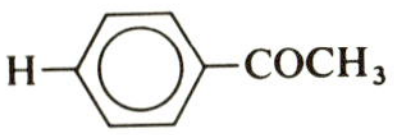 H–C$_6$H$_4$–$COCH_3$	0.35	8.66	15.9

[a] Quantum yield of photoreduction (0.2 M ketone and 2.0 M isopropanol in benzene).

It is also important to photoreduction studies that aryl pinacol can be cleaved photochemically to α-hydroxyarylmethyl radicals [207].

Irradiation of polyvinylbenzophenone leads to photoreduction [515].

Secondary isotope effects have been proposed as photochemical mechanistic criteria. Thus, a study of the benzophenone photosensitized isomerization of 1,2-dichloroethylene-d revealed that $(k_H/k_D) = 1.6$, and no detectable yield of oxetane was obtained. Since the excited carbonyl was expected to show an inverse isotope effect $(k_H/k_D) \sim 0.95$ and sensitization is clearly not diffusion-controlled in this case, it was suggested that both a Schenck and energy-transfer mechanism for sensitization are not likely. On the other hand, the benzophenone-photosensitized isomerization of the 2-butenes showed no deuterium isotope effect and was accompanied by oxetane formation. These data were taken as evidence for the existence of a precursor to the 1,3-biradical often envoked as the result of interaction of excited ketones and olefins [47, 516].

The use of t-BuNO as a trap for ketyl radicals produced by the irradiation of benzophenone in the presence of hydrogen donors has been demonstrated [28, 517] (see also Chapter 5).

Several quantitative studies of the photoreduction of benzophenone [501, 507, 503] and substituted benzophenones (511, 512) by alcohols have appeared. Reports have also been published on the photoreduction of benzophenone by ethers [504], aromatic hydrocarbons [502], amides [496], triphenyl silane [497], phenols [497], and amines [498, 500]. A quantitative study of the photoreduction of basic aqueous amine solutions of **256** has been reported [513]:

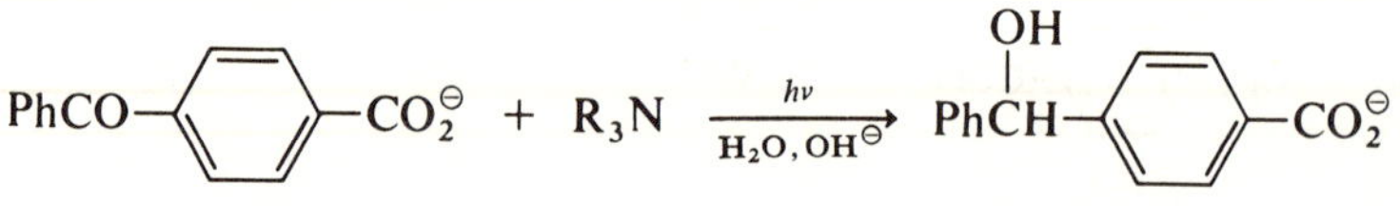

$$\text{PhCO}-\!\!\left\langle\;\right\rangle\!\!-\text{CO}_2^\ominus + R_3N \xrightarrow[H_2O, OH^\ominus]{h\nu} \text{PhCH}(\text{OH})-\!\!\left\langle\;\right\rangle\!\!-\text{CO}_2^\ominus$$

256

The use of chemically induced dynamic nuclear polarization in mechanistic photochemistry has been pointed out [502]. The ability of aminobenzophenones to sensitize stilbene *cis-trans* isomerization [514] and to emit light [518] were found to be solvent sensitive.

The rate constant of the addition of benzophenone to ketenimines has been found to be close to that for diffusion-controlled quenching [519]. The photocycloaddition of benzophenone and other ketones to ethylene carbonate has been reported [520];

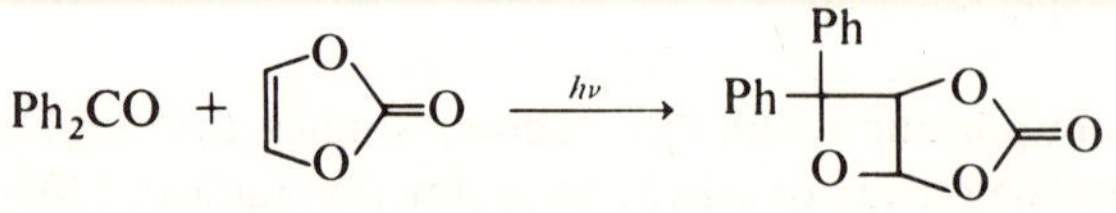

$$\text{Ph}_2\text{CO} + \underset{O}{\overset{O}{\bigcirc}}\!=\!O \xrightarrow{h\nu} \text{Ph}_2\text{C}(O\text{---})(O\text{---})C\!=\!O$$

Benzophenone has been found to photoadd smoothly to vinyl ethers [399, 401].

Although 4-hydroxyacetophenone is unreactive to photoreduction in both isopropanol and cyclohexane, 4-hydroxybenzophenone showed some photoreduction in isopropanol ($\Phi = 0.03$) and was much more reactive in cyclohexane [521]. 4-Phenylbenzophenone, which is believed to have a lowest π,π^* triplet state, is photoreduced to a diol by triethylamine [522]. Thioxanthone triplets have been found to be quenched by ground-state thioxanthone molecules at close to the diffusion-controlled rate [463].

Although **257** is photochemically inert in isopropyl alcohol, the irradiation of **257** in cyclohexane results in the loss of a *t*-butyl group, by way of its triplet state [521]:

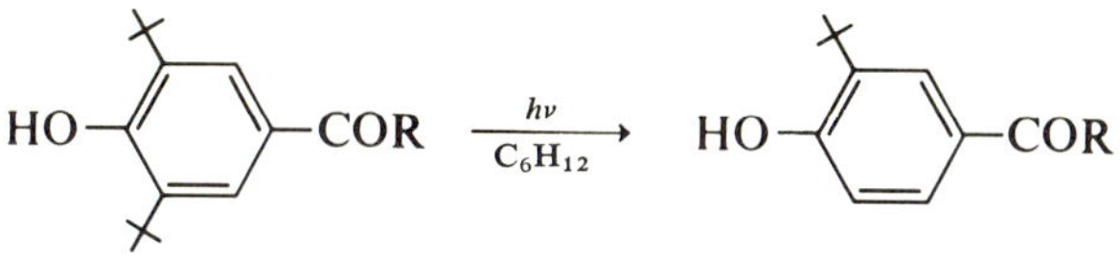

257, R = CH$_3$ or Ph

258, R = CH$_3$, 95%
259, R = Ph, 50%

In triethylamine **257** is photoreduced to a mixture of the corresponding hydrol and pinacol. Compound **260** is photoinert, as expected, under a variety of experimental conditions, and **261** is readily photoreduced to the corresponding pinacol by isopropyl alcohol [see also 523]:

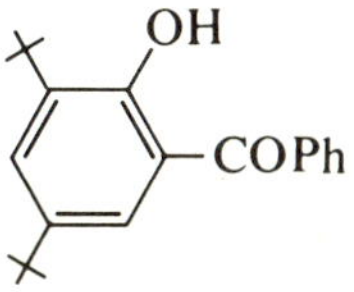

260, photoinert

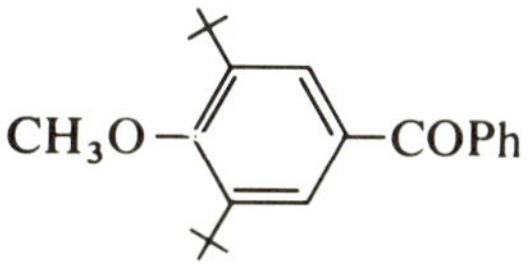

261, readily photoreduced

Irradiation of **261** in cyclohexane through which oxygen is continuously bubbled yields the quinone **262** and benzoic acid:

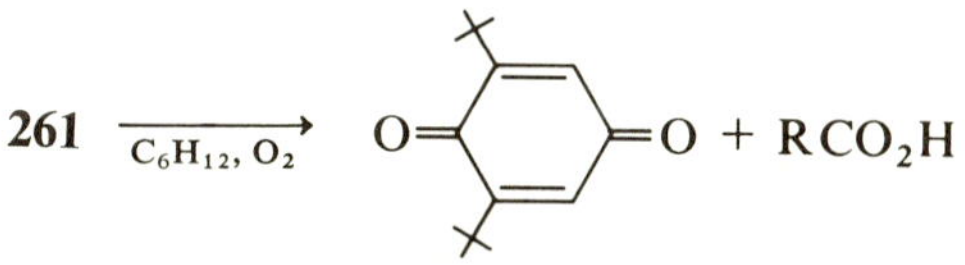

$$261 \xrightarrow{C_6H_{12},\ O_2} O=\!\!\bigcirc\!\!=O + RCO_2H$$

R = CH$_3$ or Ph

262, 12–13%

2,4,6-Trimethylbenzophenone is photoreduced to the corresponding pinacol more readily than expected by isopropyl alcohol [480]. This result speaks for a twisted benzoyl group that cannot easily achieve the required planarity for photoenolization:

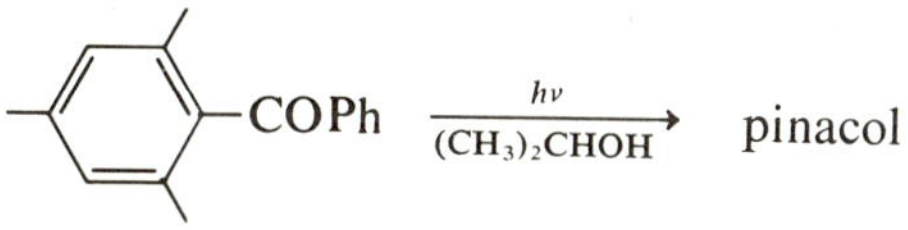

263 $\xrightarrow[(CH_3)_2CHOH]{h\nu}$ pinacol

264, 50%

Several examples of intramolecular hydrogen abstraction by ortho-substituted benzophenones have appeared [521, 524, 525, 481, 482].

Irradiation of oxygenated solutions of 2'-methylbenzophenone resulted in the formation of **268** and **269** [481]. The same products are formed by the irradiation of **270** in the presence of oxygen:

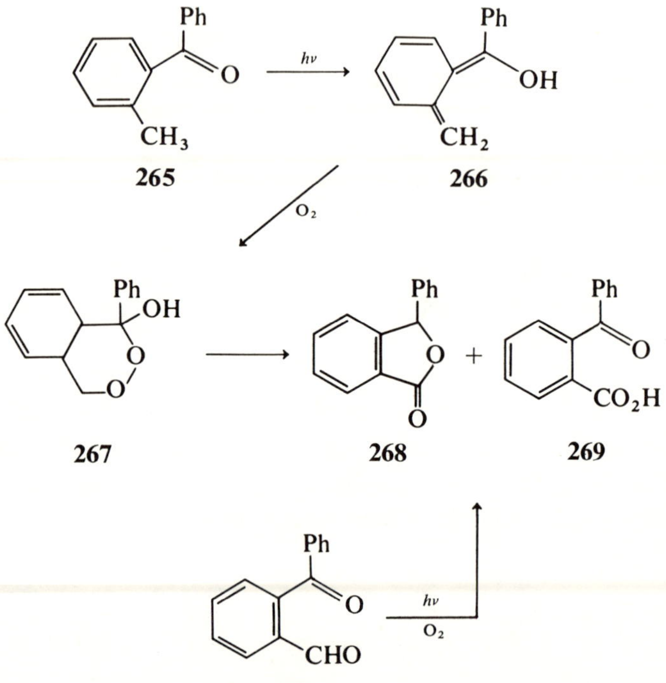

The enol **266** is a probable intermediate in this sequence.

A very interesting example of intramolecular hydrogen abstraction was found for **271**. The hydrogens on carbons C_{10} and beyond, except for the terminal CH_3 group, were abstracted by the benzophenone triplet:

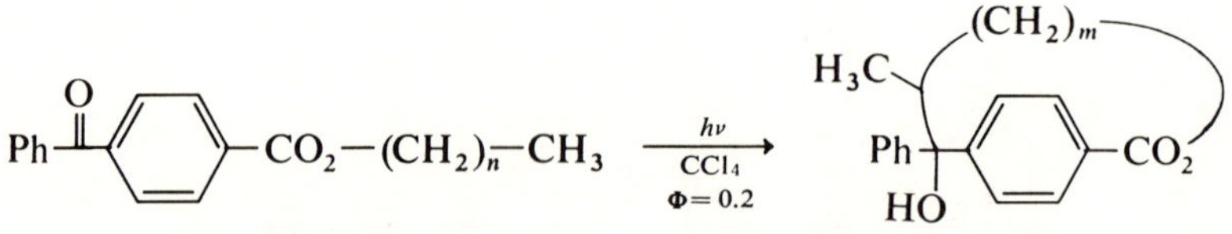

271, $n = 13-19$

272

Some selectivity was observed in the abstraction step [526].

Irradiation of **273** yields **275**, probably by way of a seven-membered cyclic transition state to yield **274** [524]:

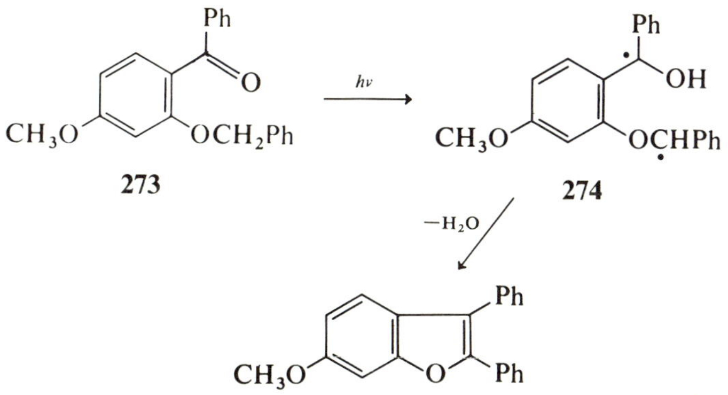

Irradiation of xanthone **277** in the presence of **276** yields **278** [527]:

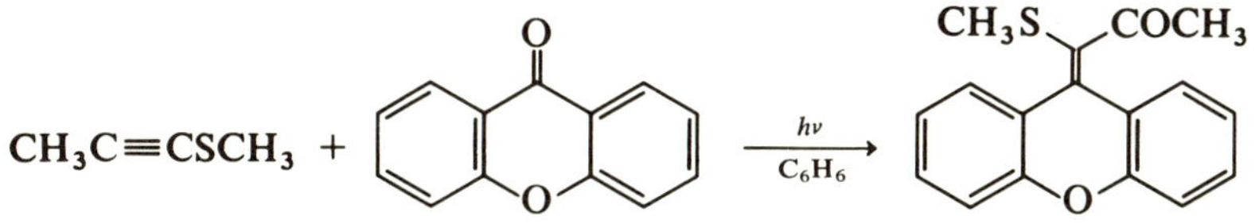

An analogous reaction also occurs when **276** is irradiated with benzophenone and thioxanthone.

Two papers on bianthrone photochemistry and its photochromic properties have appeared [528, 529; see also 530].

In contrast to the case of alkyl β-keto-phosphonates (see Section I.A.3, this chapter), phenacylphosphate is photoreduced to a pinacol [412].

Phosphines and amines have been used to quench fluorenone singlets, possibly by way of a charge-transfer mechanism [509, 522]. Production of the radical anion of fluorenone by way of the irradiation of the ketones in amines [531, 532] has been demonstrated by ESR experiments [500, 498]. Amines [531, 532] and tri-n-butylstannane [531] have been found to photoreduce fluorenone, and benzophenone promotes photoreduction in isopropanol, probably by way of chemical sensitization [533]. Intersystem crossing in fluorenone has been found to be solvent-dependent [534, 532, 533]. Fluorenone is about five times less reactive than benzophenone in cycloaddition to ketenimines [519].

Phosphorescence spectra and intramolecular triplet energy transfer have been reported for carbazole naphthyl ketones [535].

III. CARBOXYLIC ACIDS AND DERIVATIVES

A. Esters and Acids

Amino acids, such as **279**, undergo photodeamination to yield cinnamic acids [537]:

$$PhCH_2CHCO_2H \xrightarrow{h\nu} PhCH=CHCO_2H + NH_3$$
$$\underset{\textbf{279}}{\underset{|}{NH_2}}$$

Irradiation of acid **279a** in acidified methanol results in the formation of **279b** and **279c**:

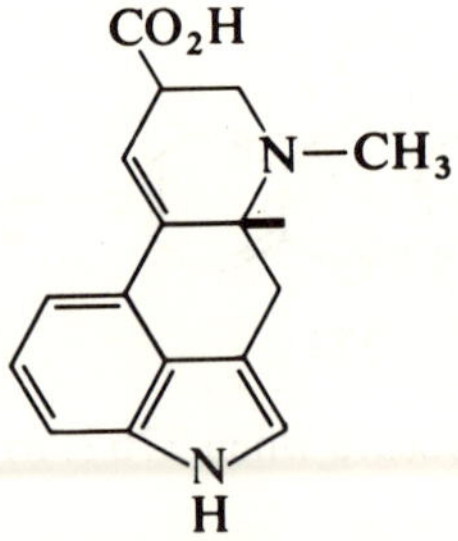

[Structure of **279a**]

Pyrex
$CH_3OH : H_2SO_4$ | $h\nu$

[Structures of **279b** and **279c**]

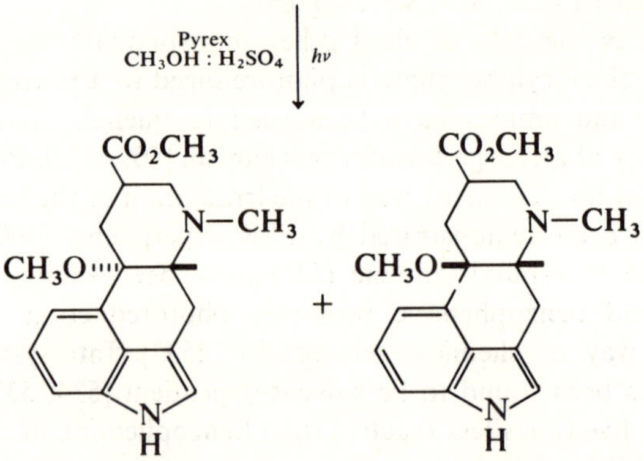

A similar result is observed for the corresponding amides [538].

Irradiation of the *threo* ester **280** or the *erythro* ester **281** results in different ratios of the ethylenes **282** and **283** [539]:

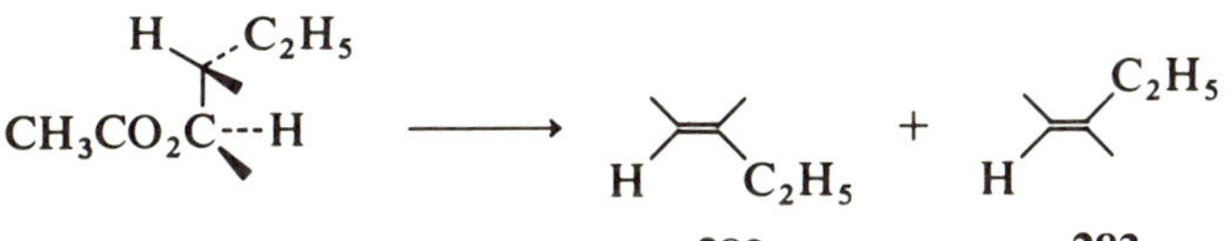

$$\text{CH}_3\text{CO}_2\overset{\text{H}\quad\text{C}_2\text{H}_5}{\underset{}{\text{C}}}\text{---H} \longrightarrow \overset{\text{H}}{\underset{\text{H}}{>}}\!=\!\overset{\text{C}_2\text{H}_5}{\underset{\text{C}_2\text{H}_5}{<}} + \overset{\text{H}}{\underset{\text{H}}{>}}\!=\!\overset{\text{C}_2\text{H}_5}{\underset{}{<}}$$

280   **282**   **283**

$\uparrow h\nu$

$$\text{CH}_3\text{CO}_2\overset{\text{H}\quad\text{C}_2\text{H}_5}{\underset{}{\text{C}}}\text{---H}$$

281

There also appears to be no photochemical interconversion of the two esters **280** and **281**; for example, no photoracemization is observed.

Quadricyclane carboxylic acid esters are reversibly converted to norbornadienes by light [540], for example, **284** ⇌ **285**:

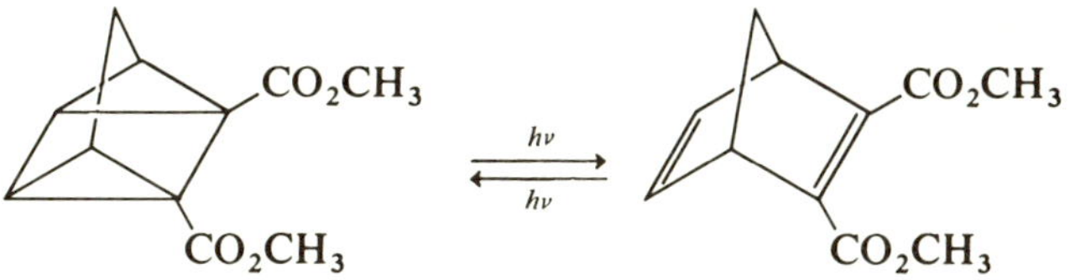

284   **285**

Neither naphthalene nor triphenylene sensitizes these $2\sigma \rightarrow 2\pi$ photorearrangements, but they do sensitize the reverse $2\pi \rightarrow 2\sigma$ transition.

A number of examples of fragmentation [541, 542, 543, 544, 544a] and rearrangement [545] reactions of alkyl esters have been reported [see also 545a].

The first examples of a benzoic acid ester undergoing oxetane formation [546, 547] and photoreduction [548] have been reported; for example, **286** → **287** [547]:

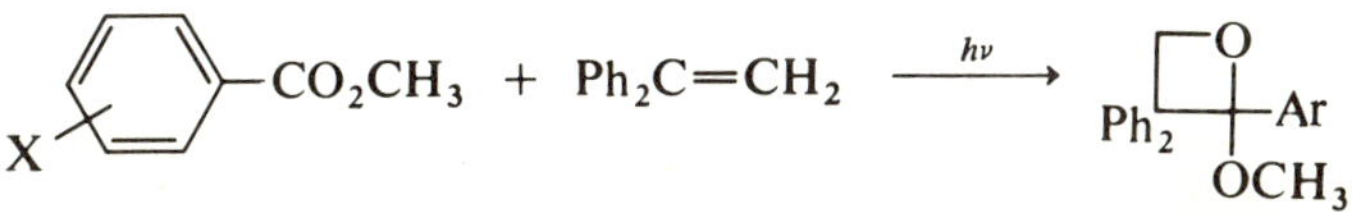

$$\text{X}\!-\!\!\underset{}{\bigcirc}\!-\!\text{CO}_2\text{CH}_3 + \text{Ph}_2\text{C}=\text{CH}_2 \xrightarrow{h\nu} \underset{\text{Ph}_2}{\overset{\text{O}}{\square}}\!\!\overset{}{\underset{\text{OCH}_3}{\text{Ar}}}$$

286   **287**

Ethyl acetate does not yield an oxetane under comparable conditions [547]. Photoaddition of **288** to tolan yields **290**, presumably by way of the oxetene **289** [546]:

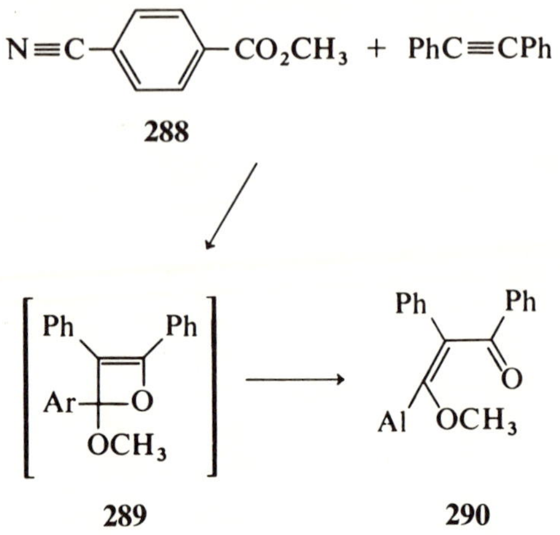

$$N \equiv C - C_6H_4 - CO_2CH_3 + PhC \equiv CPh$$

288

$$\left[\begin{array}{c} Ph \quad\quad Ph \\ \diagup\!\!\!\diagdown \\ Ar - O \\ | \\ OCH_3 \end{array} \right] \longrightarrow \begin{array}{c} Ph \quad\quad Ph \\ \diagup\!\!\!\diagdown \quad O \\ Al \quad OCH_3 \end{array}$$

289   **290**

The photodecarboxylation of pyridyl carboxylic acids has been reported [549]. Attempts to photosensitize the photo-Fries rearrangement of phenyl benzoate with triplet sensitizers have been unsuccessful [550]. The effect of nitro substituents on the photo-Fries rearrangement of phenyl benzoates has been investigated [551].

Irradiation of **291** results in the formation of **292**, **293**, and **294** [552]:

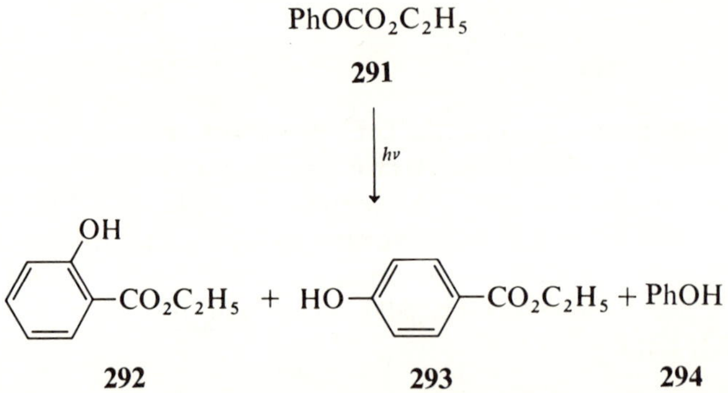

$$PhOCO_2C_2H_5$$

291

$$\downarrow h\nu$$

(2-OH)C$_6$H$_4$-CO$_2$C$_2$H$_5$ + HO-C$_6$H$_4$-CO$_2$C$_2$H$_5$ + PhOH

292   **293**   **294**

The photo-Fries rearrangement was also found to occur for carbonate esters containing 4—CH$_3$O—, 2—, 3—, and 4—CH$_3$, 4—Ph, and 4—CH$_3$CO but not for 4—Cl, 3— or 4—NO$_2$, or 4-CO$_2$H.

The phenoxy acid **295** is photorearranged to **296** [553]:

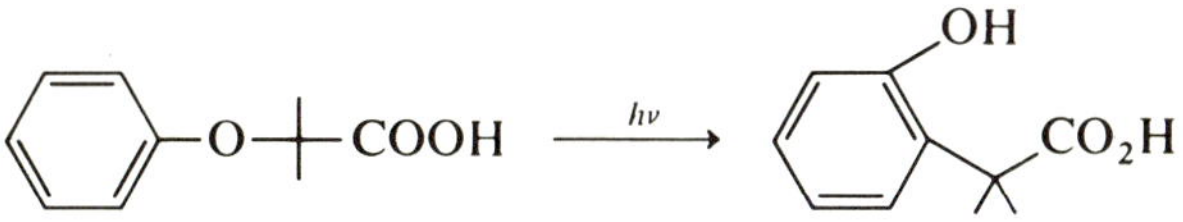

PhO—O—C(CH3)2—COOH $\xrightarrow{h\nu}$ [2-hydroxyphenyl]—C(CH3)2—CO2H

295 → **296**

The irradiation of a number of substituted phenyl acetates has been reported [554, 555]. The products observed result from (1) cleavage of the O-acyl bond leading to phenols, (2) photo-Fries rearrangement, and (3) decarboxylation. The last reaction is strongly sensitive to solvent and is enhanced by ortho or meta substitution. All the reactions above apparently occur from the S_1 state of the esters.

The diol ester **297** is photorearranged to the lactone **298** [413]:

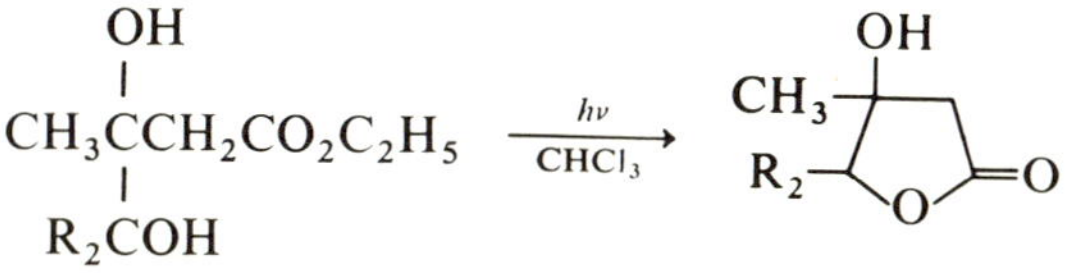

$$\underset{\underset{R_2COH}{|}}{\overset{\overset{OH}{|}}{CH_3CCH_2CO_2C_2H_5}} \xrightarrow[CHCl_3]{h\nu} \text{lactone with } CH_3, R_2, OH$$

297 → **298**, 100%

Photoester exchange has been found and interpreted as evidence for heterolytic cleavage of the diazoester **299**, which also undergoes the expected loss of N_2 to yield a ketocarbene [555]:

$$N_2CHCO_2R \xrightarrow{h\nu} N_2CHC^+O + O^-R \xrightarrow{R'OH} N_2CHCO_2R'$$

299 → **300**

This result (heterolytic cleavage) contrasts sharply with the more usual loss of nitrogen that occurs when diazocompounds are photolyzed (see Chapter 5).

Irradiation of **301** and related compounds results in products whose formation can be explained on the basis of the homolytic cleavage of bonds a, b, and c [556]:

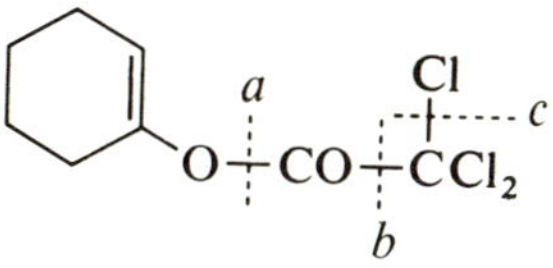

cyclohexenyl—O $\overset{a}{\vdots}$ CO $\overset{b}{\vdots}$ CCl2 $\overset{c}{\vdots}$ Cl

301

Photolysis of basic aqueous solutions of the salt **302** yields decarboxylation products **303** and **304** [557]:

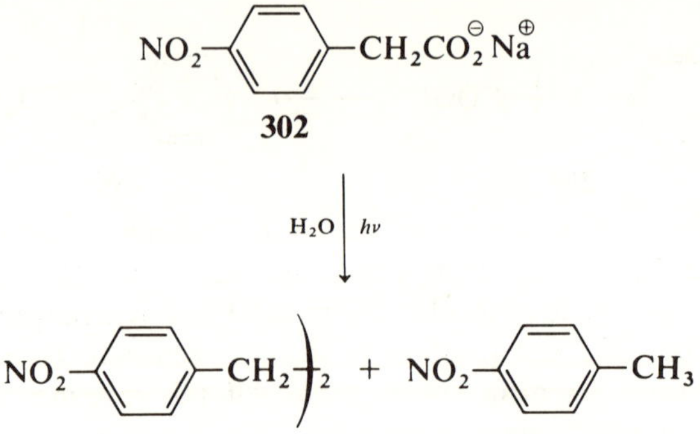

$$NO_2-\langle C_6H_4 \rangle-CH_2CO_2^\ominus\ Na^\oplus$$
302

$\xrightarrow{H_2O,\ h\nu}$

$NO_2-\langle C_6H_4 \rangle-(CH_2)_2$ + $NO_2-\langle C_6H_4 \rangle-CH_3$

303, 27% **304**, 3%

Flash spectroscopy provided evidence for the paranitrobenzyl anion as an intermediate in this photolysis.

Several reports of the photochemistry of conjugated cyclopropylcarboxylic esters have appeared [558, 559, 559a]. Cyclopropane ring opening is generally the most significant primary photochemical process for these systems.

Photolysis of P_1,P_1-diethyl-P_2-(m-nitrophenyl)pyrophosphate in cold, dilute alkali affords the unstable, unsymmetrical pyrophosphate diester, P_1,P_1-diethyl pyrophosphate [559b].

Study of Φ_{ST} for CH_3COCN as a function of 1,3-pentadiene indicates that the diene quenches both excited singlets and triplets of this cyanoketone [559c].

A number of examples of photolysis of peroxyesters have appeared [559d, 504, 505, 560, 561, 562, 563]. Of these, the conversion of **305 → 306** [561]:

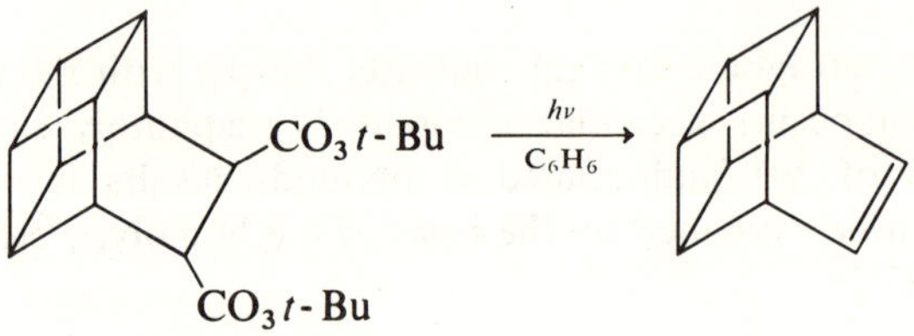

305 $\xrightarrow[C_6H_6]{h\nu}$ **306**, 40%

and the photolysis of **307** [563] merit citation:

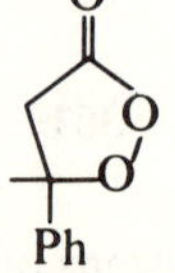

307

B. Cyclic Derivatives of Carboxylic Acids

Cyclic derivatives of α,β-unsaturated carboxylic acids are discussed in Chapter 4.

Stereochemistry of the C—O bond, relative to the enone double bond, has a marked effect on the fragmentation [review: 536] of the lactones **308** and **309**, the latter being inert to irradiation and the former being photolyzed smoothly to **310** [564]:

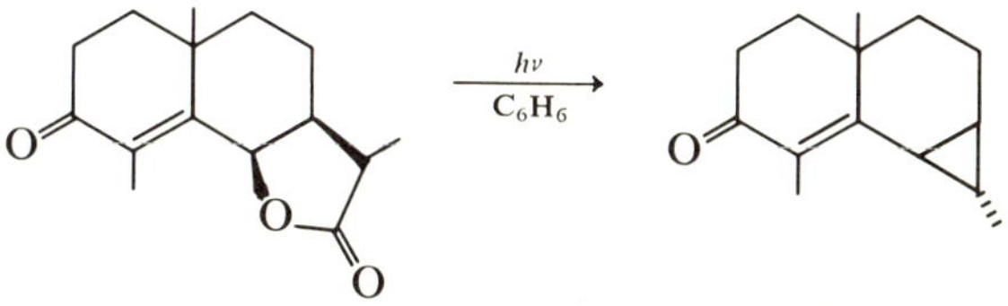

308 $\xrightarrow[C_6H_6]{h\nu}$ **310**

309 $\not\xrightarrow{h\nu}$

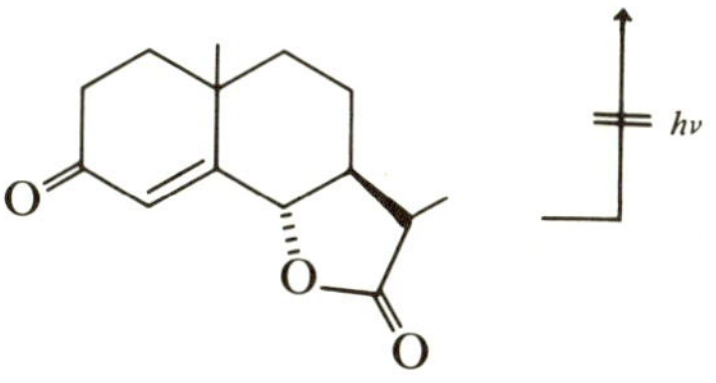

It was suggested that this difference is the result of better overlap of the enone double bond pi-system with the cleaving O—C bond of **308** relative to **309** [see also 565].

A number of photolytic fragmentations have been reported [566, 567, 568, 569]. In the case of **313**, the ketene **312** was detected by low-temperature IR when **311** was irradiated [566]:

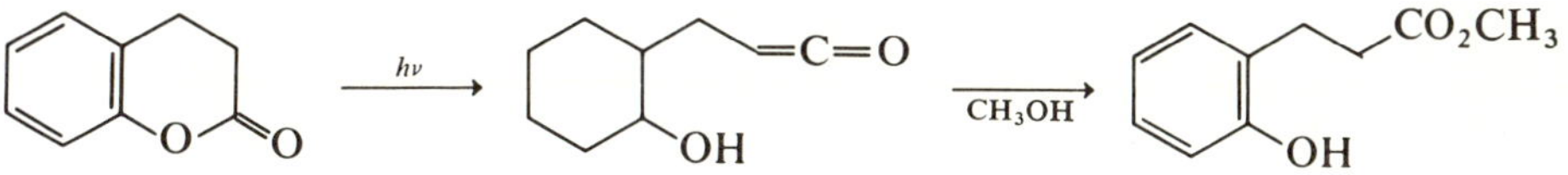

311 $\xrightarrow{h\nu}$ **312** $\xrightarrow{CH_3OH}$ **313**

Irradiation of the polycyclic ketolactone **314** leads to the loss of CO_2 and formation of **166**, which is partially photolyzed (see above Section I.A.4) [570]:

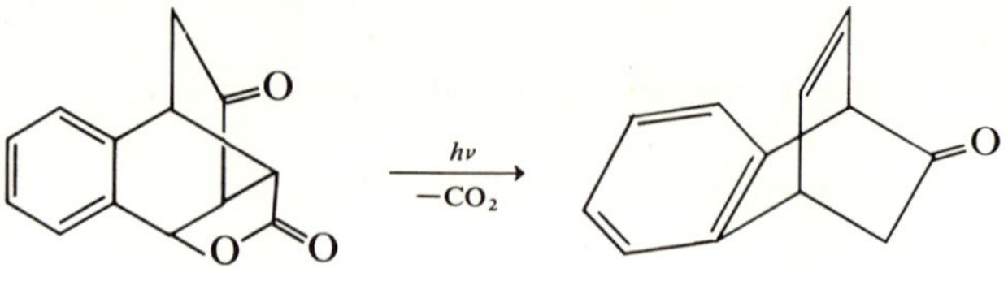

314 → **166** (hv, $-CO_2$)

In addition to the expected photocleavage rearrangement reaction of **315**, which leads to **316**, the former also suffers reduction of a double bond [571]:

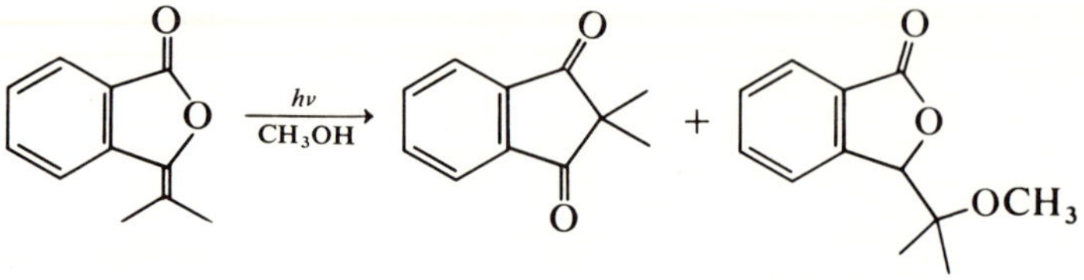

315 → **316** + **317** (hv, CH_3OH)

Lactones, such as **318**, are photorearranged to **319** [572]:

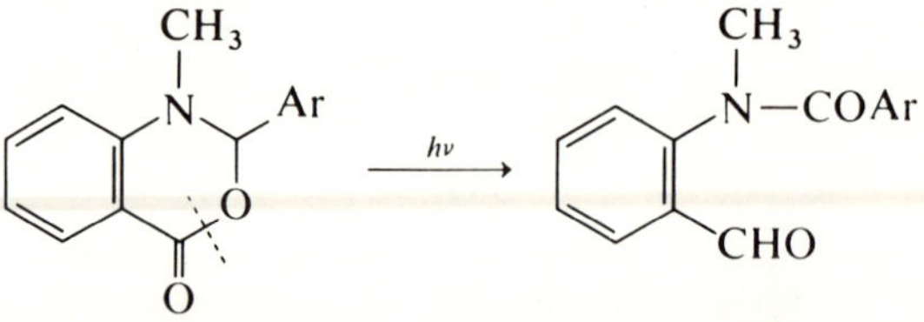

318 → **319** (hv)

A number of examples of anhydrides related to **320** that apparently photofragment to cyclobutadienes and cyclopentadienones have been discovered [573, 424, 574], for example, **320** to **321** and **322** [424]:

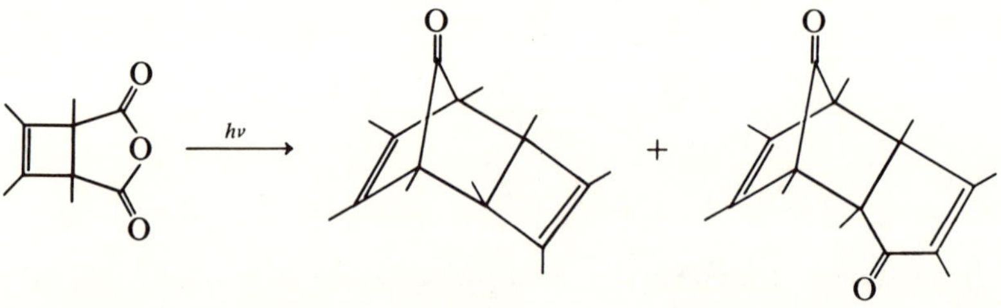

320 → **321** + **322** (hv)

Irradiation of **307** results in the loss of carbon dioxide as the major reaction pathway [563]:

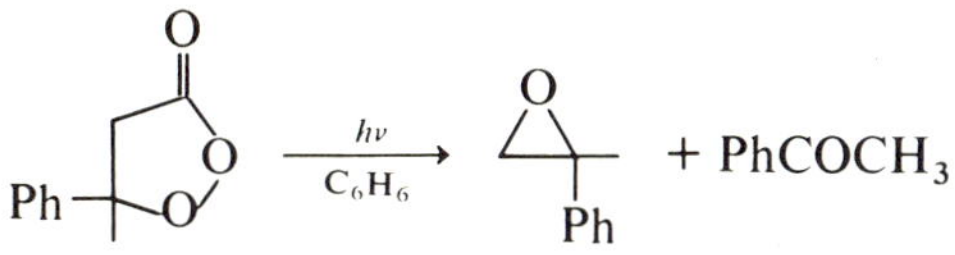

$$\text{Ph}\underset{\textbf{307}}{\overset{O}{\underset{O}{\bigtriangleup}}}\!\!\!\!\overset{O}{\underset{}{}} \xrightarrow[C_6H_6]{h\nu} \underset{\textbf{323, 50\%}}{\overset{O}{\underset{Ph}{\bigtriangleup}}} + \underset{\textbf{324, 11\%}}{\text{PhCOCH}_3}$$

$$+ \underset{\textbf{325, 8\%}}{\text{PhCOCH}_2\text{CH}_3} + \underset{\textbf{326, 8\%}}{\text{PhCH}_2\text{COCH}_3}$$

Irradiation of the cyclic carbonate **327** results in fragmentation as shown [575]:

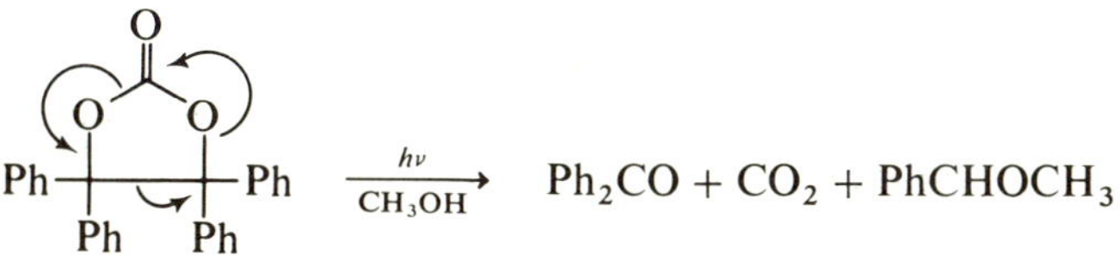

$$\underset{\textbf{327}}{\text{Ph}_2\text{C}-\text{O}-\text{C}(=O)-\text{O}-\text{CHPh}} \xrightarrow[\text{CH}_3\text{OH}]{h\nu} \text{Ph}_2\text{CO} + \text{CO}_2 + \text{PhCHOCH}_3$$

C. Amides and Related Compounds

A mechanistic study of the photo-Fries rearrangement of acetanilide to **329** and **330** suggests that the reaction proceeds by way of the $S_1(\pi,\pi^*)$ state, which crosses to a dissociative triplet σ^* state [576, 577, 577a]:

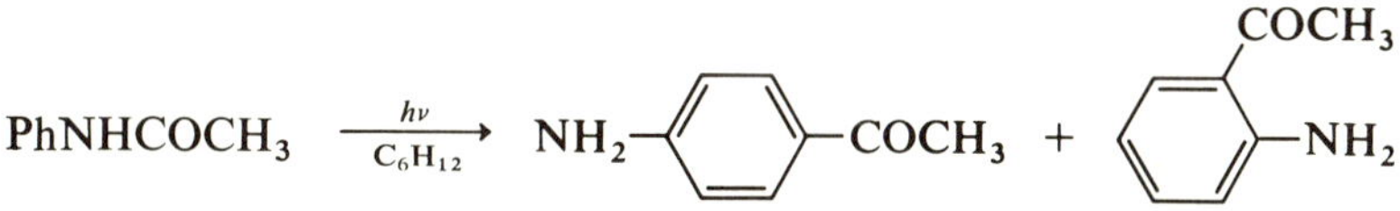

$$\underset{\textbf{328}}{\text{PhNHCOCH}_3} \xrightarrow[C_6H_{12}]{h\nu} \underset{\textbf{329, } \Phi = 0.07}{\text{NH}_2\text{-C}_6\text{H}_4\text{-COCH}_3} + \underset{\textbf{330, } \Phi = 0.06}{\text{(o-COCH}_3\text{)C}_6\text{H}_4\text{-NH}_2}$$

The quantum yields for product formation drop sharply in polar solvents [576]. The displacement of a methoxy group was noted in addition to the expected products when 2,4-dimethoxyacetanilide was irradiated [578].

Photolysis of α-chloroacetamide in the presence of anisole or phenol leads to the formation of ortho, meta, and para substitution products [578a; see also 578b]:

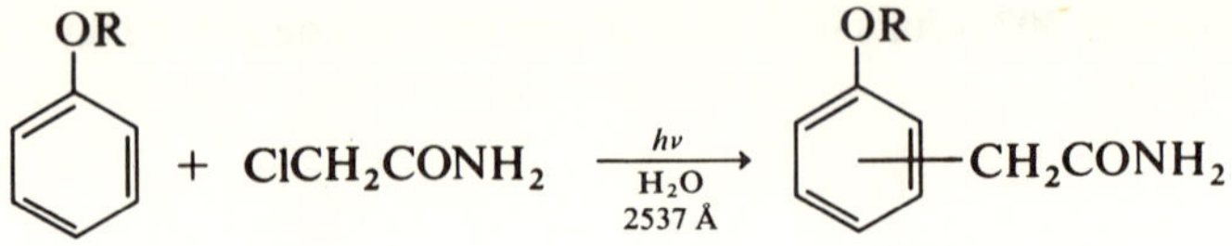

[Scheme: PhOR + ClCH$_2$CONH$_2$ $\xrightarrow[2537\ \text{Å}]{h\nu, \text{H}_2\text{O}}$ (OR)C$_6$H$_4$–CH$_2$CONH$_2$; R = H or CH$_3$; ortho, 21%; meta, 4%; para, 6%]

A number of studies of the photochemistry of amides and related compounds have appeared [579, 580, 543]. Irradiation of **331** results in acyl migration to yield **332** [581]:

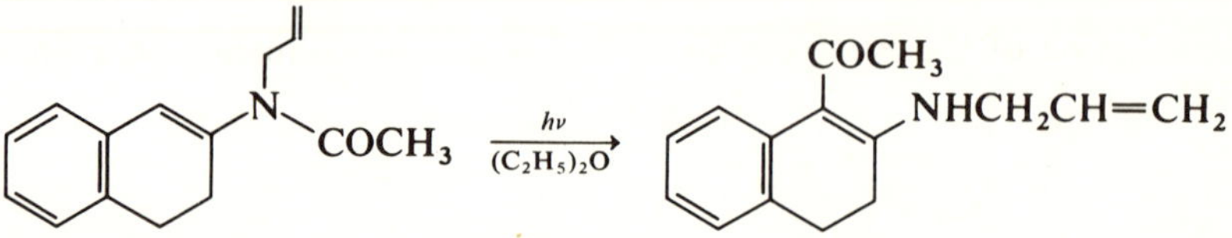

331 $\xrightarrow[(C_2H_5)_2O]{h\nu}$ **332, 60%**

Further examples of the photodecarbonylation of aziridone have been reported [582, 583, 584].

Irradiation of **333** results in the formation of **334** as the major product [585]:

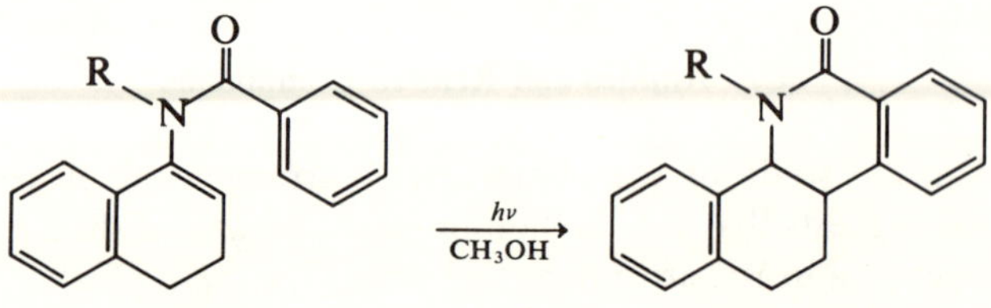

333 $\xrightarrow[CH_3OH]{h\nu}$ **334, 55%**

Irradiation of amide **334a** results in cyclization to **334b** [586]:

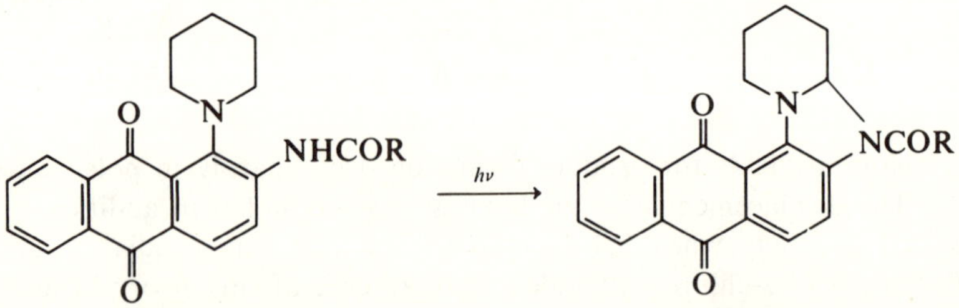

334a $\xrightarrow{h\nu}$ **334b**

The enamide **335** undergoes *cis-trans* isomerization when irradiated with > 300 nm light [587]. Excitation with shorter-wavelength light causes a 1,3-acyl shift to occur (**337**):

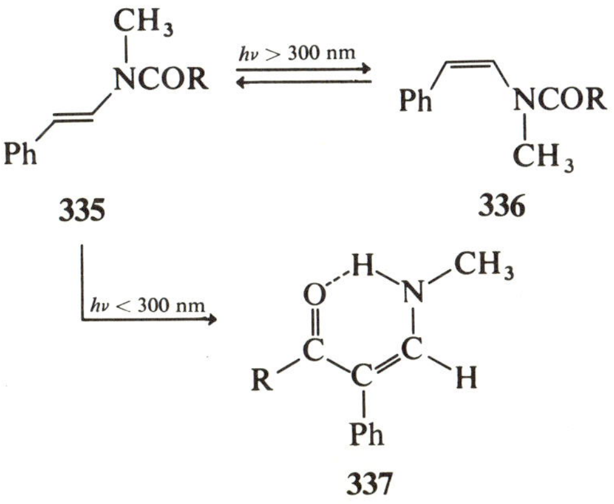

$$\underset{\textbf{335}}{\overset{Ph}{\underset{}{\diagup}}\!\!\!\!\!\overset{CH_3}{\underset{}{\diagdown}}NCOR} \xrightarrow[\hphantom{hv < 300 nm}]{h\nu > 300\ nm} \underset{\textbf{336}}{\overset{Ph}{\diagup}\!\!\!\!\!\overset{}{\diagdown}\underset{CH_3}{\overset{}{\underset{}{|}}}NCOR}$$

$$\Big\downarrow h\nu < 300\ nm$$

$$\underset{\textbf{337}}{R-\overset{O\cdots H}{\underset{}{C}}=\overset{}{\underset{Ph}{C}}-\overset{N-CH_3}{\underset{}{C}}-H}$$

Dimethyl sulfoximines are photolyzed to amino nitrenes [588].

Irradiation of **338** yields the rearranged product **339** probably by way of the cyclic intermediate **340** [589, 590]:

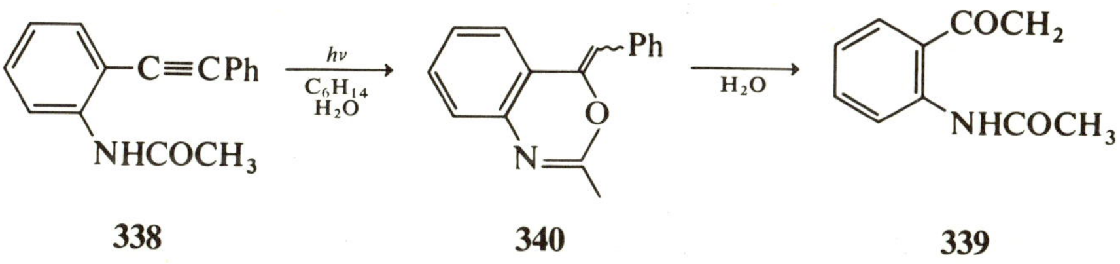

$$\underset{\textbf{338}}{\overset{}{\text{Ar}}-C\equiv CPh} \xrightarrow[\substack{C_6H_{14}\\H_2O}]{h\nu} \underset{\textbf{340}}{\text{(bicyclic)}} \xrightarrow{H_2O} \underset{\textbf{339}}{\overset{COCH_2}{\text{Ar-NHCOCH}_3}}$$

where Ar-NHCOCH₃ at **338** and product **339** has COCH₂ and NHCOCH₃ substituents.

The adamantyl lactone **341** is photodecarbonylated to the imine **342** in good yield [583]:

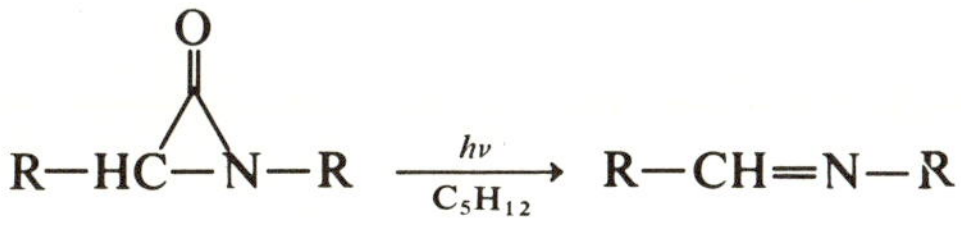

$$\underset{\textbf{341},\ R\ =\ 1\text{-adamantyl}}{R-HC\overset{\overset{O}{\|}}{\overset{}{\diagup\!\!\!\diagdown}}N-R} \xrightarrow[C_5H_{12}]{h\nu} \underset{\textbf{342}}{R-CH=N-R}$$

Derivatives of **343**, with $n \geq 7$, undergo photo-Fries rearrangement to paracyclophanes [591]. The nor-analogs of **343** cyclize by way of the ortho positions [542]:

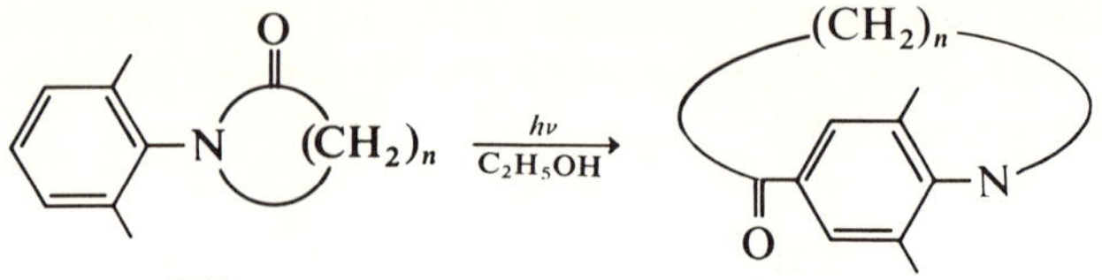

343 →(hv, C₂H₅OH)→ **344**, 30–50%

N-Phenylpyrrolidone **345** is photorearranged to N-phenylpyrrole **346** [542]:

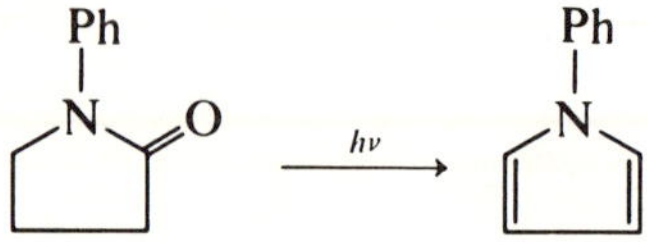

345 →(hv)→ **346**

Irradiation of amide **346a** in methanol results in the formation of a β-lactam **346b** in addition to cis and trans azo compounds **346c** [592]:

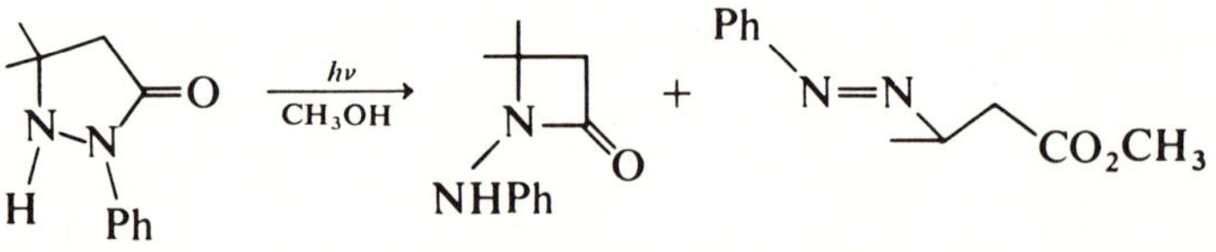

346a →(hv, CH₃OH)→ **346b** + **346c**, cis + trans

The mechanism of these reactions is undetermined.

The photochemistry of polymers, such as polyvinylpyrrolidone, continues to be of interest [593, 594] (Volume 2, page 71).

Irradiation of **347** in the presence of nucleophiles yields adducts, such as **349**, which derive from decarbonylation [595; see also 595a]:

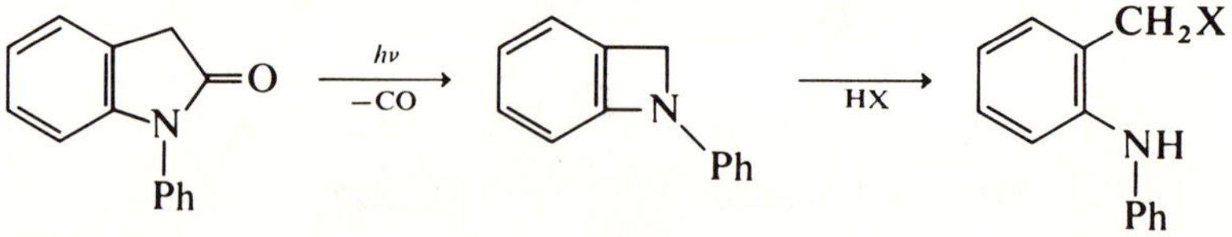

347 →(hv, −CO)→ **348** →(HX)→ **349**

Irradiation of **350** in the presence of acetylenes effects cycloaddition, for example, **350 → 352** [596, 597]:

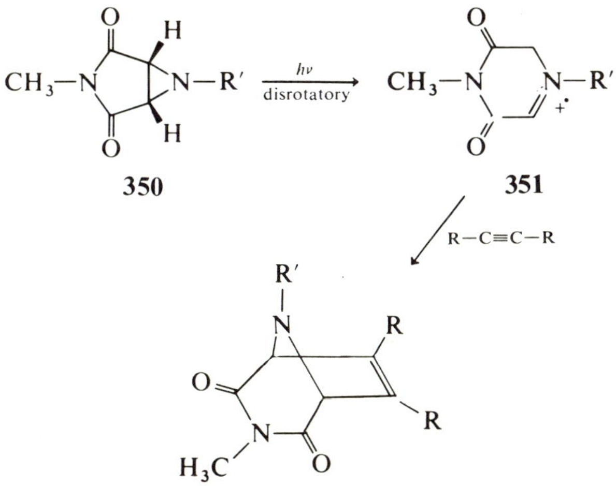

350 →(hv, disrotatory) **351**

→ R—C≡C—R

352

Cycloaddition could not be effected thermally, a result expected from orbital symmetry considerations, since conrotatory (thermal) opening of **350** would produce a highly strained cyclic structure.

In analogy to recent reports (Chapter 3, III.A.) of photocycloaddition of esters, the urethane **353** photoadds to diphenylethylene [598]:

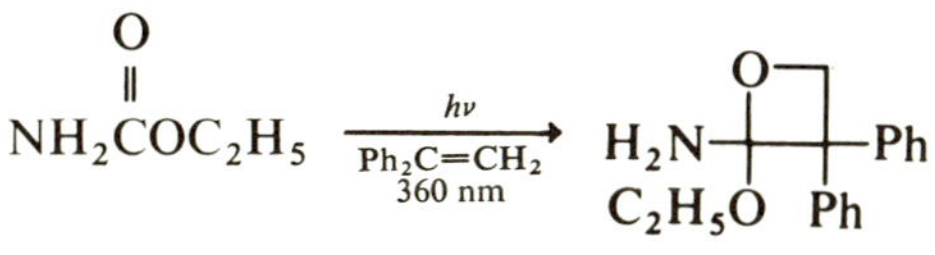

$$NH_2COC_2H_5 \xrightarrow[360 \text{ nm}]{hv, Ph_2C=CH_2}$$

353 → **354, 64%**

Since 360 nm light was used, it is possible that the diphenylethylene and not **353** is primarily responsible for light absorption.

Phenol is photoalkylated by chloroacetamide [599]. The photodecomposition of 3-(para-chlorophenyl)-1,1-dimethylurea has been reported [599a].

D. Sulfur Compounds

Electron-rich olefins have been found to add to the n,π^* triplet state of thiobenzophenone to yield either 1,4-dithiane or thietane kinds of structures [600, 600b]. Steric effects may help to determine which kind of product is formed, for example, **356 → 357** or **356 → 359**:

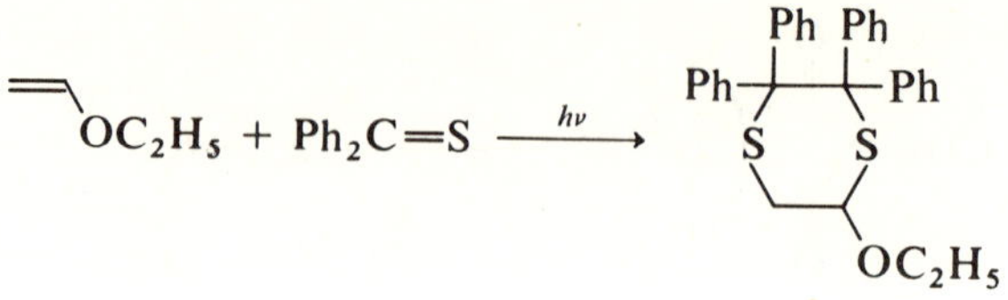

$$\text{CH}_2=\text{CHOC}_2\text{H}_5 + \text{Ph}_2\text{C}=\text{S} \xrightarrow{h\nu} \underset{\underset{\text{OC}_2\text{H}_5}{}}{\text{357}}$$

355   356   357

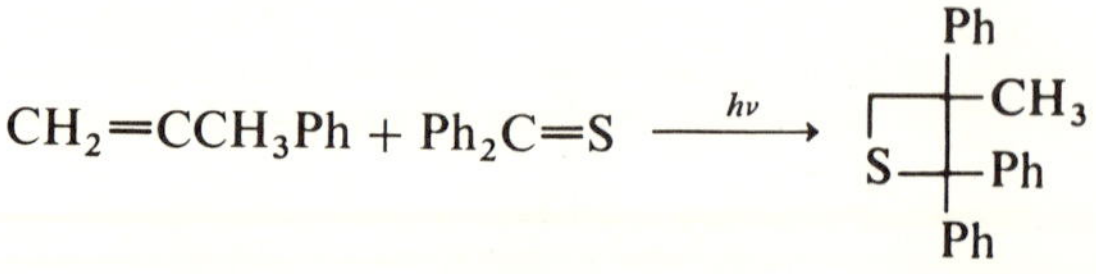

$$\text{CH}_2=\text{CCH}_3\text{Ph} + \text{Ph}_2\text{C}=\text{S} \xrightarrow{h\nu} \text{359}$$

358   356   359

Irradiation of thiobenzophenone **356** in the presence of electron-deficient olefins results in the formation of thietanes [600a]. The orientation of addition as well as the stereospecificity of the addition suggests attack by the π,π^* singlet state of **356** (see also Volume 2, page 66).

Irradiation of either **359a** or **359b** results in the formation of diphenyl disulfide [601]:

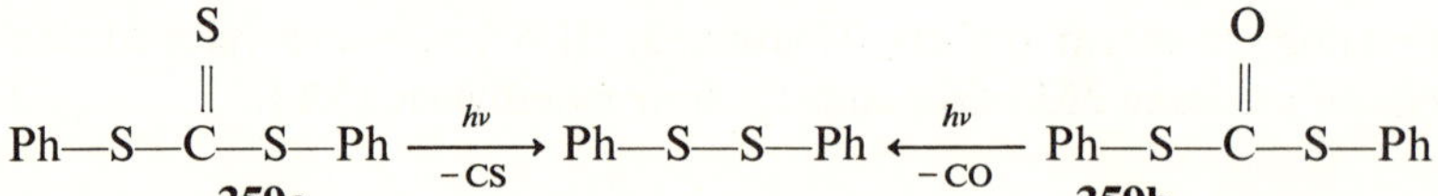

$$\underset{\textbf{359a}}{\text{Ph}-\text{S}-\overset{\text{S}}{\underset{\|}{\text{C}}}-\text{S}-\text{Ph}} \xrightarrow[-\text{CS}]{h\nu} \text{Ph}-\text{S}-\text{S}-\text{Ph} \xleftarrow[-\text{CO}]{h\nu} \underset{\textbf{359b}}{\text{Ph}-\text{S}-\overset{\text{O}}{\underset{\|}{\text{C}}}-\text{S}-\text{Ph}}$$

Irradiation of **360** in ethanol yields **361**, and in 2-methyl-2-butene, photocycloaddition to yield **362** occurs [602]:

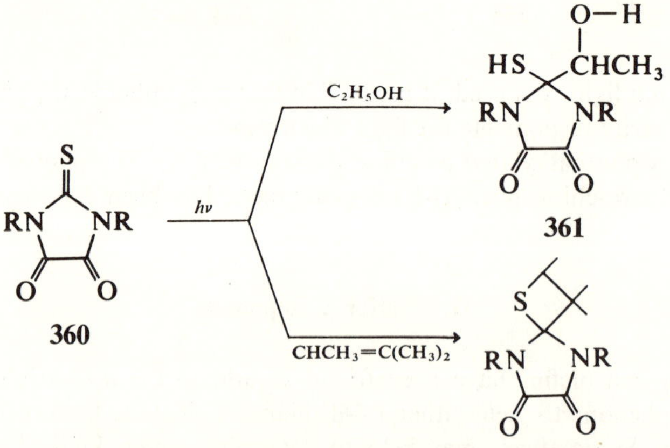

Thioesters, such as **363**, are photolyzed to yield **364** and **365**, probably by way of thioradicals produced by type I cleavage [603]:

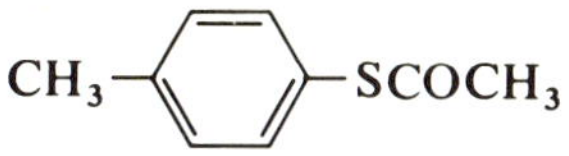

$$CH_3-\langle\rangle-SCOCH_3$$
363

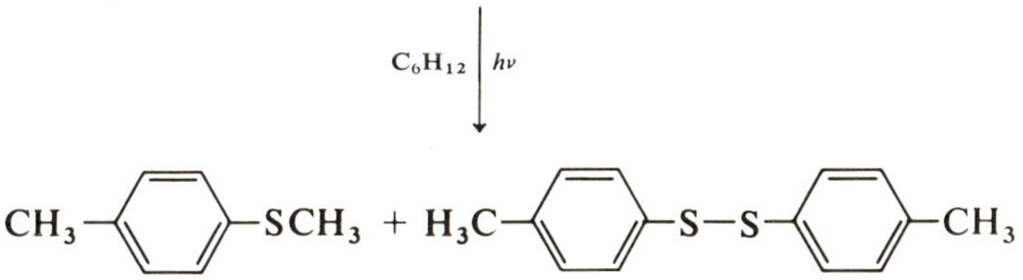

$\downarrow C_6H_{12} \mid h\nu$

$$CH_3-\langle\rangle-SCH_3 \;+\; H_3C-\langle\rangle-S-S-\langle\rangle-CH_3$$
364, 7% **365**, 77%

Several photoreactions of thioesters have also been reported [549, 604, 605]. Photochemical decomposition of sodium N,N-dimethyldithiocarbamate yields carbon disulfide [605a].

E. Ketenes and Isocyanates

Irradiation of 2-biphenyl isocyanate has been investigated under conditions of sensitized and direct irradiation [606, 606a]. The excited singlet state of the unsubstituted compound **366** affords carbazole **367**, and the triplet state yields phenanthridone **368**:

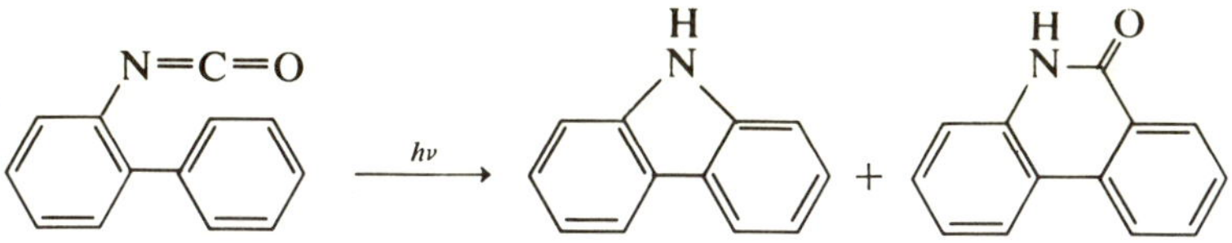

366 → **367** + **368**

CHAPTER FOUR: CONJUGATED ENONES, DIONES, AND RELATED COMPOUNDS

I. α,β-UNSATURATED CARBONYL COMPOUNDS

A. Acyclic Enones, Esters, and Amides

CNDO calculations suggest that the equilibrium configuration of the n,π^* triplet state of acraldehyde should be planar, and that of the π,π^* triplet nonplanar [607].

Several examples of *cis-trans* isomerization of α,β-unsaturated aldehydes [608, 609], ketones [610, 611, 612], and ketoacids [613, 614] have appeared. In several cases, subsequent irradiation led to cyclized products. The *trans*-dienone **1** gives **2** as the major product presumably involving initial *trans* → *cis* isomerization:

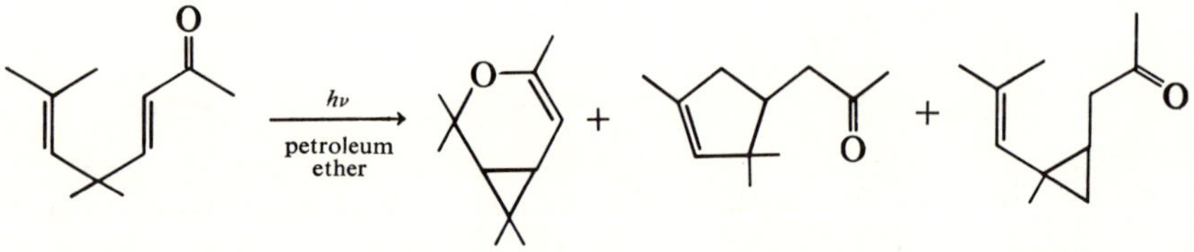

The oxetene **6** is formed in good yield by the irradiation of **5** [615]:

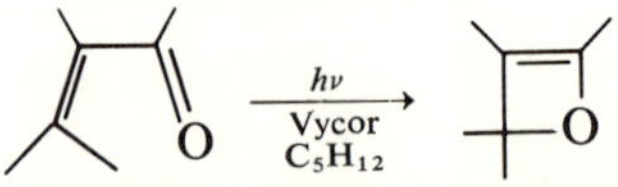

A number of examples of intermolecular photocycloaddition [616, 617] and dimerization reactions [618, 619] have appeared. Irradiation of **7** gives the dimer **8** in excellent yield with no apparent *cis-trans* isomerization of the starting material [619]:

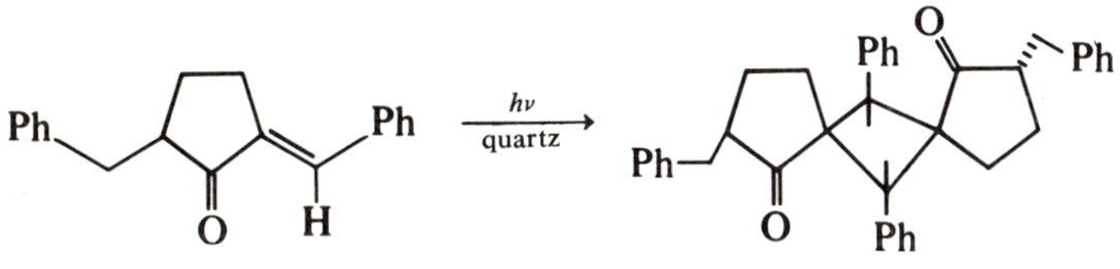

[Structure **7** → hv/quartz → **8, 95%**]

A number of cyclobutenones have been found to undergo carbene insertion reactions similar to those observed for the corresponding cyclobutanones (Volume 1, pages 77–78). Irradiation of **9** in methanol is found to give **10**. When the irradiation is carried out in pentane, an unstable compound, thought to be **11**, and which undergoes a subsequent dark reaction with methanol to give **10** is generated [620]:

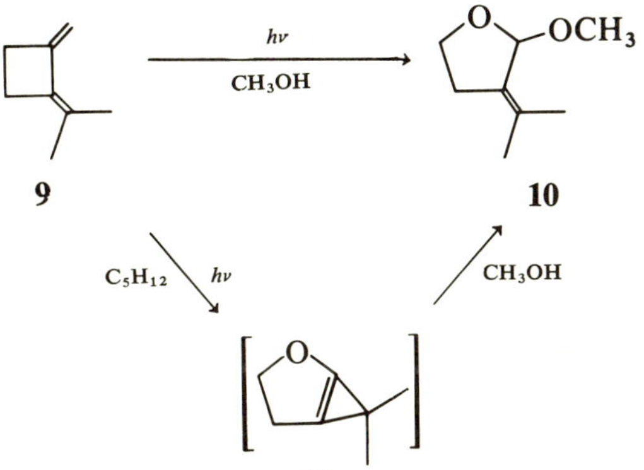

[Structure **9** → hv, CH₃OH → **10**; **9** → C₅H₁₂, hv → **11** → CH₃OH → **10**]

Irradiation of **12** in ethanol leads to **13** and **14** by reduction of the double bond [621]. The *cis* isomer predominates in a reaction that depends on the wavelength. Ketone **13** undergoes a subsequent photocyclization to give **15** in high yield, presumably by way of a seven-centered transition state:

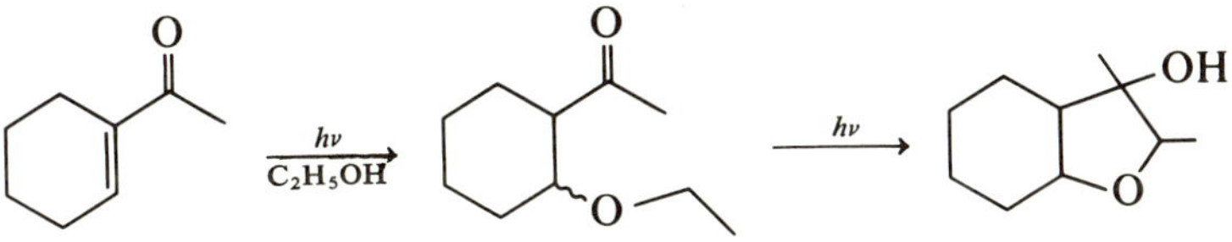

[**12** → hv, C₂H₅OH → **13, cis,** **14, trans,** → hv → **15**]

Irradiation of **16** in either the solid state or pentane solution gives **17** in quantitative yield [617]:

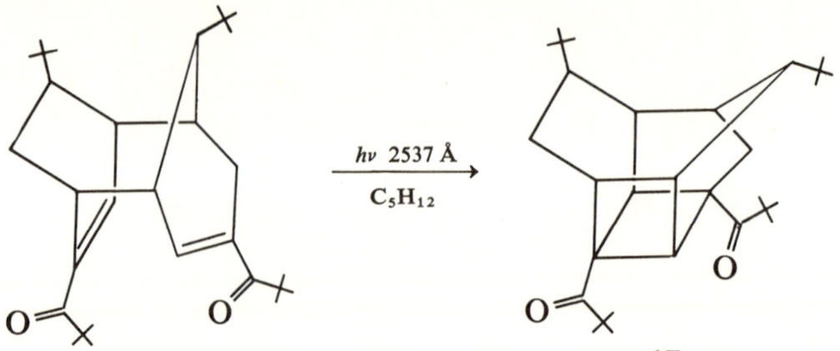

Photolysis of the *o*-hydroxyl-substituted enone **19** causes photoisomerization to **18** and **20** [622]:

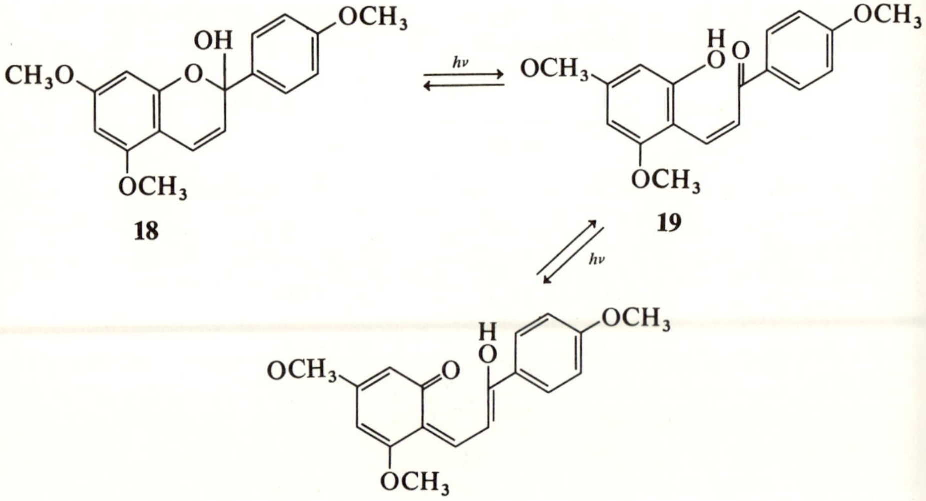

A study of the photochemical isomerization of α,β-unsaturated esters to β,γ-isomers has appeared [623]. An interesting isotope effect has been observed in the photolysis of **21** and **24** [624]. The formation of **22** from **21** is the major photorearrangement in methanol, and **23**, along with **22** is a major product in methanol-*d*:

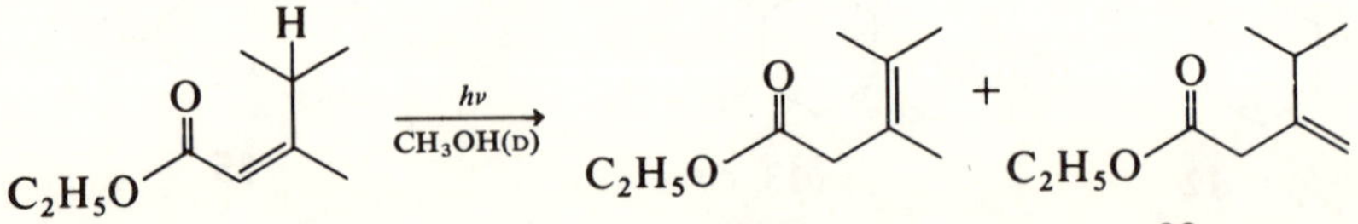

Similar results have been found for the conversion of **24** to **25** (major reaction in methanol) or **26** (major reaction in methanol-*d*):

24 **25** **26**

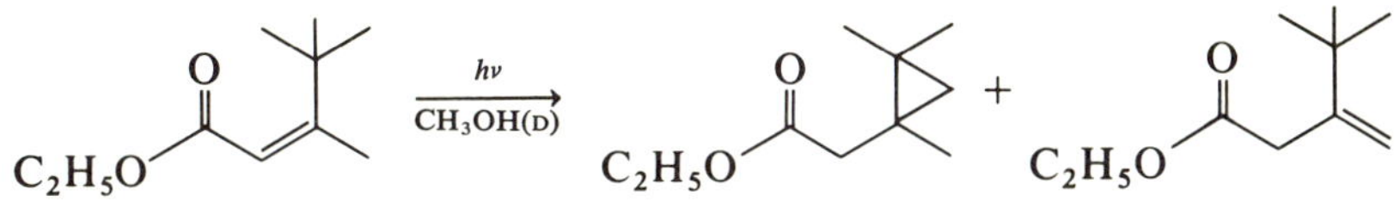

In a related case, the same author carried out an extensive study of the photochemistry of a series of methyl-substituted α,β-unsaturated esters that contain a β-cyclopropane ring [625]. For example, irradiation of **27** in hexane led to the formation of a series of products **28** to **31**. Sensitized irradiation of **27** initially gave a *cis*:*trans* photostationary mixture (50 : 50) that subsequently gave way to a single product **28**:

27 **28** **29**

30 **31**

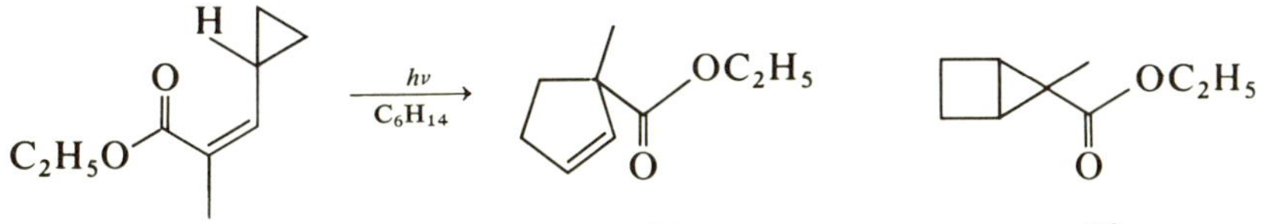

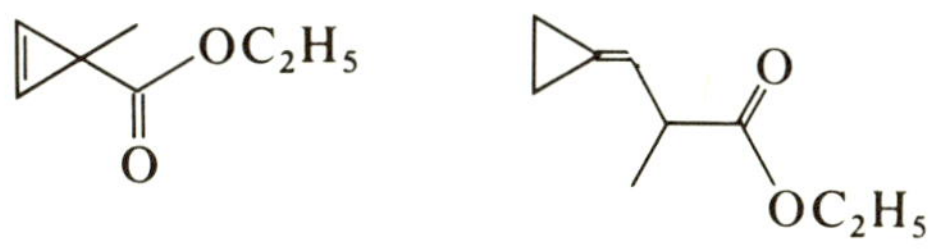

Irradiation of ethyl propiolate in ethanol and isopropanol gave **32** and **33**, respectively [626]:

32, 10% **33**, 20%

Irradiation of the ene diester **33a** results in cyclization to **33b** [627]:

33a **33b**

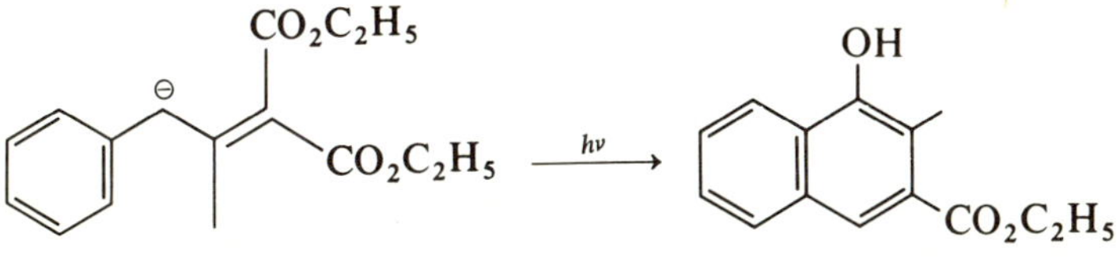

Several studies of the photo-Fries rearrangement of variously substituted phenyl cinnamates and related compounds have appeared [628, 629, 630, 631, 632]. Irradiation of **34** in benzene resulted in the formation of 2′-hydroxy- and 4′-hydroxychalcone isomers [628]:

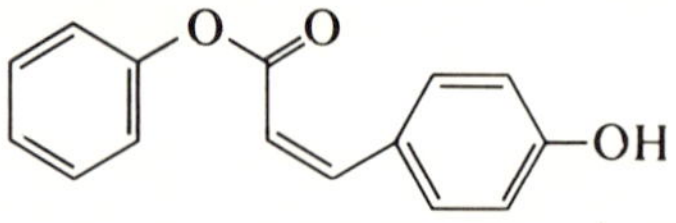

34

Irradiation of peracid **35** in benzene gives a good yield of benzaldehyde, benzoic acid, and α-phenyl acetic acid [633]:

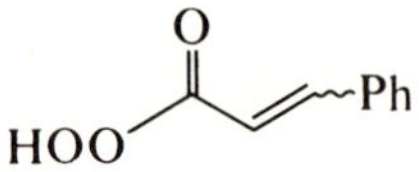

35

Benzophenone photosensitizes the addition of dioxane across the C=C bond of **36** [243]:

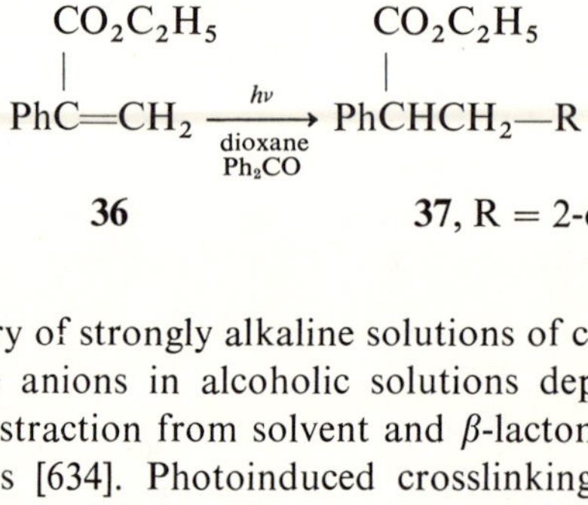

$$\underset{\textbf{36}}{\underset{|}{\text{PhC}=\text{CH}_2}\atop{\text{CO}_2\text{C}_2\text{H}_5}} \xrightarrow[\substack{\text{dioxane}\\ \text{Ph}_2\text{CO}}]{h\nu} \underset{\textbf{37, R = 2-dioxyl}}{\underset{|}{\text{PhCHCH}_2-\text{R}}\atop{\text{CO}_2\text{C}_2\text{H}_5}}$$

The photochemistry of strongly alkaline solutions of cinnamate and cyclohexene-1-carboxylate anions in alcoholic solutions depends on the wavelength. Hydrogen abstraction from solvent and β-lactone formation are the major reaction paths [634]. Photoinduced crosslinking of terpolymers of vinyl coumarin-3-carboxylate has also been reported [635].

Direct irradiation of the lactam **38**:

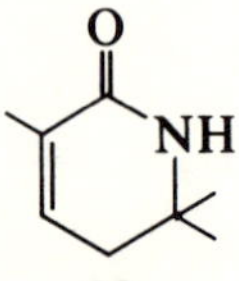

38

results mainly in the formation of cleavage products, and acetophenone sensitization yields a single cyclobutane dimer [636]. Irradiation of 4-acetoxycoumarin results in dimerization [637]. For a similar example see also [638]. The amide **39**:

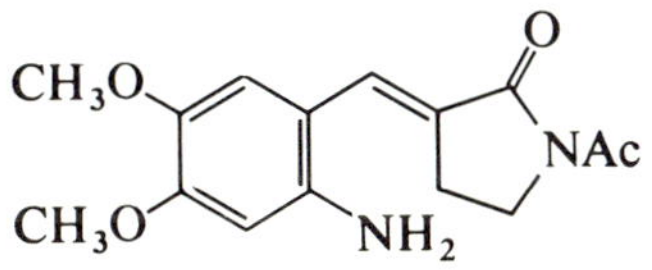

undergoes light-induced cyclizations in which the amino group attacks the pyrrolidone ring [639].

The unsaturated imide **40** was found to undergo photochemical cyclization to **41** in good yield [640, 641]:

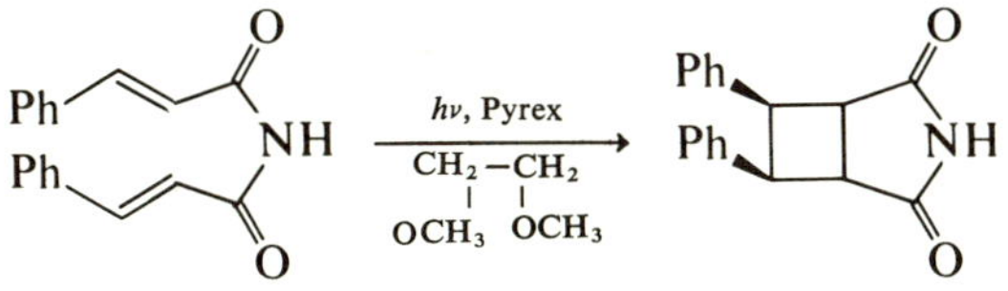

B. Conjugated Cyclic Enones

1. THREE-, FOUR-, AND FIVE-RING ENONES

Irradiation of di-t-butylcyclopropenone in anhydrous ether afforded the decarbonylation product di-t-butylacetylene [642] (Volume 1, page 123; Volume 2, page 79). Cyclobutenones undergo stereospecific ring opening to ketenes on irradiation [643, 644, 645] (Volume 1, page 124; Volume 2, page 79).

Examples of dimerizations [646, 647, 648] and photocycloadditions of cyclopentenones (Volume 2, page 79) to cyclopentene [649, 650], cyclohexene [650, 651], dichloroethylene [650, 651], diethoxyethylene [652, 653], and other olefins [654, 655] have been reported. Several important papers concerned with the mechanistic pathways of photocycloaddition and the dimerization of cyclopentenone have appeared [649, 656, 648, 657] and seem to discount earlier mechanisms involving two triplet states (Volume 2, page 79). The maximum efficiency for the photodimerization of cyclopentenone in acetonitrile (36%) is a function of ketone concentration. The photodimerization results totally from a triplet excited state, and since intersystem is 100% efficient, the dimerization apparently proceeds through a metastable dimeric intermediate that partitions between the ground-state dimer and the two ground-state monomers [656, 648]. The authors think that this dimerization occurs from π,π^* triplet states. In a similar study, evidence is presented for the existence of a 1:1 intermediate in the photocycloaddition of cyclopentenone to cyclohexene and *trans*-3-hexene [649]. The effects of temperature on the photocycloaddition of cyclopentenone to various olefins have also been studied [650].

Quantum efficiencies for the formation of cyclohexene cycloaddition products of **42** and **43** were measured by way of both direct and sensitized irradiation [651]:

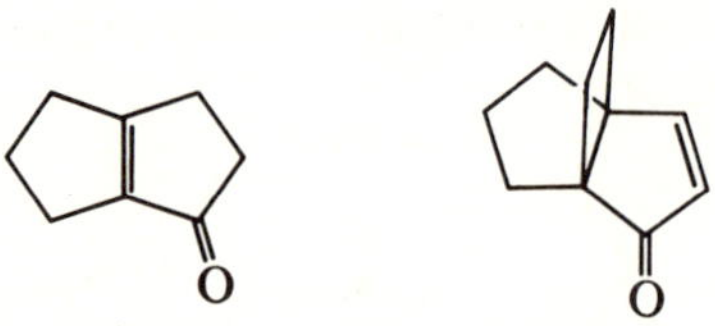

42   **43**

It was observed that although the lowest spectroscopic triplet (n,π^*) of **43** was near 71 kcal/mole, efficient sensitizers of triplet energy as high as 74 kcal/mole were not fully efficient in this case. No photochemical rearrangement products of **42** and **43** were observed.

Irradiation of **44** in methylene chloride gives **45**, with one scrambled deuterium indicating that the allylic rearrangement probably proceeds by a diradical intermediate [658]:

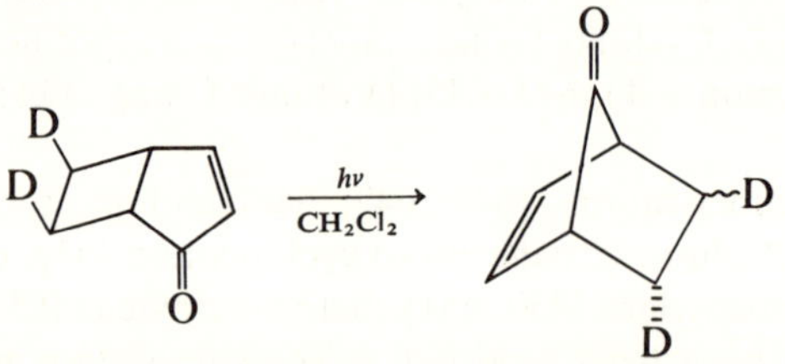

44   **45**

Photolyses of very dilute solutions (1 mg/ml) of **46** and **48** result in the formation of the cyclopropyl compounds **47** and **49**, respectively [659]:

46 **47**, 78%

48 **49**, 62%

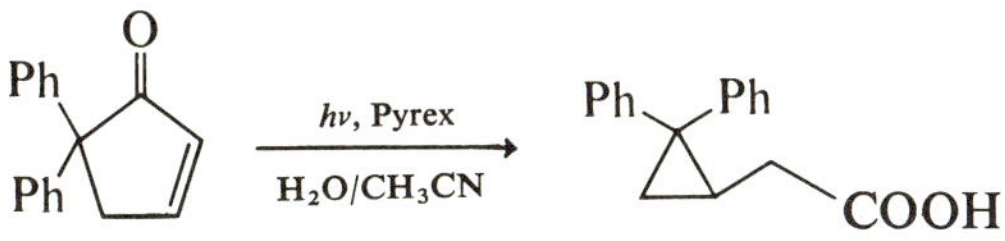

Initial cleavage on the most-substituted side followed by insertion of the alkyl radical into the double bond gives rise to an intermediary ketene. Reaction with solvent gives the observed products.

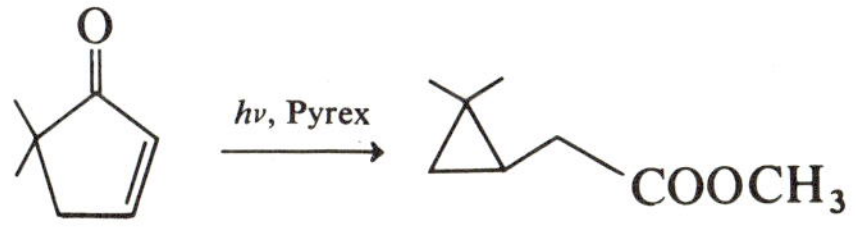

Irradiation of **50** in isopropanol afforded a 60% yield of **51** [660]:

50 **51**, 60%

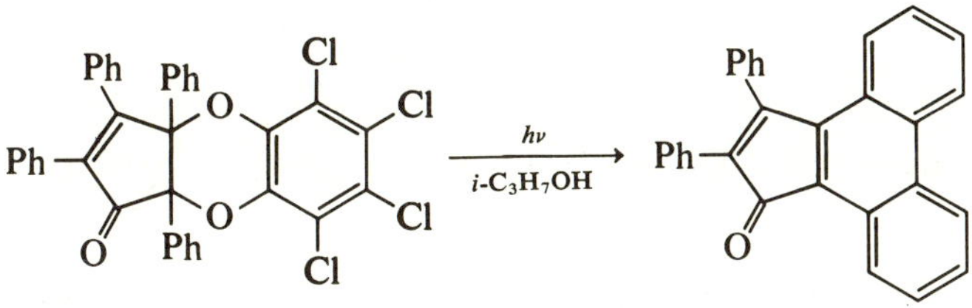

When the reaction was carried out in the presence of sensitizers or in cyclohexane solvent, no identifiable products could be detected.

A study comparing the photochemistry and the mass spectroscopy of a cyclopentenone has been reported [661]. Irradiation of either **52** or **53** gives a photostationary mixture of the two [662]:

52, 41% **53**, 59%

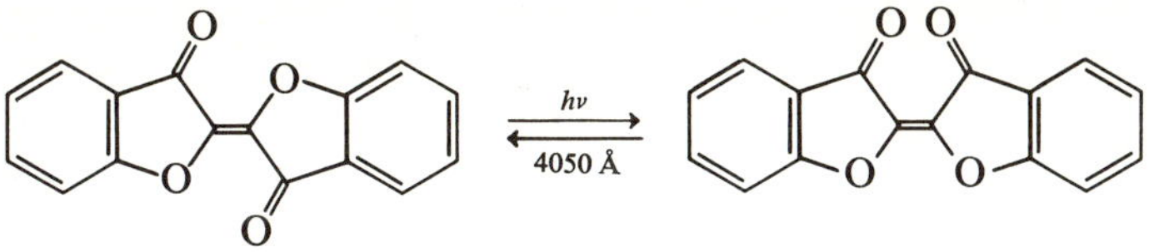

Photolysis of **54** in alcoholic solvent gives the rearranged product **55**:

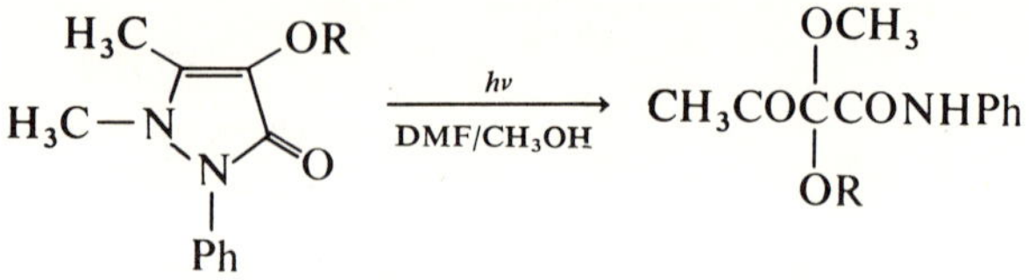
H3C
OR
H3C—N
N
O
Ph
hν
DMF/CH3OH
OCH3
CH3COCCONHPh
OR

54, R = —CH_3 or —CH_2Ph **55**

and irradiation of **56** affords two acyclic products of different structure [662a]:

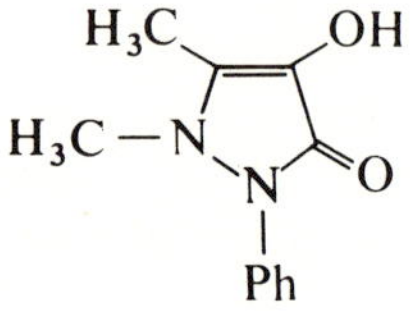
H3C
OH
H3C—N
N
O
Ph

56

2. CYCLOHEXENONES AND LARGER RING CONJUGATED ENONES

Variously substituted cyclohexenones have been found to undergo photocycloaddition with norbornadiene [663, 665], cyclohexene [666], and vinyl alkyl thioethers [667]. Several rather extensive studies have appeared involving the cycloadditions of 3-substituted cyclohexenones [666, 668, 669]. Irradiation of enone **57** in methyl acrylate gives a mixture of **58** and **59** after subsequent work-up [666]:

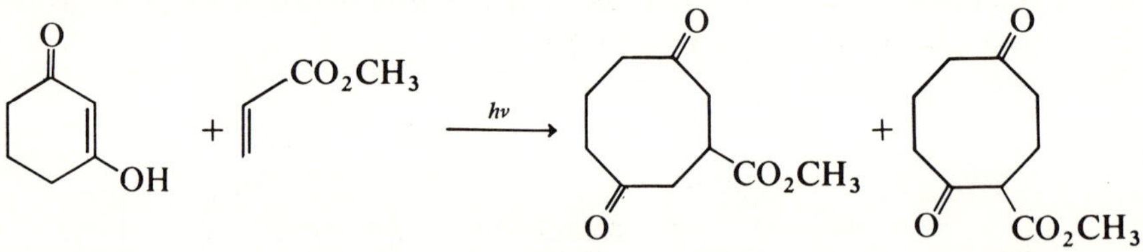
O
OH
+
CO2CH3
hν
O
CO2CH3
O
+
O
O
CO2CH3

57 **58** **59**

3-Carbethoxycyclohexenone **60** readily adds ethylene in excellent yield [668]:

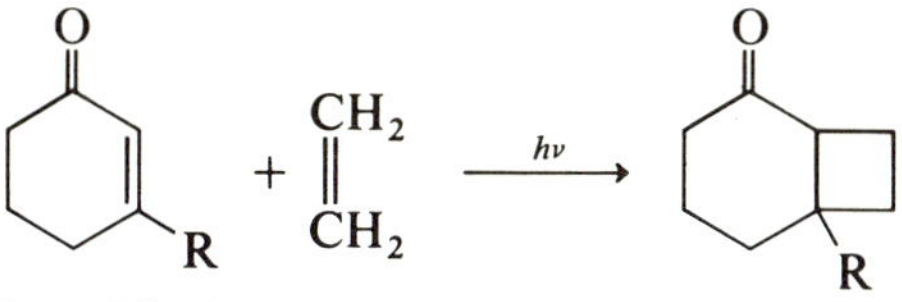

60, R = CO$_2$C$_2$H$_5$
also R = COOH, CN

>90%

trans-Dichloroethylene readily cycloadds to 3-methylcyclohexenone to give products exhibiting both *trans* and *cis* stereochemistry **61** and **62**, suggesting the intermediacy of a biradical [669]:

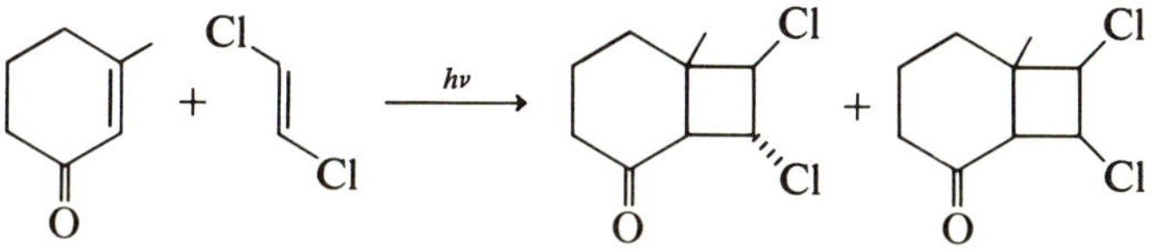

61 **62**

Photolysis of 4,4-dimethylcyclohexenone **63** [670] in tetramethylethylene gives the *trans*-fused cycloaddition product **64** [436]. A number of interesting secondary photolysis products **65** to **66** are also obtained (Chapter 3):

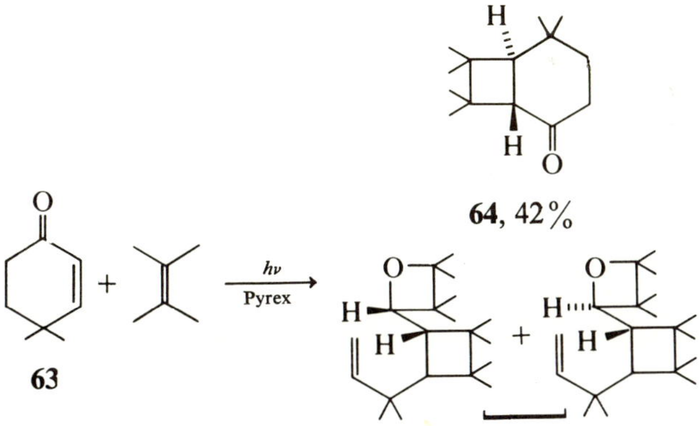

64, 42%

63 **65** 47% **66**

An example of intramolecular photocycloaddition has appeared [671].

Irradiation of (+)-isopiperitenone **67** in cyclohexane afforded **68** as the only isolated product, and the same irradiation in methanol gave **71** as well as some **68**. The ketene **70** is proposed to arise by way of biradical **69**—derived from α–cleavage on the most substituted side of **67** (see also Chapter 4):

(+)–**67** (+)–**68**, 42%

69 (±)–**70** (±)–**71**, 49%

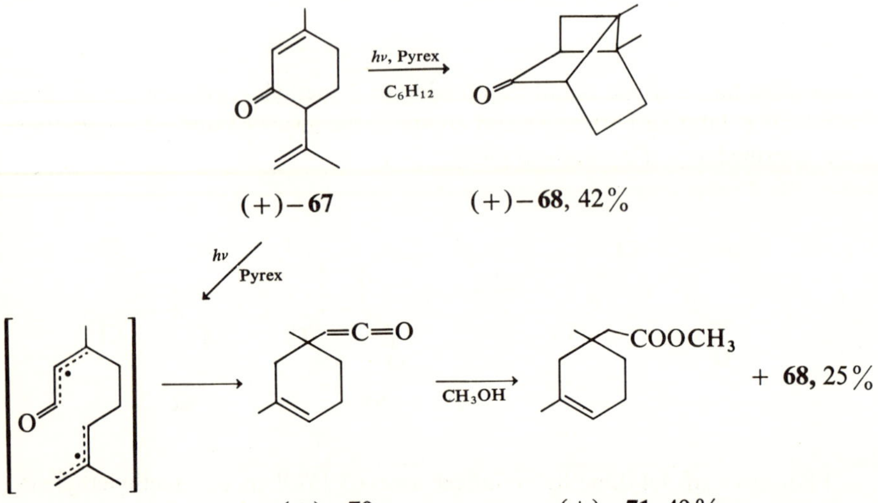

A possible alternative mechanism for the formation of **70** that would involve an initial cycloaddition **72** followed by a thermal cleavage **73** was discounted, because the ketene derived by this pathway would retain its optical activity:

72 **73** (+)– or (–)–**70**

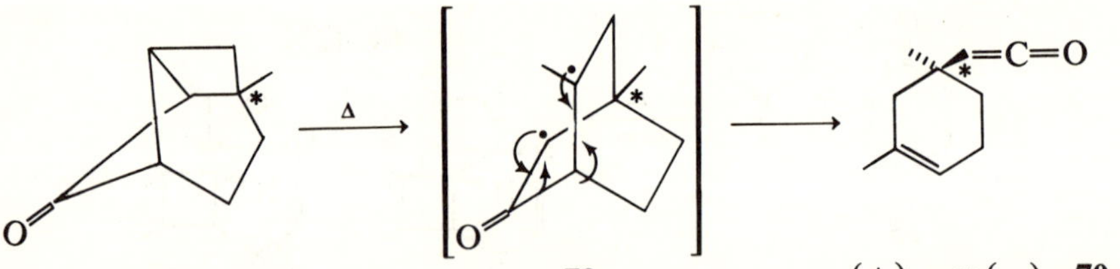

The **70**, as **71**, obtained from photolysis was racemic. 4-Acetoxyisopiperitenone undergoes analogous reactions [671].

The preferential migration of *p*-cyanophenyl relative to phenyl in the irradiation of **74** has been reported [672]:

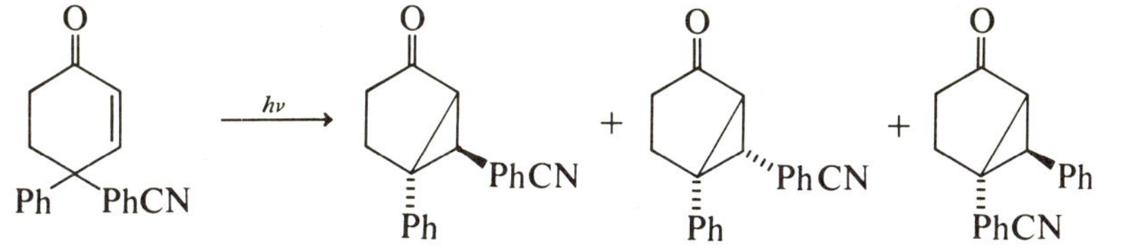

74, Φ = 0.17 → **75**, Φ = 0.17 + **76**, Φ = 0.001 + **77**, Φ = 0.0016

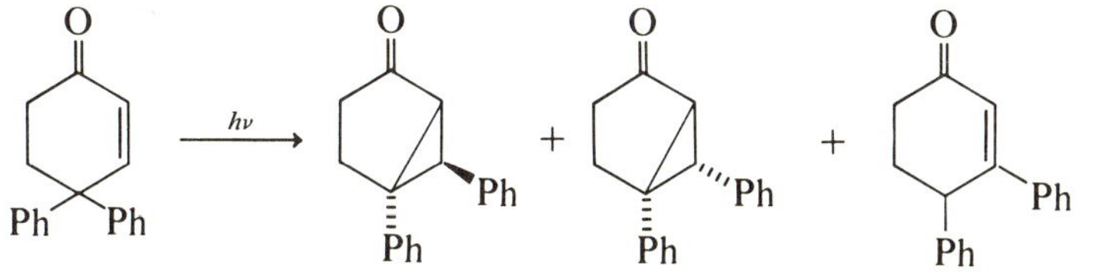

78, Φ = 0.043 → **79**, 0.043 + **80**, Φ = 0.0003 + **81**, Φ = 0.0002

Quantum efficiencies for the migration of the two groups were found to be 0.177 and 0.016, respectively. The authors conclude that α and β aryl migration involve rearrangement of the β-carbon, and their results provide evidence against electron deficiency in the migrating groups. Additional studies in the 4,4-diphenyl case **78** indicate that E_a for migration is ~ 10 kcal/mole [673].

Irradiation of the related cyclopropyl enone **82** in *t*-butanol gives the rearranged product **83** exclusively [674]:

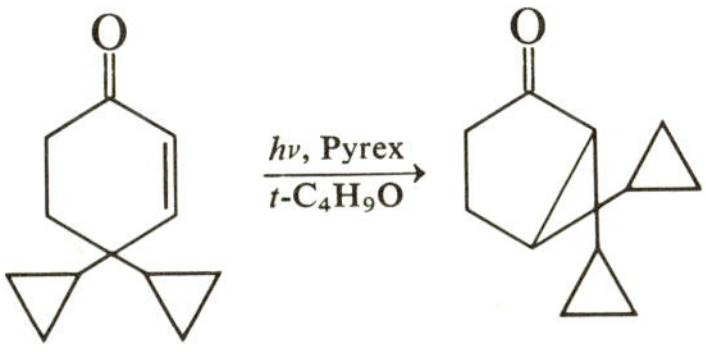

82 → **83**, 93%

A mass spectroscopic study of variously deuterated 4,4-dimethylcyclohexenones has been carried out [675]. A study of the influence of solvents on the n,π^* transitions of a number of cyclohexenones has been reported [676].

Enone **84** undergoes reduction when irradiated through Pyrex in dioxane [677]:

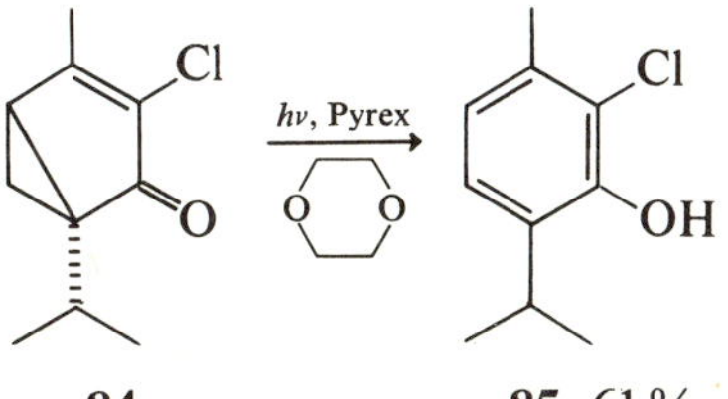

84 → **85**, 61%

Irradiation of **86** gives the cleavage product **87a** as well as the rearranged product **87b** [678]:

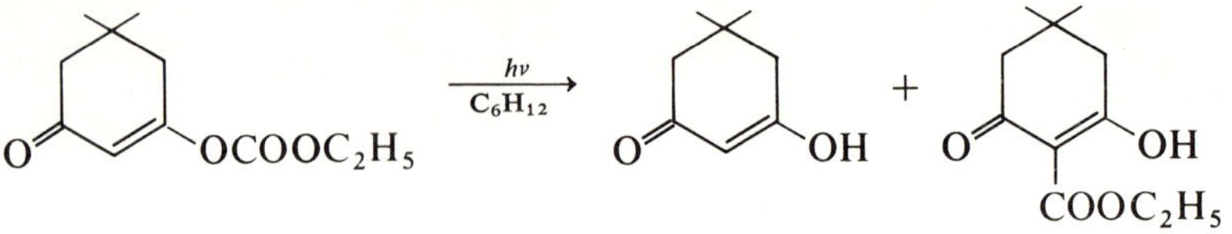

$$86 \xrightarrow{h\nu, C_6H_{12}} 87a, 3\% + 87b, 5\%$$

Similarly, irradiation of **88a** gives **88b** [679]:

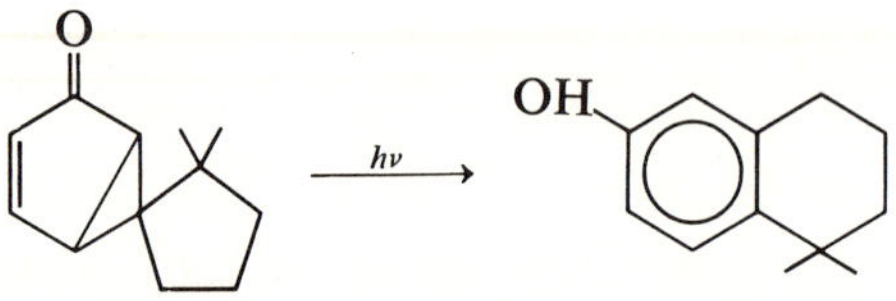

$$88a \xrightarrow{h\nu} 88b$$

The photolysis of a number of cyclohexenones containing sulfur has been reported [447]. Irradiation of **89** gives **90** as the only product:

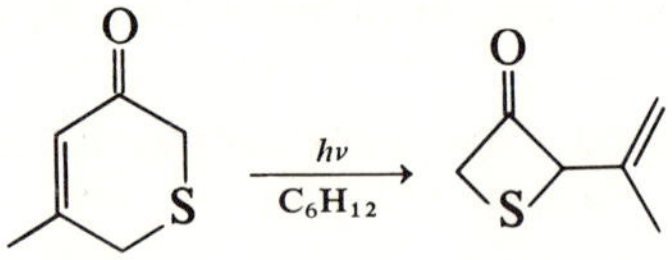

$$89 \xrightarrow{h\nu, C_6H_{12}} 90, 30\%$$

The bicyclic enone **91** gives **93** and **94**, perhaps by way of **92**:

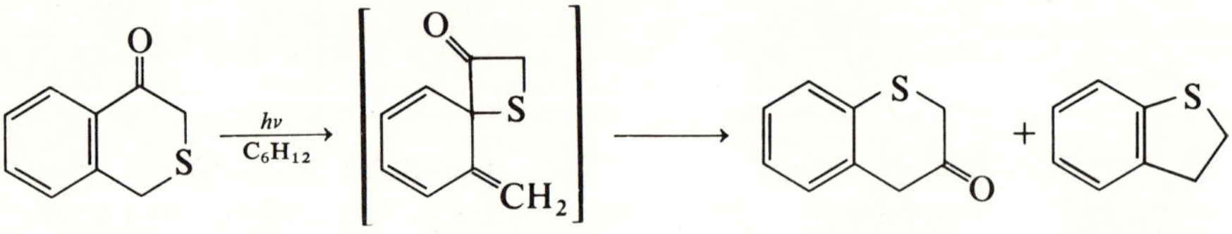

$$91 \xrightarrow{h\nu, C_6H_{12}} [92] \longrightarrow 93, 21\% + 94, 6\%$$

Irradiation of the pyrone derivative **95** results in a mixture of dimers **96** [680]:

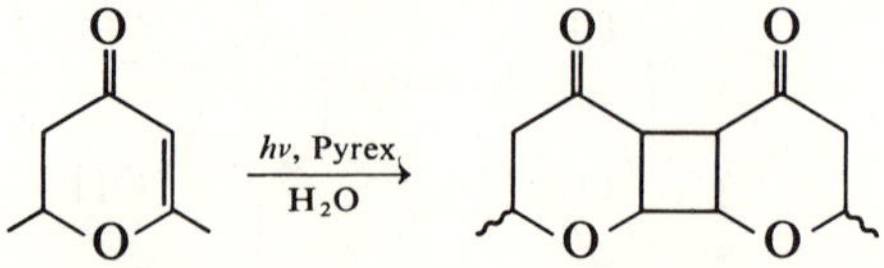

$$95 \xrightarrow{h\nu, \text{Pyrex}, H_2O} 96, 72\%$$

The photodimerization of a thiopyran-4-one has also been reported [681].

An extensive study of the photocycloaddition of various kinds of olefins to testosterone acetate **97** to give 4,5-cyclobutanosteroids has been carried out [682, 683]:

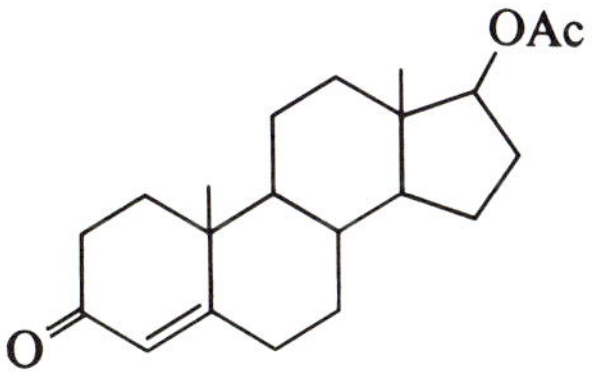

97

Vinyl acetate and acrylonitrile gave the 4α-*trans*-adducts.

Irradiation of cholest-5-en-7-one **98** in tertiary butyl alcohol through Pyrex gave **99** as the primary photoproduct [684]. Continued irradiation gave **100**, the expected allylic rearrangement product of a β,γ-unsaturated ketone **99**:

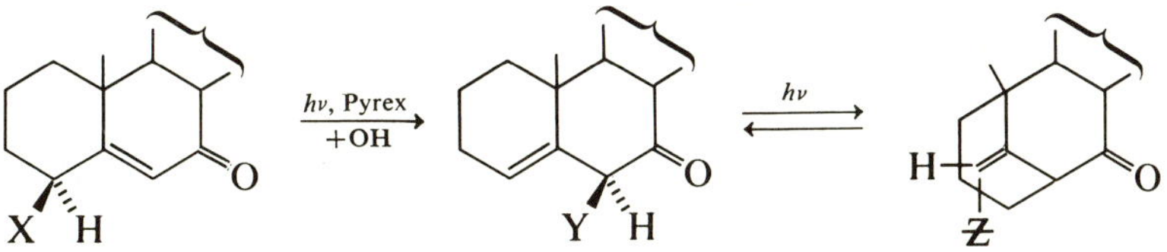

98, X = D, H   **99**, Y = H, D   **100**, Z = H, D

When the reaction was carried out by way of irradiation through quartz, small amounts of the dimeric materials **101** and **102** were formed in addition to **99** [685]:

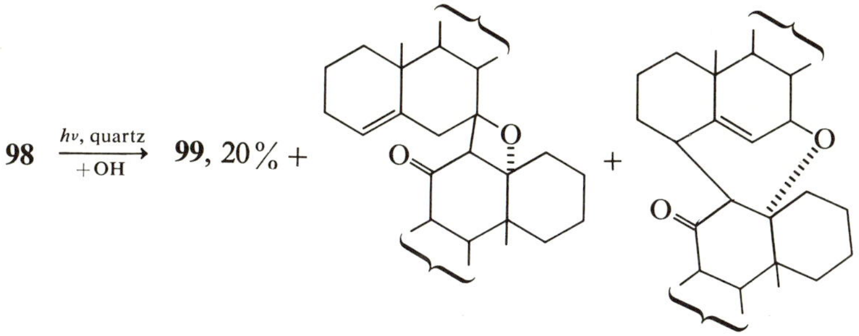

98 $\xrightarrow[+OH]{h\nu, \text{quartz}}$ **99**, 20% +   **101**, 10%   +   **102**, 7%

3β-Acetoxy-cholest-5-en-7-one afforded a 1 : 1 mixture of products analogous to **99** and **100**.

The testosterone derivative **103** undergoes a similar net deconjugation reaction **104**:

103 **104**

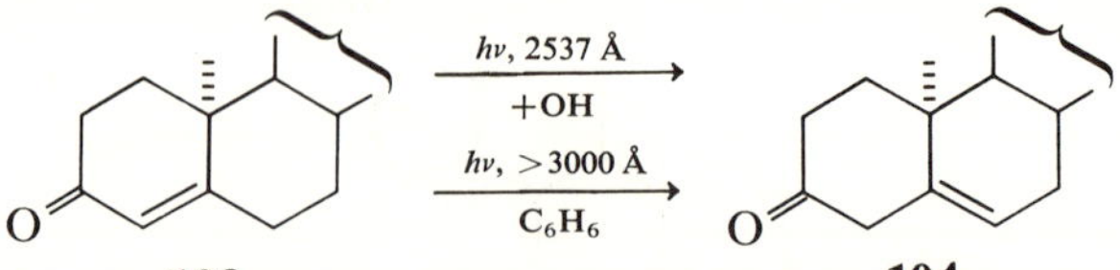

The reaction proceeds in tertiary butyl alcohol with 2537 Å light or in benzene with light of wavelength greater than 3000 Å. When **103** was irradiated (2537 Å) in *t*-butanol-*d*, deuterium incorporation occurred only at the α-carbon, thus eliminating a simple 1,3 sigmatropic shift.

Irradiation of **105D** (λ > 3270 Å) leads to a photostationary mixture with **106** that slowly gives way to **107** [686]. When the reaction is carried out with short-wavelength light (λ < 2800 Å), **108** is the sole product in what seems a bimolecular reaction [437, 428]. Photolysis of **105H** in toluene gives **109** as the primary product:

105, D, H **106** **107**

109 **108**

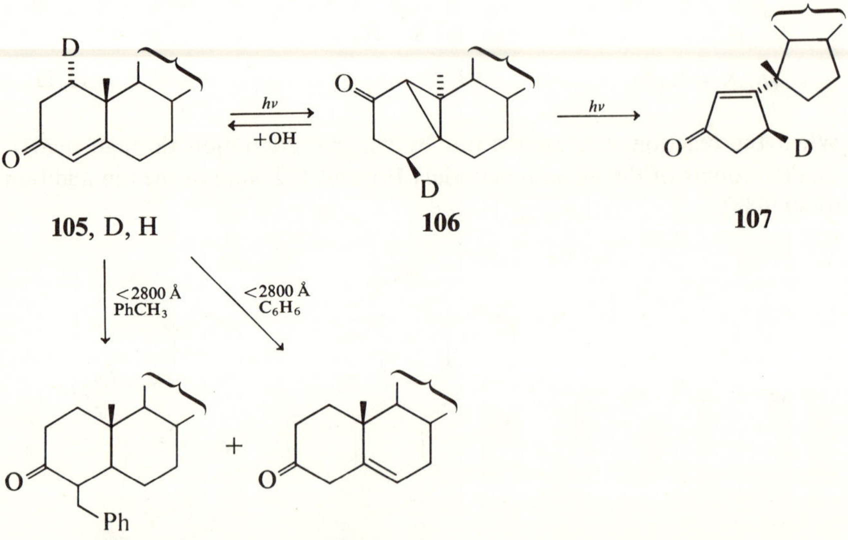

Irradiation of **110** [687] and related compounds [688] leads to the intramolecular cyclization product **111**:

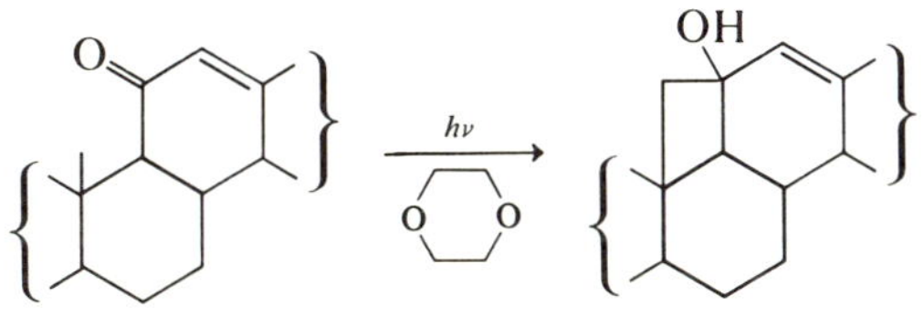

110 →(hν, dioxane) **111**, 30%

Theoretical calculations have appeared for the dimerization of isomers of **112** [689]:

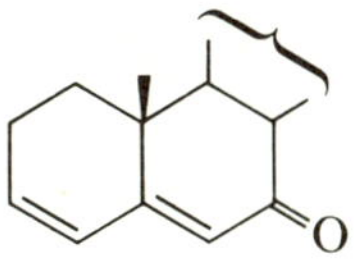

112

Evidence has been presented for the existence of two reactive triplets in the photochemistry of santonin [690].

Irradiation of androstenedione **112a** results in the formation of **112b** and **112c**, apparently by way of the triplet excited state of **112a**:

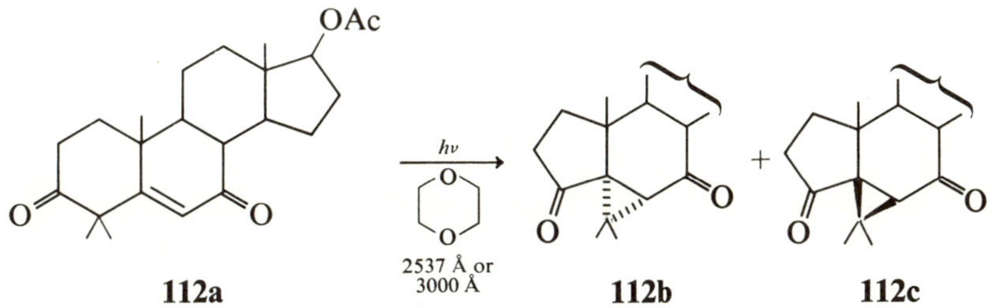

112a →(hν, dioxane, 2537 Å or 3000 Å) **112b** + **112c**

On the other hand, the 3-ethylene ketal derivative **112d** undergoes photoaddition of dioxane to the carbonyl group **112e** [691]:

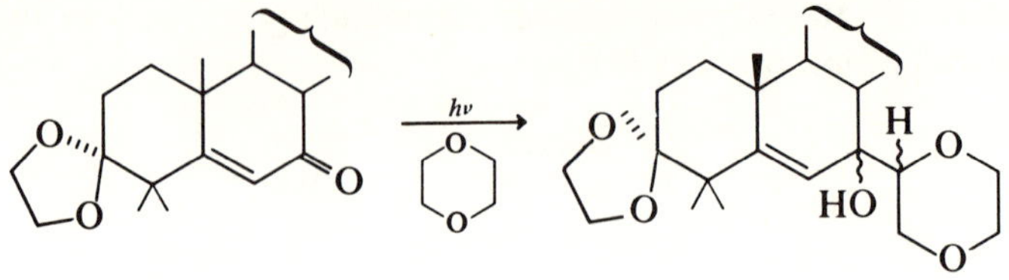

112d → **112e**, isomers

Low-temperature ($-190°$) photolysis (2537 Å) of neat **113** gave a compound exhibiting an infrared band at 1812 cm^{-1}, presumed to be **114** [414]. Irradiation of **114** ($\lambda < 3000$ Å) gave a 65% yield of **115**:

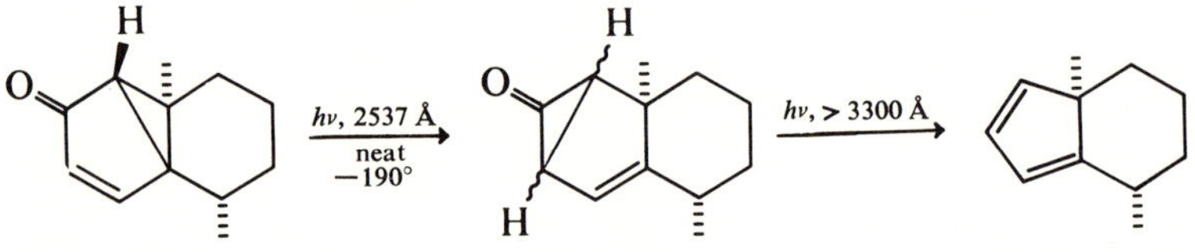

113 $\xrightarrow{h\nu, 2537 \text{ Å} \atop \text{neat} \atop -190°}$ **114**, 1812 cm^{-1} $\xrightarrow{h\nu, > 3300 \text{ Å}}$ **115**, 65%

Cyclopropanone **114** also forms an adduct with furan on warming.

An extensive study of the photochemistry of 10-hydroxymethyl-$\Delta^{1,9}$-2-octalone has appeared [692].

A flash-photolysis study of an α,β-unsaturated enone has led to the detection of two short-lived transients [693].

Photolysis of enone **116** in water gives a mixture of dimers [636]:

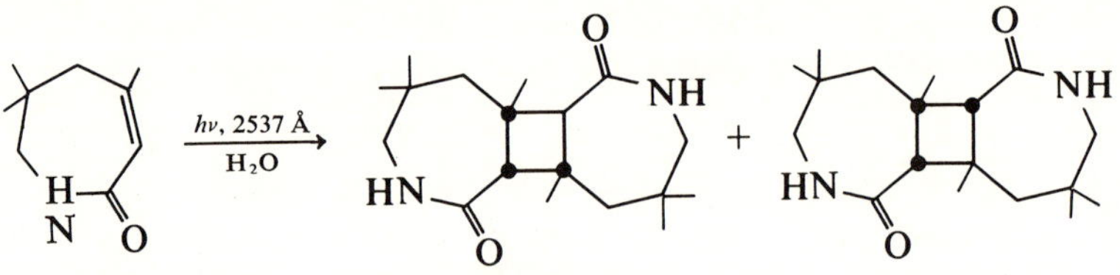

116 $\xrightarrow{h\nu, 2537 \text{ Å} \atop \text{H}_2\text{O}}$ **118**, 80% + **118**, 2%

The reaction apparently occurs by way of a triplet state, since acetone photolysis gives the same ratio of products.

Cycloheptenone **119a** undergoes photoinduced rearrangement to *cis*-fused cyclopentanone **119b** [453]:

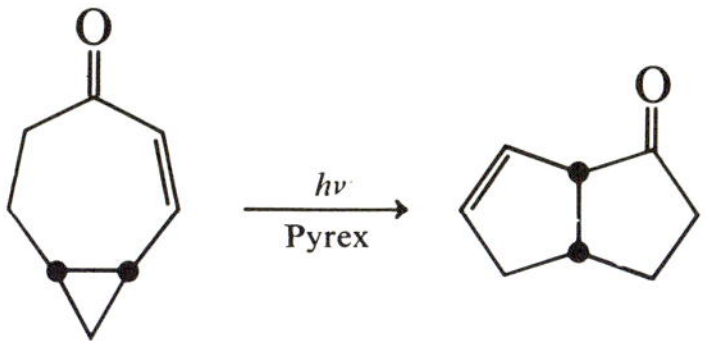

ORGANIC PHOTOCHEMISTRY

119a →(hv, Pyrex) **119b**

Although the yield of **119b** was greater when photolysis was carried out in acetone (80%) than in pentane (60%), the reaction could not be sensitized or quenched.

Irradiation of **120** in hexane gives a mixture of **121**, **122**, and **123** [694] (see Chapter 3):

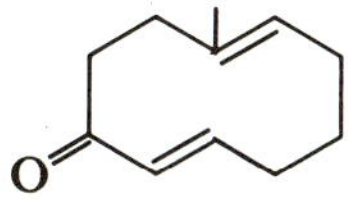

120

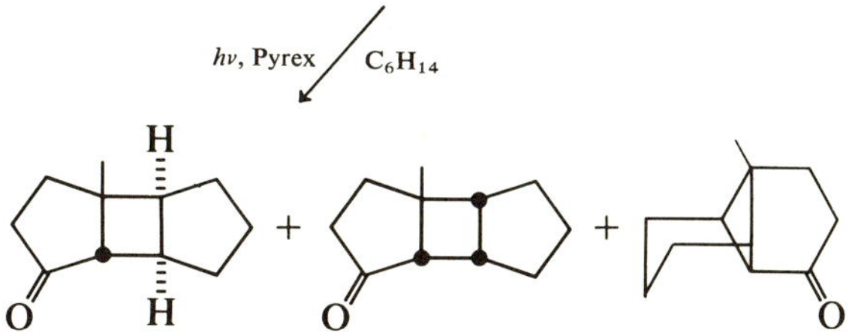

hv, Pyrex / C_6H_{14}

121 + **122** + **123**

The ratio of **121** to **123** is 2 : 1 under these conditions, and it is 9 : 1 when the irradiation is carried out in the presence of piperylene. Trienone **124** undergoes a similar cyclization to give **125** [655]:

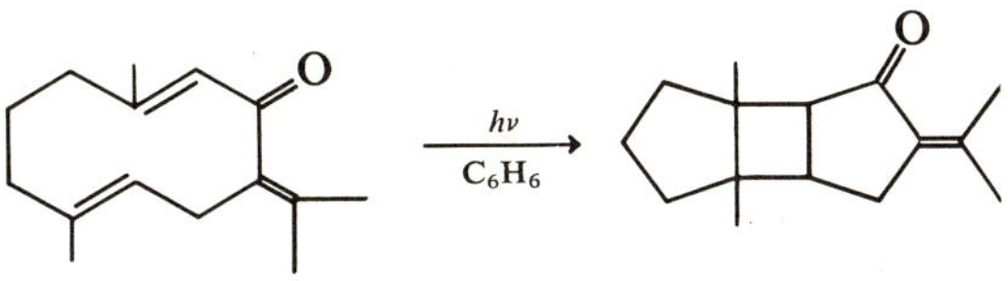

124 →(hv, C_6H_6) **125, 80%**

Allene was found to add to **126** readily to give a single isomer **127** when the irradiation was carried out at low temperature [695] (Volume 1, page 134):

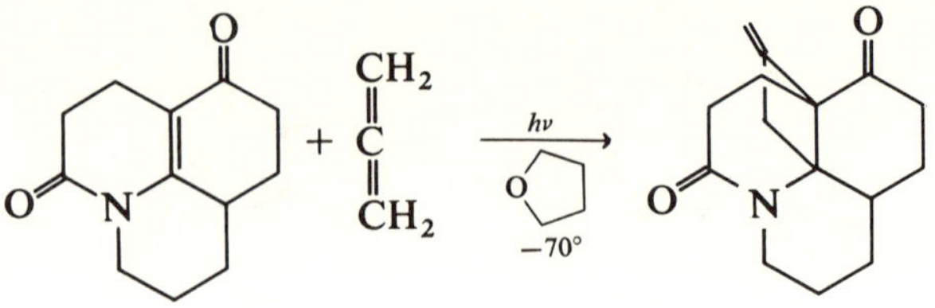

[Structures 126 + CH₂=C=CH₂ →(hν, O/−70°) 127]

126 **127**

Photolysis of **128**

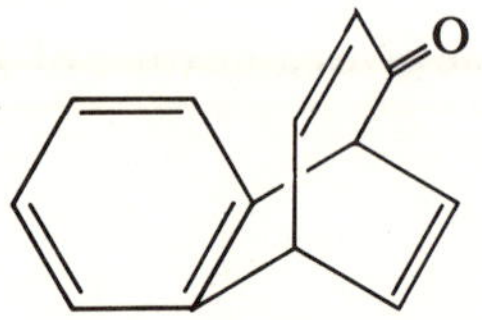

128

in acetonitrile produces the labile **129**

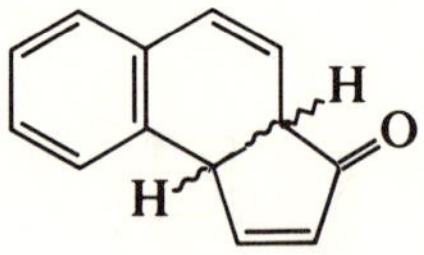

129

as the only photoproduct [696, 697]. When the reaction was carried out in water, the benzonorcaradiene **130**

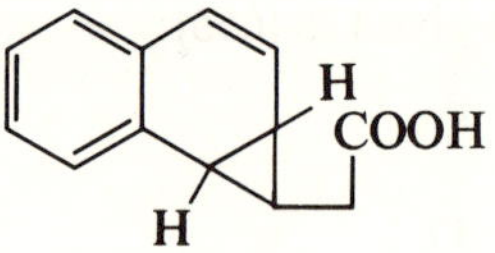

130

was obtained as the major product. Neither reaction could be sensitized by benzophenone. A ketene intermediate was postulated for the formation of **130** (Chapter 4). In a study of similar compounds, using low-temperature infrared techniques, direct evidence for ketene products was obtained [698, 699]. Thus, irradiation of **131** and **133** gave **132** and **134**, respectively:

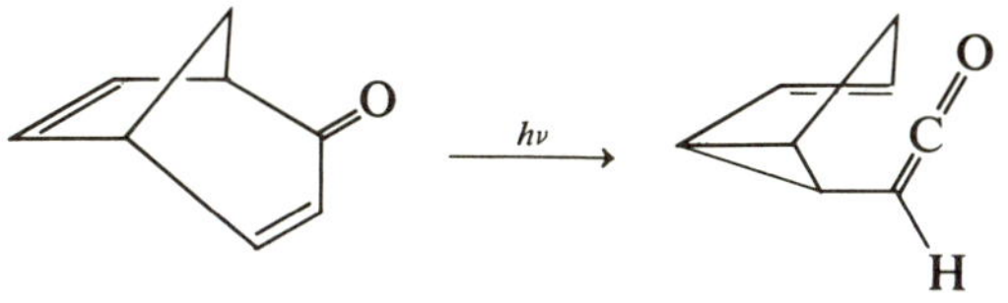

131 →hν→ **132**

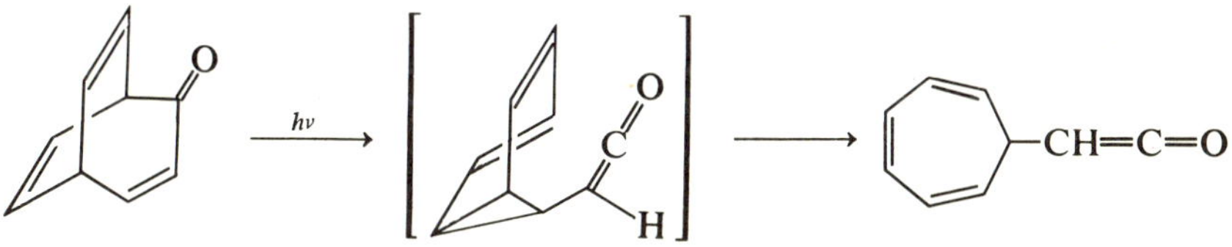

133 →hν→ [**134**] → **135** (cycloheptatrienyl-CH=C=O)

3. CYCLOHEXADIENONES AND TRIENONES

A review of 4,4-disubstituted cyclohexadienones has appeared [700]. A series of 3,5-methylsubstituted-2,4,6-trimethyl-4-*n*-propylcyclohexadienones has been irradiated in methanol, using 2537 Å light [701]. Thus, irradiation of **136** gave a mixture of **139** and **140**:

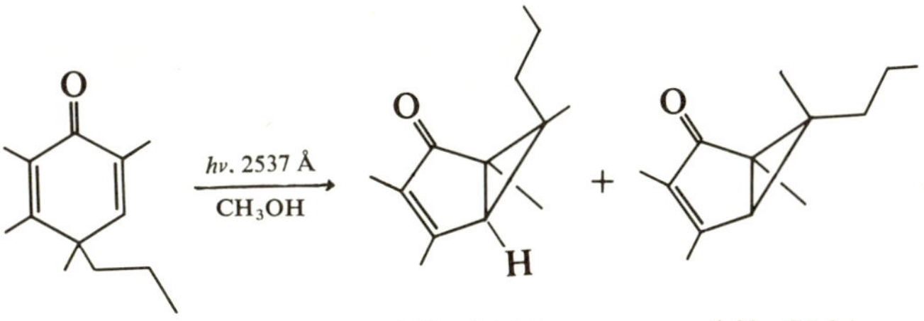

136 →hν, 2537 Å, CH$_3$OH→ **139**, 68% + **140**, 32%

The authors propose the intermediacy of species **137** and **138**:

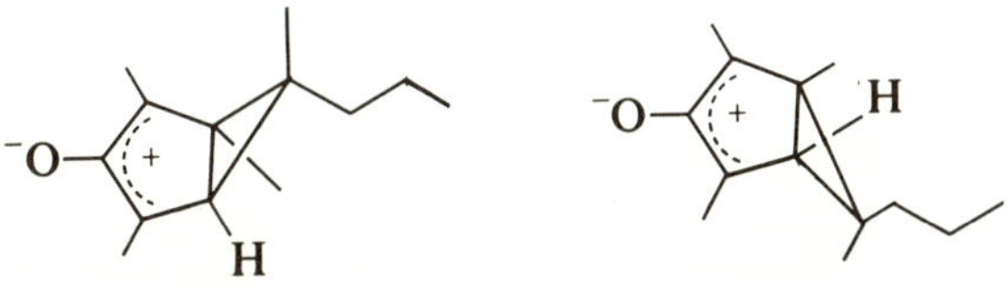

137   **138**

and point out how steric factors may control the stereochemistry of these rearrangements.

The photoreduction of 1,4-cholestadien-3-one by sodium borohydride has been reported [702]. The quantum efficiencies for the reduction of a number α-halocyclohexadienones have been measured [703, 704].

Ionic intermediates, similar to **137** and **138**, were proposed in the photolysis of **141** to give **142** and **143** [705]:

141 **142** **143**

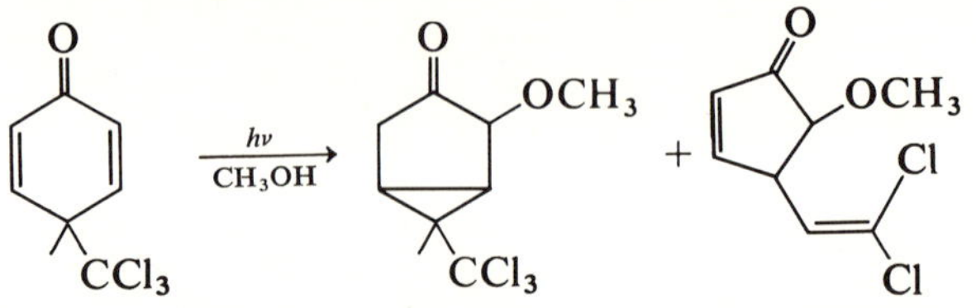

In a related paper [306] it was found that treatment of **144** or **145** with strong base gave **147** as the major product, presumably by way of the ionic intermediate **146** [706]:

144 **146** **145** **147**

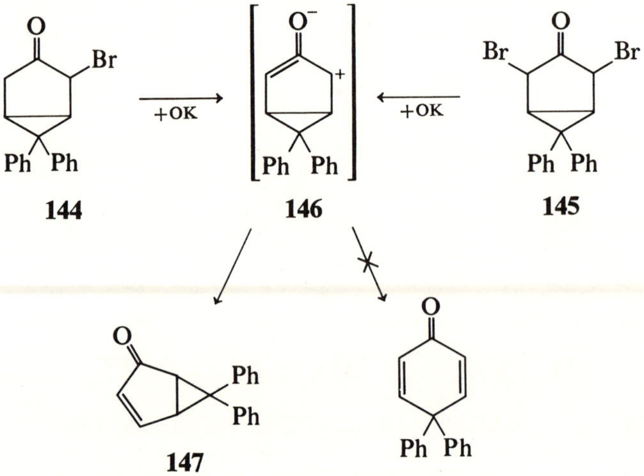

Irradiation of **148** gave the cyclobutenone **149** [707]:

148 **149**, 33%

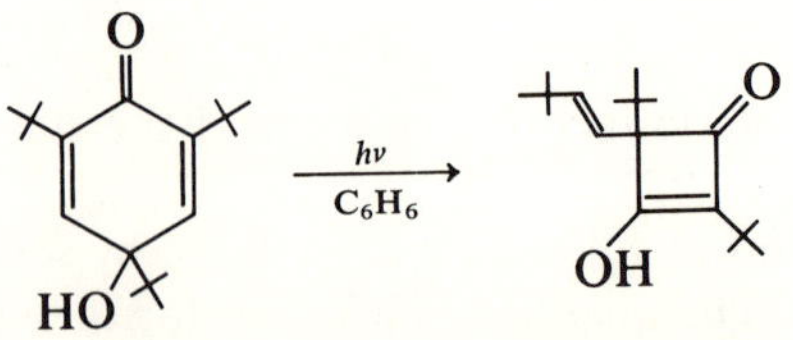

Photolysis of **150** or **151** with long-wavelength light leads to a photostationary mixture [708]. When the reaction is carried out with 2537 Å light, the rearrangement products **152** and **153** are obtained:

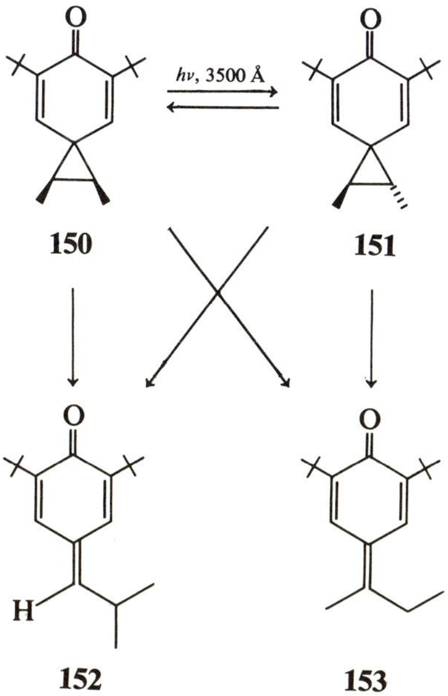

Sensitized irradiation gave only the isomerization of **150** and **151**.

Quinone methides, such as **152** and **153**, are photoreduced to phenols in the presence of hydrogen donors [709].

Dienone **154** has been found to undergo photochemical decomposition ($\Phi = 0.065$) in order to give **155** to **158**:

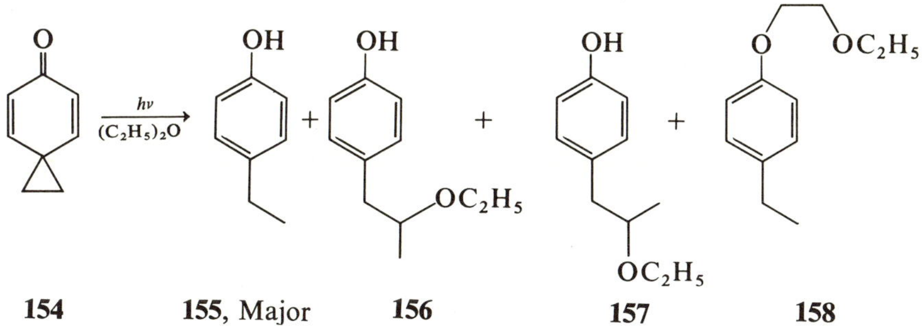

Quenching studies indicate that both singlet and triplet states are involved.

The low quantum yield is attributed to the reversal of the cyclopropyl ring opening [710, 711]:

Irradiation of **159** in the presence of benzophenone results in the formation of **160** by a mechanism that presumably involves chemical sensitization by way of the benzophenone [712]:

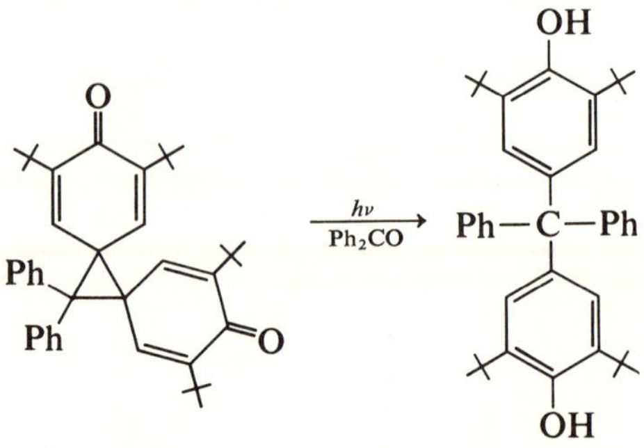

159 → (hv, Ph₂CO) → **160**

The photolysis of santonin in protic solvents has appeared [713]. Evidence for ionic intermediates was obtained for the photolysis of halogen-substituted santonins [714]. Irradiation of a number of steroidal dienones in the presence of sodium borohydride leads to a mixture of corresponding phenolic materials [702]. The photolysis of androsta-1,4-dien-3-one has appeared [715].

Photolysis of **161** in isopropanol gives **162** as the only product ($\Phi = 0.10$) [716]:

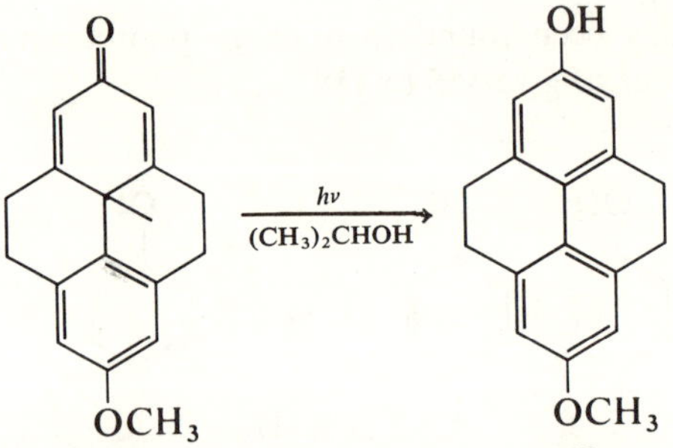

161 → (hv, (CH₃)₂CHOH) → **162**

Attempts to trap a postulated zwitterionic intermediate were unsuccessful.

Irradiation of a 10% solution of **163** in acetic acid gave a 29% yield of the ring-expansion product **164** [717, 718; see also 719, 719a]:

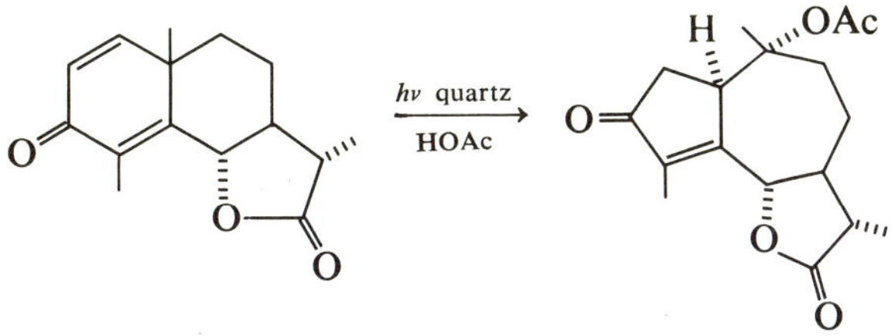

163 → **164**, 29%

hv quartz, HOAc

Several reports related to the photochemistry of Δ^1-4-alkyltestosterones have appeared [720, 679].

Photochemical dimerization [721, 722, 723] and cycloadditions [724] of a number of 4-pyrone derivatives have appeared.

The photochemistry of a number of 2,4-cyclohexadienones has been reported [725]. Irradiation of **165**

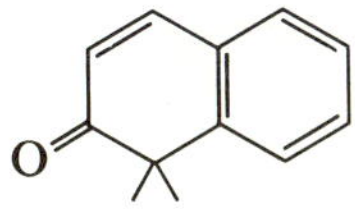

165

in ethanol gives the head-to-head dimer [726]. Dienone **166**

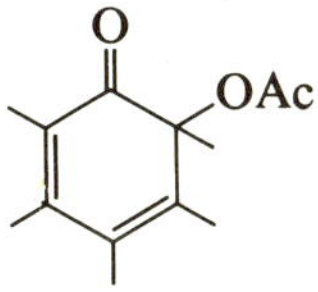

166

undergoes ring-opening and rearrangement when irradiated in the presence of cyclohexylamine-ether [727]. Photolysis of **167** in methylene chloride through Pyrex gave no product. When the reaction was carried out using methanol as solvent, a 50 to 55% yield of **168** was obtained [728]:

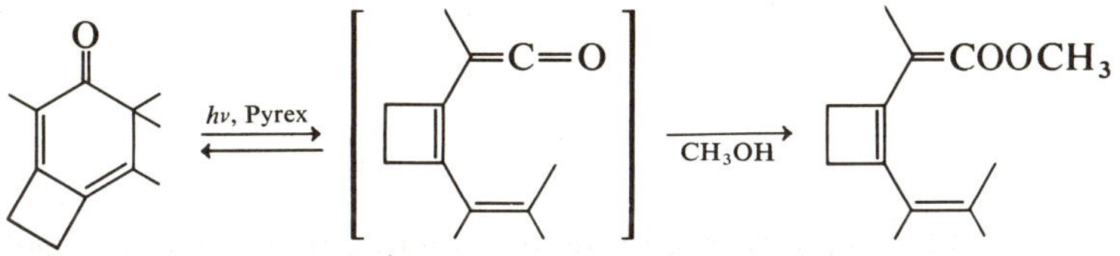

167 ⇌ [**169**] → **168**, 55%

hv, Pyrex ; CH₃OH

The presumed ketene intermediate **169** apparently reclosed in the methylene chloride solvent—thus, no apparent reaction.

A number of 2-pyrone derivatives have been irradiated [637, 729, 730]. The labile bicyclic lactone **171** is obtained from the irradiation of **170** in ether [731]:

170   **171**

In methanol-ether, the product is **172**:

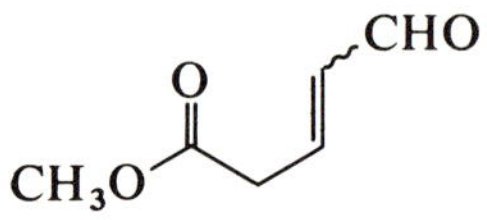

172

Photolysis of **173** in benzene affords a high yield of **175**, presumably by way of the secondary photolysis of cyclobutenone **174** [732] (Chapter 4):

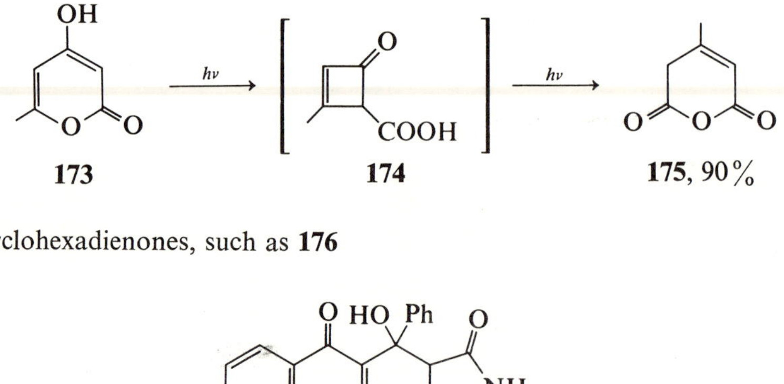

173   **174**   **175**, 90%

Cyclohexadienones, such as **176**

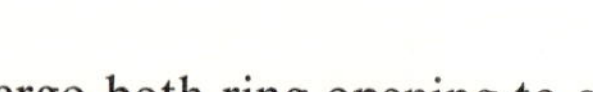

176

have been found to undergo both ring opening to a triene and di-π-methane rearrangement to an isomeric vinyl cyclopropane [733].

Some photochemical reactions of tropone and its derivatives have appeared

[734, 735, 736]. The trienone **177** undergoes (1) photochemical ring opening to the ketene **178**, which reverts back to **177** thermally, and (2) isomerization to **179** [737]:

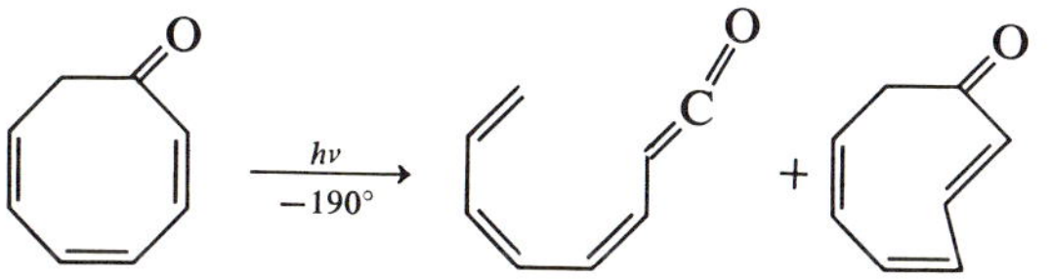

177 $\xrightarrow{h\nu}_{-190°}$ **178** + **179**

Support for the structural assignments of β- and α-lumicolchicine has appeared [738].

A mechanistic study of the photochemical intramolecular cyclization of eucarvone has appeared [739]. Quenching studies indicate that closure occurs by way of both singlet and triplet excited states.

4. ENEDIONES, ENEDIACIDS, AND RELATED COMPOUNDS

Enediones of general structure **180** add to a variety of unsaturated compounds across the C=C bond [740], for example, **180 → 182**:

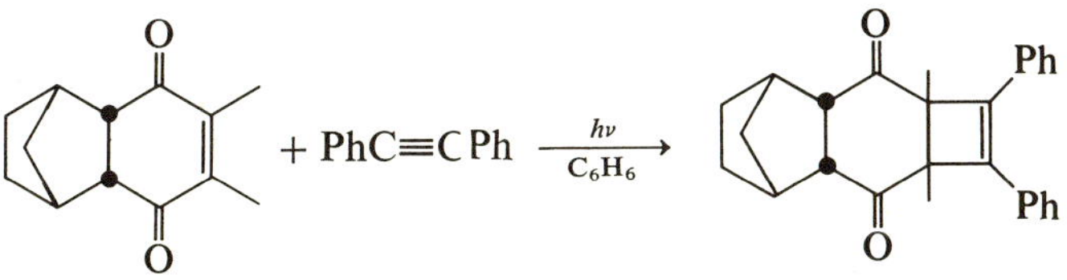

180 + PhC≡CPh **181** $\xrightarrow{h\nu}_{C_6H_6}$ **182, 70%**

Both pyridyl and quinoline α-dials are photorearranged to lactones [741], presumably by way of an intramolecular hydrogen abstraction cyclization.

The diene dione **183** undergoes photorearrangement to **184** and **185**, the latter probably arising from a type II process [742; see also 743]:

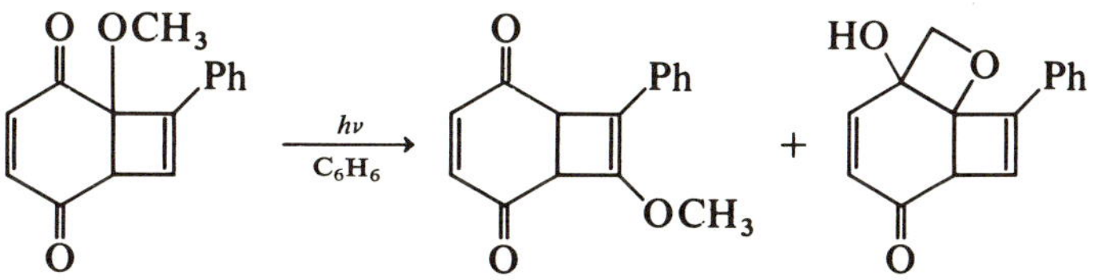

183 $\xrightarrow{h\nu}_{C_6H_6}$ **184, 55%** + **185, 20%**

Irradiation of *trans*-β-acetylacrylate yields the *cis* isomer, and the free acid photocyclizes to a lactone [613].

The enedione **186** is photocyclized to the cage dione **187** [427]:

186 **187**

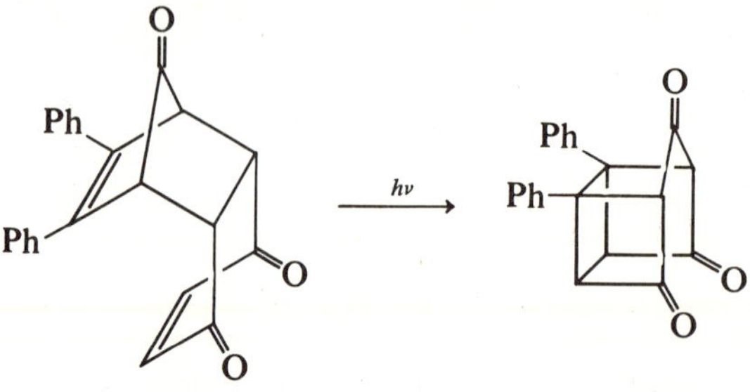

Related structures yield decarbonylation products (Chapter 3).

A number of reports concerning intramolecular photocycloadditions of the C═C bond of maleic acid derivatives to C═C [540, 744, 745, 725, 746, 747, 63] and C—C [748, 749, 750] have appeared (Volume 1, page 142). Although direct excitation of **188** leads to the formation of the internal $2\pi + 2$ adduct **191** and some C═C bond reduction, acetone sensitization leads also to external C═C bond additions [748]:

188, X = CO_2CH_3 **189** **190** **193** **191** **192**

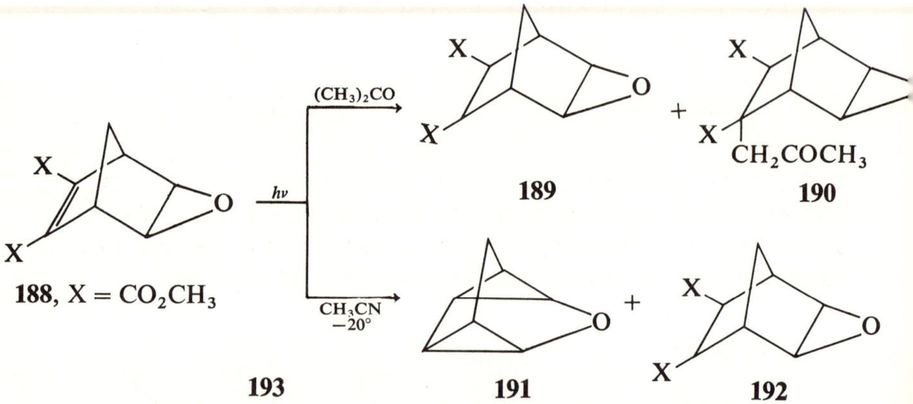

The tricyclic ene diester **193** (X = O or $CHCO_2CH_3$) is photorearranged to **194** in good yield [750]:

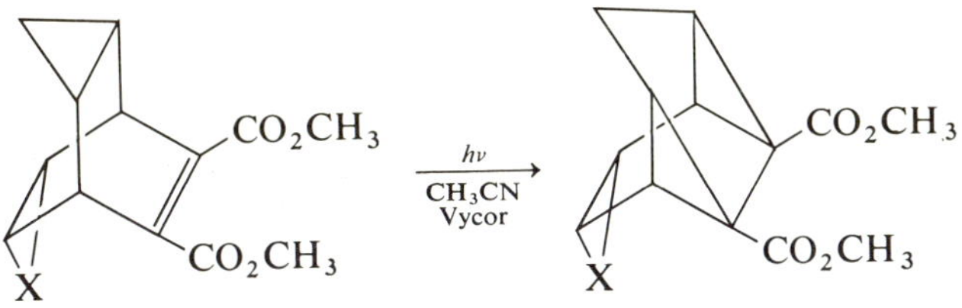

$X = CHCO_2CH_3, O$
193

194, 86–96%

and the polycyclic system **195** photorearranges to **196**, possibly by way of the suggested mechanism [749]:

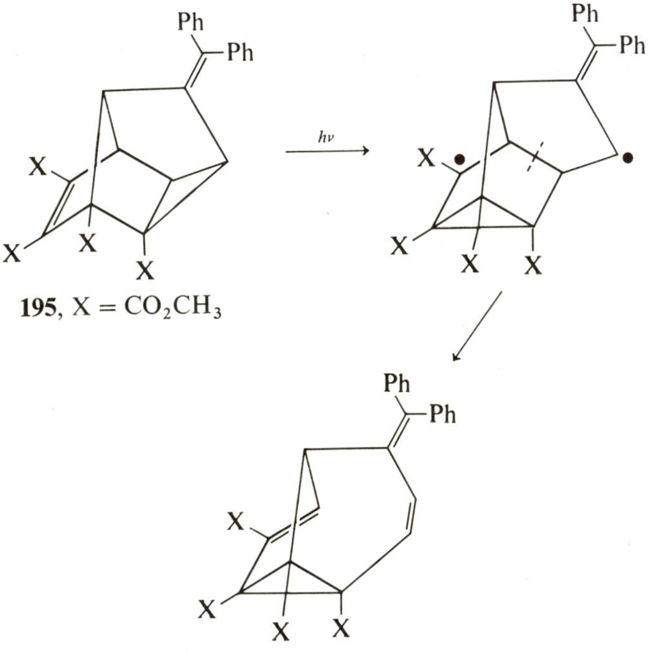

195, $X = CO_2CH_3$

196, 95%

In contrast to the intramolecular ring closures that are well known for norbornenes [746] and for systems related to **197** (Volume 1, page 142) [751], the latter is apparently photolyzed by 258 nm light to cyclobutadiene and dimethylphthalate [752]. Interestingly, irradiation of **197** with 313 nm light yields the ring-closed isomer **198** [753]:

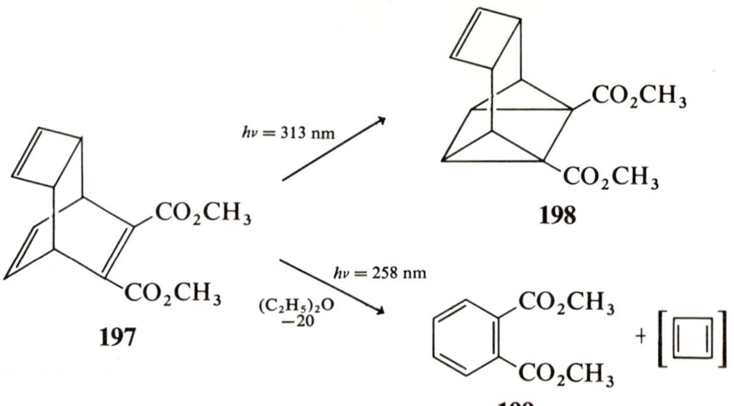

197 198 199

Irradiation of quinolizine ester **200a** in alcoholic solvents results in pyrroloazepine formation **200b** [754; see also 755]:

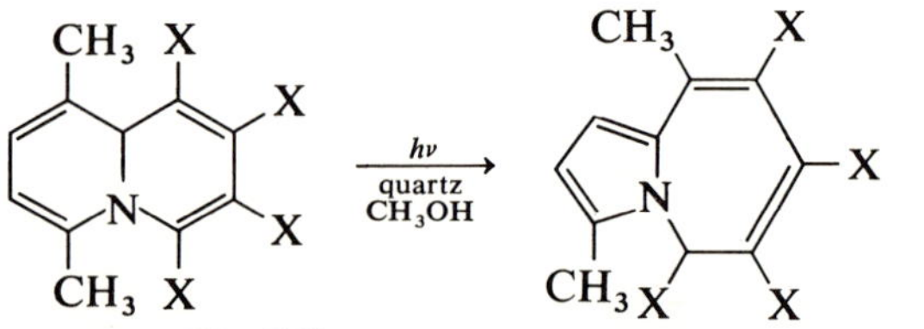

200a, X = CO_2CH_3 **200b**, 5%

On the other hand, dihydro pyrridine **200c** yields the ring-opened product **200d** [756]:

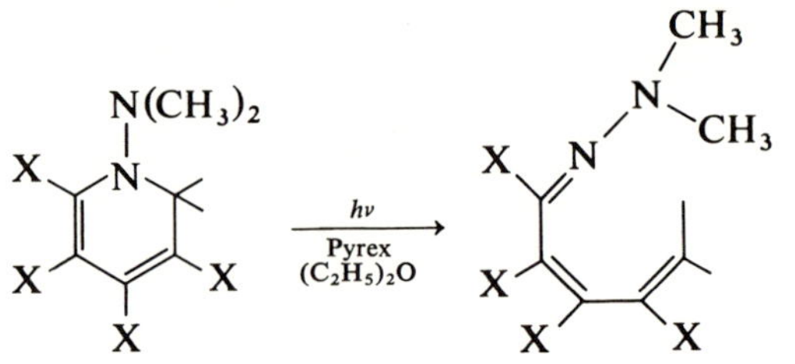

200c, X = CO_2CH_3 **200d**

The photocycloaddition of maleic anhydride and its derivatives to unsaturated systems is well known (Volume 1, page 144) [757]. The photoaddition of maleic anhydride to acetylene yields the interesting bicyclopropyl structure **202** in addition to the expected adduct **201** [758]:

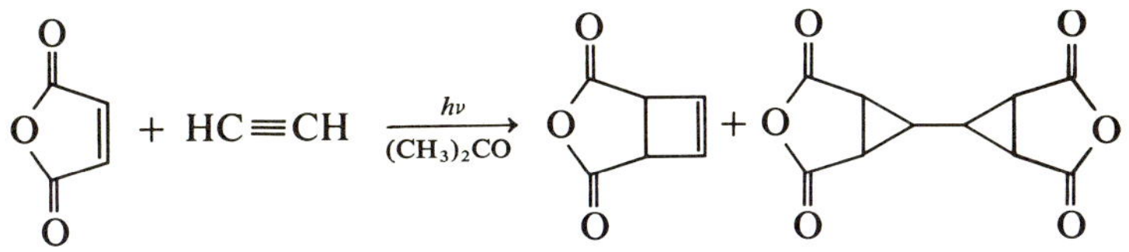

201, 36%   202, 48%

Alkyl acetylenes give higher yields of cyclobutene products. Although the mechanism of the formation of **202** is unclear, the trapping of the biradical **203** by maleic anhydride seems plausible:

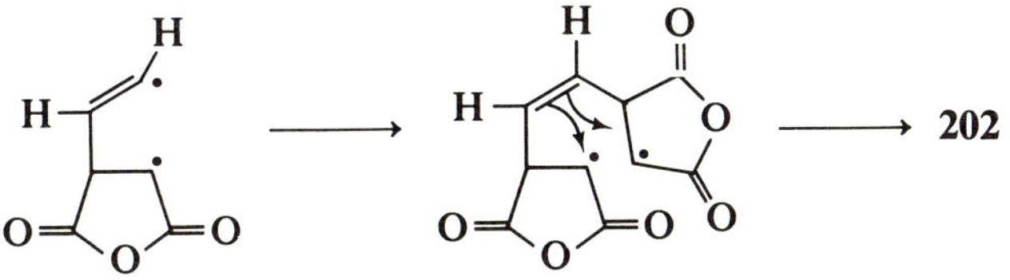

203   204   → 202

Intramolecular examples of maleimide photodimerization [759] are provided by the series **205** [760]:

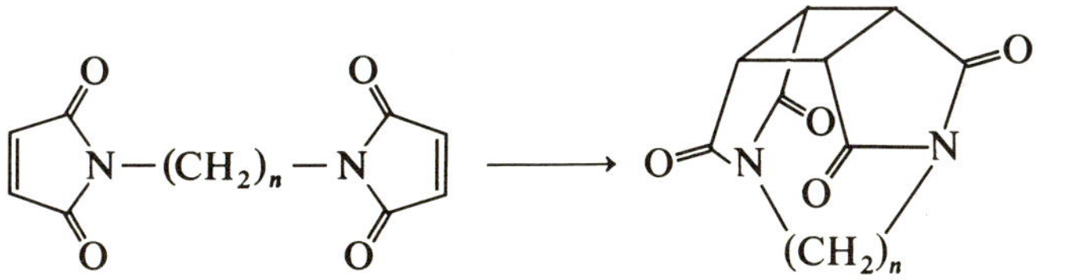

205, $n = 3–7$ **206**

Irradiation of phenylmaleimide and dimethylaniline yields the cycloadduct **207** [761]:

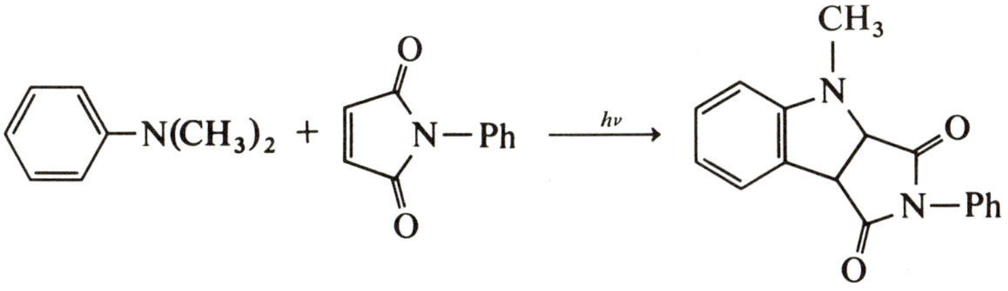

207

Thiomaleic anhydride, in analogy to maleic anhydride, yields a 2:1 photoadduct with benzene [762].

Irradiation of enaminonitriles results in *cis-trans* isomerization and cyclization [763].

Alcoholic solutions of pyrazolinone **207a** on irradiation yield imidazolinone **207b** and *N*-acetonylphenylcarbamate **207c** in addition to other minor acyclic products [592]. Irradiation in the presence of acetone, however, yields aldehyde **207d** as the major product:

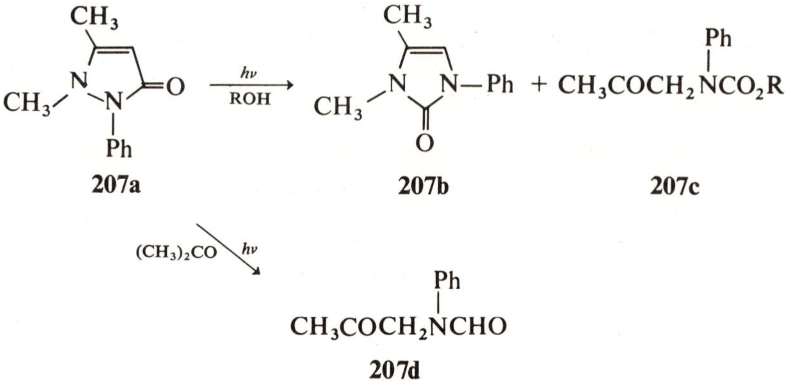

$$\text{207a} \xrightarrow{h\nu, \text{ROH}} \text{207b} + CH_3COCH_2NCO_2R \text{ (207c, N-Ph)}$$

207a **207b** **207c**

$$\xrightarrow{(CH_3)_2CO, h\nu} CH_3COCH_2\overset{Ph}{N}CHO$$

207d

5. BENZOQUINONE AND RELATED COMPOUNDS

A mathematical model of the mechanism of photoreduction of quinones has been proposed [764].

Photoaddition of acetylenes to methoxybenzoquinones and methoxynaphthoquinones yields adducts of types **208** and **209** [765, 766]:

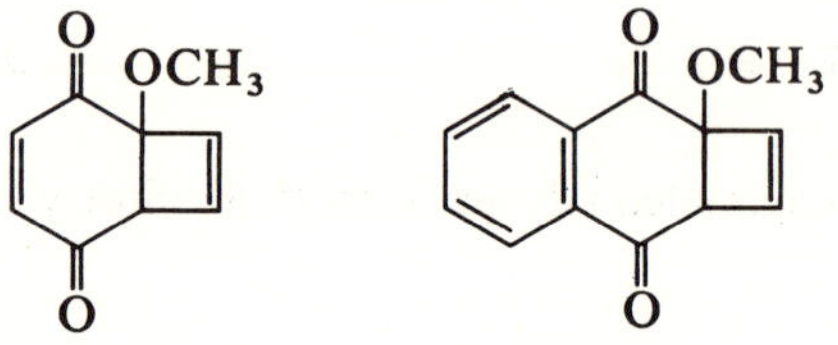

208 **209**

The quinone **210** is photorearranged to **211** [767; see also 767a]:

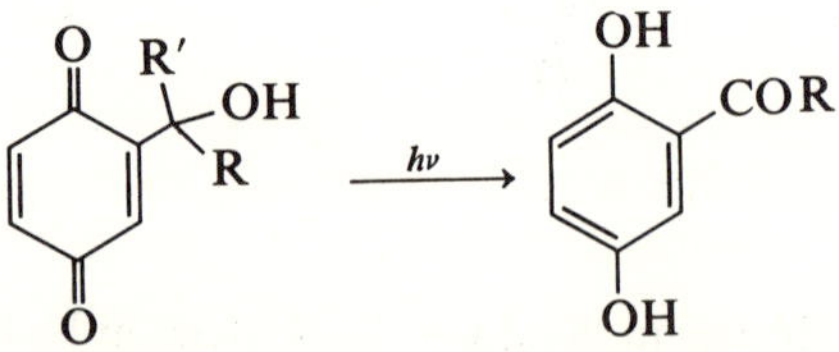

210 **211**

The diphenylbenzoquinone **212** photodimerizes in benzene but photorearranges to **213** in acetonitrile [768]:

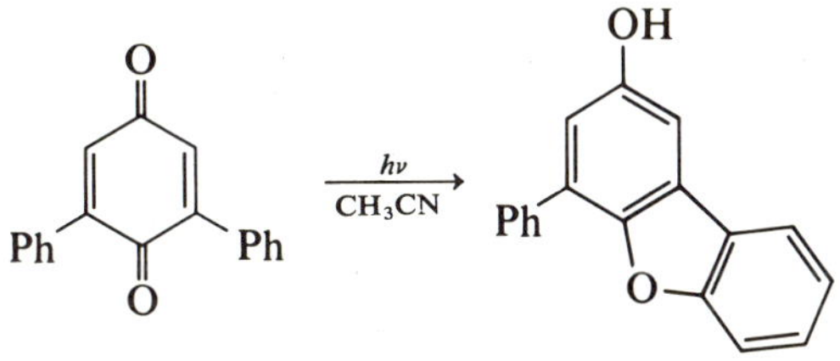

212 → **213, 80%**

Studies of hydrogen abstraction by duroquinone [769], perfluorobenzoquinone [770], and tetrachlorobenzoquinone [771] have been reported.

Irradiation of naphthoquinone and dioxane yields the adduct **216**, presumably by way of photoreduction-addition, followed by thermal oxidation [772]:

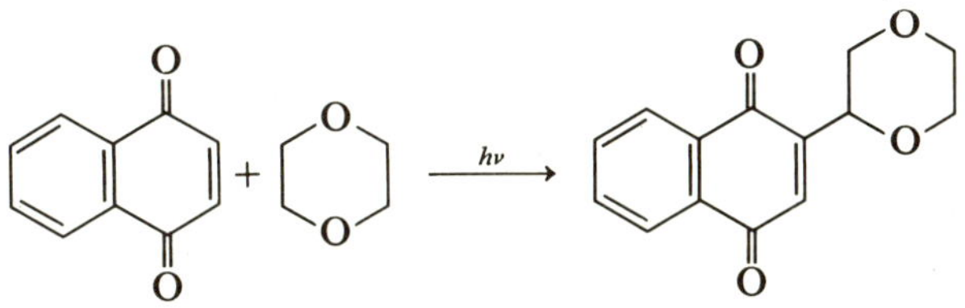

214 + **215** → **216, 50%**

Photoacylation of **217** by acetic anhydride yields **218** and **219** [773]:

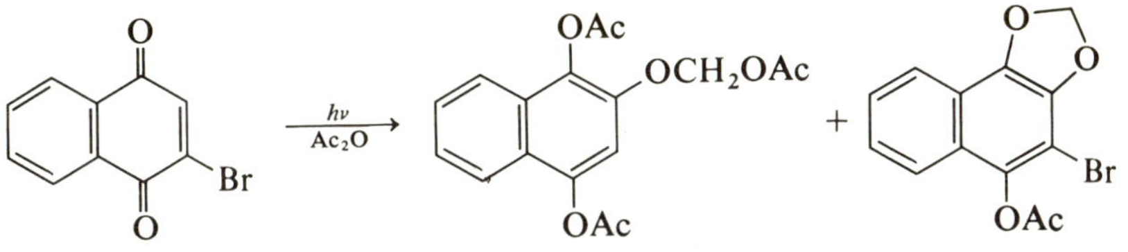

217 → **218** + **219**

Irradiation of the 2-piperidino and 2-morpholino isomers of **217** have appeared [774].

Studies of the mechanism of the photoreduction of 9,10-anthraquinones [775] and an unusual photodimerization of an anthraquinone derivative have appeared [777].

6. α-DICARBONYL COMPOUNDS

An extensive review of the photochemistry of orthoquinones and α-diketones has been published [778]. Biacetyl continues to be a favorite molecule for use in probing photochemical mechanisms. Triplet quantum yields for aromatic compounds in cyclohexane solution from measurements of the intensity and lifetime of sensitized biacetyl phosphorescence have been published [779]. Flash-photolysis studies of biacetyl and camphorquinone reveal that the ketyl radicals derived from hydrogen abstraction have spectra similar to α-diketone triplets [780, 780a]. The quenching of biacetyl triplets by a stable triplet, free radicals, a paramagnetic complex [781], amines, phenols, anilines, alcohols [782, 783, 784], and oxygen [785] have been studied. Both biacetyl fluorescence and phosphorescence are quenched by phenols, anilines, and amines [782, 783, 784]; the mechanism of this luminescence quenching has been discussed [784].

An attempt to detect excited triplet ketene molecules by sensitized biacetyl emission was unsuccessful [786]. Guest-host interactions in benzil-benzoin mixed crystals have been studied [787].

Irradiation of biacetyl and the unsaturated phosphate **220** yields **222** [405, 788], presumably by way of the intermediate oxetane **221** (Chapter 3):

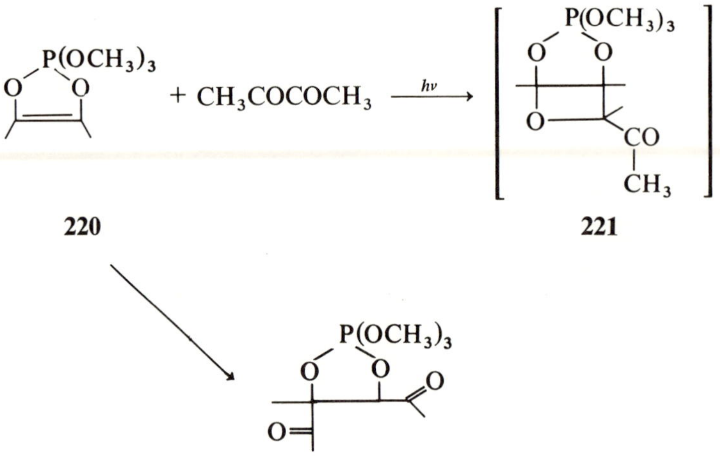

Irradiation of branched α-diketones results in cyclization to α-hydroxycyclobutanones [789]. The reactivity toward hydrogen abstraction increases as the γ-hydrogen goes from primary ($R_1 = R_2 = H$, $k_r = 2.5 \times 10^3$ sec^{-1}) to secondary ($R_1 = H$, $R_2 = CH_3$, $k_r = 1.3 \times 10^5$ sec^{-1}), to tertiary ($R_1 = R_2 = CH_3$, $k_r = 8.5 \times 10^5$ sec^{-1}).

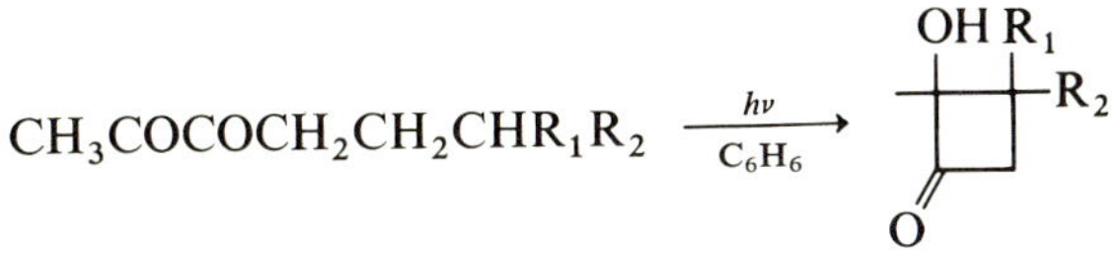

$$CH_3COCOCH_2CH_2CHR_1R_2 \xrightarrow[C_6H_6]{h\nu}$$

223 → **224**

Irradiation of unsaturated α-diketones, such as **225**, results in intramolecular cycloaddition [790]:

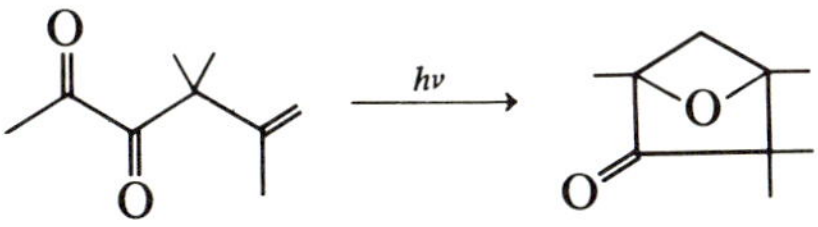

225 $\xrightarrow{h\nu}$ **226, 100%**

Although the unmethylated analog of **225** is also photocyclized to an oxetane, **227** undergoes photocyclization to an α-hydroxycyclobutanone:

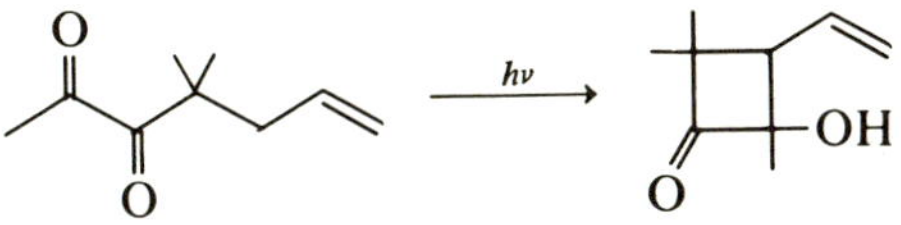

227 $\xrightarrow{h\nu}$ **228**

The unsaturated α-diketone **229** is photocyclized to the cyclopentenone **230** [791]:

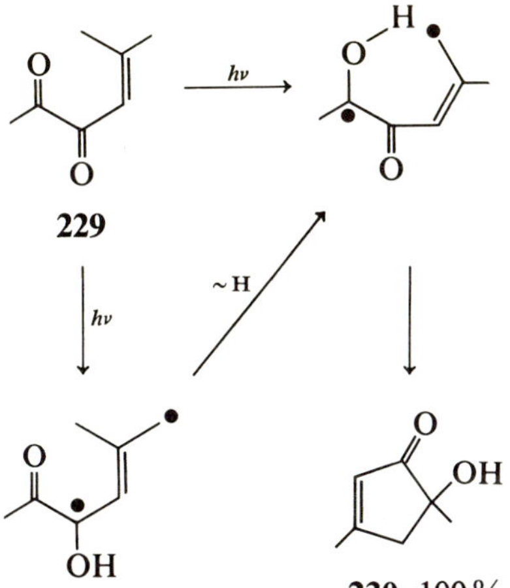

229 $\xrightarrow{h\nu}$ ... $\xrightarrow{\sim H}$ ... → **230, 100%**

This rearrangement apparently requires either an unusual seven-membered cyclic transition state or hydrogen abstraction by the "near" carbonyl followed by rearrangement of the ensuing biradical. An analogous reaction occurs when **231** [791] and **232** [792] are irradiated [see also 793]:

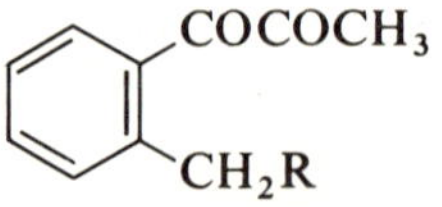

        COCOCH$_3$
Ar $\Big\langle$
        CH$_2$R

231, R = H or CH$_3$ 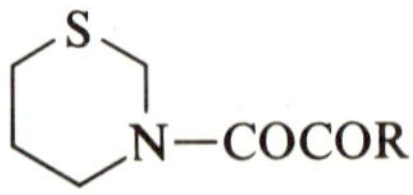   **232**, R = Ph or CH$_3$

(morpholine-like ring with S and N—COCOR)

In the case of **232**, some α-hydroxycyclobutanone is also produced. The thioester **233** is photodecarbonylated to PhSSPh [601]:

PhSCOCOSPh
233

The photoreduction of benzil by isopropanol, cumene, and cyclohexane [794, 795] (Volume 2, page 94) and its photooxydation by O_2 have been reported [796], as has a flash-photolysis study of the photoreduction of phenanthraquinone by isopropanol [797].

Photocycloaddition reactions of benzil [798], orthoquinones [799], and phenanthraquinone [495, 800, 801, 798] have been studied (Volume 2, pages 96, 97). Irradiation of phenanthraquinone in the presence of *cis*- or *trans*-β-phenoxystyrene results in the formation of both *cis* and *trans* dioxenes **235** [495]. A kinetic analysis of the photoreduction of **234** has appeared [801a]:

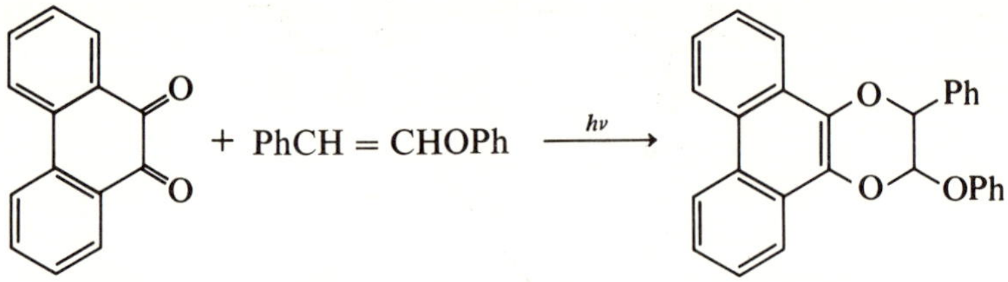

phenanthraquinone + PhCH=CHOPh $\xrightarrow{h\nu}$ dioxene adduct

234   **235**

Perturbational MO calculations indicate that both thermal and photochemical Diels-Alder reactions of **234** can be concerted [799].

Irradiation of the cyclobutanedione **236** yields the cyclopropanone **237** and polymer [802]:

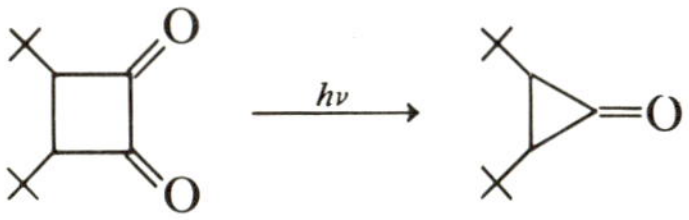

The photochemical behavior of the dione **238**

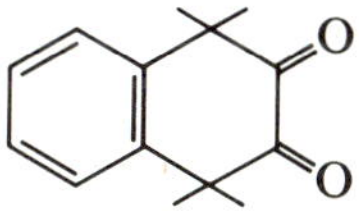

in the presence [795] and absence [803] of oxygen has been investigated. As in the case of benzil [796], peroxidized intermediates may be involved in the aerated systems.

Studies of the photoreduction of camphorquinone have appeared [804, 805, 788]. Both ketol and solvent adducts are produced. The quantum efficiency for the reduction of **239**

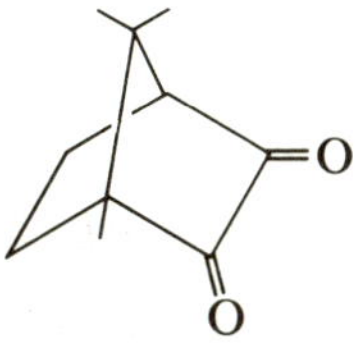

is low (0.06 in isopropanol). Benzophenone sensitizes these photoreductions, probably by "chemical sensitization" in which benzophenone ketyl radical formed by hydrogen abstraction from solvent, transfers a hydrogen atom to **239**. The cyclic α-diketone **240**

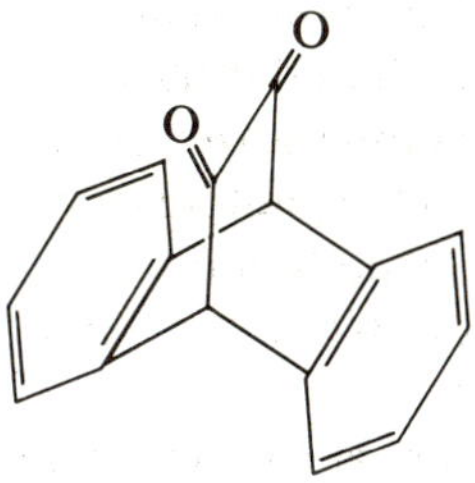

and related compounds undergo bisphotodecarbonylation [806].

Irradiation of **241** in the presence of oxygen and water results in the formation of **242** and **243** [806a] (see Volume 2, page 96):

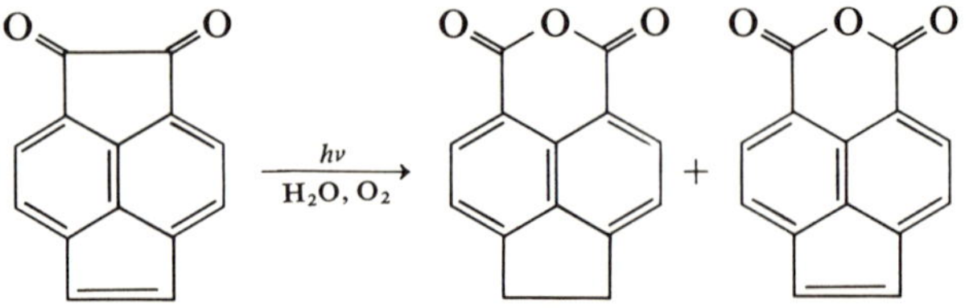

$$\mathbf{241} \xrightarrow[H_2O, O_2]{h\nu} \mathbf{242} + \mathbf{243}$$

An interesting intramolecular hydrogen abstraction cyclization [807] has been reported for a ginkgolide (partial structure **244**):

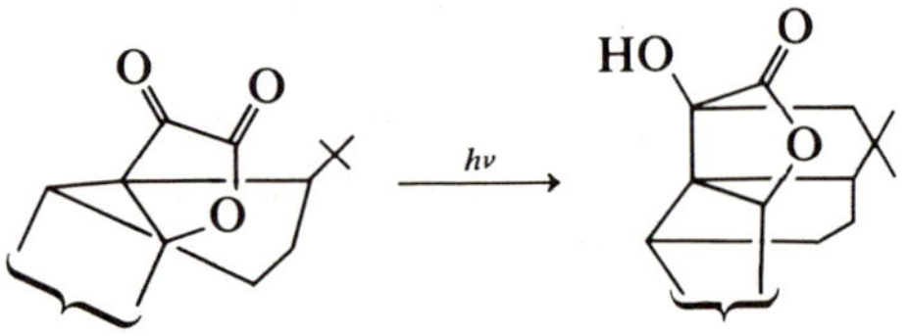

$$\mathbf{244} \xrightarrow{h\nu} \mathbf{245}$$

7. PURINES, PYRIMIDINES, AND RELATED COMPOUNDS

The photochemistry of purines and pyrimidines is of considerable interest, because these molecules are the basic units of DNA [Reviews: 808, 809]. This section is intended to be only a brief survey of the recent photochemistry of these biologically important molecules.

The specific formation of thymine dimers in DNA is possible by acetophenone photosensitization [810]. The photochemical properties [811] of thymine dimers and the mechanism of dimer formation [812, 813, 814, 815, 599, 657] and cleavage [812] have been studied. X-ray analysis has established the structure of one of the photodimers of dimethylthymine [815a]. The dimerization of ortic acid is photosensitized by benzophenone and quenched by 1,3-pentadiene [816]. Studies of the photoreaction of furocoumarins and DNA indicate that a photo-C_4-cycloaddition occurs in which the 5-6 double bond of the pyrimidine bases and the 3-4 or 4'-5' double bond of the furocoumarin are involved [817, 818, 819, 820].

The excited triplet state of uracil has been observed in solution [821]. Some pyrimidine bases are photoreduced across the C═C bond by sodium borohydride [822]. The ratio of hydration to dimerization of pyrimidines is different in D_2O and H_2O [823]. The azauracil **246**

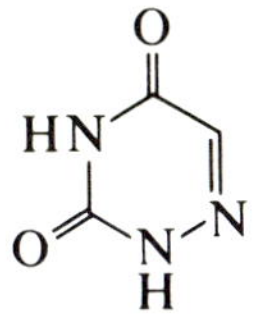

246

is photohydrated across the C=N bond [824]. The thiouracil **247** is photoreduced by sodium borohydride to **248** [825; see also 826]:

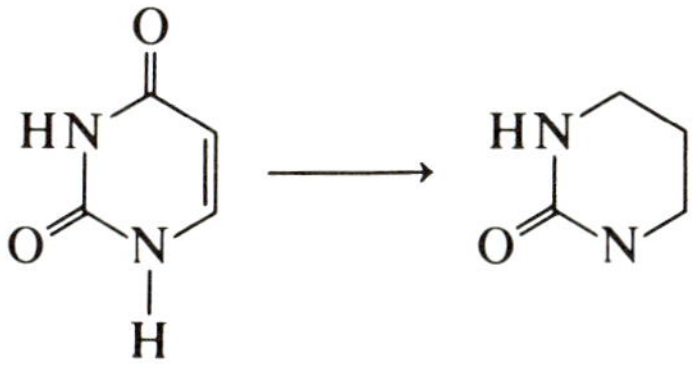

247 → **248**

The addition of alcohols [827, 828] (Volume 2, page 99) and amino acids [828, 829] to caffeine and other purines as well as their oxidation [830] has been reported. The photodimerization of 5-ethyl-2'-deoxyuridine has been reported [831].

CHAPTER FIVE: NITROGEN-CONTAINING CHROMOPHORES

I. UNSATURATED NITROGEN COMPOUNDS

A. Azines, Oximes, and Related Compounds

Irradiation of 1,1-diphenylmethylenimine in isopropanol results in hydrogen abstraction (Volume 2, page 100) [831a]. The role of chemical sensitization in the photoreduction of N-alkylimines has been elucidated (Vol. 2, page 100) [832, 833, 834]. Benzaldehyde N-alkylimines in 95% ethanol yield dihydrophotodimers, and benzophenone N-alkylimines yield benzhydryl-alkylamines. The imine excited states do not react, but ketyl radicals derived from carbonyl impurities "sensitize" the reduction [832].

The recently reported photoreactions of imines include an azabullvalene rearrangement [835], an intramolecular hydrogen abstraction [836], a photoreduction of an adjacent hydroxyl group [837], and cyclization, for example, 1 → 2 [838]:

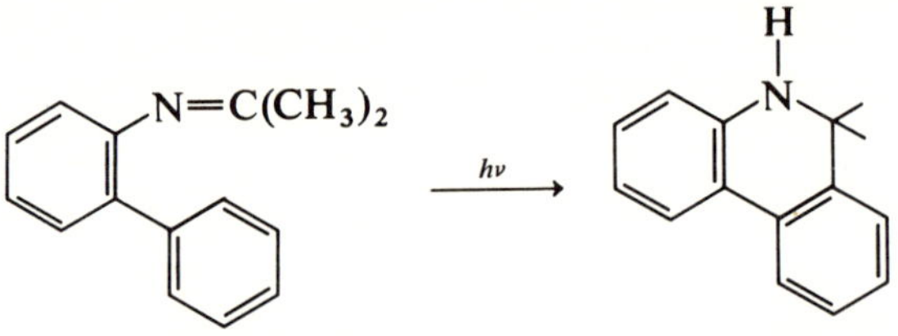

Further examples (Volume 2, page 101) of hydrazone photochemistry [839, 840, 841, 842, 843] and oxime photochemistry [844, 845, 846, 847, 848, 849] have appeared. In addition to the known *syn-anti* photoisomerization of oximes [844, 847], photorearrangements of oximes involving oxaziridines [849,

845, 846, 848] were reported. Direct evidence for the formation of oxaziridine [849, 846] has been provided, for example, **3 → 4** [849]:

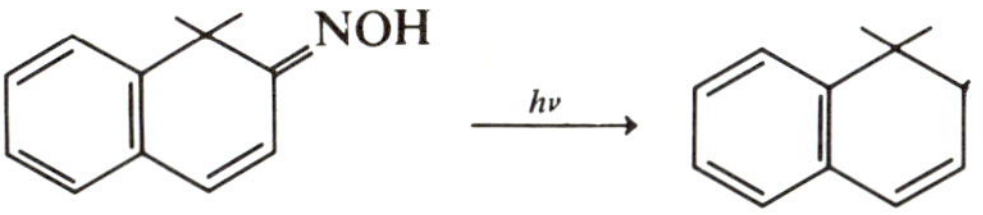

[Structure: 3 (dihydronaphthalene with NOH substituent) → hν → 4 (oxaziridine-fused dihydronaphthalene)]

Irradiation of the *N*-acylketimine **5** in isopropanol yields the reduced product **6** [851], probably by way of a triplet mechanism:

$$Ph_2C=NCOR \xrightarrow[(CH_3)_2CHOH]{h\nu} Ph_2CHNHCOR$$

5, R=Ph, CH$_3$ **6**, 100%

It is suggested that hydrogen abstraction by the carbonyl is the first step in the photoreduction.

Irradiation of the salts of tosylhydrazones [852, 853, 854] results in the formation of carbenes (Volume 1, page 176); the sulfoximine **7** yields a nitrene [588]:

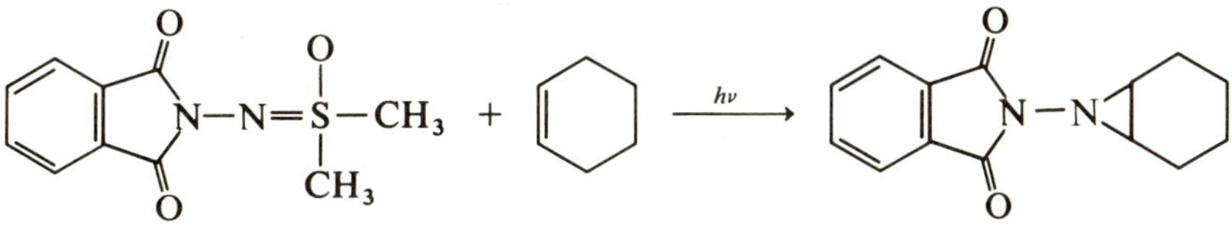

[Structure: phthalimide-N-N=S(CH$_3$)$_2$ (**7**) + cyclohexene → hν → phthalimide-N-N-cyclohexyl aziridine (**8**, 20%)]

and **9** yields a number of products [855], including diphenyl acetylene:

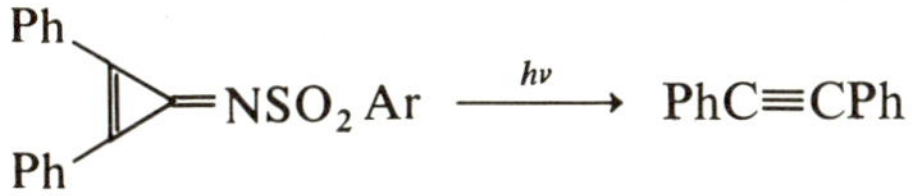

[Structure: Ph$_2$C=C=NSO$_2$Ar (**9**) → hν → PhC≡CPh (**10**)]

The azirine **11** is photorearranged to the cyanide **12** and the isocyanate **13** [856]:

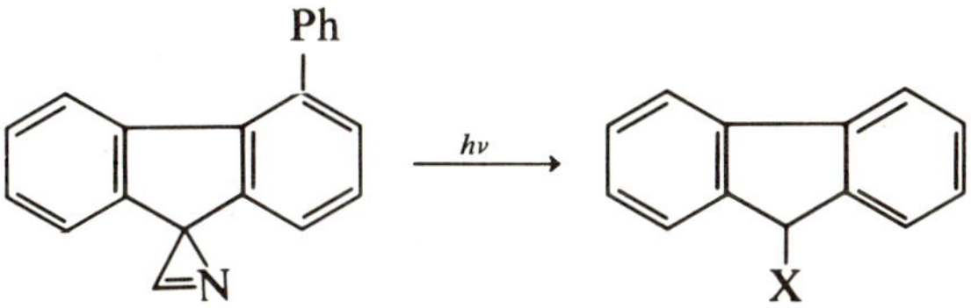

[Structure: Ph-substituted fluorene-azirine (**11**) → hν → fluorene-CHX (**12**, X = CN; **13**, X = N=C=O)]

The photochemistry of oxazoles and imidazoles has been studied. The 1,2,5-oxadiazole **14** is photofragmented to **15** and **16**, which then presumably combine to yield **17** and **18** [857]:

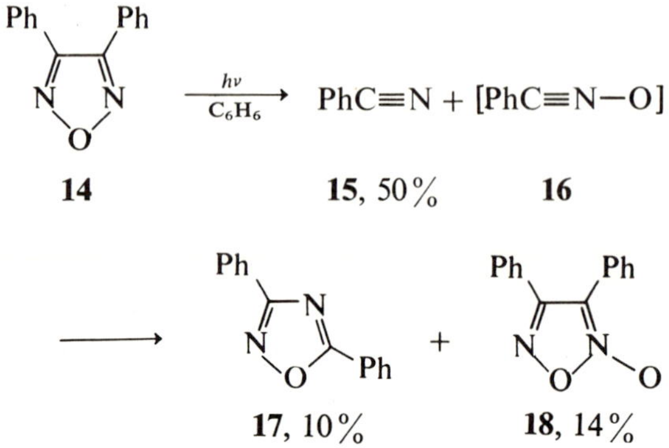

$$\text{14} \xrightarrow[C_6H_6]{h\nu} \text{PhC}\equiv\text{N} + [\text{PhC}\equiv\text{N}-\text{O}]$$

14   **15**, 50%   **16**

$$\longrightarrow \text{17} + \text{18}$$

17, 10%   **18**, 14%

The photorearrangements of a 2,5-oxazole [858], an imidazole [859], and certain diazoles [860] have been reported. An unusual photooxidation of a methyl group of benzoimidazole has been observed (861).

Further examples of the ring expansion of anthranils to azepines, possibly by way of a nitrene, have appeared [862]. However, some anthranils (for example, **19**) yield acridanones [862]:

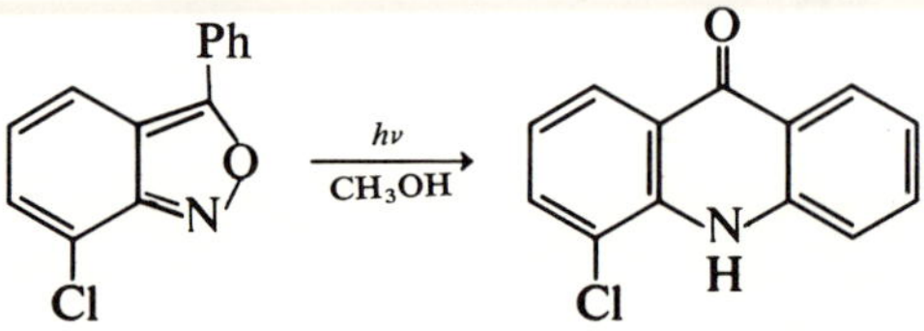

$$\text{19} \xrightarrow{h\nu,\ CH_3OH} \text{20}$$

Stilbene-phenanthrene photocyclizations of 4,5-diphenyloxazoles, thiazoles, imidazoles [863], and tetrazoles [864] have been observed.

Irradiation of **21** in benzene-diethyl amine allows isolation of **22** [865]:

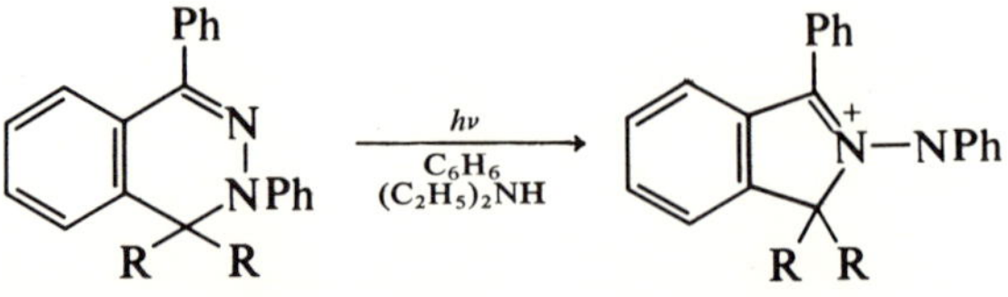

$$\text{21, R} = \text{PhCH}_2 \xrightarrow[(C_2H_5)_2NH]{h\nu,\ C_6H_6} \text{22, 70\%}$$

Irradiation of **23** with 254 nm light results in photofragmentation to **24**, and >300 nm light causes cyclization to **25** [866]:

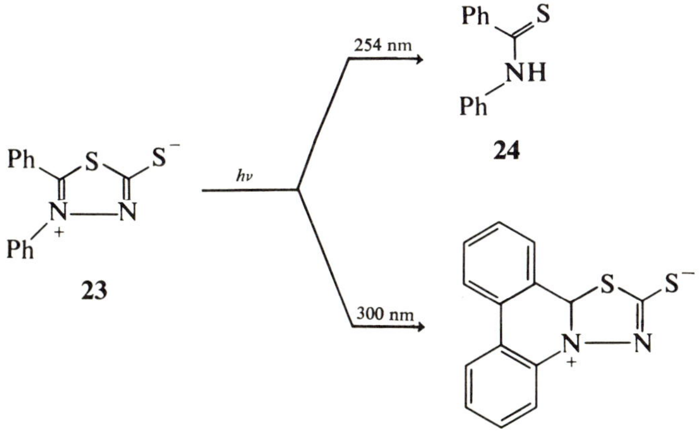

The tosylhydrazone dianion **26** is photorearranged to **27**, which is then photofragmented to 2-butyne [867]:

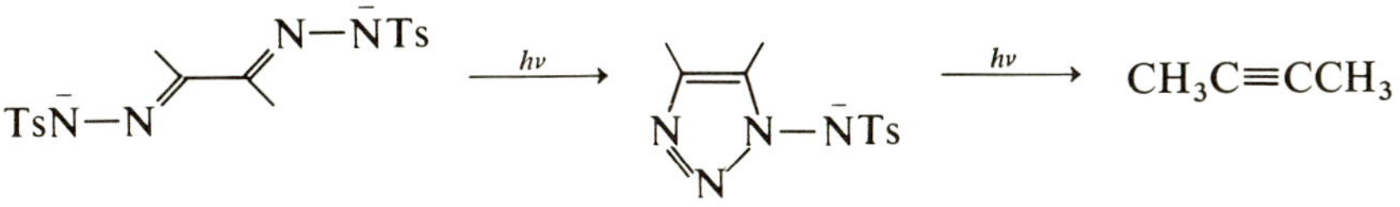

A reexamination [868] of the effect of benzophenone on the photolysis of benzalazine [869, 870] (Volume 2, page 101) indicates that oxygen is required for benzonitrile formation.

Irradiation of **28** results in photofragmentation to benzonitrile and **29** [871]:

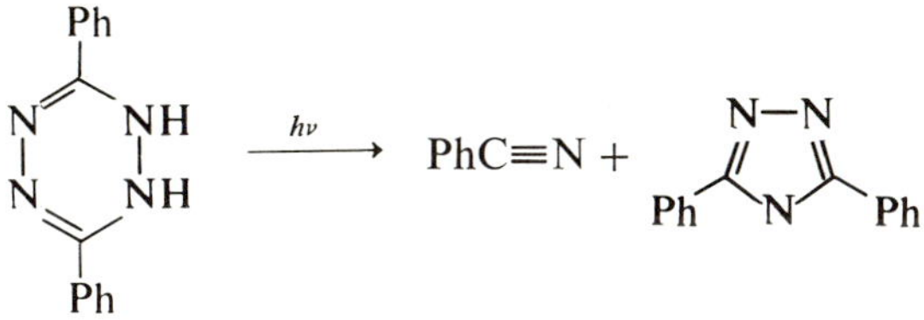

The effect of various side chains in photobleaching of riboflavins has been studied [872].

The azepine 30 is photorearranged to 31, 32, and 33 [873]:

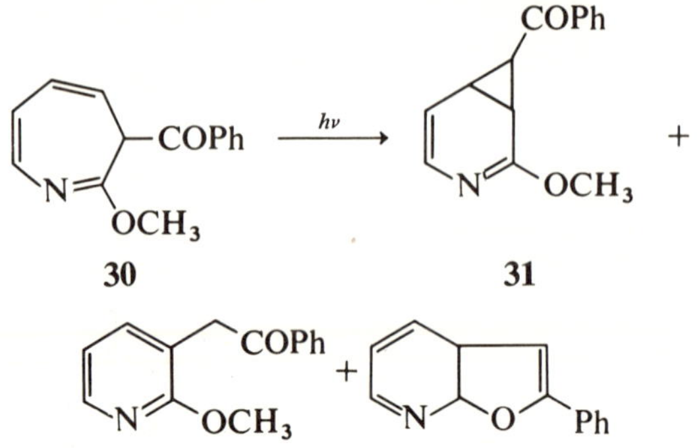

30 →(hν) 31 + 32 + 33

2-Substituted-3H-azepines undergo photocyclization [874], for example, 34 → 35:

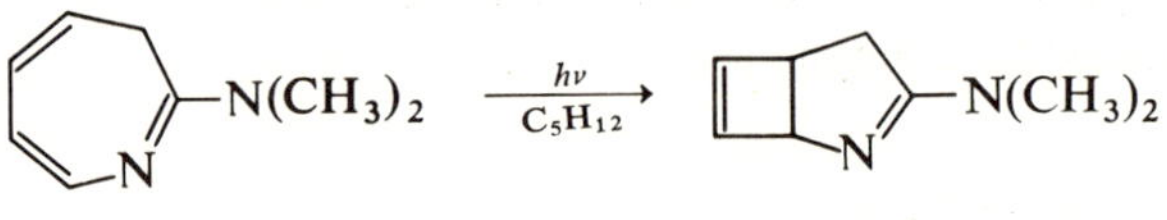

34 →(hν, C_5H_{12}) 35, 70%

B. Azo Compounds

The reactions of alkyl radicals produced by photolyses of azo compounds continue to receive attention [875, 876, 877]. The fate of an initially optically active radical [878] and the cage effect of cumyl radicals [879] have been investigated by way of azo photolyses.

The photo *cis-trans* isomerization [880, 881, 882] of azo alkanes is thought to proceed by way of a $^3(\pi,\pi^*)$ state [883].

The technique of chemically induced nuclear-spin polarization has been used as a probe for the determination of spin multiplicities of radical-pair precursors in the photolyses of azoalkanes.

The photochemical conversion of pyrazolines to cyclopropanes (Volume 2, page 105) continues to be of synthetic [884, 885, 886, 887, 888] and mechanistic [889, 890, 891, 892, 893, 893a] interest.

Aromatic hydrocarbons whose singlet excited states lie higher than that of 36

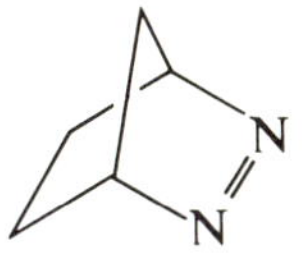

36

cause photosensitized decomposition. It seems that singlet **37** does not cross to triplet **37**

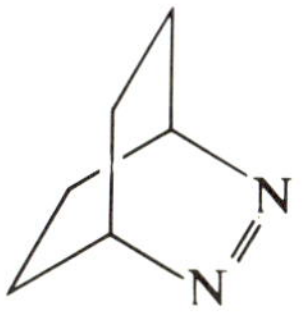

37

before decomposition. The energy of the latter is estimated to be about 60 kcal/mole [890]. The quenching of the fluorescence of **37** by conjugated dienes has been studied [892].

The same ratio of products is obtained from the direct and sensitized irradiation of **38** [893]:

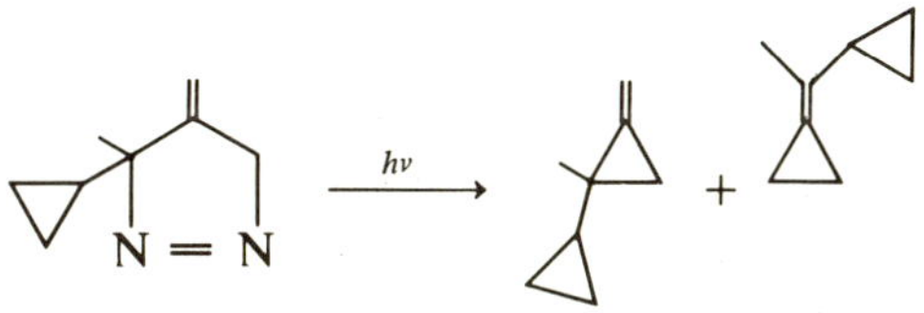

38   **39**   **40**

Although a biradical intermediate is implicated, no cyclopropyl ring-opened products are observed.

Irradiation of **41**, on the other hand, results in ring opening to **42** [894]:

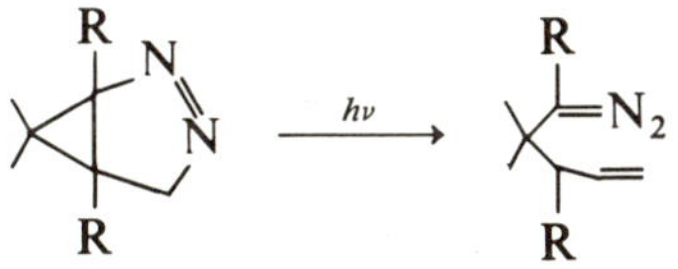

41, R = CO_2CH_3 **42**

The photorearrangement of a cyclic azo compound to a cyclic hydrazone has been reported [895].

Irradiation of **43** results in stereospecific ring inversion [896]:

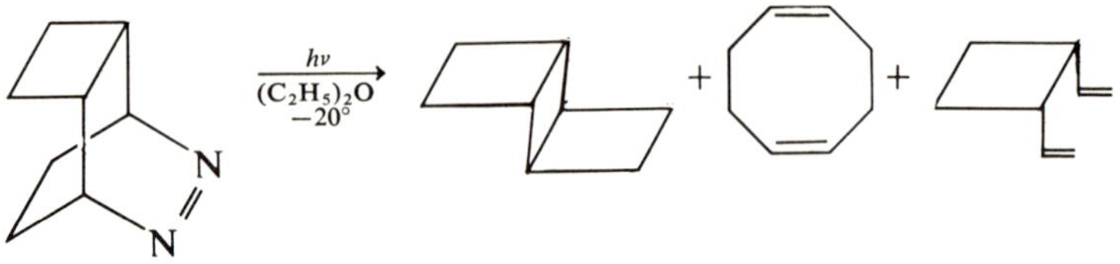

43 → **44**, 18%   +   **45**, 9%   +   **46**, 5%

(conditions: $h\nu$, $(C_2H_5)_2O$, $-20°$)

The homolog **47** undergoes an analogous photolysis [897]:

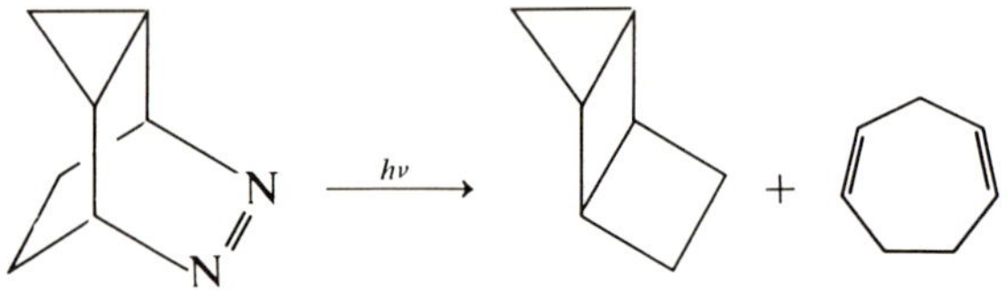

47 $\xrightarrow{h\nu}$ **48**, 20%   +   **49**, 60%

Irradiation of **50** yields bicyclopentane **51** [898]:

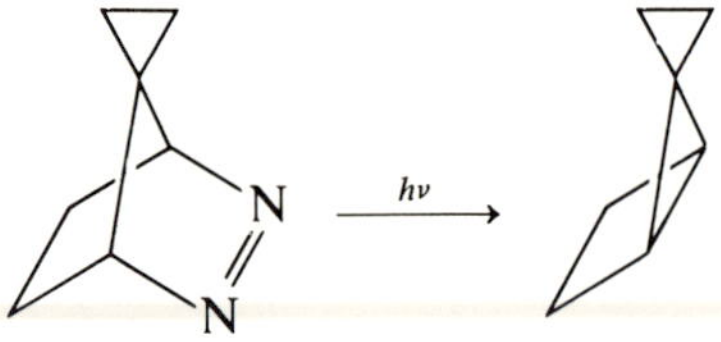

50 $\xrightarrow{h\nu}$ **51**

Further examples of the photolyses of pyrazoles to cyclopropenes, generally by way of open-chain diazo compounds (Volume 2, page 102), have appeared [899, 900, 901, 902]. Irradiation of **52**, however, results in the unexpected formation of **53** and **54** [902a]:

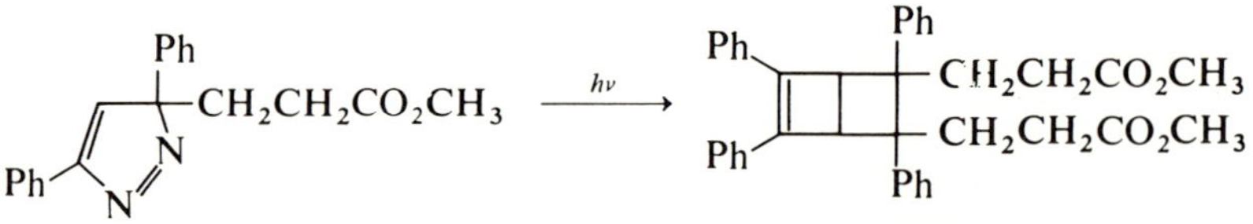

52 $\xrightarrow{h\nu}$ **53**, 8% 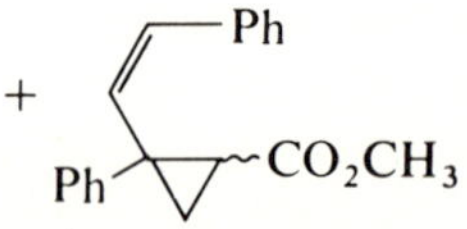   +   **54**

Irradiation of 3H-indazoles at 77°K produces molecules believed to be triplet 1,3-biradicals [903]. In the case of **55**, and some of its derivatives, these intermediates may be trapped by ethylenes, and 1,5-triplet biradicals are produced (**57**):

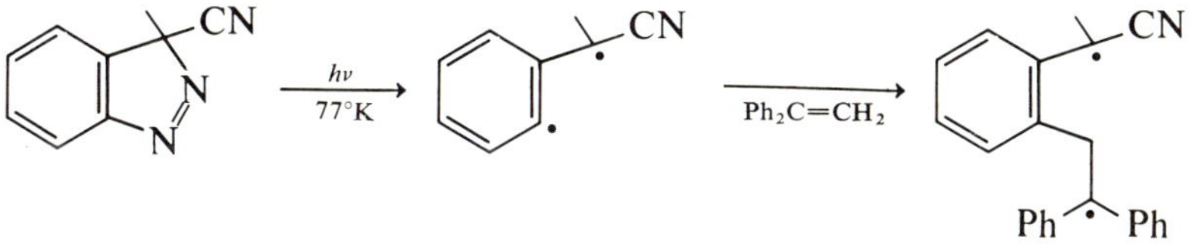

$$\underset{\mathbf{55}}{\text{3H-indazole-CN}} \xrightarrow[77°K]{h\nu} \underset{\mathbf{56}}{\text{biradical}} \xrightarrow{Ph_2C=CH_2} \underset{\mathbf{57}}{\text{1,5-biradical}}$$

Studies of the photoreduction [904, 905, 906] and phototropy [907] of azobenzene derivatives have been published. In the case of **58**, the N=N bond is photoreduced in alcohol solvents, but the NO_2 group is photoreduced in amines [906]:

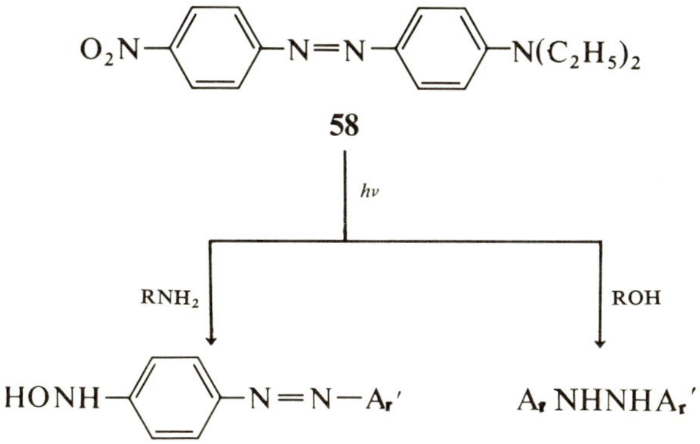

$$O_2N-C_6H_4-N=N-C_6H_4-N(C_2H_5)_2$$
58

$\xrightarrow{h\nu}$ with RNH_2 → HONH-C_6H_4-N=N-A_r' (**60**)

with ROH → $A_r NHNHA_r'$ (**59**)

Benzophenone seems to act as a "chemical sensitizer" in these reactions.

The chemically induced nuclear-spin polarizations (CINSP) obtained in two reactions which yield the same products but which differ in the multiplicity of the precursor of a radical pair has been studied [877]. The formation of 1,1,2-triphenylethane from diphenylcarbene and toluene and the photochemical decomposition of **60a** were studied. The acyclic azo compound was presumed to decompose by way of the singlet state, and diphenylcarbene is believed to react in its triplet state [908]. The ratio of unsymmetrical **60b** to symmetrical products **60c** and **60d** is higher in the case of **60a** than in the methylene reaction, possibly as the result of a spin-correlation effect. The CINSP spectra for the two methods of decomposition showed opposite polarizations:

$$Ph_2CHN=NCH_2Ph \longrightarrow Ph_2CHCH_2Ph + (Ph_2CH)_2 + (PhCH_2)_2$$
$$\quad\quad \textbf{60a} \quad\quad\quad\quad\quad\quad \textbf{60b} \quad\quad\quad \textbf{60c} \quad\quad \textbf{60d}$$

$$Ph_2C: + PhCH_3 \longrightarrow Ph_2\dot{C}H + Ph\dot{C}H_2$$

C. Azides, Triazolines, and Related Compounds

A review of the photochemical decomposition of organic azides has appeared [909], and a number of reports of the photolysis of alkyl azides have been published [910, 911, 912, 856, 913, 914, 838]. In general, the loss of nitrogen occurs to produce a nitrene that then undergoes reactions characteristic of monovalent nitrogen.

The azide **61**, like other unsaturated azides (Volume 2, page 110) [913], is photolyzed to an azirine [856], which is photorearranged (see Section A, this chapter):

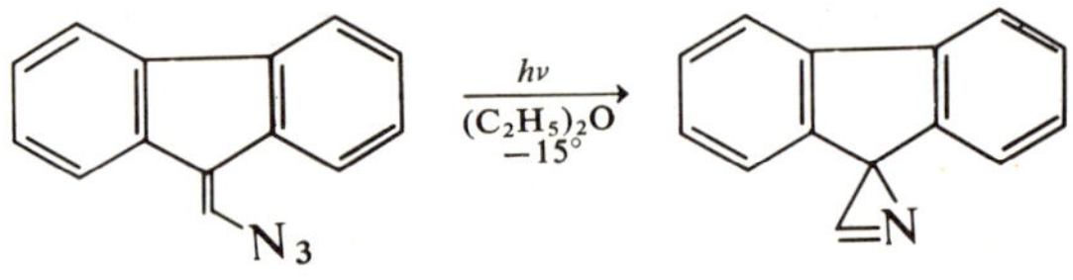

$$\underset{\textbf{61}}{\text{fluorenylidene-N}_3} \xrightarrow[-15°]{h\nu \atop (C_2H_5)_2O} \underset{\textbf{11}}{\text{azirine}}$$

Alkyl azides were found to be efficient quenchers of aromatic hydrocarbon fluorescence [915]. Energy transfer from donors with low singlet energies to alkyl azides is more efficient than that previously observed for triplet sensitization.

Ring substituents have been found to have little effect on the sensitivity of aryl azides toward photolysis [916, 917]. Markedly different results were found for the ketone- and aromatic hydrocarbon-sensitized decomposition of 2-azobiphenyl [918] (Volume 2, page 111). It seems that ketones transfer triplet excitation to the azide, and aromatic hydrocarbons transfer singlet excitation.

A study of the stereochemistry of the intermolecular singlet C—H insertion reactions of carbethoxynitrene reveals that selectivity does not change over a hundredfold variation in solvent concentration and is totally stereospecific [919]. No heavy atom effect on selectivity was observed. Similarly, **62** is photolyzed to **63** in low yield but with high retention of optical purity [920; see also 921]:

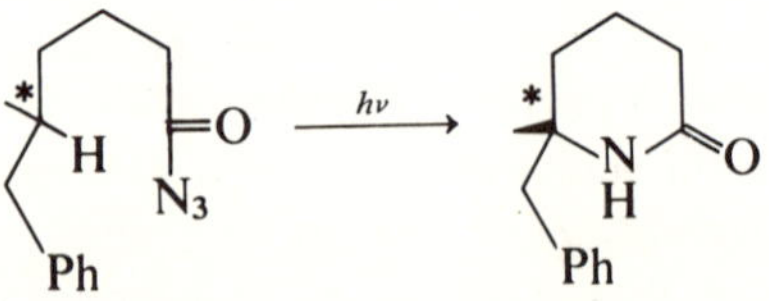

62 $\xrightarrow{h\nu}$ **63**, 4%

Related intramolecular insertions [922] are the conversion of **64** → **65** [923, 20], **66** → **67** [924], and **68** → **69** [922, 925]:

64 **65**

66 **67**, 45%

68 **69**, 24%

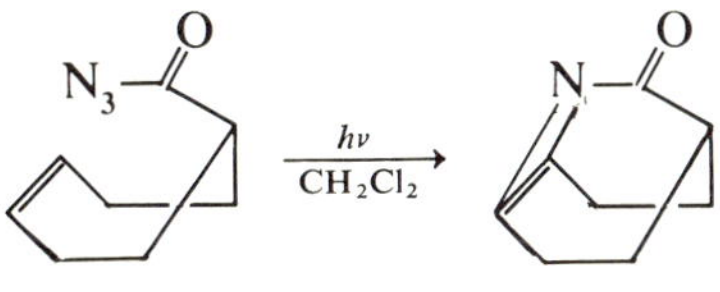

The effect of substituents on the acyl azide to isocyanate photorearrangement has been reported [926].

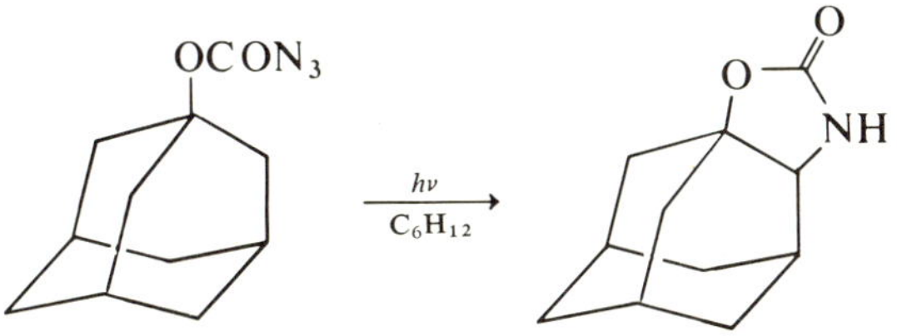

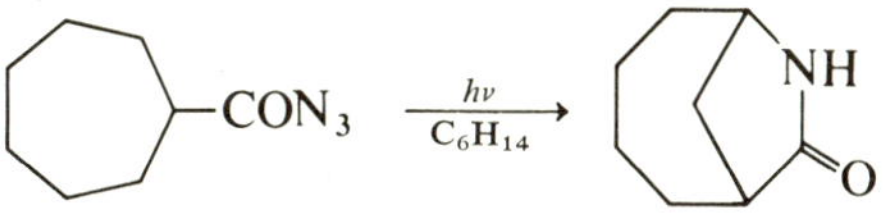

Aryl azides also undergo photocyclization reactions [927, 928].

The photolyses of triazolines to aziridines continues to receive attention [929, 930, 931, 932, 933, 934; see also 934a]. The conversion of **70** to **71** represents an interesting application of this photoreaction [929]:

70 **71**, 56%

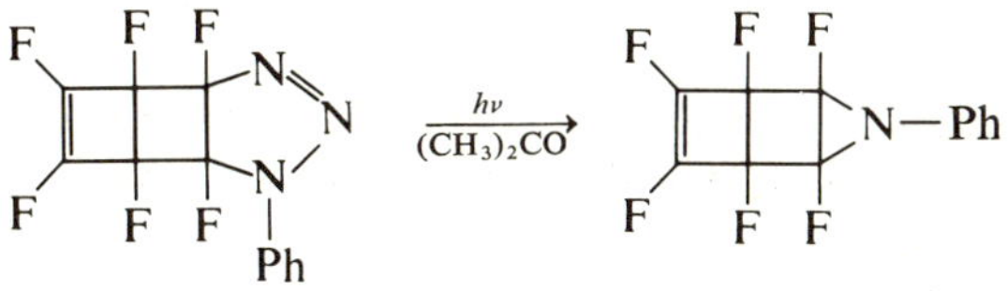

Photolysis of benzotriazole and related compounds leads to rearranged products **73** and **74** [928, 927] or adducts **76** [935]; both cases in which the expected aziridines are unstable:

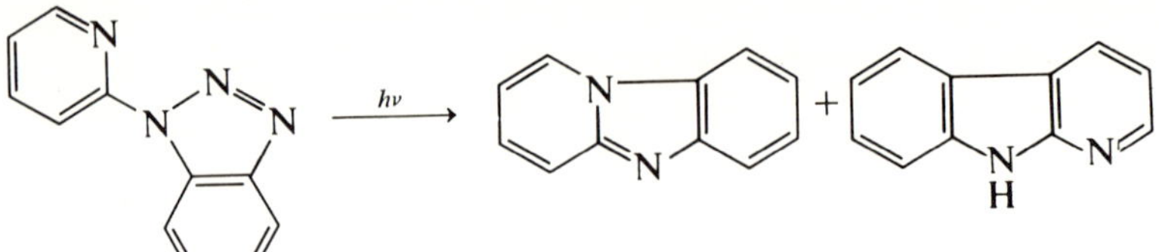

72 → **73, 70%** + **74, 10%**

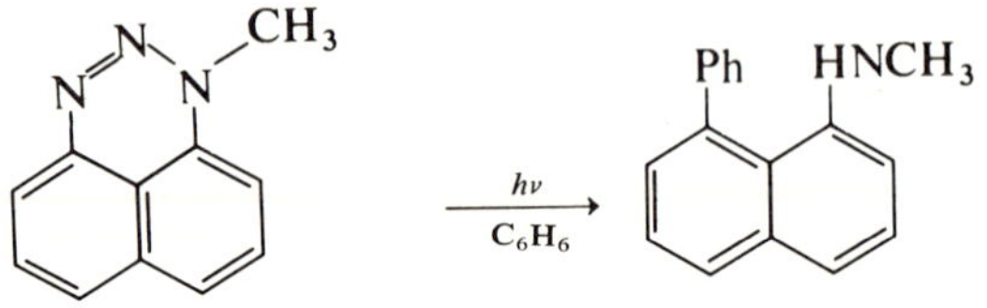

75 → **76**

Similarly, **77** is photolyzed to **78** and **79** [936]:

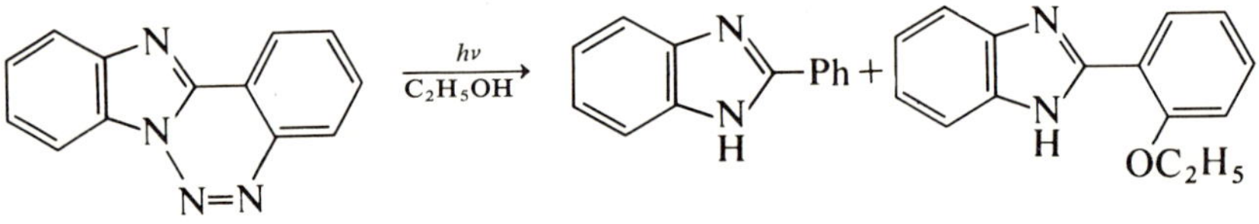

77 → **78, 50%** + **79, 16%**

In the case of 1-vinylbenzotriazole **80**, an internal trap is available, and indole is formed [937]:

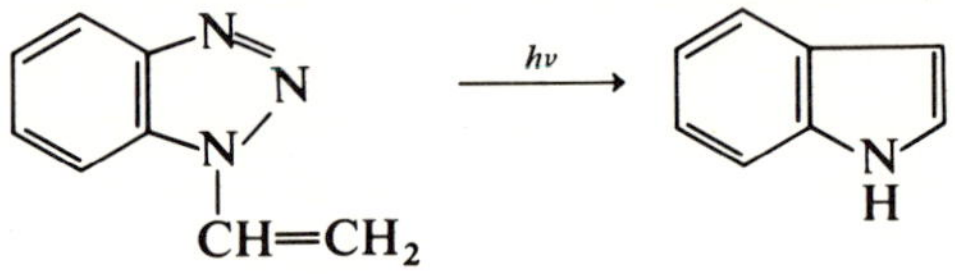

80 → **81**

The same products, derived from the benzoazetine **83**, are produced by the photodecarbonylation of **84** or the photolysis of **82** [595a]:

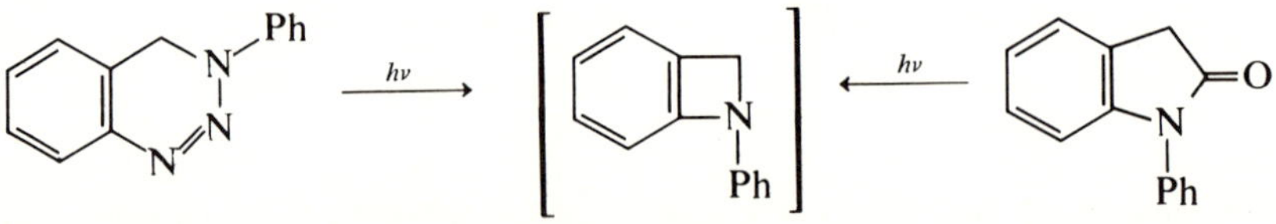

82 → [**83**] ← **84**

Irradiation of tetrazoles [938, 871] and tetrazolides [939] has been found to result in the loss of N_2 with a concomitant formation of trappable unstable intermediates, for example, **85** → **86** [871; see also 939a] and **87** → **88** [939]:

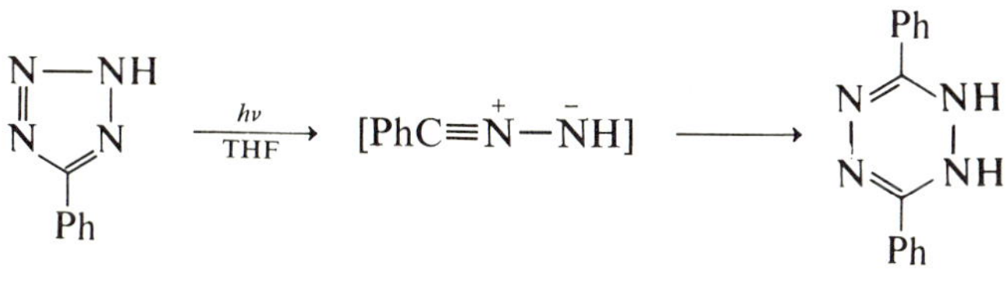

The use of **87** as a photosensitizer ($E_3 \sim 75$ kcal/mole) has been demonstrated. The loss of nitrogen serves as a probe for sensitization efficiency.

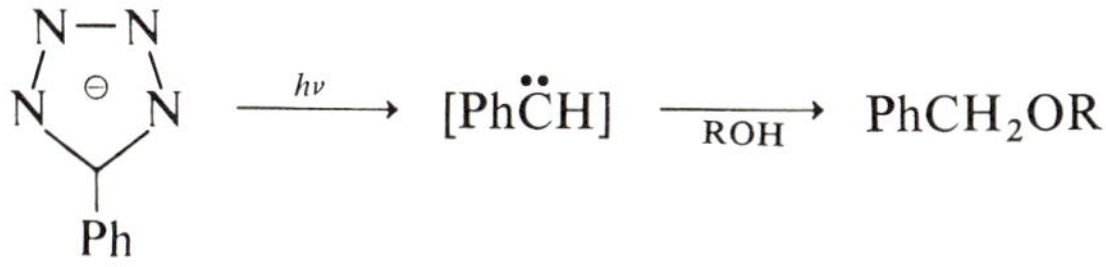

Benzocyclopropenone is implicated as an intermediate in the photolysis of **89** to **91** [940; see also 941]:

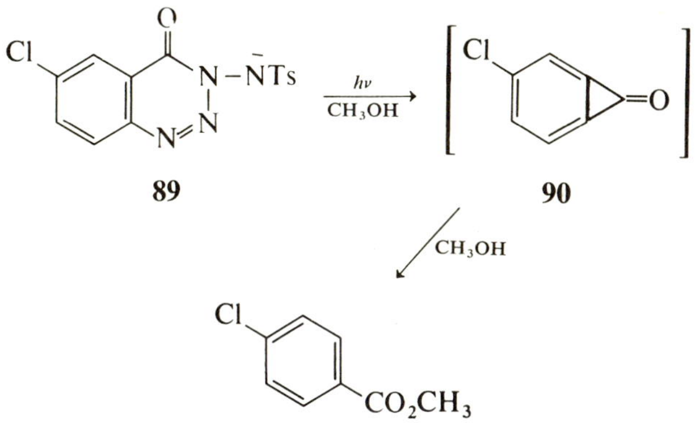

D. Diazo and Diazonium Compounds

Photolysis of diazo compounds generally results in the loss of nitrogen followed by the formation of carbenes.

In addition to methylene [942, 943], a number of alkyl carbenes [944, 902] and cyclic divinyl carbenes [947, 948, 949, 945, 946, 950] may be generated.

Stable norcaradienes may be synthesized by the irradiation of cyclic divinylcarbenes in benzene [945, 946, 947], for example, **92 → 93** [946]:

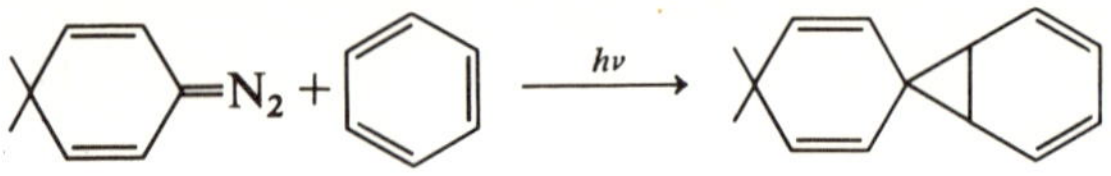

$$\text{92} =N_2 + \bigcirc \xrightarrow{h\nu} \text{93}$$

Photolyses of acyclic α-diazoketones [951, 952], esters [953, 944, 902a], and Wolff rearrangements of acyclic [953a] and cyclic α-diazoketones [954, 955, 956, 957, 958, 959, 960, 961] have been reported.

Irradiation of biscarbomethoxymethylene in ethers [962] or sulfides [963, 964] results in rearranged products believed to be derived from ylids, such as **96** [962], **99**, [963], and **101** [965]:

$$\underset{94}{CH_3OCH_3} + \underset{95}{N_2C(CO_2CH_3)_2} \xrightarrow{h\nu} \underset{96}{(CH_3)_2O^+-C^-(CO_2CH_3)_2}$$

$$\xrightarrow{} \underset{97}{CH_3OC(CO_2CH_3)_2} \overset{CH_3}{|}$$

$$\underset{98}{CH_3SCH_3} + \underset{95}{N_2C(CO_2CH_3)_2} \xrightarrow{h\nu} \underset{99}{(CH_3)_2S^+-C^-(CO_2CH_3)_2}$$

$$\underset{100}{RSCH_2CH=CH_2} + \underset{95}{N_2C(CO_2CH_3)_2} \longrightarrow \underset{101}{\underset{|}{RS^+CH_2-CH=CH_2} \atop ^-C(CO_2CH_3)_2}$$

$$\xrightarrow{} \underset{102}{R-S-C(CO_2CH_3)_2CH_2CH=CH_2}$$

Irradiation of **95** in the presence of ethylenes [966] and thiobenzophenone [967] has been reported.

Irradiation of **103** results in the formation of tolan **104** [968]:

$$\underset{103}{\underset{\parallel}{\text{N}_2}\ \underset{\parallel}{\text{N}_2} \atop PhCCOCPh} \xrightarrow[C_6H_5CH_2]{h\nu} \underset{104}{PhC\equiv CPh}$$

Ketenimines are produced by the photolysis of diphenyldiazomethane in the presence of isonitriles [969, 970]. Other examples of photolyses of diaryl diazomethanes [971, 972], including a chemically induced nuclear polarization study of the products from diphenyldiazomethane [908], have been reported.

A common intermediate does not appear in the photolysis and thermolysis of aryl diazonium salts [973].

Irradiation of bis(benzenesulfonyl)diazomethane in the presence of oxetanes results in carbene addition and ring-expansion products [974].

Irradiation of diazoketone **105** in the presence of excess primary or secondary amines yields the corresponding amide **106** [975]:

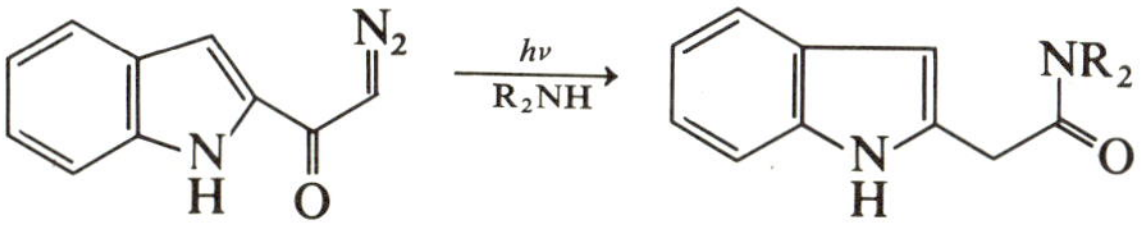

$$\text{105} \xrightarrow[R_2NH]{h\nu} \text{106}$$

A study of the electronic effects in rearrangements to electron-deficient species derived from 2-diazo-1,1,1-triarylethanes has appeared [976].

II. ORGANIC COMPOUNDS HAVING NO, NO$_2$, AND ONO FUNCTIONS

A. Nitrocompounds and Organic Nitrites

Irradiation of trimethylnitrosomethane in cyclohexane yields **107** [977]:

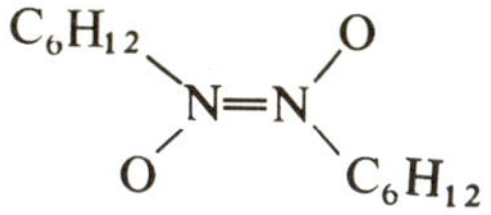

$$C_6H_{12}\text{-N(O)=N(O)-}C_6H_{12}$$
107

A helium-neon laser has been used to study the kinetics and mechanism of the photolysis of 2-chloro-2-nitrosobutane [978]. Irradiation of nitrosobenzene leads to a number of products, all of which seem derived from an initial primary photochemical formation of a phenyl radical and nitric oxide [979]. It has been pointed out that the possibility exists that some or all of the photochemistry of **108** may arise from its dimeric form:

$$\text{PhNO} \xrightarrow{h\nu} \text{Ph}^\cdot + \text{NO}^\cdot \longrightarrow \text{products}$$
108

The use of nitroso compounds as scavengers for the study (ESR) of short-lived free radicals has found wide applicability [980, 981, 982, 983, 984, 985, 986, 987, 988, 988a; see also 988b]. For example, the radical produced in the Pb(OAc)$_4$ α-acetoxylation of propionic acid has been trapped by 1,1-dimethyl-1-nitrosoethane **109** [980]:

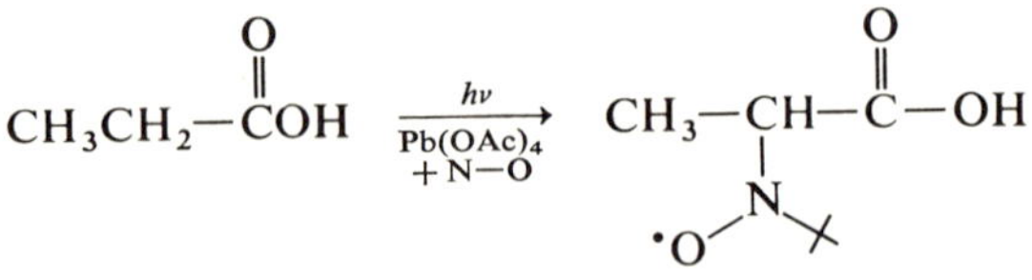

$$CH_3CH_2-\overset{O}{\underset{\|}{C}}OH \xrightarrow[+N-O]{\underset{Pb(OAc)_4}{h\nu}} CH_3-\underset{\underset{\cdot O}{\overset{|}{N}}\diagdown}{\overset{|}{C}H}-\overset{O}{\underset{\|}{C}}-OH$$

109

Irradiation of disulfide **110** in the presence of a suitable nitroso compound gives rise to nitroxide radical **111** [986]:

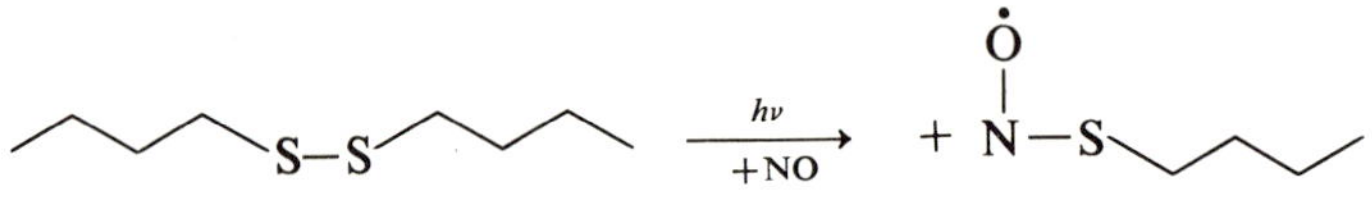

110 $\xrightarrow[+NO]{h\nu}$ **111**

In general, the nitroso compounds used exist as dimeric species at ambient temperature and are photochemically "activated" to their monomeric forms.

Irradiation of N-nitrosocompounds generally results in homolytic cleavage of the N—NO bond [989, 990, 991, 992] (Volume 2, page 119).

A study of the reversible intramolecular photocyclization of **112** has been reported [993; see also 994, 995] Volume 2, page 117):

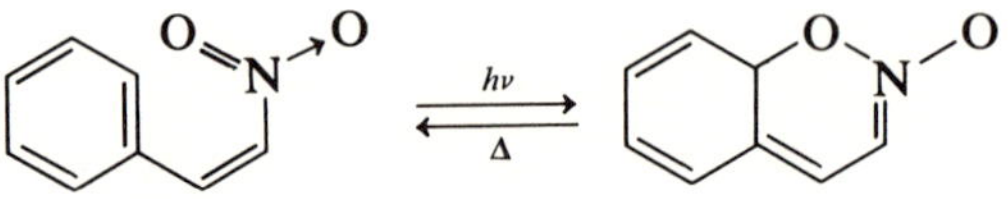

112 $\underset{\Delta}{\overset{h\nu}{\rightleftarrows}}$ **113**

Studies of the photoreduction of nitrobenzene (Volume 2, page 117) and its primary photoproducts have appeared [996, 997, 998, 999, 578a, 1000, 1001, 1002, 1003, 1004]. Intramolecular photoreductions of ortho-substituted nitrobenzenes [1005, 1006, 1007] have also been studied, for example, **114 → 115** [1006]:

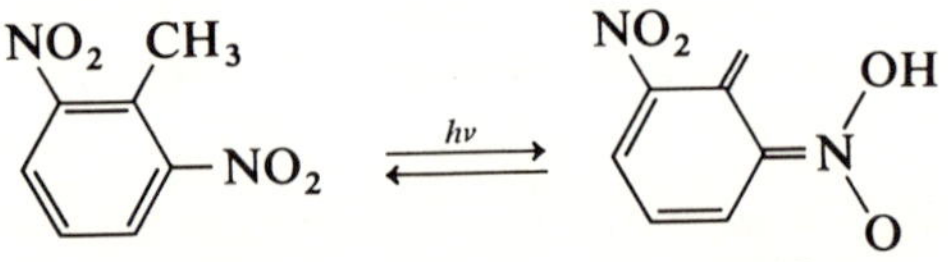

114 $\overset{h\nu}{\rightleftarrows}$ **115**

The photoreduction of α-nitronaphthalene by isopropanol has been reported [1008].

A full report on the photolyses of nitrobenzene in triethylphosphite has appeared [1009, 1010] (Volume 2, page 117).

Aspects of the photoamination [1011] and other nucleophilic substitution reactions of nitrobenzenes [1012, 1013, 1014, 1015] have appeared. It has been suggested that photoexcited 4-nitroanisole reacts with cyanide to form a short-lived intermediate that is oxidized by molecular oxygen [1015: see also 1014].

A review of the synthetic aspects of the Barton reaction of nitrites [1016] as well as a number of further examples of this photolysis [1017, 1018, 1019, 1020, 1021, 1022, 1023, 1024, 1025] have appeared.

Although the photochemical transformation of nitrite esters of fused five-membered ring alcohols usually leads to cyclic hydroxamic acids, a steroid case has been discovered in which a cyclic nitrone **117** is formed [1026]

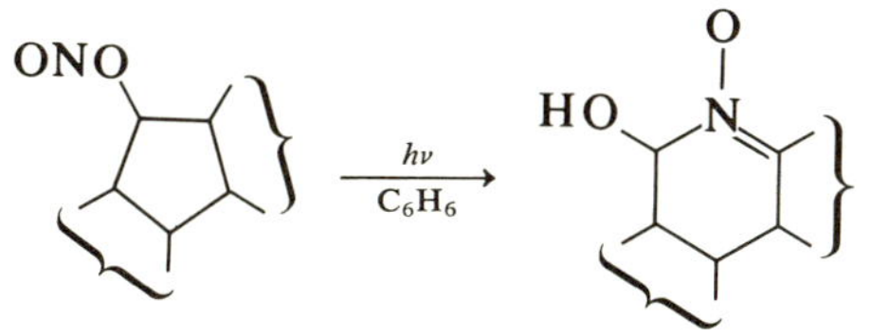

Irradiation of **118** results in initial N—O cleavage followed by rearrangement to γ-amino ketone **120**:

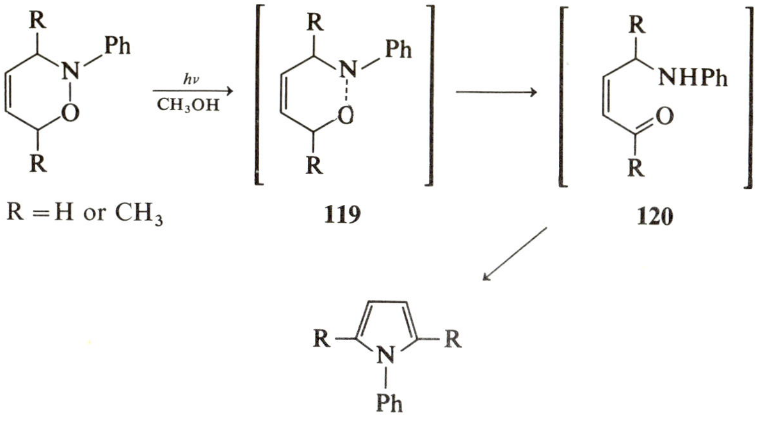

118, R = H or CH$_3$

121, 68%

Ketone **120** subsequently rearranges to *N*-phenylpyrrole **121** [1027]. Evidence for intermediate **120** was provided by low-temperature infrared spectroscopy.

Photolysis of **122** generates 4-pentenoxy radicals that undergo intramolecular cyclization to form α-substituted tetrahydrofuran derivatives preferentially [1028, 1028a]:

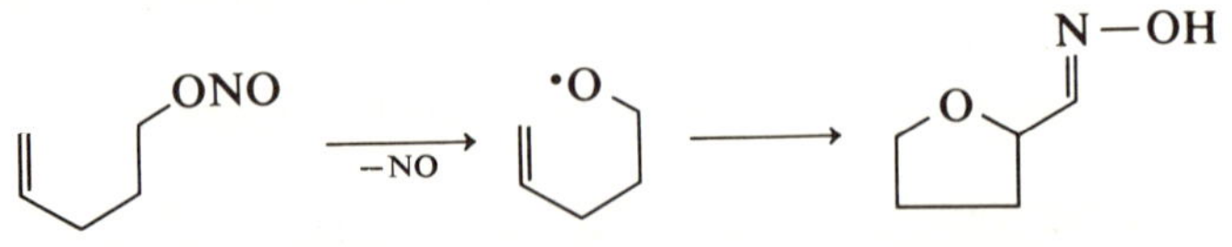

122 → **123** → **124**, 68%

An ESR study of methoxy radicals generated by the photolysis of nitrite ester **125** [1029] has appeared:

$$CH_3NO_2 \xrightarrow[77°K]{h\nu} CH_3^{\cdot} + NO_2^{\cdot} \longrightarrow CH_3ONO \xrightarrow{h\nu} CH_3O^{\cdot} + \dot{N}O$$
$$\mathbf{125}$$

The NOCl photochemical preparation of oximes from cycloalkanes [1030, 1031, 1032] and *n*-alkyl carboxylic acids [1033] has been reported. Photolysis of perfluorodiaminomethanes in the presence of N_2O produces fluoroimino compounds [1034].

B. *N*-Oxides and Related Compounds

The *N*-oxide **126** photoisomerizes smoothly to **128**, presumably by way of an oxaziridine [1035]:

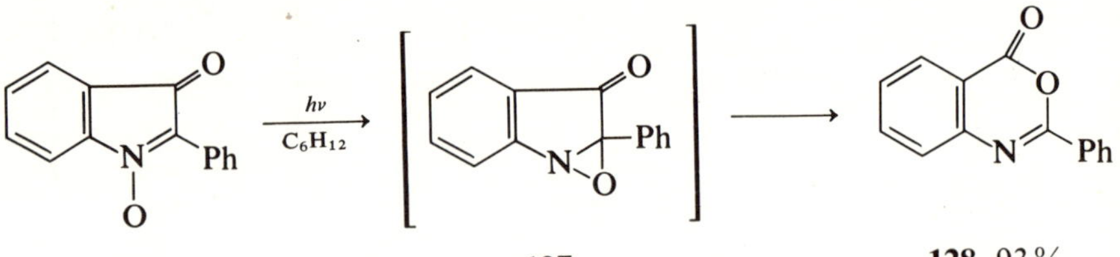

126   **127**   **128**, 93%

The latter kind of intermediate has been isolated in some cases [1036, 1037, 1037a, 1026] (Volume 2, page 120). *N*-oxide photorearrangements to nitrosoalcohols have been suggested to occur in the photolysis of poly(nitrobenzyl)-vinylpyridines [1038].

The azine monoxide **129** is photoisomerized to **131** and the interesting heterocyclic compound **132** [1039], the former possibly arising from the diazoketone **130**:

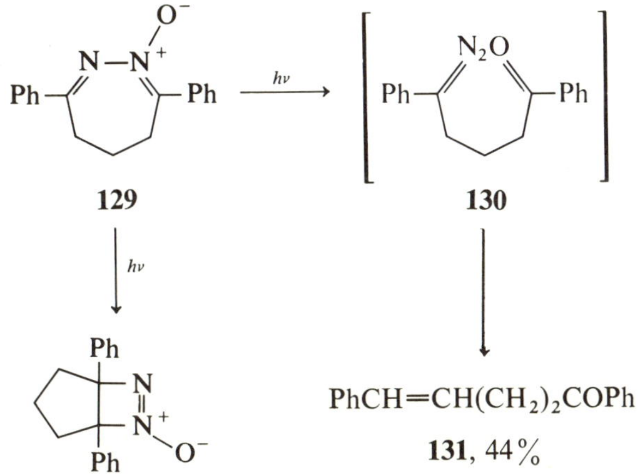

129 → [**130**]

129 →(hv) **132**, 24%

130 → PhCH=CH(CH$_2$)$_2$COPh
131, 44%

An investigation of the photochemical reactions of 3,5-di-*t*-butylbenzene-1,4-diazoxide has been reported [711].

Irradiation of pyridine *N*-oxides generally leads to the formation of pyridines, pyrroles, and azaoxipines (Volume 2, page 120) [1040, 1041]. In the case of **133**, the acyclic isomer **134** is formed [1042]:

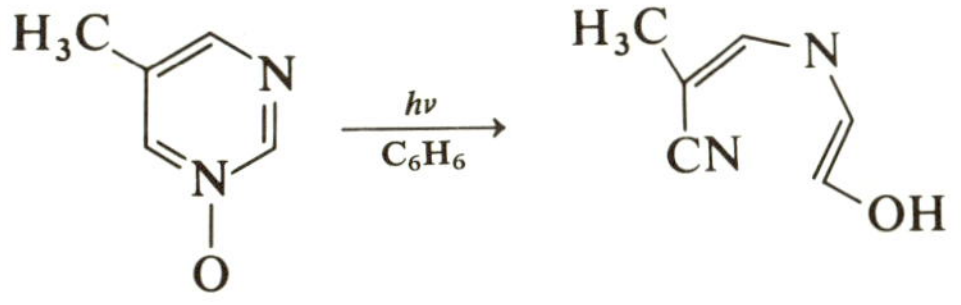

133 →(hv, C$_6$H$_6$) **134**, 22%

Photolysis of **136** in cyclohexene causes oxygenation of the solvent [1043]. Irradiation of **136** in aromatic hydrocarbon solvents yields phenols:

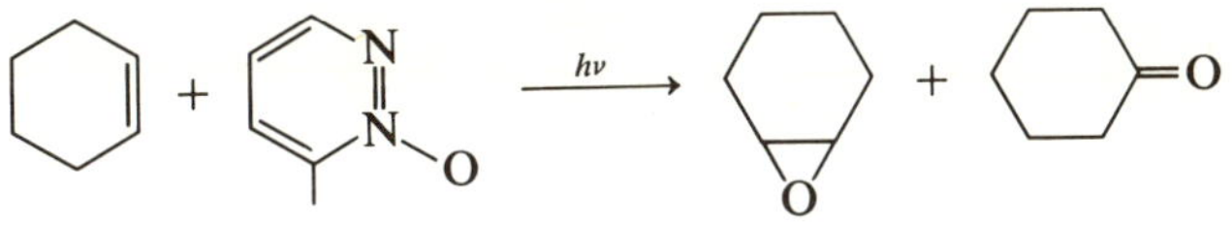

135 + **136** →(hv) **137**, 25% + **138**, 5%

The photorearrangement of the quinoline *N*-oxide **139** yields expected aldehyde **140** [1044, 1045], possibly by way of a nitrene intermediate [1046] (Volume 2, page 121):

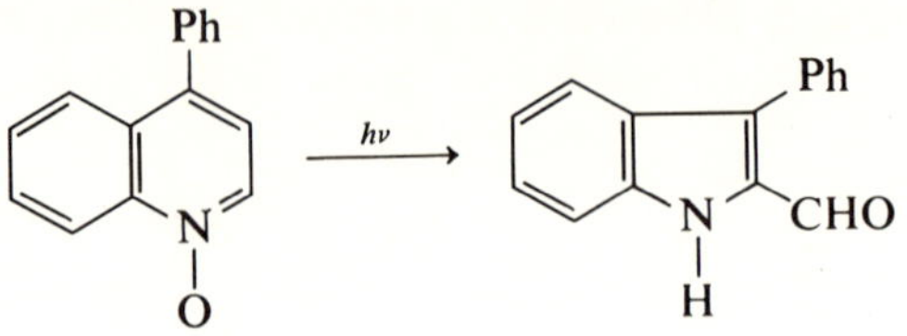

139 →^{hv} **140**

The photorearrangements of quinoline N-oxides have been found to depend on the solvent. In protonic solvents carbostyrils **142** are major products, and N-formyl-2-indolinols **143** are the major products in aprotonic solvents [1044]:

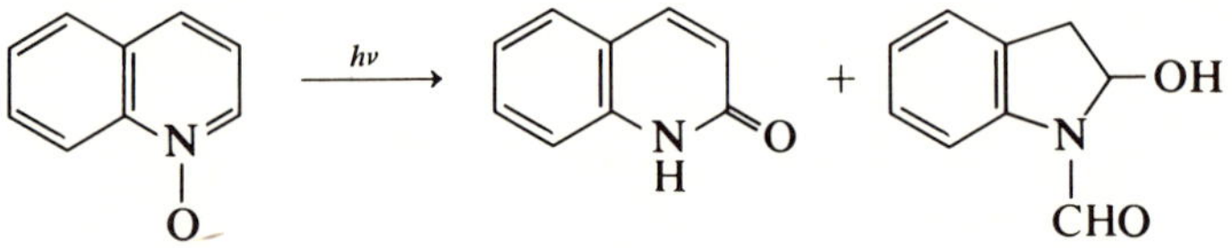

141 →^{hv} **142** + **143**

Irradiation of N-ethoxyquinolinium perchlorate yields 2- and 3-hydroxymethyl quinoline and quinoline itself [1047].

Although acridine N-oxide **147** photorearranges to a tropolone derivative (Volume 2, page 122), 9-cyanoacridine N-oxide **144** also yields oxepines and [10]annulenes [1048]:

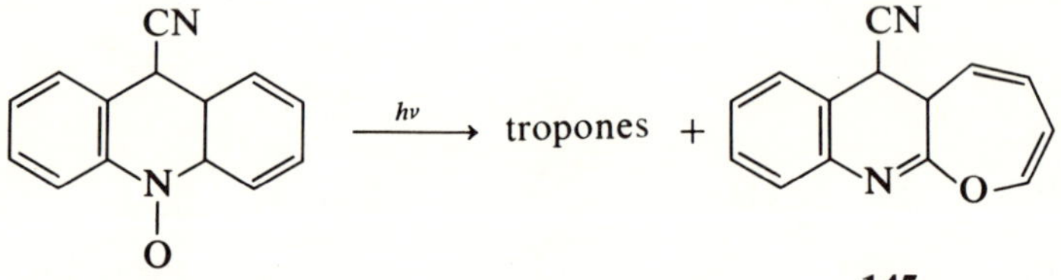

144 →^{hv} tropones + **145**

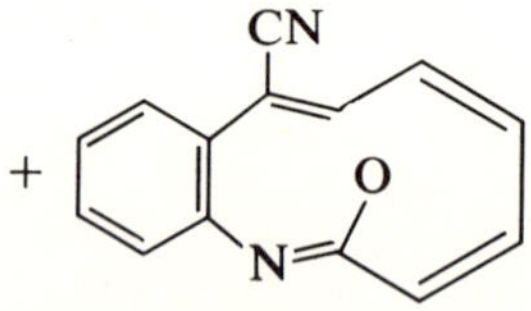

+ **146**

Irradiation of **147** in alcohol results in an oxygen amine rearrangement and solvent addition [1049]:

147 148

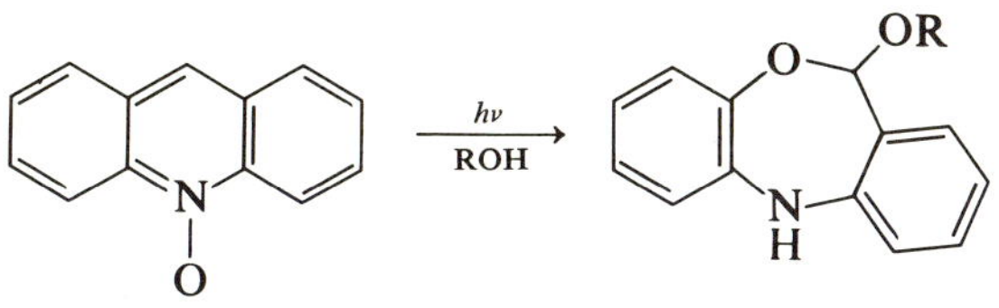

III. N-HETEROAROMATICS

A. Pyridine Derivatives

Pyridine is photohydrated with low quantum efficiency [1050]. Ring-substituted pyridines, such as **149**, undergo photoring contraction and photoreduction [1051]:

149 **150**, 16% **151**, 30%

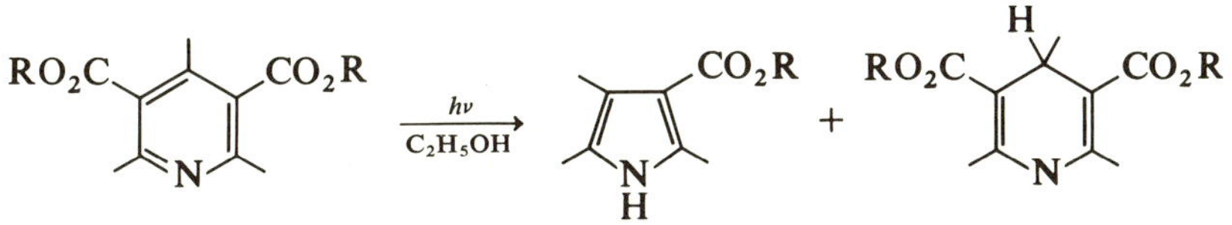

Irradiation of 2- and 3-bromopyridines results in reduction to pyridine [1052]. Photodecarboxylation has been reported for 2,3-, 2,4-, 2,6-, and 3,5-pyridinedicarboxylic acids [1053].

The 1,2-diazobenzene **152** is photorearranged to **153** and **154** [1054]:

152 **153** **154**

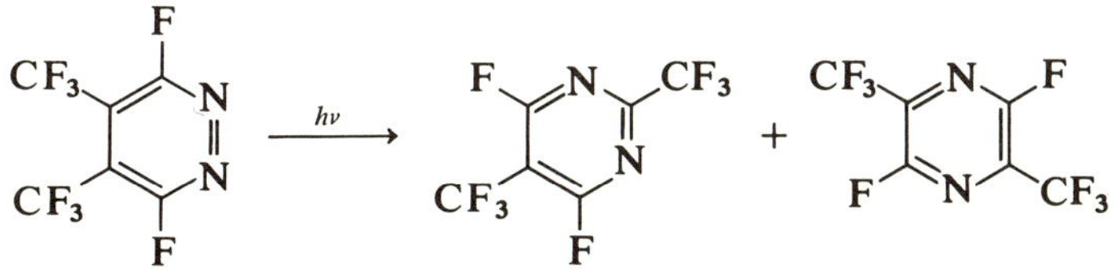

Irradiation of methylthio-S-triazines results in desulfurization [1055].

A number of further examples of photorearrangements of *N*-ethoxycarbonyl pyridines (Volume 2, page 123) have appeared [1056, 1057, 1040]. In addition to **156**, irradiation of **155** yields some pyridine [1058]:

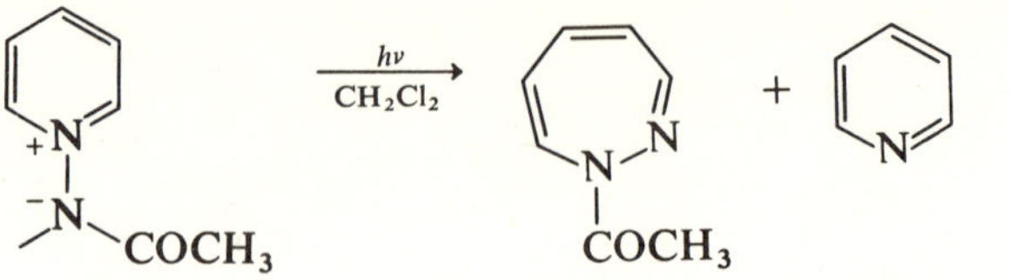

155 →(hv, CH₂Cl₂)→ 156, 55% + 157, 37%

The photochemistry of bromopyridines has been investigated [1059]. The base strengths of various 2- and 4-styrlpyridines in the ground and first excited singlet states have been determined [1060].

B. Purines, Quinolines, Acridines, and Related Compounds

7-Azaindole is believed to undergo a biprotonic photoisomerization to yield the tautomer **159** [1061]:

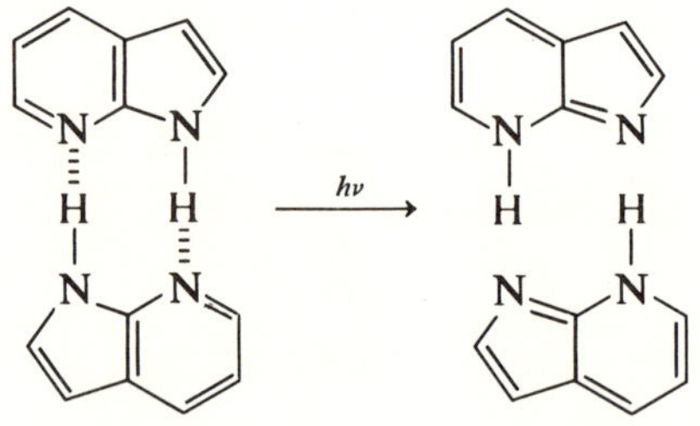

158 →hv→ **159**

A review of the role of the photochemistry of purines and nicotinamides in chemical evolution has appeared [1062].

Photoalkylations of caffeine with amino acids [828] and alcohols [827] have been reported. Photoalkylation of related compounds by alcohols [1063, 1064] have also appeared.

Irradiation of 2-alkyl quinolines results in a type II reaction to yield methyl quinolines [1065; see also 1066]:

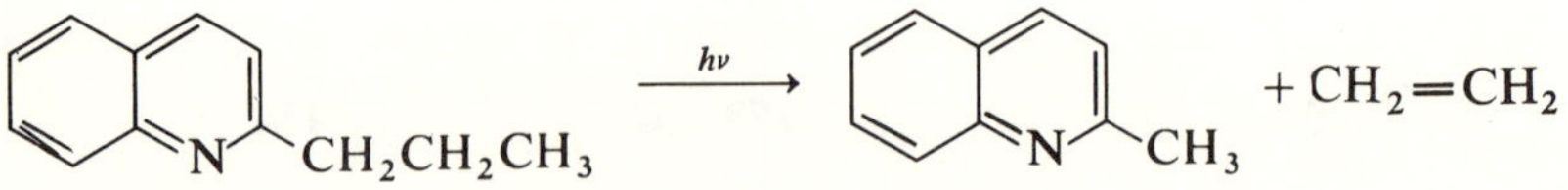

160 →hv→ **161** + $CH_2{=}CH_2$

Further studies of the photoreduction of acridine (Volume 2, page 123) by alcohols [1067, 213; see also 1068] and its alkylation and reduction by acids

[1069] have appeared. Energy transfer from triplet-state acridine molecules to paramagnetic complex ions has been reported [1070]. The quantum yields for intersystem crossing for proflavin ($\phi = 0.47$), acriflavin ($\phi = 0.47$), and acridine orange ($\phi = 0.30$) have been published [704]. Irradiation of **162** results in the loss of SO_2 [1071]:

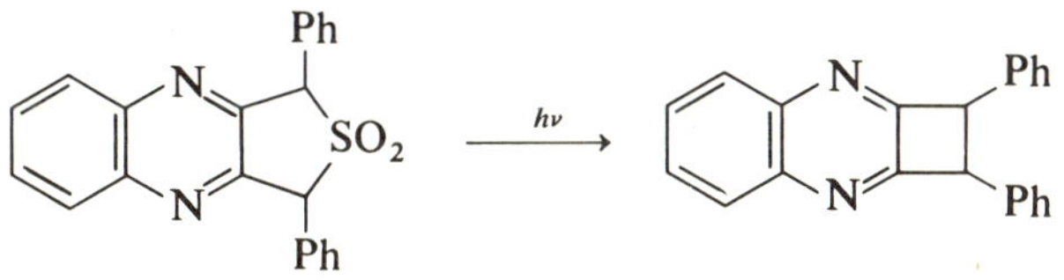

162 $\xrightarrow{h\nu}$ **163**

REFERENCES

1. R. A. Holroyd, *J. Amer. Chem. Soc.* **91**, 2208 (1969).
2. W. Zich and N. Getoff, *Monatsh. Chem.*, **100**, 1745 (1969).
3. W. W. Binkley and R. W. Binkley, *Carbohyd. Res.*, **11**, 1 (1969).
4. D. T. Carty, *Tetrahedron Lett.*, 4753 (1969).
5. R. Criegee and R. Huber, *Angew. Chem. Int. Ed. Engl.*, **8**, 759 (1969).
6. P. R. Story, W. H. Morrison, and J. M. Butler, *J. Amer. Chem. Soc.*, **91**, 2398 (1969).
7. J. Rigaudy, R. DuPont, and N. Y. Cuong, *C. R. Acad. Sci., Paris, Ser. C*, **269**, 416 (1969).
8. J. J. Basselier and J. P. LeRoux, *C. R. Acad. Sci., Paris, Ser. C*, **268**, 970 (1969).
9. N. Suguyama, M. Iwata, M. Yoshioka, K. Yamada, and H. Aoyama, *Bull. Chem. Soc. Japan*, **42**, 1377 (1969).
10. D. J. Carlsson and D. M. Miles, *Macromolecules*, **2**, 597 (1969).
11. J. M. Surzur and M. P. Crozet, *C. R. Acad. Sci., Paris, Ser. C*, **268**, 2109 (1969).
12. J. M. Surzur, C. Dupuy, M.-P. Crozet, and N. Aimar, *C. R. Acad. Sci., Paris, Ser. C*, **269**, 849 (1969).
13. T. Laird and H. Williams, *Chem. Commun.*, 561 (1969).
14. R. A. Archer and P. V. DeMarco, *J. Amer. Chem. Soc.*, **91**, 1530 (1969).
15. P. R. Brown and J. O. Edwards, *J. Org. Chem.*, **34**, 3131 (1969).
16. E. J. Corey and E. Block, *J. Org. Chem.*, **34**, 1233 (1969).
17. D. Herlens and J. Khuong-Huu, *C. R. Acad. Sci., Paris, Ser. C*, **269**, 1405 (1969).
18. G. Adam and K. Schreiber, *Chem. Ber.*, **102**, 878 (1969).
19. O. Yonemitsu, H. Nakai, Y. Kanaoka, I. L. Karle, and B. Witkop, *J. Amer. Chem. Soc.*, **91**, 4591 (1969).
20. B. Umezawa, O. Hoshino, and S. Sawaki, *Chem. Pharm. Bull. Japan*, **17**, 1115 (1969).
21. B. Umezawa, O. Hoshino, and S. Sawaki, *Chem. Pharm. Bull. Japan*, **17**, 1120 (1969).
22. K. Wiesner and T. Inaba, *J. Amer. Chem. Soc.*, **91**, 1036 (1969).
23. R. Beugelmans and H. Compaignon de Marcheville, *Chem. Commun.*, 241 (1969).
24. B. W. Finucane and J. B. Thomson, *Chem. Commun.*, 1220 (1969).
25. R. W. Phillips and D. H. Volman, *J. Amer. Chem. Soc.*, **91**, 3418 (1969).
26. J. G. Traynham, A. G. Lane, and N. S. Bhacca, *J. Org. Chem.*, **34**, 1302 (1969).
27. P. Boldt, W. Thielecke, and J. Etzemuller, *Chem. Ber.*, **102**, 4157 (1969).

28. T. Ohashi, M. Sugie, M. Okahara, and S. Komori, *Tetrahedron*, **25**, 5349 (1969).
29. R. Mews and O. Glemser, *Chem. Ber.*, **102**, 4188 (1969).
30. M. A. Golub, Abstracts of the 158th Meeting of the American Chemical Society, New York, September, 1969, PHYS-156.
31. R. Salovey, J. P. Luongo, and W. A. Yager, *Macromolecules*, **2**, 198 (1969).
32. H. R. Gloria and R. F. Reinisch, Abstracts of the 5th Western Regional Meeting of the American Chemical Society, Disneyland, Calif., October, 1969, no. 171.
33. D. D. Tanner and N. J. Bunce, *J. Amer. Chem. Soc.*, **91**, 3028 (1969).
34. K. Terauchi and H. Sakurai, *Kogyo Kagaku Zasshi*, **72**, 215 (1969).
35. D. G. Pobedimskii and V. A. Belyakov, *Kinet. Katal.*, **10**, 64 (1969).
36. C. von Sonntag, *Tetrahedron*, **23**, 5853 (1969).
37. A. J. Merer and R. S. Mulliken, *Chem. Rev.*, **69**, 639 (1969).
38. H. D. Scharf, *Fortschr. Chem. Forschung*, **11**, 216 (1969).
39. J. S. Swenton, *J. Chem. Ed.*, **46**, 7 (1969).
40. W. Dilling, *Chem. Rev.*, **69**, 845 (1969).
41. D. Seebach, *Fortschr. Chem. Forschung*, **11**, 177 (1969).
42. N. J. Turro, *Photochem. Photobiol.*, **9**, 555 (1969).
43. R. B. Cundall and A. J. R. Voss, *Chem. Commun.*, **116** (1969).
44. K. Fukano and S. Sato, *J. Chem. Soc. Japan*, **72**, 213 (1969).
45. K. Fukano and S. Sato, *Kogyo Kagaku Zasshi*, **72**, 213 (1969).
46. J. Saltiel, K. R. Neuberger, and M. Wrighton, *J. Amer. Chem. Soc.*, **91**, 3658 (1969).
47. R. A. Caldwell and S. P. James, *J. Am. Chem. Soc.*, **91**, 5184 (1969).
48. A. A. Zimmerman et al., *J. Org. Chem.*, **34**, 73 (1969).
49. M. A. Golub, *Macromolecules*, **2**, 550 (1969).
50. H. Yamazaki and R. J. Cvetanovic, *J. Amer. Chem. Soc.*, **91**, 520 (1969).
51. R. A. Carruthers, R. A. Crellin, and A. Ledwith, *Chem. Commun.*, 252 (1969).
52. T. Mori, K. Kimoto, M. Kawanisi, and H. Nozaki, *Tetrahedron Lett.*, 3653 (1969).
53. S. J. Cristol and G. A. Lee, *J. Amer. Chem. Soc.*, **91**, 7554 (1969).
54. S. J. Cristol and G. O. Mayo, *J. Org. Chem.*, **34**, 2363 (1969).
55. J. K. Crandall and D. R. Paulson, *Tetrahedron Lett.*, 2751 (1969).
56. F. T. Bond, *Appl. Spectrosc.*, **23**, 637 (1969).
57. F. Weiss, *Angew. Chem., Int. Ed. Engl.*, **8**, 218 (1969).
58. E. Montandon, H. Francois, and R. Lalande, *Bull. Soc. Chim. France*, 2773 (1969).
59. S. Tazuke and S. Okamura, *J. Polymer Sci., A-1*, **7**, 715 (1969).
60. J. D. Park, S. K. Choi, and H. E. Romine, *J. Org. Chem.*, **34**, 2521 (1969).
61. J. Saltiel and L. S. Ng Lim, *J. Amer. Chem. Soc.*, **91**, 5404 (1969).
62. P. J. Kropp and H. J. Krauss, *J. Amer. Chem. Soc.*, **91**, 7466 (1969).
63. A. A. Gorman and J. B. Sheridan, *Tetrahedron Lett.*, 2569 (1969).
64. E. Payo, L. Cortes, J. Mantecon, and C. Rivas, *Acta Cient. Venez.*, **18**, 130 (1967); *Chem Abstracts*, **70**, 3681d (1969).
65. J. M. Landesberg and J. Sieczkowski, *J. Amer. Chem. Soc.*, **91**, 2120 (1969).
66. G. R. Ziegler, *J. Amer. Chem. Soc.*, **91**, 446 (1969).
67. J. M. Holovka, *Diss. Abstr.*, **29**, 3260B (1969).
68. B. M. Trost, *J. Org. Chem.*, **34**, 3644 (1969).
69. J. R. Edman, *J. Amer. Chem. Soc.*, **91**, 7103 (1969).
70. D. Bienick and F. Korte, *Tetrahedron Lett.*, 4059 (1969).
71. L. Vollner, W. Klein, and F. Korte, *Tetrahedron Lett.*, 2967 (1969).
72. H. M. Fischler and F. Korte, *Tetrahedron Lett.*, 2793 (1969).
73. J. D. Rosen, D. J. Sutherland, and M. A. O. Khan, *J. Agr. Food Chem.*, **17**, 404 (1969).
74. M. A. O. Khan, J. D. Rosen, and D. J. Sutherland, *Science*, **164**, 318 (1969).

75. T. Sugimura, Y. Takeda, T. Osawa, and T. Ukita, *Eisei Kagaku*, **14**, 221 (1968).
76. K. L. Williamson, Y. F. L. Hsu, R. Lacko, and C. H. Youn, *J. Amer. Chem. Soc.*, **91**, 6129 (1969).
77. W. L. Dilling, C. E. Reineke, and R. A. Plepys, *J. Org. Chem.*, **34**, 2605 (1969).
78. J. A. Marshall, *Accounts Chem. Res.*, **2**, 33 (1969).
79. P. J. Kropp, *J. Amer. Chem. Soc.*, **91**, 5783 (1969).
80. J. A. Marshall and A. E. Greene, *Tetrahedron*, **25**, 4183 (1969).
81. J. A. Waters and B. Witkop, *J. Org. Chem.*, **34**, 3774 (1969).
82. J. A. Marshall and A. R. Hochstetler, *J. Amer. Chem. Soc.*, **91**, 648 (1969).
83. H. C. deMarcheville and R. Beugelmans, *Tetrahedron Lett.*, 1901 (1969).
84. A. R. Hochstetler, *Diss. Abstr.*, **29**, 3678B (1969).
85. E. LeGoff and S. Oka, *J. Amer. Chem. Soc.*, **91**, 5665 (1969).
86. E. LeGoff, S. Oka, and W. Deadman, 158th National Meeting of the American Chemical Society, New York, September, 1969, ORGN-57.
87. R. S. H. Liu, *Tetrahedron Lett.*, 1409 (1969).
88. H. E. Zimmerman, R. W. Binkley, R. S. Givens, G. L. Grunewald, and M. A. Sherwin, *J. Amer. Chem. Soc.*, **91**, 3317 (1969).
89. R. S. H. Liu and C. G. Krespan, *J. Org. Chem.*, **34**, 1271 (1969).
90. R. S. H. Liu and J. R. Edman, *J. Amer. Chem. Soc.* **91**, 1492 (1969).
91. M. A. Sherwin, *Diss. Abstr.*, **30**, 1053B (1969).
92. I. F. Eckhard, H. Heaney, and B. A. Marples. *Tetrahedron Lett.*, 3273 (1969).
93. R. K. Murray and H. Hart, 157th National Meeting of the American Chemical Society, Minneapolis, Minn. April, 1969, ORGN-141.
94. H. Hart and R. K. Murray, Jr., *J. Amer. Chem. Soc.*, **91**, 2183 (1969).
95. R. M. Pagni, *Diss. Abstr.*, **30**, 3267B (1969).
96. H. E. Zimmerman and C. O. Bender, *J. Amer. Chem. Soc.*, **91**, 7516 (1969).
97. R. C. Hahn and L. J. Rothman, *J. Amer. Chem. Soc.*, **91**, 2409 (1969).
98. R. C. Hahn and L. J. Rothman, 157th National Meeting of the American Chemical Society, Minneapolis, Minn. April, 1969, ORGN-139.
99. S. J. Cristol, G. O. Mayo, and G. A. Lee, *J. Amer. Chem. Soc.*, **91**, 214 (1969).
100. J. S. Swenton, *J. Org. Chem.*, **34**, 3217 (1969).
101. J. G. Traynham and H. H. Hsieh, *Tetrahedron Lett.*, 3905 (1969).
102. G. M. Whitesides, G. L. Goe, and A. C. Cope, *J. Amer. Chem. Soc.*, **91**, 2608 (1969).
103. R. B. Bates, G. D. Forsythe, G. A. Wolfe, G. Ohloff, and K. H. Schulte-Elte, *J. Org. Chem.*, **34**, 1059 (1969).
104. L. A. Paquette, G. V. Meehan, and R. F. Eizember, *Tetrahedron Lett.*, 999 (1969).
105. H. R. Ward and E. Karafiath, *J. Amer. Chem. Soc.*, **91**, 522 (1969).
106. H. R. Ward and E. Karafiath, *J. Amer. Chem. Soc.*, **91**, 7475 (1969).
107. A. M. Braun, W. B. Hammond, and H. G. Cassidy, *J. Amer. Chem. Soc.*, **91**, 6196 (1969).
108. P. Scribe, D. Hourdin, and J. Wiesmann, *C. R. Acad. Sci., Paris, Ser. C*, **268C**, 178 (1969).
109. A. C. Waiss, Jr., and M. Wiley, *Chem. Commun.*, 512 (1969).
110. J. M. Holovka, R. R. Grabbe, P. D. Gardner, C. B. Strow, M. L. Hill, and T. V. Van Auken, *Chem. Commun.*, 1522 (1969).
111. T. Sato and I. Moritani, *Tetrahedron Lett.*, 3181 (1969).
112. G. Feler, *Theoret. Chim. Acta.*, **12**, 412 (1968).
113. W. T. A. M. van der Lugt and L. J. Oosterhoff, *J. Amer. Chem. Soc.*, **91**, 6042 (1969).
114. W. G. Dauben, J. H. Smith, and J. Saltiel, *J. Org. Chem.*, **34**, 261 (1969).
115. D. M. Gale, *U.S. Patent* 3,459,647 (1969).

116. J. Saltiel, L. Metts, and M. Wrighton, *J. Amer. Chem. Soc.*, **91**, 5684 (1969).
117. H. L. Hyndman, B. M. Monroe, and G. S. Hammond, *J. Amer. Chem. Soc.*, **91**, 2852 (1969).
118. P. A. Leermakers, J. P. Montillier, and R. D. Rauh, *Mol. Photochem.*, **1**, 57 (1969).
119. J. Saltiel and O. C. Zafirious, *Mol. Photochem.*, **1**, 319 (1969)
120. S. Kita and K. Fukui, *Bull. Chem. Soc. Japan*, **42**, 66 (1969).
121. J. D. White and D. N. Gupta, *Tetrahedron*, **25**, 3331 (1969).
122. J. L. Courtney and S. McDonald, *Aust. J. Chem.*, **22**, 2411 (1969).
123. E. C. Sanford and G. S. Hammond, *Appl. Spectrosc.*, **23**, 640 (1969).
124. J. M. Garrett and G. J. Fonken, *Tetrahedron Lett.*, 191 (1969).
125. E. F. Kiefer and C. H. Tanna, *J. Amer. Chem. Soc.*, **91**, 4478 (1969).
126. E. F. Kiefer and J. Y. Fukunaga, *Tetrahedron Lett.*, 993 (1969).
127. M. S. Toy, 5th Western Regional Meeting of the American Chemical Society, Disneyland, Calif., December, 1969, no. 241.
128. P. H. Mazzocchi, *Tetrahedron Lett.*, 989 (1969).
129. J. E. Baldwin and S. M. Krueger, *J. Amer. Chem. Soc.*, **91**, 6444 (1969).
130. G. R. deMare, P. Goldfinger, G. Huybrechts, E. Jonas, and M. Toth, *Ber. Bunsenges.*, **73**, 867 (1969).
131. M. Miyashita, H. Uda, and A. Yoshikoshi, *Chem. Commun.*, 1396 (1969).
132. R. A. Caldwell, *J. Org. Chem.*, **34**, 1886 (1969).
133. J. S. McConaghy and J. J. Bloomfield, *Tetrahedron Lett.*, 1121 (1969).
134. J. Altman, E. Babad, D. Ginsburg, and M. B. Rubin, *Israel J. Chem.*, **7**, 435 (1969).
135. J. Altman, E. Babad, M. B. Rubin, and D. Ginsburg, *Tetrahedron Lett.*, 1125 (1969).
136. S. Masamune and R. T. Seidner, *Chem. Commun.*, 542 (1969).
137. L. A. Paquette and J. C. Phillips, *J. Amer. Chem. Soc.* **91**, 3973 (1969).
138. L. A. Paquette and J. C. Stowell, *Tetrahedron Lett.*, 4159 (1969).
139. W. J. Nebe and G. J. Fonken, *J. Amer. Chem. Soc.*, **91**, 1249 (1969).
140. J. C. Sircar and G. S. Fisher, *J. Org. Chem.*, **34**, 404 (1969).
141. H. E. Zimmerman and G. E. Samuelson, *J. Amer. Chem. Soc.*, **91**, 5307 (1969).
142. G. E. Petrowski, *Diss. Abstr.*, **30**, 1603B (1969).
143. K. Yoshihara, Y. Ohta, T. Sakai, and Y. Hirose, *Tetrahedron Lett.*, 2263 (1969).
144. T. Tabata and H. Hart, *Tetrahedron Lett.*, 4929 (1969).
145. J. D. DeVrieze, *Diss. Abstr.*, **30**, 1586B (1969).
146. E. K. von Gustorf, F. W. Grevels, and G. O. Schenck, *Justus Liebigs Ann. Chem.* **719**, 1 (1969).
147. M. Mousseron-Canet and J. P. Chabaud, *Bull. Soc. Chim. France*, 308 (1969).
148. D. A. Seeley, *Diss. Abstr.*, **30**, 1052B (1969).
149. L. B. Jones and V. K. Jones, *Fortschr. Chem. Forsch.*, **13**, 307 (1969).
150. L. B. Jones and V. K. Jones, *J. Org. Chem.*, **34**, 1298 (1969).
151. A. P. Ter Borg and H. Kloosterziel, *Rec. Trav. Chem.*, **88**, 266 (1969).
152. K. Shen, W. E. McEwen, and A. P. Wolf, *Tetrahedron Lett.*, 827 (1969).
153. R. S. Givens, *Tetrahedron Lett.*, 663 (1969).
154. P. Datta, T. D. Goldfarb, and R. S. Boikess, *J. Amer. Chem. Soc.*, **91**, 5429 (1969).
155. S. Masamune, P. M. Baker, and K. Hojo, *Chem. Commun.*, 1203 (1969).
156. J. Schwartz, *Chem. Commun.*, 833 (1969).
157. G. Schroder, G. Kirsch, and J. F. M. Oth, *Tetrahedron Lett.*, 4575 (1969).
158. A. G. Anastassiou and J. H. Gebrian, 158th National Meeting of the American Chemical Society, September, 1969, New York, ORGN-14.
159. A. G. Anastassiou and J. H. Gebrian, *J. Amer. Chem. Soc.*, **91**, 4011 (1969).
160. S. Masamune, K. Hojo, and S. Takada, *Chem. Commun.*, 1204 (1969).

161. A. G. Anastassiou and J. H. Gebrian, *Tetrahedron Lett.*, 5239 (1969).
162. A. G. Anastassiou and R. P. Cellura, *Chem. Commun.*, 903 (1969).
163. H. Rottele, P. Nikoloff, J. F. M. Oth, and G. Schroder, *Chem. Ber.*, **102**, 3367 (1969).
164. G. Schroder, W. Martin, and H. Rottele, *Angew. Chem. Int. Ed. Eng.*, **8**, 69 (1969).
165. H. Rottele, W. Martin, J. F. M. Oth, and G. Schroder, *Chem. Ber.*, **102**, 3985 (1969).
166. A. G. Anastassiou, V. Orfanos, and J. H. Gebrian, *Tetrahedron Lett.*, 4491 (1969).
167. J. M. Patterson and L. T. Burka, *Tetrahedron Lett.*, 2215 (1969).
168. R. M. Kellogg, M. B. Groen, and H. Wynberg, 157th National Meeting of the American Chemical Society, Minneapolis, Minn., April, 1969, ORGN-24.
169. A. Couture and A. Lablance-Combier, *Chem. Commun.*, 524 (1969).
169a. G. Wittig and M. Rings, *Justus Liebigs Ann. Chem.*, **719**, 127 (1969).
170. R. Srinivasan and H. Hiraoka, *Tetrahedron Lett.*, 2767 (1969).
171. A. P. Bindra, J. A. Elix, and M. V. Sargent, *Aust. J. Chem.*, **22**, 1449 (1969).
172. A. G. Anastassiou and R. P. Cellura, *Chem. Commun.*, 1521 (1969).
173. L. A. Paquette and D. E. Kuhla, *J. Org. Chem.*, **34**, 2885 (1969).
174. I. G. Bolesov, *Russ. Chem. Rev.*, **37**, 666 (1968).
175. S. Sato, H. Kobayashi, and K. Fukano, *Kogyo Kagaku Zasshi*, **72**, 209 (1969).
176. D. Bryce-Smith, *Chem. Commun.*, 806 (1969).
177. H. Morrison and W. I. Ferree, *Chem. Commun.*, 268 (1969).
178. K. H. Grellmann and W. Kuhnle, *Tetrahedron Lett.*, 1537 (1969).
179. H. Nozaki, H. Kato, and R. Noyori, *Tetrahedron*, **25**, 1661 (1969).
180. T. Sato, M. Wakabayashi, and K. Hata, *Bull. Chem. Soc. Japan*, **42**, 773 (1969).
181. J. Szmuszkovicz, *Org. Prep.*, **1**, 105 (1969).
182. D. J. Cram, 21st National Organic Symposium, Salt Lake City, Utah, June, 1969, p. 8.
183. T. D. Walsh, *J. Amer. Chem. Soc.*, **91**, 515 (1969).
184. N. J. Turro, M. Tobin, L. Friedman, and J. B. Hamilton, *J. Amer. Chem. Soc.*, **91**, 516 (1969).
185. E. Grovenstein, T. C. Campbell, and T. Shibata, *J. Org. Chem.*, **34**, 2418 (1969).
186. N. C. Perrins and J. P. Simons, *Trans. Faraday Soc.*, **65**, 390 (1969).
187. Y. Ogata, Y. Izawa, H. Tomioka, and T. Ukigai, *Tetrahedron*, **25**, 1817 (1969).
188. E. T. Enisor and D. I. Metelitsa, *Russ. Chem. Rev.*, **37**, 656 (1968).
189. E. A. Fitzgerald, *Diss. Abstr.*, **29**, 3652B (1969).
189a. G. Quinkert, M. Finke, J. Palmowski, and W. W. Wiersdorff, *Mol. Photochem.*, **2**, 433 (1969).
190. G. Camaggi and F. Gozzo, *J. Chem. Soc., C*, 489 (1969).
191. D. Bryce-Smith, A. Gilbert, and B. H. Orger, *Chem. Commun.*, 800 (1969).
192. A. E. Pedler, J. C. Tatlow, and A. J. Uff, *Tetrahedron*, **25**, 1597 (1969).
193. N. Kharasch and J. L. Day, *Quart. Reports Sulfur Chem.*, **3**, 177 (1968).
194. G. E. Robinson and J. M. Vernon, *Chem. Commun.*, 977 (1969).
195. T. Sato, *Yuki Gosei Kagaku Kyokai Shi.*, **27**, 715 (1969); *Chem. Abstr.*, **71**, 112667v (1969).
196. G. R. Lappin and J. S. Zannucci, *Tetrahedron Lett.*, 5085 (1969).
197. T. Tosa, C. Poc, and H. Sakurai, *Tetrahedron Lett.*, 3635 (1969).
198. J. R. Plimmer and B. E. Hummer, *J. Agr. Food Chem.*, **17**, 83 (1969).
199. J. T. Pinhey and R. D. G. Rigby, *Tetrahedron Lett.*, 1267 (1969).
200. J. T. Pinhey and R. D. G. Rigby, *Tetrahedron Lett.*, 1271 (1969).
201. T. Latowski, *Zesz. Nauk., Mat., Fiz., Chem., Wyzsza Szk. Pedagog. Gdansku*, **8**, 189 (1968); *Chem. Abstracts*, **71**, 105683y (1969).
202. J. L. Neumeyer, K. H. Oh, K. K. Weinhardt, and B. R. Neustadt, *J. Org. Chem.*, **34**, 3786 (1969).

203. F. Pietra, *Quart. Rev.*, **23**, 504 (1969).
204. K. Omura and T. Matsuura, *Chem. Commun.*, 1394 (1969).
205. Y. Ogata and T. Ukigai, *J. Chem. Soc.*, *C*, 2413 (1969).
206. N. Yamamoto, T. Kojikawa, H. Sato, and H. Tsubomura, *J. Amer. Chem. Soc.*, **91**, 265 (1969).
207. R. S. Davidson, F. A. Younis, and R. Wilson, *Chem. Commun.*, 826 (1969).
208. R. C. Petterson, C. S. Irving, A. M. Khan, G. W. Griffin, and I. M. Sarkar, 158th National Meeting of the American Chemical Society, September, 1969, New York, ORGN-10.
209. G. W. Griffin, 157th National Meeting of the American Chemical Society, April, 1969, Minneapolis, Minn., S-17.
210. R. L. Smith, A. Manmade, and G. W. Griffin, *J. Heterocyclic Chem.*, **6**, 443 (1969).
211. G. W. Griffin and H. Kristinsson, U.S. Patent 3,445,357 (1969); *Chem. Abstr.*, **71**, 21847n (1969).
212. V. Zanker and D. Benicke, *Z. Physik. Chem. (Frankfurt)*, **66**, 34 (1969).
213. A. Kira and M. Koizumi, *Bull. Chem. Soc. Japan*, **42**, 625 (1969).
214. Y. I. Kiryukhin and K. S. Bagdasaryan, *Khim. Vys. Energ.*, **3**, 179 (1969).
215. O. H. Iles and A. Ledwith, *Chem. Commun.*, 364 (1969).
216. M. Zander, *Z. Naturforsch.*, **24a**, 254 (1969).
217. M. Zander and W. H. Franke, *Chem. Ber.*, **102**, 2728 (1969).
218. C. Pac and H. Sakurai, *Kogyo Kagaku Zasshi*, **72**, 230 (1969).
219. T. Hinohara, *Nippon Kag. Zasshi*, **90**, 860 (1969).
220. C. Pac and H. Sakurai, *Chem. Commun.*, 20 (1969).
221. N. V. Khromov-Borisov and V. E. Gmiro, *Doklady Chem.* (*USSR*), **185**, 217 (1969).
222. B. E. Smith, *J. Chem. Soc.*, *A*, 2673 (1969).
222a. H. J. Shine, C. M. Baldwin, and J. F. Sullivan, 25th Southwest Regional Meeting of the American Chemical Society, ORGN-22.
223. I. Avigal, J. Feitelson, and M. Ottolenghi, *J. Chem. Phys.*, **50**, 2614 (1969).
224. V. D. Pokhodenko, V. A. Khizhnyi, and V. A. Bidzilya, *Russ. Chem. Rev.*, **37**, 435 (1968).
224a. M. Kuwahara, N. Shindo, N. Kato, and K. Munakata, *Agr. Biol. Chem.*, **33**, 892 (1969).
225. W. Rundel, *Chem. Ber.*, **102**, 359 (1969).
226. W. Rundel, *Chem. Ber.*, **102**, 359 (1969).
227. W. Rundel, *Chem. Ber.*, **102**, 1649 (1969).
228. W. D. Weringa, *Tetrahedron Lett.*, 273 (1969).
229. C. R. Olander, R. J. Warnet, D. M. S. Wheeler, and G. A. Sim, 157th National Meeting of the American Chemical Society, April, 1969, Minneapolis, Minn., ORGN-92.
229a. Y. Saburi, T. Yoshimoto, and K. Minami, *Nippon Kagaku Zasshi*, **90**, 587 (1969).
230. D. C. Neckers and J. deZwaan, *Chem. Commun.*, 813 (1969).
231. D. M. Lemal, J. V. Starvos, and V. Austel, *J. Amer. Chem. Soc.*, **91**, 3373 (1969).
232. M. G. Barlow, R. N. Haszeldine, and R. Hubbard, *Chem. Commun.*, 202 (1969).
233. M. A. Pak, D. N. Shigorin, and G. A. Ozerova, *Doklady Phys. Chem.*, **186**, 311 (1969).
234. K. Okumura, S. Takamuku, and H. Sakurai, *Kogyo Kagaku Zasshi*, **72**, 200 (1969).
235. K. Omura and T. Matsuura, *Chem. Commun.*, 1516 (1969).
236. R. A. Abramovitch and T. Takaya, *Chem. Commun.*, 1369 (1969).
237. Y. Nagao, K. Shima, and H. Sakurai, *Kogyo Kagaku Zasshi*, **72**, 236 (1969).
238. T. Sato, S. Shimada, and K. Hata, *Bull. Chem. Soc. Japan*, **42**, 766 (1969).

239. F. Bolletta, P. G. DiMarco, and S. Pietra, *Z. Naturforsch.*, **24a**, 482 (1969).
240. E. Vander Donckt, J. Nasielski, and P. Thiry, *Chem. Commun.*, 1249 (1969).
241. J. R. Bunting and N. Filipescu, 158th National Meeting of the American Chemical Society, September, 1969, New York, PHYS-183.
242. R. C. Cookson, S. M. deB. Costa, and J. Hudec, *Chem. Commun.*, 753 (1969).
243. W. Droste, H.-D. Scharf, and F. Korte, *Justus Liebigs Ann. Chem.*, **724**, 71 (1969).
244. W. Metzner and D. Wendisch, *Justus Liebigs Ann. Chem.*, **730**, 111 (1969).
245. C. DeBoer, *J. Amer. Chem. Soc.*, **91**, 1855 (1969).
246. J. J. McCullough and C. W. Huang, *Can. J. Chem.*, **47**, 757 (1969).
247. R. M. Bowman, J. J. McCullough, and J. S. Swenton, *Can. J. Chem.*, **47**, 4503 (1969).
248. S. Fujita, T. Nomi, and H. Nozaki, *Tetrahedron Lett.*, 3557 (1969).
249. G. Dumenil, P. Maldonado, R. Guglielmetti, and J. Metzger, *Bull. Soc. Chim. France*, 817 (1969).
250. R. S. Becker, E. Dolan, and D. E. Balke, *J. Chem. Phys.*, **50**, 239 (1969).
251. T. Bercovici, R. Heiligman-Rim, and E. Fischer, *Mol. Photochem.*, **1**, 23 (1969).
252. T. Bercovici, R. Heiligman-Rim, and E. Fischer, *Mol. Photochem.*, **1**, 189 (1969).
253. C. Balny, P. Douzon, T. Bercovici, and E. Fischer, *Mol. Photochem.*, **1**, 225 (1969).
254. P. H. Vandewyer, J. Hoefnagels, and G. Smets, *Tetrahedron*, **25**, 3251 (1969).
255. J. Kale and R. S. Becker, *J. Amer. Chem. Soc.*, **91**, 6513 (1969).
256. I. Shimizu, H. Kokado, and E. Inone, *Bull. Chem. Soc. Japan*, **42**, 1726 (1969).
257. P. Serve, H. M. Rosenberg, and R. Rondeau, *Can. J. Chem.*, **47**, 4295 (1969).
258. R. C. Cookson, S. M. deB. Costa, and J. Hudec, *Chem. Commun.*, 1272 (1969).
259. G. W. Gruber and M. Pomerantz, *J. Amer. Chem. Soc.*, **91**, 4004 (1969).
260. D. L. Horrocks, *J. Chem. Phys.*, **50**, 4151 (1969).
261. D. L. Horrocks, *J. Chem. Phys.*, **50**, 4151 (1969).
262. C. M. Chopin and J. H. Wharton, *Chem. Phys. Lett.*, **3**, 552 (1969).
263. L. Benati, G. Martelli, P. Spagnolo, and M. Tiacco, *J. Chem. Soc., B*, 472 (1969).
264. D. C. Neckers, J. H. Dopper, and H. Wynberg, *Tetrahedron Lett.*, 2913 (1969).
265. W. H. F. Sasse, P. J. Collin, and D. B. Roberts, *Tetrahedron Lett.*, 4791 (1969).
266. D. C. Neckers, J. DeZwaan, J. H. Dopper, and H. Wynberg, 158th National Meeting of the American Chemical Society, September, 1969, New York, ORGN-11.
267. T. J. Barton and A. J. Nelson, *Tetrahedron Lett.*, 5037 (1969).
268. T. Sata, I. Moritani, and M. Matsuyama, *Tetrahedron Lett.*, 5113 (1969).
269. T. J. Barton and A. J. Nelson, *Tetrahedron Lett.*, 5037 (1969).
270. G. Jori, G. Galiazzo, and G. Gennari, *Photochem. Photobiol.*, **9**, 179 (1969).
271. A. R. Mosier, W. D. Guenzi, and L. L. Miller, *Science*, **164**, 1083 (1969).
272. J. R. Plimmer and U. I. Klingbell, *Chem. Commun.*, 648 (1969).
273. H. E. Zimmerman and P. S. Mariano, *J. Amer. Chem. Soc.*, **91**, 1718 (1969).
274. E. Block and E. J. Corey, *J. Org. Chem.*, **34**, 896 (1969).
274a. M. B. Hocking, *Can. J. Chem.*, **47**, 4567 (1969).
275. J.-J. Basselier and J.-C. Cherton, *C. R. Acad. Sci., Paris, Ser. C*, **269**, 1412 (1969).
276. P. Courtot and R. Rumin, *Bull. Soc. Chim. France*, 3665 (1969).
277. T. Miyadera and R. Tachikawa, *Tetrahedron*, **25**, 837 (1969).
278. T. Toda, M. Nitta, and T. Mukai, *Tetrahedron Lett.*, 4401 (1969).
279. G. Martelli, P. Spagnolo, and M. Tiecco, *Chem. Commun.*, 282 (1969).
280. E. Muller, M. Sauerbier, and G. Zountsas, *Tetrahedron Lett.*, 3003 (1969).
281. E. Muller, J. Heiss, and M. Sauerbier, *Justus Liebigs Ann. Chem.*, **723**, 61 (1969).
282. W. H. F. Sasse, *Aust. J. Chem.*, **22**, 1257 (1969).
283. E. H. White, E. W. Friend, Jr., R. L. Stern, and H. Maskill, *J. Amer. Chem. Soc.*, **91**, 523 (1969).

284. A. Schonberg, U. Sodtke, and K. Praefcke, *Chem. Ber.*, **102**, 1453 (1969).
285. A. Bylina and Z. R. Grabowski, *Trans. Faraday Soc.*, **65**, 458 (1969).
286. D. G. Whitten and M. T. McCall, *J. Amer. Chem. Soc.*, **91**, 5097 (1969).
287. H. A. Hammond, D. E. DeMeyer, and J. L. R. Williams, *J. Amer. Chem. Soc.*, **91**, 5180 (1969).
288. K. Krueger and E. Lippert, *Z. Physick. Chem. (Frankfurt)*, **66**, 293 (1969).
289. W. M. Gelbart and S. A. Rice, Jr., *J. Chem. Phys.*, **50**, 4775 (1969).
290. L. Pedersen, D. G. Whitten, and M. T. McCall, *Chem. Phys. Lett.*, **3**, 569 (1969).
291. G. Fischer, K. A. Muszkat, and E. Fischer, *Israel J. Chem.*, **6**, 965 (1968).
292. J. Saltiel and E. D. Megarity, *J. Amer. Chem. Soc.*, **91**, 1265 (1969).
293. E. V. Blackburn and C. J. Timmons, *Quart. Rev.*, **23**, 482 (1969).
294. R. N. Nurmukametov and G. I. Grishina, *Zh. Fiz. Khim.*, **43**, 2676 (1969).
295. G. P. DeGunst, *Rec. Trav. Chem.*, **88**, 801 (1969).
296. A. Bromberg and K. A. Muszkat, *J. Amer. Chem. Soc.*, **91**, 2860 (1969).
297. S. Farid, W. Kothe, and G. Pfundt, *Tetrahedron Lett.*, 4147 (1968).
298. P. Bortolus and G. Cauzzo, *Z. Physik. Chem. (Frankfurt)*, **63**, 29 (1969).
299. G. Galiazzo, P. Bortolus, G. Cauzzo, and U. Mazzucato, *J. Heterocyclic Chem.*, **6**, 465 (1969).
300. B. Chauncy and E. Gellert, *Aust. J. Chem.*, **22**, 993 (1969).
301. G. Kortum and P. Krieg, *Chem. Ber.*, **102**, 3033 (1969).
302. R. H. Martin, G. Morren, and J. J. Schurter, *Tetrahedron Lett.*, 3683 (1969).
303. R. H. Martin and J. J. Schurter, *Tetrahedron Lett.*, 3679 (1969).
304. H. M. Rosenberg, R. Rondeau, and P. Serve, *J. Org. Chem.*, **34**, 471 (1969).
305. H. Durr, *Justus Liebigs Ann. Chem.*, **723**, 102 (1969).
306. M. Hasegawa, Y. Suzuki, F. Suzuki, and H. Nakanishi, *J. Polymer Sci.*, **7A**, 743 (1969).
307. M. Hasegawa, Y. Suzuki, *J. Polymer Sci.*, **5B**, 813 (1967).
308. P. J. Grisdale and J. L. R. Williams, *J. Org. Chem.*, **34**, 1675 (1969).
309. D. G. Whitten, Y. J. Lee, and M. T. McCall, 158th National Meeting of the American Chemical Society, September, 1969, New York, ORGN-16.
310. M. T. McCall and D. G. Whitten, *J. Amer. Chem. Soc.*, **91**, 5681 (1969).
311. J. Nasielski, M. Jauquet, E. Vander Donckt, and A. Van Sinoy, *Tetrahedron Lett.*, 4859 (1969).
312. J. Nasielski, M. Jauquet, E. Vander Donckt, and A. Van Sinoy, *Tetrahedron Lett.*, 4859 (1969).
313. D. G. Whitten, P. D. Wildes, and J. G. Lopp, *J. Amer. Chem. Soc.*, **91**, 3393 (1969).
314. J. C. Dotty, J. L. R. Williams, and P. J. Grisdale, *Can. J. Chem.*, **47**, 2355 (1969).
315. B. K. Selinger and M. Sterns, *Chem. Commun.*, 978 (1969).
316. T. W. Mattingly and A. Zweig, *Tetrahedron Lett.*, 621 (1969).
317. K. Sandros, *Acta Chem. Scand.*, **23**, 2815 (1969).
318. J. Christie and B. Selinger, *Photochem. Photobiol.*, **9**, 471 (1969).
319. B. S. Solomon, C. Steel, and A. Weller, *Chem. Commun.*, 927 (1969).
320. R. B. Hurley, *Diss. Abstr.*, **30**, 1069B (1969).
321. M. Schwoerer and H. Sixl, *Z. Naturforsch.*, **24a**, 952 (1969).
322. H. H. Wasserman and P. M. Keehn, *J. Amer. Chem. Soc.*, **91**, 2374 (1969).
323. R. J. McDonald and B. K. Selinger, *Ber. Bunsenges.*, **73**, 789 (1969).
324. J. Blum, F. Grauer, and E. D. Bergmann, *Tetrahedron*, **25**, 3501 (1969).
325. G. S. Hammond, S. C. Shim, and S. P. Van, *Mol. Photochem.*, **1**, 89 (1969).
326. W. A. Henderson and A. Zweig, *Tetrahedron Lett.*, 625 (1969).
327. W. A. Henderson, Jr., R. Lopresti, and A. Zweig, *J. Amer. Chem. Soc.*, **91**, 6049 (1969).

328. J. Lugtenburg and E. Havinga, *Tetrahedron Lett.*, 1505 (1969).
329. C. Goedicke and H. Stegemeyer, *Ber. Bensenges.*, **73**, 782 (1969).
330. R. L. E. Drisko, *Diss. Abstr.*, **30**, 4585B (1969).
331. G. P. Petrenko and A. S. Kovbasyuk, *Ukrain. Khim. Zhur.*, **35**, 989 (1969).
332. N. Mataga and Y. Torihashi, *J. Chem. Soc. Japan*, **72**, 120 (1969).
333. R. S. Davidson, *Chem. Commun.*, 1450 (1969).
334. A. G. Schultz and R. H. Schlessinger, *Chem. Commun.*, 1483 (1969).
335. R. H. Schlessinger and A. G. Schultz, *Tetrahedron Lett.*, 4513 (1969).
336. J. Meinwald, 21st National Organic Symposium, Salt Lake City, Utah, June, 1969, p. 61.
337. B. Bossenbroek, D. C. Sanders, H. M. Curry, and H. Schechter, *J. Amer. Chem. Soc.*, **91**, 371 (1969).
338. R. B. Fox and R. F. Cozzens, *Macromolecules*, **2**, 181 (1969).
339. M. G. Kuz'min, V. L. Ivanov, and Y. Y. Kulis, *Khim. Vys. Energ.*, **2**, 228 (1969).
340. M. D. Cohen, Z. Lumer, J. M. Thomas, and J. O. Wiliams, *Chem. Commun.*, 1172 (1969).
341. O. L. Chapman and K. Lee, *J. Org. Chem.*, **34**, 4166 (1969).
342. S. Singh and C. Sandorfy, *Can. J. Chem.*, **47**, 257 (1969).
343. D. T. Wilson and B. K. Selinger, *Photochem. Photobiol.*, **9**, 171 (1969).
344. E. Baum, *Appl. Spectrosc.*, **23**, 649 (1969).
345. R. Lapouyade, A. Castellan, and H. Bouas-Laurent, *C. R. Acad. Sci. Paris, Ser. C*, **268C**, 217 (1969).
346. H. Bouas-Laurent and R. Lapouyade, *Chem. Commun.*, 817 (1969).
347. R. Lapouyade, A. Castellan, and H. Bouas-Laurent, *Tetrahedron Lett.*, 3537 (1969).
348. C. Pac and H. Sakurai, *Tetrahedron Lett.*, 3829 (1969).
349. T. J. Kemp and J. P. Roberts, *Trans. Faraday Soc.*, **65**, 725 (1969).
350. G. Meyer, *Bull. Soc. Chim. France*, 3629 (1969).
351. R. H. Martin and M. Deblecker, *Tetrahedron Lett.*, 3597 (1969).
352. W. Jenny and R. Paioni, *Chimia*, **23**, 41 (1969).
353. G. Sugowdz, P. J. Collin, and W. H. F. Sasse, *Tetrahedron Lett.*, 3843 (1969).
354. J. Olmsted, *Mol. Photochem.*, **1**, 331 (1969).
355. D. R. Maulding and B. G. Roberts, *J. Org. Chem.*, **34**, 1734 (1969).
356. R. Speed and B. Selinger, *Aust. J. Chem.*, **22**, 9 (1969).
357. T. Forster, *Angew. Chem. Int. Ed. Engl.*, **8**, 333 (1969).
358. G. Reske, *Z. Naturforsch.*, **24a**, 17 (1969).
359. A. Muller and U. Sommer, *Ber. Bunsenges.*, **73**, 819 (1969).
360. J. L. Kropp and W. R. Dawson, *J. Phys. Chem.*, **73**, 1747 (1969).
361. W. R. Dawson and J. L. Kropp, *J. Phys. Chem.*, **73**, 693 (1969).
362. H. Tezuka, *Yuki Gosei Kagaku Koyokai Shi.*, **27**, 430 (1969); *Chem. Abstr.*, **71**, 100948z (1969).
363. J. S. Swenton, *J. Chem. Ed.*, **46**, 217 (1969).
364. L. M. Stephenson and G. S. Hammond, *Angew. Chem. Int. Ed. Engl.*, **8**, 261 (1969).
365. T. Tezuka, *Yuki Gosei Kagaku Koyokai Shi*, **27**, 309 (1969).
365a. K. Shima and H. Sakuri, *Bull. Chem. Soc. Japan*, **42**, 849 (1969).
366. T. Kubota, K. Shima, S. Toki, and H. Sakurai, *Chem. Commun.*, 1462 (1969).
367. K. Kojima, K. Sakai, and K. Tanabe, *Tetrahedron Lett.*, 1925 (1969).
368. E. Baggiolini, H. P. Hamlow, K. Schaffner, and O. Jeger, *Chimia*, **23**, 181 (1969).
368a. R. G. Tonkyn and R. J. Cotter, *J. Polym. Sci.*, A-1, **7**, 2744 (1969).
369. W. C. Agosta, D. K. Herron, and W. W. Lowrance, *Tetrahedron Lett.*, 4521 (1969).
370. R. Srinivasan and S. E. Cremer, *J. Amer. Chem. Soc.*, **87**, 1647 (1965).

370a. J. Meinwald and A. T. Hamner, *Chem. Commun.*, 1302 (1969).
371. R. R. Sauers and J. A. Whittle, *J. Org. Chem.*, **34**, 3579 (1969).
372. R. R. Sauers, W. Schinski, and M. M. Mason, *Tetrahedron Lett.*, 79 (1969).
373. W. G. Dauben and G. W. Shaffer, *J. Org. Chem.*, **34**, 2301 (1969).
374. H. E. O'Neal and C. W. Larson, *J. Phys. Chem.*, **73**, 1011 (1969).
375. G. D. Renkes and F. S. Wettack, *J. Amer. Chem. Soc.*, **91**, 7514 (1969).
376. K. R. Kopecky and C. Mumford, *Can. J. Chem.*, **47**, 709 (1969).
377. E. H. White, J. Wiecko, and D. F. Roswell, *J. Amer. Chem. Soc.*, **91**, 5194 (1969).
378. F. McCapra and R. A. Hann, *Chem. Commun.*, 442 (1969).
379. G. Golde, K. Mobius, and W. Kaminski, *Z. Naturforsch.*, **24a**, 1214 (1969).
380. S. G. Cohen, S. Aktipis, and H. Rubinstein, *Photochem. Photobiol.*, **10**, 45 (1969).
381. H. E. Seyfarth, A. Hesse, and H. Pastohr, *Z. Chem.*, **9**, 150 (1969).
382. J. S. Jewel and W. A. Szarek, *Tetrahedron Lett.*, 43 (1969).
383. M. Julia and M. Maumy, *Bull. Soc. Chim. France*, 2427 (1969).
384. J. Sperling, *J. Amer. Chem. Soc.*, **91**, 5389 (1969).
385. D. Elad and J. Sperling, *Chem. Commun.*, 34 (1969).
386. D. Elad and J. Sperling, *J. Chem. Soc.*, **C**, 1579 (1969).
387. S. H. Schroeter, *Tetrahedron Lett.*, 1591 (1969).
388. M. A. Golub, *J. Amer. Chem. Soc.*, **91**, 4925 (1969).
389. R. F. Borkman and D. R. Kearns, *J. Amer. Chem. Soc.*, **88**, 3467 (1966).
390. K. Kawaoka and D. R. Kearns, *J. Chem. Phys.*, **45**, 147 (1966).
391. J. T. Przybytek, S. P. Singh, and J. Kagan, *Chem. Commun.*, 1224 (1969).
392. C. H. Bibart, M. G. Rockley, and F. S. Wettack, *J. Amer. Chem. Soc.*, **91**, 2802 (1969).
392a. C. H. Bibart, M. G. Rockley, G. D. Renkes, and F. S. Wettack, 158th National Meeting of the American Chemical Society, September, 1969, New York, PHYS-182.
393. N. C. Yang and S. P. Elliott, *J. Amer. Chem. Soc.*, **91**, 7550 (1969).
394. N. C. Yang, S. P. Elliott, and B. Kim, *J. Amer. Chem. Soc.*, **91**, 7551 (1969).
395. A. Heller, *Mol. Photochem.*, **1**, 257 (1969).
396. M. D. Shetter, *Appl. Spectrosc.*, **23**, 651 (1969).
397. D. J. Carlsson and D. M. Wiles, *Macromolecules*, **2**, 587 (1969).
398. E. Baggliolini, K. Schaffner, and O. Jeger, *Chem. Commun.*, 1103 (1969).
398a. A. G. Brook and J. M. Duff, *J. Amer. Chem. Soc.*, **91**, 2118 (1969).
399. S. H. Schroeter, *Chem. Commun.*, 12 (1969).
400. N. J. Turro and P. A. Wriede, *J. Org. Chem.*, **34**, 3562 (1969).
401. S. H. Schroeter and C. M. Orlando, Jr., *J. Org. Chem.*, **34**, 1181 (1969).
402. E. W. Abrahamson and S. W. Jasper, 158th National Meeting of the American Chemical Society, September, 1969, New York, PHYS-181.
403. N. J. Turro, J. C. Dalton, D. M. Pond, and P. A. Wriede, 158th National Meeting of the American Chemical Society, September, 1969, New York, PHYS-150.
404. N. J. Turro, *Photochem. Photobiol.*, **9**, 555 (1969).
405. W. G. Bentrude and K. R. Darnall, *Chem. Commun.*, 862 (1969).
406. J. K. Crandall and C. F. Mayer, *J. Org. Chem.*, **34**, 2814 (1969).
407. W. G. Dauben, L. Schutte, and R. E. Wolf, *J. Org. Chem.*, **34**, 1849 (1969).
408. W. G. Dauben, L. Schutte, R. E. Wolf, and E. J. Deviny, *J. Org. Chem.*, **34**, 2512 (1969).
409. K. E. Davis, *Diss. Abstr.*, **30**, 4585b (1969).
410. E. J. Baum, L. D. Hess, J. R. Wyatt, and J. N. Pitts, Jr., *J. Amer. Chem. Soc.*, **91**, 2461 (1969).
411. C. E. Griffin, W. G. Bentrude, and G. M. Johnson, *Tetrahedron Lett.*, 969 (1969).

412. H. Tomeoka, Y. Izawa, and Y. Ogata, *Tetrahedron*, **25**, 1501 (1969).
413. S. P. Singh and J. Kagan, *Chem. Commun.*, 1121 (1969).
414. L. L. Barber, O. L. Chapman, and J. D. Lassila, *J. Amer. Chem. Soc.*, **91**, 3664 (1969).
414a. N. J. Turro and D. M. McDaniel, 158th National Meeting of the American Chemical Society, September, 1969, New York, PHYS-151.
415. N. J. Turro and T. Cole, *Tetrahedron Lett.*, 3451 (1969).
415a. G. B. Vermont, *Def. Publ., U.S. Patent Office,* 755, 456 (1969).
416. E. J. Volker and J. A. Moore, *J. Org. Chem.*, **34**, 3639 (1969).
416a. H. U. Hostettler, *Chem. Abstracts*, **71**, 123692a (1969).
417. J. G. Pacifici and C. Diebert, *J. Amer. Chem. Soc.*, **91**, 4595 (1969).
418. B. E. Kaplan and L. T. Turner, 158th National Meeting of the American Chemical Society, September, 1969, New York, ORGN-40.
419. P. J. Wagner and R. W. Spoerke, *J. Amer. Chem. Soc.*, **91**, 4437 (1969).
420. P. Yates and G. Hagen, *Tetrahedron Lett.*, 3623 (1969).
421. A. G. Fallis, *Diss. Abstr.*, **30**, 2348b (1969).
421a. D. M. Pond, D. S. Weiss, F. D. Lewis, and N. J. Turro, 21st Southeastern Regional Meeting of the American Chemical Society, November, 1969, Richmond, Va., no. 30.
422. P. S. Engel and H. Ziffer, *Tetrahedron Lett.*, 5181 (1969).
423. J. Ipaktschi, *Tetrahedron Lett.*, 2153 (1969).
424. G. Maier and U. Mende, *Tetrahedron Lett.*, 3155 (1969).
425. B. Fuchs and S. Yankelevich, *Israel J. Chem.*, **6**, 511 (1968).
426. B. Fuchs, *Israel J. Chem.*, **6**, 517 (1968).
427. C. M. Anderson, J. B. Bremner, H. H. Westberg, and R. N. Warrener, *Tetrahedron Lett.*, 1585 (1969).
428. D. Bellus, D. R. Kearns, and K. Schaffner, *Helv. Chim. Acta*, **52**, 971 (1969).
428a. R. Srinivasan, *J. Amer. Chem. Soc.*, **81**, 2601 (1959).
429. C. C. Badcock, M. J. Perona, G. O. Pritchard, and B. Richborn, *J. Amer. Chem. Soc.*, **91**, 543 (1969).
430. J. A. Barltrop and J. D. Coyle, *Chem. Commun.*, 1081 (1969).
431. R. Imhof, W. Graf, H. Wehrli, and K. Schaffner, *Chem. Commun.*, 852 (1969).
432. R. G. Carlson and E. L. Biersmith, *Chem. Commun.*, 1049 (1969).
433. A. Sonoda, I. Moritani, F. Miki, and T. Tsuji, *Tetrahedron Lett.*, 3187 (1969).
434. M. S. Carson, W. Cocker, S. M. Evans, and P. U. R. Shannon, *Chem. Commun.*, 726 (1969).
435. J. P. Pete and M. L. Villaume, *Tetrahedron Lett.*, 3753 (1969).
436. P. J. Nelson, D. Ostrem, J. D. Lassila, and O. L. Chapman, *J. Org. Chem.*, **34**, 811 (1969).
437. S. Kuwata and K. Schaffner, *Helv. Chim. Acta*, **52**, 173 (1969).
438. L. A. Paquette and G. V. Meehan, *J. Org. Chem.*, **34**, 450 (1969).
439. K. Kojima, K. Sakai, and K. Tanabe, *Tetrahedron Lett.*, 3399 (1969).
440. J. E. Baldwin and S. M. Krueger, *J. Amer. Chem. Soc.*, **91**, 2396 (1969).
441. J. Ipaktschi, *Tetrahedron Lett.*, 215 (1969).
442. H. Hart and R. K. Murray, *Tetrahedron Lett.*, 379 (1969).
442a. H. Hart, R. K. Murray, and G. D. Appleyard, *Tetrahedron Lett.*, 4785 (1969).
443. P. K. Grover and N. Anand, *Chem. Commun.*, 982 (1969).
444. D. R. Graber, M. W. Grimes, and A. Haug, *J. Chem. Phys.*, **50**, 1623 (1969).
445. J. H. Stocker and D. H. Kern, *Chem. Commun.*, 204 (1969).
445a. P. M. Collins and P. Gupta, *Chem. Commun.*, 90 (1969).
446. P. M. Collins and P. Gupta, *Chem. Commun.*, 1288 (1969).

447. W. C. Lumma, Jr., and G. A. Berchtold, *J. Org. Chem.*, **34**, 1566 (1969).
448. K. K. Maheshwari and G. A. Berchtold, *Chem. Commun.*, 13 (1969).
449. C. Ganter and J.-F. Moser, *Chimia*, **23**, 180 (1969).
450. C. Ganter and J.-F. Moser, *Helv. Chim. Acta*, **52**, 725 (1969).
451. C. Ganter and J.-F. Moser, *Helv. Chim. Acta*, **52**, 967 (1969).
452. A. Padwa, A. Battisti, and E. Shefter, *J. Amer. Chem. Soc.*, **91**, 4000 (1969).
453. L. A. Paquette, G. V. Meehan, and R. F. Eizember, *Tetrahedron Lett.*, 995 (1969).
454. T. Miwa, M. Kato, and T. Tamano, *Tetrahedron Lett.*, 1761 (1969).
455. R. G. Carlson and D. E. Henton, *Chem. Commun.*, 674 (1969).
456. O. Cox, *Diss. Abstr.*, **30**, 118b (1969).
456a. L. A. Paquette and R. F. Eizember, *J. Amer. Chem. Soc.*, **91**, 7108 (1969).
457. J. R. Scheffer and M. L. Lungle, 24th Northwest Regional Meeting of the American Chemical Society, June, 1969, Salt Lake City, Utah, no. 184.
458. J. R. Scheffer and M. L. Lungle, *Tetrahedron Lett.*, 845 (1969).
458a. J. W. Stankorb and K. Conrow, *Tetrahedron Lett.*, 2395 (1969).
458b. H. Nozaki, H. Yamamoto, and T. Mori, *Can. J. Chem.*, **47**, 1107 (1969).
459. D. A. Warwick and C. H. J. Wells, *Tetrahedron Lett.*, 4955 (1969).
459a. D. A. M. Watkins, *Phytochemistry*, **8**, 979 (1969).
459b. D. G. Crosby and C. S. Tang, *J. Agr. Food Chem.*, **17**, 1291 (1969).
460. W. D. K. Clark, A. D. Litt, and C. Steel, *J. Amer. Chem. Soc.*, **91**, 5413 (1969).
461. W. D. K. Clark, A. D. Litt, and C. Steel, *Chem. Commun.*, 1087 (1969).
462. C. A. Parker and T. A. Joyce, *Trans. Faraday Soc.*, **65**, 2823 (1969).
462a. M. Hofert, *Photochem. Photobiol.*, **9**, 427 (1969).
463. O. L. Chapman and G. Wampfler, *J. Amer. Chem. Soc.*, **91**, 5390 (1969).
464. B. D. Challand, *Can. J. Chem.*, **47**, 685 (1969).
465. J. H. Stocker and D. H. Kern, 21st Southeast Regional Meeting of the American Chemical Society, November, 1969, Richmond, Va., no. 32.
466. S. G. Cohen and B. Green, *J. Amer. Chem. Soc.*, **91**, 6824 (1969).
467. N. C. Yang and R. L. Dusenbery, *Mol. Photochem.*, **1**, 159 (1969).
468. D. C. Nicodem, F. R. Stermitz, V. P. Muralidharan, and C. M. O'Donnell, 24th Northwest Regional Meeting of the American Chemical Society, June, 1969, Salt Lake City, Utah, no. 186.
469. J. H. Stocker, R. M. Jenevein, and D. H. Kern, *J. Org. Chem.*, **34**, 2810 (1969).
470. A. Padwa, W. Eisenhardt, R. Gruber, and D. Pashayan, *J. Amer. Chem. Soc.*, **91**, 1851 (1969).
471. D. Bellus and K. Schaffner, *Helv. Chim. Acta*, **52**, 1010 (1969).
472. P. J. Wagner and I. Kochevar, 158th National Meeting of the American Chemical Society, September, 1969, New York, PHYS-149.
473. P. J. Wagner and H. N. Schott, *J. Amer. Chem. Soc.*, **91**, 5383 (1969).
474. P. J. Wagner and H. N. Schott, 158th National Meeting of the American Chemical Society, September, 1969, New York, PHYS-43.
475. P. J. Wagner and G. Copen, *Mol. Photochem.*, **1**, 173 (1969).
476. R. A. Caldwell and P. M. Fink, *Tetrahedron Lett.*, 2987 (1969).
477. P. J. Wagner and P. A. Kelso, *Tetrahedron Lett.*, 4151 (1969).
478. P. J. Wagner, A. E. Kemppainen, and P. A. Kelso, 158th National Meeting of the American Chemical Society, September, 1969, New York, PHYS-148.
479. P. J. Wagner and A. E. Kemppainen, *J. Amer. Chem. Soc.*, **91**, 3085 (1969).
480. T. Matsuura and V. Kitaura, *Tetrahedron*, **25**, 4487 (1969).
481. M. Pfau, E. W. Sarver, and N. D. Heindel, *C. R. Acad. Sci., Paris, Ser. C*, **268c**, 1167 (1969).

482. T. Matsuura and V. Kitaura, *Tetrahedron*, **25**, 4487 (1969).
483. A. Padwa, E. Alexander, and M. Niemcyzk, *J. Amer. Chem. Soc.*, **91**, 456 (1969).
484. A. Padwa and D. Eastman, *J. Amer. Chem. Soc.*, **91**, 462 (1969).
485. Y. Kanda, J. Stanislaus, and E. C. Lim, *J. Amer. Chem. Soc.*, **91**, 5085 (1969).
486. N. Filipescu and J. M. Menter, *J. Chem. Soc.*, **B**, 616 (1969).
487. D. O. Cowan and A. A. Baum, 157th National Meeting of the American Chemical Society, April, 1969, Minneapolis, Minn., ORGN-117.
488. J. R. Collier and J. Hill, *Chem. Commun.*, 640 (1969).
489. T. Laird and N. Williams, *Chem. Commun.*, 561 (1969).
491. A. G. Szabo, *Diss. Abstr.*, **30**, 2369b (1969).
492. R. H. Fisch, L. C. Chow, and M. C. Caserio, *Tetrahedron Lett.*, 1259 (1969).
493. D. R. Arnold, L. B. Gillis, and E. B. Whipple, *Chem. Commun.*, 918 (1969).
493a. S. K. Chakrabarti, *Mol. Phys.*, **16**, 467 (1969).
494. C. H. J. Wells, A. Horsfield, and J. Paxton, *Chem. Commun.*, 393 (1969).
495. S. Farid and K.-H. Scholz, *Chem. Commun.*, 572 (1969).
496. M. Ogata, H. Matsumoto, and H. Kano, *Kogyo Kagaku Zasshi*, **72**, 195 (1969).
497. H.-D. Becker, *J. Org. Chem.*, **34**, 2469 (1969).
498. R. S. Davidson, P. F. Lambeth, F. A. Younis, and R. Wilson, *J. Chem. Soc.*, C, 2203 (1969).
499. G. O. Schenck, G. Koltzenburg, and E. Roselius, *Z. Naturforsh.*, **24b**, 222 (1969).
500. R. S. Davidson, P. F. Lamberth, J. F. McKellar, P. H. Turner, and R. Wilson, *Chem. Commun.*, 732 (1969).
501. J. Guttenplan and S. G. Cohen, *Chem. Commun.*, 247 (1969).
502. G. L. Closs and L. E. Closs, *J. Amer. Chem. Soc.*, **91**, 4550 (1969).
503. P. J. Wagner, *Mol. Photochem.*, **1**, 71 (1969).
504. W. F. Smith and B. W. Rossiter, *Tetrahedron*, **25**, 2059 (1969).
505. W. F. Smith, *Tetrahedron*, **25**, 2071 (1069).
506. G. O. Schenck, G. Matthias, M. Pape, M. Cziesla, G. van Bunau, E. Roselius, and G. Koltzenburg, *Justus Liebigs Ann. Chem.*, **719**, 80 (1968).
507. N. Ogata, H. Matsumoto, and H. Kano, *Kogyo Kagaku Zasshi*, **72**, 195 (1969).
508. S. G. Cohen and J. I. Cohen, *Israel J. Chem.*, **6**, 757 (1968).
509. R. S. Davidson and P. F. Lambeth, *Chem. Commun.*, 1098 (1969).
510. N. Filipescu and F. L. Minn, *J. Chem. Soc.*, **B**, 84 (1969).
511. D. H. Nelson, G. D. Norton, and W. H. Stanley, 157th National Meeting of the American Chemical Society, April, 1969, Minneapolis, Minn., ORGN-146.
512. P. Traynard and J. P. Blanchi, *C. R. Acad. Sci., Paris, Ser. C*, **267**, 1381 (1968).
513. S. G. Cohen and N. Stein, *J. Amer. Chem. Soc.*, **91**, 3690 (1969).
514. S. G. Cohen, M. D. Saltzman, and J. B. Guttenplan, *Tetrahedron Lett.*, 4321 (1969).
515. C. David, W. Demarteau, and G. Geushens, *Polymer*, **10**, 21 (1969).
516. J. E. Baldwin and J. A. Kopecki, *J. Amer. Chem. Soc.*, **91**, 3106 (1969).
517. I. H. Leaver and G. Caird Ramsay, *Tetrahedron*, **25**, 5669 (1969).
518. E. J. O'Connell, *Chem. Commun.*, 571 (1969).
519. L. A. Singer, G. A. Davis, and V. P. Muralidharan, *J. Amer. Chem. Soc.*, **91**, 897 (1969).
520. A. G. Kalle, *Chem. Abstracts*, **71**, 101692y (1969).
521. T. Matsuura and Y. Kitaurd, *Tetrahedron*, **25**, 4501 (1969).
522. R. S. Davidson, P. F. Lambeth, F. A. Younis, and R. Wilson, *J. Chem. Soc.*, C, 2203 (1969).
523. B. N. Tripathi and C. L. Gare, *Indian J. Chem.*, **7**, 778 (1969).
524. G. R. Lappin and J. S. Zannucci, *Chem. Commun.*, 1113 (1969).

525. Y. B. Zimin, Z. B. Chelnokova, P. I. Levin, and Y. A. Gurvich, *Izv. Akad. Nauk. SSSR, Ser. Khim.*, 412 (1969).
526. R. Breslow and M. A. Winnik, *J. Amer. Chem. Soc.*, **91**, 3083 (1969).
527. H. J. T. Bos and J. Boleij, *Rec. Trav. Chim.*, **88**, 465 (1969).
528. T. Berkovici and E. Fischer, *Israel J. Chem.*, **7**, 127 (1969).
529. L. J. Dombrowski, C. L. Groncki, R. L. Strong, and H. H. Richtol, *J. Phys. Chem.*, **73**, 3481 (1969).
530. L. J. Dobrowski, *Diss. Abstr.*, **29**, 3651B (1969).
531. G. A. Davis, P. A. Carapelluci, K. Szoc, and J. D. Gresser, *J. Amer. Chem. Soc.*, **91**, 2264 (1969).
532. R. A. Caldwell, *Tetrahedron Lett.*, 2121 (1969).
533. J. B. Guttenplan and S. G. Cohen, *Tetrahedron Lett.*, 2125 (1969).
534. L. A. Singer, *Tetrahedron Lett.*, 923 (1969).
535. M. Zander, *Z. Naturforsch.*, **24a**, 1387 (1969).
536. N. P. Shusheriva and R. Ya. Levina, *Russ. Chem. Rev.*, **37**, 198 (1968).
537. B. Monties, *C. R. Acad. Sci., Paris, Ser. C*, **269**, 1069 (1969).
538. W. Barbieri, L. Bernardi, G. Bosisio, and A. Temperilli, *Tetrahedron*, **25**, 2401 (1969).
539. J. E. Gano, *Tetrahedron Lett.*, 2549 (1969).
540. G. Kaupp and H. Prinzbach, *Justus Liebigs Ann. Chem.*, **725**, 52 (1969).
541. R. Bengelmans, M.-T. LeGoff, and H. C. de Marcheville, *C. R. Acad. Sci., Paris, Ser. C*, **269**, 1309 (1969).
542. M. Fischer, *Chem. Ber.*, **102**, 342 (1969).
543. B. Danieli, P. Manitto, and G. Russo, *Chem. Ind.*, 329 (1969).
544. Y. Ogata, T. Itoh, and Y. Izawa, *Bull. Chem. Soc. Japan*, **42**, 794 (1969).
544a. A. A. Scala, 158th National Meeting of the American Chemical Society, September, 1969, New York, PHYS-98.
545. H. Shizuka, T. Morita, Y. Mori, and I. Tanaka, *Bull. Chem. Soc. Japan.* **42**, 1831 (1969).
545a. A. Haas and D. Y. Oh, *Chem. Ber.*, **102**, 77 (1969).
546. T. Mujamoto, Y. Shigemitsu, and Y. Odaira, *Chem. Commun.*, 1410 (1969).
547. Y. Shigemitsu, F. Nakai, and Y. Odaira, *Tetrahedron*, **25**, 3039 (1969).
548. K. Fukui and Y. Odaira, *Tetrahedron Lett.*, 5255 (1969).
549. C. Azuma and A. Sugimori, *Kogyo Kagaku Zasshi*, **72**, 239 (1969).
549a. T. Yamase, H. Kokado, and E. Inoue, *Kogyo Kagaku Zasshi*, **72**, 162 (1969).
550. D. A. Plank, *Tetrahedron Lett.*, 4365 (1969).
551. C. D. Pande, B. N. Tripathi, and B. Venkataramani, *Indian J. Chem.*, **6**, 542 (1968).
552. C. Pac, S. Tsutsumi, and H. Sakurai, *Kogyo Kagaku Zasshi*, **72**, 224 (1969).
553. D. P. Kelly, J. T. Pinhey, and R. D. G. Rigby, *Aust. J. Chem.*, **22**, 977 (1969).
554. H. J. Hageman, *Tetrahedron*, **25**, 6015 (1969).
555. T. DoMinh, O. P. Strausz, and H. E. Gunning, *J. Amer. Chem. Soc.*, **91**, 1261 (1969).
556. J. Libman, M. Sprecher, and Y. Mazur, *J. Amer. Chem. Soc.*, **91**, 2062 (1969).
557. J. D. Margerum and C. T. Petrusis, *J. Amer. Chem. Soc.*, **91**, 2467 (1969); see *J. Amer. Chem. Soc.*, **87**, 3772 (1965), *ibid.*, **88**, 4733 (1966); J. D. Margerum and C. T. Petrusis, 157th National Meeting of the American Chemical Society, April, 1969, Minneapolis, Minn., ORGN-91.
558. J. S. Swenton and A. J. Krubsack, *J. Amer. Chem. Soc.*, **91**, 786 (1969).
559. D. L. Garin and K. O. Henderson, 24th Northwest Regional Meeting of the American Chemical Society, June, 1969, Salt Lake City, Utah, No. 181.
559a. Y. L. Chen and J. E. Casida, *J. Agr. Food Chem.*, **17**, 208 (1969).
559b. D. L. Miller and T. Ukena, *J. Amer. Chem. Soc.*, **91**, 3050 (1969).

559c. T. R. Evans and P. A. Leermakers, *J. Amer. Chem. Soc.*, **91**, 5898 (1969).
559d. M. Kobayashi, H. Minato, and Y. Ogi, *Bull. Chem. Soc. Japan*, **42**, 2737 (1969).
560. A. T. Koritskii and A. V. Zubkov, *Dokl. Akad. Nauk SSSR*, **185**, 1312 (1969).
561. E. N. Cain, R. Vukov, and S. Masamune, *Chem. Commun.*, 98 (1969).
562. J. K. Kochi and P. J. Krusic, *J. Amer. Chem. Soc.*, **91**, 3940 (1969).
563. W. Adam and G. S. Aponte, 158th National Meeting of the American Chemical Society, September, 1969, New York, ORGN-42.
564. G. W. Perold and G. Ourisson, *Tetrahedron Lett.*, 3871 (1969).
565. R. Simonaitis, *Diss. Abstr.*, **30**, 2834B (1969).
566. O. L. Chapman and C. L. McIntosh, *J. Amer. Chem. Soc.*, **91**, 4309 (1969).
567. I. S. Krull and D. R. Arnold, *Tetrahedron Lett.*, 1247 (1969).
568. J. R. Throckmorton, *J. Org. Chem.*, **34**, 3438 (1969).
569. R. Simonaitis and J. N. Pitts, Jr., *J. Amer. Chem. Soc.*, **91**, 108 (1969).
570. R. S. Givens and W. F. Oettle, *Chem. Commun.*, 1164 (1969).
571. S. F. Nelsen and P. J. Hintz, *J. Amer. Chem. Soc.*, **91**, 6190 (1969).
572. M. A. Hems, *Tetrahedron Lett.*, 375 (1969).
573. G. Maier, U. Mende, and G. Fritschi, *Angew. Chem. Int. Ed. Engl.*, **8**, 912 (1969).
574. G. Maier and U. Mende, *Angew. Chem. Int. Ed. Engl.*, **8**, 132 (1969).
575. R. L. Smith, A. Manmade, and G. W. Griffin, *J. Heterocyclic Chem.*, **6**, 443 (1969).
576. H. Shizuka, *Bull. Chem. Soc., Japan*, **42**, 52 (1969).
577. H. Shizuka, *Bull. Chem. Soc., Japan*, **42**, 57 (1969).
577a. H. Shizuka and J. Tanaka, *Bull. Chem. Soc. Japan*, **42**, 909 (1969).
578. J. S. Bradshaw, R. Knudsen, and E. L. Loveridge, 24th Northwest Regional Meeting of the American Chemical Society, June, 1969, Salt Lake City, Utah, no. 188.
578a. O. Yonemitsu and S. Naruto, *Tetrahedron Lett.*, 2387 (1969).
578b. I. L. Karle, J. W. Gibson, and J. Karle, *Acta Crystallogr., Sect B.*, **25**, 2034 (1969).
579. S. R. Bosco, A. Cirillo, and R. B. Timmons, *J. Amer. Chem. Soc.*, **91**, 3140 (1969).
580. H. Shizuka, E. Okutsu, Y. Mori, and I. Tanaka, *Mol. Photochem.*, **1**, 135 (1969).
581. I. Ninomiya, T. Naito, and T. Mori, *Tetrahedron Lett.*, 2259 (1969).
582. J. C. Sheehan and M. M. Nafissi -V., *J. Amer. Chem. Soc.*, **91**, 1176 (1969).
583. E. R. Talaty, A. E. Dupuy, Jr., and T. H. Golson, *Chem. Commun.*, 49 (1969).
584. E. R. Talaty and A. E. Dupuy, Jr., *J. Medicinal Chem.*, **12**, 195 (1969).
585. I. Ninomiya, T. Naito, and T. Mori, *Tetrahedron Lett.*, 3643 (1969).
586. J. Lynch, and O. Meth-Cohn, *Tetrahedron Lett.*, 161 (1969).
587. R. W. Hoffman and K. R. Eicken, *Chem. Ber.*, **102**, 2987 (1969).
588. D. J. Anderson, T. L. Gilchrist, D. C. Horwell, and C. W. Rees, *Chem. Commun.*, 146 (1969).
589. T. D. Roberts, L. Ardemagni, and H. Schechter, *J. Amer. Chem. Soc.*, **91**, 6185 (1969).
590. T. D. Roberts, L. Ardemagni, L. Munchausen, and R. Staudenmayer, 25th Southwest Regional Meeting of the American Chemical Society, December, 1969, Oklahoma City, Okla., ORGN-23.
591. M. Fischer, *Tetrahedron Lett.*, 2281 (1969).
592. S. N. Ege, *J. Chem. Soc.*, C, 2624 (1969).
593. N. K. Frunze, *Rev. Roum. Chim.*, **14**, 1309 (1969).
594. H. H. G. Jelliner and J. F. Kryman, 5th Western Regional Meeting of the American Chemical Society, October, 1969, Disneyland, Calif., no. 172.
595. M. Fischer, *Chem. Ber.*, **102**, 3495 (1969).
595a. M. Fischer and F. Wagner, *Chem. Ber.*, **102**, 3486 (1969).
596. R. Huisgen, W. Schier, H. Mader, and E. Brunn, *Angew. Chem. Int. Ed. Engl.*, **8**, 604 (1969).

597. S. Oida and E. Ohki, *Chem. Pharm. Bull. Tokyo*, **17**, 2461 (1969).
598. T. Tominaga and S. Tsutsumi, *Tetrahedron Lett.*, 3175 (1969).
599. J. C. Nnadi and S. Y. Wang, *Tetrahedron Lett.*, 2211 (1969).
599a. D. G. Crosby and C. S. Tang, *J. Agr. Food Chem.*, **17**, 1041 (1969).
600. A. Ohno, Y. Ohnishi, and G. Tsuchihashi, *J. Amer. Chem. Soc.*, **91**, 5038 (1969).
600a. A. Ohno, Y. Ohnishi, and G. Tsuchihashi, *Tetrahedron Lett.*, 161 (1969).
600b. A. Ohno, Y. Ohnishi, and G. Tsuchihashi, *Tetrahedron Lett.*, 283 (1969).
601. H. G. Heine and W. Metzner, *Justus Liebigs Ann. Chem.*, **724**, 223 (1969).
602. T. Yonezawa, M. Matsumoto, Y. Matsumura, and H. Kato, *Bull. Chem. Soc. Japan*, **42**, 2323 (1969).
603. J. R. Grunwell, *Chem. Commun.*, 1437 (1969).
604. U. Schmidt and C. Osterroht, *Chem. Ber.*, **102**, 2140 (1969).
605. E. L. Loveridge and J. S. Bradshaw, 24th Northwest Regional Meeting of the American Chemical Society, June, 1969, Salt Lake City, Utah, no. 187.
605a. T. Yamase, H. Kokado, et al., *J. Chem. Soc. Japan*, **72**, 162 (1969).
606. J. Swenton, *Appl. Spectrosc.*, **23**, 648 (1969).
606a. J. Swenton, 5th Western Regional Meeting of the American Chemical Society, October, 1969, Disneyland, Calif., no. 104.
607. J. J. McCullough, H. Ohoronyk, and D. P. Santry, *Chem. Commun.*, 570 (1969).
608. J.-L. Olive, M. Mousseron-Canet, and J. Dornand, *Bull. Soc. Chim. France*, 3247 (1969).
609. D. E. McGreer and B. D. Page, *Can. J. Chem.*, **47**, 866 (1969).
610. M. Mousseron-Canet and J.-L. Olive, *Bull. Soc. Chim. France*, 3242 (1969).
611. J. Meinwald and J. W. Kobzina, *J. Amer. Chem. Soc.*, **91**, 5177 (1969).
612. J. Meinwald, 21st National Organic Symposium, Salt Lake City, Utah, June, 1969, p. 61.
613. N. Sugiyama, H. Kataoka, C. Kashima, and K. Yamada, *Bull. Chem. Soc. Japan*, **42**, 1098 (1969).
614. N. Sugiyama, H. Kataoka, C. Kashima, and K. Yamada, *Bull. Chem. Soc. Japan*, **42**, 1353 (1969).
615. L. E. Friedrich and G. B. Schuster, *J. Amer. Chem. Soc.*, **91**, 7204 (1969).
616. P. Sunder-Plassman, P. H. Nelson, P. H. Boyle, A. Cruz, J. Iriarte, P. Crabbe, J. A. Zderic, J. A. Edwards, and J. H. Fried, *J. Org. Chem.*, **34**, 3779 (1969).
617. C. Cottrell, R. C. Dougherty, G. Fraenkel, and E. Pecchold, *J. Amer. Chem. Soc., Soc.*, **91**, 7545 (1969).
618. N. Sugiyama, Y. Sato, M. Yoshioka, K. Yamada, and H. Kataoka, *Bull. Chem. Soc. Japan*, **42**, 1153 (1969).
619. G. C. Forward and D. A. Whiting, *J. Chem. Soc.*, C, 1868 (1969).
620. N. J. Turro, E. Lee-Ruff, D. R. Morton, and J. M. Conia, *Tetrahedron Lett.*, 2991 (1969).
621. M. B. Rubin, *Israel J. Chem.*, **7**, 49 (1969).
622. L. Jurd, *Tetrahedron*, **25**, 2367 (1969).
623. M. Itoh, M. Tokuda, K. Seguchi, K. Taniguchi, and A. Suzuki, *Kogyo Kagaku Zasshi*, **72**, 219 (1969).
624. M. J. Jorgenson, *J. Amer. Chem. Soc.*, **91**, 198 (1969).
625. M. J. Jorgenson, *J. Amer. Chem. Soc.*, **91**, 6432 (1969).
626. G. Buchi and S. H. Feairheller, *J. Org. Chem.*, **34**, 609 (1969).
627. N. C. Yang, L. C. Lin, A. Shani, and S. S. Yang, *J. Org. Chem.*, **34**, 1845 (1969).
628. H. Obara, H. Takahashi, and J. Onodera, *Kogyo Kagaku Zasshi*, **72**, 309 (1969).
629. H. Obara, H. Takahashi, and J. Onodera, *J. Chem. Soc. Japan*, **72**, 309 (1969).

630. H. Obara, H. Takahashi, and H. Hirano, *Bull. Chem. Soc. Japan*, **42**, 560 (1969).
631. M. Tsuda, *Bull. Chem. Soc. Japan*, **42**, 905 (1969).
632. M. B. Hocking, *Can. J. Chem.*, **47**, 4567 (1969).
633. K. Tokumaru, *Chem. Ind. (London)*, 297 (1969).
634. E. F. Ullman, E. Babad, and M. Sung, *J. Amer. Chem. Soc.*, **91**, 5792 (1969).
635. Y. Inukai, *Kogyo Kagaku Zasshi*, **72**, 300 (1969).
636. E. Cavalieri and S. Horoupian, *Can. J. Chem.*, **47**, 2781 (1969).
637. N. Matzat, H. Wamhoff, and F. Korte, *Chem. Ber.*, **102**, 3122 (1969).
638. C. Kashima, M. Yamamoto, Y. Sato, and N. Sugiyama, *Bull. Chem. Soc. Japan*, **42**, 3596 (1969).
639. H. Zimmer, D. C. Armbruster, S. P. Kharidia, and D. C. Laukin, *Tetrahedron Lett.*, 4053 (1969).
640. R. T. LaLonde and C. B. Davis, *Can. J. Chem.*, **47**, 3250 (1969).
641. P. H. Vandewyer, J. Hoefnagels, and G. Smets, *Tetrahedron*, **25**, 3251 (1969).
642. J. Ciabattoni and E. C. Nathan, *J. Amer. Chem. Soc.*, **91**, 4766 (1969).
643. S. M. Krueger, J. A. Kapecki, J. E. Baldwin, and J. C. Paul, *J. Chem. Soc.*, B, 796 (1969).
644. C. Trindle, *J. Amer. Chem. Soc.*, **91**, 4936 (1969).
645. T. C. Floyd, D. A. Plank, and W. H. Starnes, *Chem. Commun.*, 1237 (1969).
646. G. Mark, F. Mark, and O. E. Polansky, *Justus Liebigs Ann. Chem.*, **719**, 151 (1969).
647. A. J. Bellamy, *J. Chem. Soc.*, B, 449 (1969).
648. P. J. Wagner and D. J. Bucheck, *J. Amer. Chem. Soc.*, **91**, 5090 (1969).
649. P. de Mayo, A. A. Nicholson, and M. F. Tchir, *Can. J. Chem.*, **47**, 711 (1969).
650. R. O. Loutfy, P. de Mayo, and M. F. Tchir, *J. Amer. Chem. Soc.*, **91**, 3984 (1969).
651. R. L. Cargill, A. C. Miller, D. M. Pond, P. deMayo, M. F. Tchir, K. R. Neuberger, and J. Saltiel, *Mol. Photochem.*, **1**, 301 (1969).
652. T. Matsumoto, *Tetrahedron Lett.*, 4103 (1969).
653. S. Kogawa, S. Matsumoto, S. Nishida, S. Yu, *Tetrahedron Lett.*, 3913 (1969).
654. J. F. Bagli and T. Bogri, *Tetrahedron Lett.*, 1639 (1969).
655. J. R. Scheffer and B. A. Boire, *Tetrahedron Lett.*, 4005 (1969).
656. P. J. Wagner and D. J. Bucheck, *Can. J. Chem.*, **47**, 713 (1969).
657. D. J. Bucheck, *Diss. Abstr.*, **30**, 1583B (1969).
658. A. B. Sears and R. L. Cargill, 21st Southeastern Regional Meeting of the American Chemical Society, November, 1969, Richmond, Va., no. 28.
659. W. C. Agosta, A. B. Smith, A. S. Kende, R. G. Eilerman, and J. Benham, *Tetrahedron Lett.*, 4517 (1969).
660. W. M. Horspool, *Chem. Commun.*, 467 (1969).
661. H. Durr and P. Heitkamper, *Z. Naturforsch.*, B, 779 (1969).
662. H. Gusten, *Chem. Commun.*, 133 (1969).
662a. V. J. Reisch and A. Fitzek, *Tetrahedron Lett.*, 271 (1969).
663. J. J. McCullough and R. W. W. Rasmussen, *Chem. Commun.*, 387 (1969).
665. J. J. McCullough, J. M. Kelly, and R. W. W. Rasmussen, *J. Org. Chem.*, **34**, 2933 (1969).
666. B. D. Challard, H. Hikino, G. Kornis, G. Lange, and P. deMayo, *J. Org. Chem.*, **34**, 794 (1969).
667. J. Y. Vanderhoek, *J. Org. Chem.*, **34**, 4184 (1969).
668. W. C. Agosta, and W. W. Lowrance, *Tetrahedron Lett.*, 3053 (1969).
669. T. S. Cantrell, W. S. Haller, and J. C. Williams, *J. Org. Chem.*, **34**, 509 (1969).
670. T. H. Koch, *Diss. Abstr.*, **29**, 3263B (1969).
671. W. F. Erman and T. W. Gibson, *Tetrahedron*, **25**, 2493 (1969).

672. H. E. Zimmerman and N. Lewin, *J. Amer. Chem. Soc.*, **91**, 879 (1969).
673. H. E. Zimmerman and W. R. Elser, *J. Amer. Chem. Soc.*, **91**, 887 (1969).
674. R. C. Hahn, G. W. Jones, and D. W. Kurtz, 158th National Meeting of the American Chemical Society, September, 1969, New York, ORGN-41.
675. C. Fenselau, W. G. Dauben, G. W. Shaffer, and N. D. Vietmeyer, *J. Amer. Chem. Soc.*, **91**, 112 (1969).
676. A. Bienvenue and J.-E. Dubois, *Bull. Soc. Chim. France*, 391 (1969).
677. H. E. Smith, R. T. Gray, T. J. Schaffner, and P. G. Lenhert, *J. Org. Chem.*, **34**, 136 (1969).
678. J. Libman, M. Sprecher, and Y. Mazur, *Israel J. Chem.*, **6**, 833 (1968).
679. W. V. Curran, *Diss. Abstr.*, **30**, 119B (1969).
680. P. Yates and D. J. MacGregor, *Tetrahedron Lett.*, 453 (1969).
681. N. Sugiyama, Y. Sato, H. Kataoka, C. Kashima, and K. Yamada, *Bull. Chem. Soc. Japan*, **42**, 3005 (1969).
682. S. Terao, S. Tsushima, I. Agata, and T. Miki, *Kogyo Kagaku Zasshi*, **72**, 203 (1969).
683. S. Terao, S. Tsushima, I. Agata, and T. Miki, *J. Chem. Soc. Japan, Ind. Chem. Sect.*, **72**, 203 (1969).
684. N. Furutachi, Y. Nakadaira, and K. Nakanishi, *J. Amer. Chem. Soc.*, **91**, 1028 (1969).
685. J. Hayashi, N. Furutachi, Y. Nakadaira, and K. Nakanishi, *Tetrahedron Lett.*, 4589 (1969).
686. D. Bellus and K. Schaffner, *Chimia*, **23**, 182 (1969).
687. B. W. Finucane and J. B. Thomson, *Chem. Commun.*, 380 (1969).
688. M. Mousseron-Canet and J. P. Chabaud, *Bull. Chim. Soc. France*, 239 (1969).
689. A. Devaquet and L. Salem, *J. Amer. Chem. Soc.*, **91**, 3793 (1969).
690. M. Fisch and R. Nonnenmacher, 157th National Meeting of the American Chemical Society, April, 1969, Minneapolis, Minn., ORGN-88.
691. S. Domb, G. Bozzato, J. A. Saboz, and K. Schaffner, *Helv. Chim. Acta*, **52**, 2436 (1969).
692. D. F. Brizzolara, *Diss. Abstr.*, **30**, 1582B (1969).
693. G. Ramme, R. L. Strong, and H. H. Richtol, *J. Amer. Chem. Soc.*, **91**, 5711 (1969).
694. C. H. Heathcock, R. H. Starkey, and R. A. Badger, 158th National Meeting of American Chemical Society, September, 1969, New York, ORGN-45.
695. K. Wiesner, L. Poon, I. Jirkovski, and M. Fishman, *Can. J. Chem.*, **47**, 433 (1969).
696. A. S. Kende, Z. Goldschmidt, and P. T. Izzo, 158th National Meeting of the American Chemical Society, September, 1969, New York, ORGN-37.
697. A. S. Kende, Z. Goldschmidt, and P. T. Izzo, *J. Amer. Chem. Soc.*, **91**, 6858 (1969).
698. O. L. Chapman, M. Kane, J. D. Lassila, R. L. Loeschen, and H. E. Wright, 158th National Meeting of the American Chemical Society, September, 1969, New York, ORGN-38.
699. O. L. Chapman, M. Kane, J. D. Lassila, R. L. Loeschen, and H. E. Wright, *J. Amer. Chem. Soc.*, **91**, 6856 (1969).
700. H. E. Zimmerman, *Angew. Chem. Int. Ed. Engl.*, **8**, 1 (1969).
701. T. R. Rodgers and H. Hart, *Tetrahedron Lett.*, 4845 (1969).
702. J. A. Waters and B. Witkop, *J. Org. Chem.*, **34**, 1601 (1969).
703. M. Nemoto, K. Kokubun, and M. Koizumi, *Bull. Chem. Soc. Japan*, **42**, 1223 (1969).
704. M. Nemoto, H. Kokubun, and M. Koizumi, *Chem. Commun.*, 1095 (1969).
705. D. I. Schuster and V. Y. Abraitys, *Chem. Commun.*, 419 (1969).
706. H. E. Zimmerman, D. S. Crumrine, D. Dopp, and P. S. Huyffer, *J. Amer. Chem. Soc.*, **91**, 434 (1969).
707. D. A. Plank, J. C. Floyd, and W. H. Starnes, *Chem. Commun.*, 1003 (1969).

708. W. H. Pirkle, S. G. Smith, and G. K. Koser, *J. Amer. Chem. Soc.*, **91**, 1580 (1969).
709. T. Matsuura and K. Ogura, *Bull. Chem. Soc. Japan*, **42**, 2970 (1969).
710. D. I. Schuster and I. S. Krull, *Mol. Photochem.*, **1**, 107 (1969).
711. G. F. Koser, *Diss. Abstr.*, **30**, 127B (1969).
712. H.-D. Becker, *J. Org. Chem.*, **34**, 2472 (1969).
713. K. Schaffner-Sabba, *Helv. Chim. Acta*, **52**, 1237 (1969).
714. M. Fisch, *Chem. Commun.*, 1472 (1969).
715. G. Habermehl and A. Haaf, *Justus Liebigs Ann. Chem.*, **723**, 181 (1969).
716. H. E. Zimmerman and G. Jones, *J. Amer. Chem. Soc.*, **91**, 5678 (1969).
717. K. Schaffner, *Helv. Chim. Acta*, **52**, 1237 (1969).
718. E. H. White, S. Eguchi, and J. N. Marx, *Tetrahedron*, **25**, 2099 (1969).
719. K. F. Cheng, *Chem. Commun.*, 562 (1969).
719a. D. I. Schuster and N. K. Lau, *Mol. Photochem.*, **1**, 415 (1969).
720. W. C. Barringer, *Diss. Abstr.*, **30**, 114B (1969).
721. P. Yates, E. S. Hand, P. Singh, S. K. Roy, and I. W. J. Still, *J. Org. Chem.*, **34**, 4046 (1969).
722. P. Yates and P. Singh, *J. Org. Chem.*, **34**, 4052 (1969).
723. P. Yates, M. J. Jorgenson, and P. Singh, *J. Amer. Chem. Soc.*, **91**, 4739 (1969).
724. J. W. Hanifin and E. Cohen, *J. Amer. Chem. Soc.*, **91**, 4494 (1969).
725. P. Vogel, B. Willhalm, and H. Prinzbach, *Helv. Chim. Acta*, **52**, 584 (1969).
726. J. Carnduff, D. G. Leppard, and J. N. Low, *Chem. Commun.*, 1218 (1969).
727. M. R. Morris and A. J. Waring, *Chem. Commun.*, 526 (1969).
728. R. J. Bastiani, D. J. Hart, and H. Hart, *Tetrahedron Lett.*, 4841 (1969).
729. Y. Kamano and M. Komatsu, *Chem. Pharm. Bull. Japan*, **17**, 1698 (1969).
730. Y. Kamano, Y. Tanaka, and M. Komatsu, *Chem. Pharm. Bull. Japan*, **17**, 1706 (1969).
731. W. H. Pirkle and L. H. McKendry, *J. Amer. Chem. Soc.*, **91**, 1179 (1969).
732. C. T. Bedford and T. Money, *Chem. Commun.*, 685 (1969).
733. K. R. Huffman, M. Burger, W. A. Henderson, Jr., M. Loy, and E. F. Ullman, *J. Org. Chem.*, **34**, 2407 (1969).
734. J. J. Looker, *Org. Chem. Bull.*, **40**, 1 (1968); *Chem. Abstr.*, **70**, 37302 (1969).
735. L. Canonica, B. Danieli, P. Manitto, and G. Russo, *Tetrahedron Lett.*, 607 (1969).
736. S. Seto, H. Sugiyama, S. Takenada, and H. Watanabe, *J. Chem. Soc.*, C, 1625 (1969).
737. L. L. Barber, O. L. Chapman, and J. D. Lassila, *J. Amer. Chem. Soc.*, **91**, 531 (1969).
738. L. Canonica, B. Danieli, P. Manitto, and G. Russo, *Tetrahedron Lett.*, 607 (1969).
739. D. H. Sussman, *Diss. Abstr.*, **29**, 142B (1969).
740. J. A. Barltrop and D. Giles, *J. Chem. Soc.*, C, 105 (1969).
741. G. Queguiner and A. Godard, *C. R. Acad. Sci., Paris, Ser. C*, **269**, 1646 (1969).
742. F. A. L. Anet and D. P. Mullis, *Tetrahedron Lett.*, 737 (1969).
743. W. Kothe, *Tetrahedron Lett.*, 5201 (1969).
744. W. Eberbach and H. Prinzbach, *Chem. Ber.*, **102**, 4164 (1969).
745. H. Prinzbach and P. Vogel, *Helv. Chim. Acta*, **52**, 396 (1969).
746. G. Kaupp and H. Prinzbach, *Helv. Chim. Acta*, **52**, 956 (1969).
747. R. C. Bansal, A. W. McCulloch, and A. G. McInnes, *Can. J. Chem.*, **47**, 2391 (1969).
748. H. Prinzbach and M. Klaus, *Angew. Chem. Int. Ed. Eng.*, **8**, 276 (1969).
749. H. Prinzbach and W. Auge, *Angew. Chem. Int. Ed. Eng.*, **8**, 209 (1969).
750. H. Prinzbach, M. Klaus, and W. Mayer, *Angew. Chem. Int. Ed. Engl.*, **8**, 833 (1969).
751. S. F. Nelsen and J. P. Gillespie, *Tetrahedron Lett.*, 3259 (1969).
752. E. D. Miller and E. Hedaya, *J. Amer. Chem. Soc.*, **91**, 5401 (1969).

753. S. F. Nelsen and J. P. Gillespie, *Tetrahedron Lett.*, 5059 (1969).
754. R. M. Acheson and J. K. Stubbs, *J. Chem. Soc., C*, 2316 (1969).
755. A. O. Plunkett, *Chem. Commun.*, 1044 (1969).
756. S. F. Nelsen, *J. Org. Chem.*, **34**, 2248 (1969).
757. P. C. Ford, D. H. Stuermer, and D. P. McDonald, *J. Amer. Chem. Soc.*, **91**, 6209 (1969).
758. W. Hartmann, *Chem. Ber.*, **102**, 3974 (1969).
759. D. Bryce-Smith and M. A. Hems, *Tetrahedron*, **25**, 247 (1969).
760. F. C. DeSchryver, I. Bhardwaj, and J. Put, *Angew. Chem. Int. Ed. Engl.*, **8**, 213 (1969).
761. J. M. Fayadh and G. A. Swan, *J. Chem. Soc., C*, 1781 (1969).
762. M. Verbeek, H. D. Scharf, and F. Korte, *Chem. Ber.*, **102**, 2471 (1969).
763. J. E. Kuder and J. P. Ferris, 158th National Meeting of the American Chemical Society, September, 1969, New York, ORGN-12.
764. J. Beutel, R. J. Ruszkay, and J. F. Brennan, *J. Phys. Chem.*, **73**, 3240, 3245 (1969).
765. S. P. Pappas, B. C. Pappas, and N. A. Portnoy, *J. Org. Chem.*, **34**, 520 (1969).
766. S. P. Pappas and N. A. Portnoy, *Chem. Commun.*, 597 (1969).
767. J. M. Bruce, D. Creed, and K. Dawes, *Chem. Commun.*, 594 (1969).
767a. E. P. Fokin and A. M. Detsina, *Izv. Sib. Akad. Nauk SSSR*, 95 (1969); *Index Chem.*, **36**, 133573 (1969).
768. H. J. Hageman and W. G. B. Huysmans, *Chem. Commun.*, 837 (1969).
769. H. Hermann and G. O. Schenck, *Photochem. Photobiol.*, **8**, 255 (1968).
770. A. Hydson and J. W. Lewis, *J. Chem. Soc.*, **B**, 531 (1969).
771. D. R. Kemp and G. Porter, *Chem. Commun.*, 1029 (1969).
772. H. J. Piek, *Tetrahedron Lett.*, 1169 (1969).
773. J. E. Baldwin and J. E. Brown, *Chem. Commun.*, 167 (1969).
774. E. P. Fokin and A. M. Detsina, *Izv. Sib. Otd. Akad. Nauk SSSR, Ser. Khim. Nauk*, 95 (1969); *Index Chem.*, **36**, 133573 (1969).
775. G. O. Phillips, N. W. Worthington, J. F. McKellar, and R. R. Sharpe, *J. Chem. Soc.*, **A**, 767 (1969).
777. S. Seo, U. Sankawa, Y. Ogihara, and S. Shibata, *Tetrahedron Lett.*, 767 (1969).
778. M. B. Rubin, *Fortsch. Chem. Forsch.*, **13**, 251 (1969).
779. K. Sandros, *Acta Chem. Scand.*, **23**, 2815 (1969).
780. A. Singh, A. R. Scott, and F. Sopchyshyn, *J. Phys. Chem.*, **73**, 2633 (1969).
780a. L. Tsai and E. Charney, *J. Phys. Chem.*, **73**, 2463 (1969).
781. R. B. Cundall, G. B. Evans, and E. J. Land, *J. Phys. Chem.*, **73**, 3982 (1969).
782. N. J. Turro and R. Engel, *Mol. Photochem.*, **1**, 143 (1969).
783. N. J. Turro and R. Engel, *Mol. Photochem.*, **1**, 235 (1969).
784. N. J. Turro and R. Engel, *J. Amer. Chem. Soc.*, **91**, 7113 (1969).
785. A. Cicolella, X. Deglise, M. Bouchy, J.-C. Andre, J. Lemaire, and M. Niclause, *C. R. Acad. Sci., Paris, Ser. C*, **268**, 1929 (1969).
786. M. Grossman, G. P. Semeluk, and I. Unger, *Can. J. Chem.*, **47**, 3079 (1969).
787. J. C. Mackie, *Aust. J. Chem.*, **22**, 255 (1969).
788. K. R. Darnall, *Diss. Abstr.*, **30**, 555B (1969).
789. N. J. Turro and T. J. Lee, *J. Amer. Chem. Soc.*, **91**, 5651 (1969).
790. R. Bishop and N. K. Hamer, *Chem. Commun.*, 804 (1969).
791. R. Bishop and N. K. Hamer, *Chem. Commun.*, 804 (1969).
792. B. Akermark, N. G. Johansson, and B. Sjoberg, *Tetrahedron Lett.*, 371 (1969).
793. J. E. Alexander and S. P. Pappas, 157th National Meeting of the American Chemical Society, April, 1969, Minneapolis, Minn., ORGN-87.
794. D. L. Bunbury and T. T. Chuang, *Can. J. Chem.*, **47**, 2045 (1969).

795. G. E. Gream, J. C. Paice, and C. C. R. Ramsay, *Aust. J. Chem.*, **22**, 1229 (1969).
796. J. Saltiel and H. C. Curtis, *Mol. Photochem.*, **1**, 239 (1969).
797. P. A. Carapellucci, H. P. Wolf, and K. Weiss, *J. Amer. Chem. Soc.*, **91**, 4635 (1969).
798. K. R. Eichen, *Justus Liebigs Ann. Chem.*, **724**, 66 (1969).
799. W. C. Herndon and W. B. Giles, *Chem. Commun.*, 497 (1969).
800. S. Farid and D. Hess, *Chem. Ber.*, **102**, 3747 (1969).
801. W. Friedricksen, *Tetrahedron Lett.*, 1219 (1969).
801a. K. Kyoto, K. Ono and, J. Osugi, *Bull. Chem. Soc. Japan*, **42**, 3357 (1969).
802. A. deGroot, D. Oudman, and H. Wynberg, *Tetrahedron Lett.*, 1529 (1969).
803. G. E. Gream, and J. C. Paice, *Aust. J. Chem.*, **22**, 1249 (1969).
804. B. M. Monroe and S. A. Weiner, *J. Amer. Chem. Soc.*, **91**, 450 (1969).
805. M. B. Rubin, *Tetrahedron Lett.*, 3931 (1969).
806. J. Strating, B. Zwanenburg, A. Wagenaar, and A. C. Udding, *Tetrahedron Lett.*, 125 (1969).
806a. B. M. Trost, *J. Amer. Chem. Soc.*, **91**, 918 (1969).
807. Y. Nakadaira, Y. Hirota, and K. Nakanishi, *Chem. Commun.*, 1469 (1969).
808. E. Fahr, *Ann. Chim.*, **8**, 578 (1969).
809. T. Goto, *Pure Appl. Chem.*, **17**, 421 (1968).
810. A. A. Lamola, *Photochem. Photobiol.*, **9**, 291 (1969).
811. D. Herbert, *Photochem. Photobiol.*, **9**, 33 (1969).
812. J. Eisinger and A. A. Lamola, *Mol. Photochem.*, **1**, 209 (1969).
813. N. Camerman and S. C. Nyburg, *Theoret. Chim. Acta*, **13**, 162 (1969).
814. R. Ben-Ishai, E. Ben-Hur, and Y. Hornfeld, *Israel J. Chem.*, **6**, 769 (1968).
815. T. Lisewski and K. L. Wierzchowski, *Chem. Commun.*, 348 (1969).
815a. N. Camerman, D. Weinblum, and S. C. Nyburg, *J. Amer. Chem. Soc.*, **91**, 982 (1969).
816. M. Charlier, C. Helene, and M. Dourlent, *J. Chim. Phys.*, **66**, 700 (1969).
817. F. Dall'Acqua, S. Marciani, and G. Rodighiero, *Z. Naturforsch.*, **24b**, 307 (1969).
818. F. Bordin, L. Musajo, and R. Bevilacqua, *Z. Naturforsch.*, **24b**, 691 (1969).
819. R. O. Rahn, 5th Western Regional Meeting of the American Chemical Society, October, 1969, Disneyland, Calif., no. 170.
820. K. C. Smith, 5th Western Regional Meeting of the American Chemical Society, October, 1969, Disneyland, Calif., no. 245.
821. E. Hayon, *J. Amer. Chem. Soc.*, **91**, 5397 (1969).
822. Y. Kondo and B. Witkop, *J. Amer. Chem. Soc.*, **91**, 5264 (1969).
823. J. C. Nnadi and S. Y. Wang, *Tetrahedron Lett.*, 2211 (1969).
824. L. Kittler and G. Lober, *Photochem. Photobiol.*, **10**, 35 (1969).
825. E. Sato and Y. Kanaoka, *Tetrahedron Lett.*, 3547 (1969).
826. M. G. Pleiss, H. Ochiai, P. A. Cerutti, *Biochem. Biophys. Res. Commun.*, **34**, 70 (1969).
827. D. Elad, I. Rosenthal, and H. Steinmaus, *Chem. Commun.*, 305 (1969).
828. D. Elad and I. Rosenthal, *Chem. Commun.*, 905 (1967).
829. K. C. Smith, *Biochem. Biophys. Res. Commun.*, **34**, 354 (1969).
830. Y. LeRoux, J.-C. Ginisty, and C. Nofre, *C. R. Acad. Sci., Paris, Ser. C.*, **269C**, 744 (1969).
831. K. K. Gauri, K. H. W. Pflughaupt, and R. Muller, *Z. Naturforsch.*, **24b**, 833 (1969).
831a. N. Toshima, H. Hirai, and S. Makishima, *Kogyo Kagaku Zasshi*, **72**, 184 (1969).
832. A. Padwa, W. Bergmark, and D. Pashayan, *J. Amer. Chem. Soc.*, **91**, 2653 (1969).
833. G. Balogh and F. C. deSchryver, *Tetrahedron Lett.*, 1371 (1969).
834. B. Fraser-Reid, A. McLean, and E. W. Usherwood, *Can. J. Chem.*, **47**, 4511 (1969).

835. L. A. Paquette, J. R. Malpass, G. R. Krow, and T. J. Barton, *J. Amer. Chem. Soc.*, **91**, 5296 (1969).
836. R. Potashnik and M. Ottolenghi, *J. Chem. Phys.*, **51**, 3671 (1969).
837. V. I. Stenberg, E. F. Travecedo, and W. E. Musa, *Tetrahedron Lett.*, 2031 (1969).
838. R. A. Abramovitch and E. P. Kyba, *Chem. Commun.*, 265 (1969).
839. R. W. Binkley, *Tetrahedron Lett.*, 1893 (1969).
840. R. W. Binkley and A. S. Kushner, 158th National Meeting of the American Chemical Society, September, 1969, New York, ORGN-15.
841. G. Condorelli, L. L. Castanzo, *Boll. Sedute Accad. Gioenia*, **8**, 775 (1966); *Chem. Abstracts*, **70**, 10815d (1969).
842. Z. Raciszewski and J. F. Stephen, *J. Amer. Chem. Soc.*, **91**, 4338 (1969).
843. J. P. Dubois and H. Labhard, *Chimia*, **23**, 109 (1969).
844. T. Sasaki and M. Takahashi, *Yuki Gosei Kagaku Kyokai Shi*, **26**, 899 (1968).
845. B. L. Fox and H. M. Rosenberg, *Chem. Commun.*, 1115 (1969).
846. H. Izawa, P. deMayo, and T. Tabata, *Can. J. Chem.*, **47**, 51 (1969).
847. R. P. Gandhi and V. K. Chadha, *Indian J. Chem.*, **7**, 633 (1969).
848. J. P. Vermes and R. Beugelmans, *Tetrahedron Lett.*, 2091 (1969).
849. T. Oine and T. Mukai, *Tetrahedron Lett.*, 157 (1969).
851. T. Okada, M. Kawanisi, H. Nozaki, N. Toshima, and H. Hirai, *Tetrahedron Lett.*, 927 (1969).
852. A. T. Blomquist and C. F. Heins, *J. Org. Chem.*, **34**, 2906 (1969).
853. P. K. Freeman and R. C. Johnson, *J. Org. Chem.*, **34**, 1751 (1969).
854. A. T. Blomquist and C. F. Heins, *J. Org. Chem.*, **34**, 2906 (1969).
855. N. Obata, A. Hamada, and T. Takizawa, *Tetrahedron Lett.*, 3917 (1969).
856. W. Bauer and K. Hafner, *Angew. Chem. Int. Ed. Engl.*, **8**, 772 (1969).
857. T. Mukai, T. Oine, and A. Matsubara, *Bull. Chem. Soc. Japan*, **42**, 581 (1969).
858. M. Kojima and M. Maeda, *Tetrahedron Lett.*, 2379 (1969).
859. P. Beak and W. Messer, *Tetrahedron*, **25**, 3287 (1969).
860. W. R. Messer, *Diss. Abstr.*, **30**, 132B (1969).
861. H. B. Land and A. R. Frasca, *Chem. Ind.*, 1594 (1969).
862. M. Ogata, H. Matsumoto, and H. Kano, *Tetrahedron*, **25**, 5205 (1969).
863. J. L. Cooper, and H. H. Wassermann, *Chem. Commun.*, 200 (1969).
864. F. A. Neugebauer, *Chem. Ber.*, **102**, 1339 (1969).
865. B. Singh, *J. Amer. Chem. Soc.*, **91**, 3670 (1969).
866. R. M. Moriarty and R. Mukherjee, *Tetrahedron Lett.*, 4627 (1969).
867. P. K. Freeman and R. C. Johnson, *J. Org. Chem.*, **34**, 1746 (1969).
868. R. W. Binkley, *J. Org. Chem.*, **34**, 3218 (1969).
869. R. W. Binkley, *J. Org. Chem.*, **34**, 931 (1969).
870. R. W. Binkley, *J. Org. Chem.*, **34**, 2072 (1969).
871. P. Scheiner, *J. Org. Chem.*, **34**, 199 (1969).
872. W. M. Moore and C. Baylor, *J. Amer. Chem. Soc.*, **91**, 7170 (1969).
873. M. Ogata, H. Matsumoto, and H. Kano, *Tetrahedron*, **25**, 5217 (1969).
874. R. A. Odum and B. Schmall, *Chem. Commun.*, 1299 (1969).
875. R. S. Davidson, *Tetrahedron*, **25**, 3383 (1969).
876. G. Fraenkel and E. Pecchold, *Tetrahedron Lett.*, 4821 (1969).
877. G. L. Closs and A. D. Trifunac, *J. Amer. Chem. Soc.*, **91**, 4554 (1969).
878. K. R. Kopecky and T. Gillan, *Can. J. Chem.*, **47**, 2371 (1969).
879. S. A. Weiner and G. S. Hammond, *J. Amer. Chem. Soc.*, **91**, 986 (1969).
880. E. M. Kosower, P. C. Huang, and T. Tsuji, *J. Amer. Chem. Soc.*, **91**, 2325 (1969).
881. T. Mill, and R. S. Stringham, *Tetrahedron Lett.*, 1853 (1969).

882. J. F. Janssen, *J. Chem. Ed.*, **46**, 117 (1969).
883. I. I. Abram, G. S. Milne, B. S. Solomon, and C. Steel, *J. Amer. Chem. Soc.*, **91**, 1220 (1969).
884. H. Prinzbach and H. D. Martin, *Chimia*, **23**, 37 (1969).
885. E. L. Allred, J. C. Hinshaw, and A. L. Johnson, *J. Amer. Chem. Soc.*, **91**, 3382 (1969).
886. M. Franck-Neuman, *Angew. Chem. Int. Ed. Engl.*, **8**, 210 (1969).
887. K. B. Wiberg and A. deMeijere, *Tetrahedron Lett.*, 59 (1969).
888. K. Mackenzie, W. P. Lay, J. R. Telford, and D. L. Williams-Smith, *Chem. Commun.*, 761 (1969).
889. E. L. Allred and R. L. Smith, *J. Amer. Chem. Soc.*, **91**, 6766 (1969).
890. P. S. Engel, *J. Amer. Chem. Soc.*, **91**, 6903 (1969).
891. D. R. Arnold, A. B. Evnin, and P. H. Kasai, *J. Amer. Chem. Soc.*, **91**, 784 (1969).
892. A. C. Day and T. R. Wright, *Tetrahedron Lett.*, 1067 (1969).
893. T. Sanjiki, M. Ohta, and H. Kato, *Chem. Commun.*, 638 (1969).
893a. F. H. Dorer, *Appl. Spectrosc.*, **23**, 651 (1969).
894. M. Franck-Neuman and C. Buchecker, *Tetrahedron Lett.*, 2659 (1969).
895. K. R. Kopecky and S. Evani, *Can, J. Chem.*, **47**, 4041 (1969).
896. H. Tanida, S. Teratake, Y. Hata, and M. Watanabe, *Tetrahedron Lett.*, 5341 (1969).
897. H. Tanida, S. Teratake, Y. Hata, and M. Watanabe, *Tetrahedron Lett.*, 5345 (1969).
898. W. R. Roth and K. Enderer, *Justus Liebigs Ann. Chem.*, **730**, 82 (1969).
899. G. Snatzke and H. Langer, *Chem. Ber.*, **102**, 1865 (1969).
900. A. C. Day and R. N. Inwood, *J. Chem. Soc., C*, 1065 (1969).
901. S. D. Andrews, A. C. Day, and R. N. Inwood, *J. Chem. Soc., C*, 2443 (1969).
902. M. Franck-Neuman and C. Buchecker, *Tetrahedron Lett.*, 15 (1969).
902a. I. Moritani, T. Hosokawa, and N. Obata, *J. Org. Chem.*, **34**, 670 (1969).
903. C. L. Closs and L. R. Kaplan, *J. Amer. Chem. Soc.*, **91**, 2168 (1969).
904. G. Irick and J. G. Pacifici, *Tetrahedron Lett.*, 1303 (1969).
905. J. G. Pacifici and G. Irick, Jr., *Tetrahedron Lett.*, 2207 (1969).
906. J. G. Pacifici, G. Irick, Jr., C. G. Anderson, *J. Amer. Chem. Soc.*, **91**, 5654 (1969).
907. T. Kameo, T. Hirashima, O. Manabe, and H. Hiyama, *Kogyo Kagaku Zasshi*, **72**, 1327 (1969).
908. G. L. Closs and L. E. Closs, *J. Amer. Chem. Soc.*, **91**, 4549 (1969).
909. G. L'Abbé, *Chem. Rev.*, **69**, 345 (1969).
910. N. Koga, G. Koga, and J.-P. Anselme, *Can. J. Chem.*, **47**, 1143 (1969).
911. R. L. Whistler and A. K. M. Anisuzzaman, *J. Org. Chem.*, **34**, 3823 (1969).
912. J. H. Boyer and R. Selvarajan, *J. Amer. Chem. Soc.*, **91**, 6122 (1969).
913. K. Isomura, M. Okada, and H. Taniguchi, *Tetrahedron Lett.*, 4073 (1969).
914. A. M. Farid, J. McKenna, J. M. McKenna, and E. N. Wall, *Chem. Commun.*, 1222 (1969).
915. F. D. Lewis and J. C. Dalton, *J. Amer. Chem. Soc.*, **91**, 5260 (1969).
916. R. A. Odum, A. M. Aaronson, *J. Amer. Chem. Soc.*, **91**, 5680 (1969).
917. T. Tsunoda, T. Yamaoka, and K. Ikari, *Kogyo Kagaku Zasshi*, **72**, 156 (1969).
918. J. S. Swenton, T. J. Ikeler, and B. H. Williams, *Chem. Commun.*, 1263 (1969).
919. J. M. Simson and W. Lwowski, *J. Amer. Chem. Soc.*, **91**, 5107 (1969).
920. S. Yamada and S. Terashima, *Chem. Commun.*, 511 (1969).
921. S. Terashima and S. Yamada, *Chem. Pharm. Bull. Tokyo*, **16**, 1953 (1968).
922. K. Kawashima and I. Agata, *J. Pharm. Soc. Japan*, **89**, 1426 (1969).
923. I. Brown, O. E. Edwards, J. M. McIntosh, and D. Vocelle, *Can. J. Chem.* **47** 2751 (1969).

924. W. V. Curran and R. A. Angier, *J. Org. Chem.*, **34**, 3668 (1969).
925. K. Kawashima and I. Agata, *Yakugaku Zasshi*, **89**, 1426 (1969).
926. S. Munekata and S. Kikuchi, *Seisan-Kenkyn*, **20**, 519 (1968); *Chem. Abstracts*; **70**, 67362h (1969).
927. J. Hubert, *J. Chem. Soc.*, *C*, 1334 (1969).
928. A. J. Hubert, *Chem. Commun.*, 328 (1969).
929. M. G. Barlow, R. N. Haszeldine, and W. D. Morton, *Chem. Commun.*, 931 (1969).
930. P. Scheiner, U.S. Patent 3,428,538 (1969).
931. R. G. Kostyanovsky et al., *Tetrahedron Lett.*, 4021 (1969).
932. S. McLean and D. M. Findlay, *Tetrahedron Lett.*, 2219 (1969).
933. R. L. Hale and L. H. Zalkow, *Tetrahedron*, **25**, 1393 (1969).
934. J. H. Boyer and R. Selvarajan, *J. Heterocyclic Chem.*, **6**, 503 (1969).
934a. J. H. Boyer and R. Selvarajan, *Tetrahedron Lett.*, 47 (1969).
935. P. Flowerday, M. J. Perkins, *J. Amer. Chem. Soc.*, **91**, 1035 (1969).
936. R. H. Spector and M. M. Joullie, *J. Heterocyclic Chem.*, **6**, 605 (1969).
937. M. Marky, T. Doppler, H.-J. Hansen, and H. Schmid, *Chimia*, **23**, 230 (1969).
938. F. L. Bach, J. Karliner, and G. E. VanLear, *Chem. Commun.*, 1110 (1969).
939. P. Sheiner, *Tetrahedron Lett.*, 4863 (1969).
939a. R. R. Fraser, Gurudata, and K. E. Haque, *J. Org. Chem.*, **34**, 4118 (1969).
940. M. S. Ao, E. M. Burgess, A. Schauer, and E. A. Taylor, *Chem. Commun.*, 220 (1969).
941. J. Adamson, D. L. Forster, T. L. Gilchrist, and C. W. Rees, *Chem. Commun.*, 221 (1969).
942. W. Kirmse and M. Buschhoff, *Chem. Ber.*, **102**, 1087 (1969).
943. W. Kirmse and M. Buschhoff, *Chem. Ber.*, **102**, 1098 (1969).
944. M. Vidal, F. Massot, and P. Arnaud, *C. R. Acad. Sci.*, Paris, Ser. *C.*, **268C**, 423 (1969).
945. D. Schonleber, *Angew. Chem. Int. Ed. Engl.*, **8**, 76 (1969).
946. M. Jones, *Angew. Chem. Int. Ed. Engl.*, **8**, 76 (1969).
947. M. Jones, A. M. Harrison, and K. R. Rettig., *J. Amer. Chem. Soc.*, **91**, 7462 (1969).
948. H. Durr, G. Scheppers, and L. Schrader, *Chem. Commun.*, 257 (1969).
949. H. Durr and L. Schrader, *Chem. Ber.*, **102**, 2026 (1969).
950. H. Durr and L. Schrader, *Angew. Chem. Int. Ed. Engl.*, **8**, 446 (1969).
951. C. A. Converse and F. F. Richards, *Biochem.*, **8**, 4431 (1969).
952. N. R. Ghosh, C. R. Ghoshal, and S. Hah, *Chem. Ind.*, 493 (1969).
953. U. Schollkopf and M. Reetz, *Tetrahedron Lett.*, 1541 (1969).
953a. G. Snatzke, B. Ehrig, and H. Klein, *Tetrahedron*, **25**, 5601 (1969).
954. Y. Hata and H. Tanida, *J. Amer. Chem. Soc.*, **91**, 1170 (1969).
955. M. Regitz and J. Rutter, *Chem. Ber.*, **102**, 3877 (1969).
956. M. Avaro and J. Levisalles, *Bull. Chim. Soc. France*, 3180 (1969).
957. P. R. Brook and B. Brophy, *Tetrahedron Lett.*, 4187 (1969).
958. K. Wiberg and A. deMeijere, *Tetrahedron Lett.*, 519 (1969); A.-J. Ashe, *Tetrahedron Lett.*, 523 (1969).
959. W. D. Barker, R. Gilbert, J.-P. Lapointe, H. Veschambre, and D. Vocelle, *Can. J. Chem.*, **47**, 2853 (1969).
960. H. Veschambre and D. Vocelle, *Can. J. Chem.*, **47**, 1981 (1969).
961. N. R. Ghosh, C. R. Ghoshal, and S. Shah, *Chem. Commun.*, 151 (1969).
962. W. Ando, T. Yagihara, and T. Migita, *Tetrahedron Lett.*, 1983 (1969).
963. W. Ando, T. Yagihara, S. Tozune, and T. Migita, *J. Amer. Chem. Soc.*, **91,** 2786 (1969).

964. W. Ando, T. Yagihara, S. Tozune, S. Nakaido, and T. Migita, *Tetrahedron Lett.*, 1979 (1969).
965. W. Ando, K. Nakayama, K. Ichibori, and T. Migita, *J. Amer. Chem. Soc.*, **91**, 5164 (1969).
966. W. Ando, S. Kondo, and T. Migita, *J. Amer. Chem. Soc.*, **91**, 6516 (1969).
967. J. A. Kaufman and S. J. Weininger, *Chem. Commun.*, 593 (1969).
968. P. J. Whitman and B. M. Trost, *J. Amer. Chem. Soc.*, **91**, 7534 (1969).
969. J. A. Green and L. A. Singer, *Tetrahedron Lett.*, 5093 (1969).
970. J. H. Boyer and W. Beverung, *Chem. Commun.*, 1377 (1969).
971. J. C. Fleming and H. Schechter, *J. Org. Chem.*, **34**, 3962 (1969).
972. D. R. Dalton and S. A. Liebman, *Tetrahedron*, **25**, 3321 (1969).
973. E. S. Lewis, R. E. Holliday, and L. D. Hartung, *J. Amer. Chem. Soc.*, **91**, 430 (1969).
974. H. Nozaki, T. Sakai, H. Takaya, and R. Noyori, *Kogyo Kagaku Zasshi*, **72**, 280 (1969).
975. V. Snieckus and K. S. Bhandari, *Tetrahedron Lett.*, 3375 (1969).
976. J. A. Cotruvo, *Diss. Abstr.*, **30**, 118B (1969).
977. A. Mackor, J. U. Veenland, and Th. J. DeBoer, *Rec. Trav. Chim.*, **88**, 1249 (1969).
978. L. Creagh and I. Trachtenber, *J. Org. Chem.*, **34**, 1307 (1969).
979. R. Tanikaga, *Bull. Chem. Soc. Japan*, **42**, 210 (1969).
980. S. Forshult and C. Lagercrantz, *Acta Chem. Scand.*, **23**, 522 (1969).
981. C. Lagercrantz and S. Forshult, *Acta Chem. Scand.*, **23**, 708 (1969).
982. C. Lagercrantz and S. Forshult, *Acta Chem. Scand.*, **23**, 811 (1969).
983. C. Lagercrantz, *Acta Chem. Scand.*, **23**, 3259 (1969).
984. Th. A. J. W. Wajer, A. Mackor, and Th. J. deBoer, *Tetrahedron*, **25**, 175 (1969).
985. G. Barbarella and A. Rassat, *Bull. Soc. Chim. France*, 2378 (1969).
986. I. H. Leaver, G. C. Ramsay, and E. Suzuki, *Aust. J. Chem.*, **22**, 1891 (1969).
987. I. H. Leaver and G. C. Ramsay, *Aust. J. Chem.*, **22**, 1899 (1969).
988. A. L. Buchachenko, A. M. Wasserman et al., *Intern. J. Chem. Kinetics*, **1**, 361 (1969).
988a. I. H. Leaver and G. C. Ramsay, *Tetrahedron*, **25**, 5669 (1969).
988b. E. F. Ullman and D. G. B. Boocock, *Chem. Commun.*, 1161 (1969).
989. Y. L. Chow and J. N. S. Tam, *Chem. Commun.*, 747 (1969).
990. W. B. Watkins and R. N. Seelye, *Can. J. Chem.*, **47**, 497 (1969).
991. Y. L. Chow, N. S. Tam, and A. C. H. Lee, *Can. J. Chem.*, **47**, 2441 (1969).
992. E. E. J. Dekker, J. B. F. N. Engbuts, and T. J. deBoer, *Tetrahedron Lett.*, 2651 (1969).
993. J. A. Sousa, J. Weinstein, and A. L. Bluhn, *J. Org. Chem.*, **34**, 3320 (1969).
994. J. S. Cridland and S. T. Reid, *Chem. Commun.*, 125 (1969).
995. G. Howarth, D. G. Lance, W. A. Szarek, and J. K. N. Jones, *Can. J. Chem.*, **47**, 81 (1969).
996. S. Hashimoto and K. Kano, *Kogyo Kagaku Zasshi*, **72**, 191 (1969).
997. E. G. Janzen and J. L. Gerlock, *J. Amer. Chem. Soc.*, **91**, 3108 (1969).
998. G. R. Seely, *J. Phys. Chem.*, **73**, 117, 125 (1969).
999. H. Van Beek and P. M. Heertjes, *Ind. Chim. Belge*, **32**, 136 (1967).
1000. H. J. Roth and M. Adomeit, *Tetrahedron Lett.*, 3201 (1969).
1001. S. Hashimoto, K. Kano, and K. Ueda, *Tetrahedron Lett.*, 2733 (1969).
1002. H. Hart and J. W. Link, *J. Org. Chem.*, **34**, 758 (1969).
1003. G. A. Varvoglis, *Chim. Chron.*, *A.*, **33**, 121 (1968).
1004. C. Mercier and J.-P. Dubose, *Bull. Soc. Chim. France*, 4425 (1969).
1005. D. J. Needle and R. L. Pollitt, *J. Chem. Soc.*, C, 2127 (1969).
1006. M. E. Langmuir, L. Dogliotti, E. D. Black, and G. Wettermark, *J. Amer. Chem. Soc.*, **91**, 2204 (1969).

1007. P. H. McFarlane and D. W. Russell, *Chem. Commun.*, 475 (1969).
1008. S. Hashimoto and K. Kano, *Kogyo Kagaku Zasshi*, **72**, 188 (1969).
1009. R. J. Sundberg, B. P. Das, and R. H. Smith, *J. Amer. Chem. Soc.*, **91**, 658 (1969).
1010. R. J. Sundberg, R. H. Smith, Jr., and J. E. Bloor, *J. Amer. Chem. Soc.*, **91**, 3392 (1969).
1011. A. Van Vliet, J. Cornelisse, and E. Havinga, *Rec. Trav. Chim.*, **88**, 1339 (1969).
1012. R. L. Letsinger and K. E. Steller, *Tetrahedron Lett.*, 1401 (1969).
1013. R. L. Letsinger and R. R. Hautala, *Tetrahedron Lett.*, 4205 (1969).
1014. L. B. Jones, J. C. Kudrna, and J. P. Foster, *Tetrahedron Lett.*, 3263 (1969).
1015. R. L. Letsinger and J. H. McCain, *J. Amer. Chem. Soc.*, **91**, 6425 (1969).
1016. R. H. Hesse, *Adv. Free Radical Chem.*, **3**, 83 (1969).
1017. H. Suginome, I. Yamazaki, et al., *J. Chem. Soc. Japan*, **72**, 243 (1969).
1018. W. Mehrhof, K. Irmscher, R. Erb, and L. Pohl, *Chem. Ber.*, **102**, 643 (1969).
1019. H. Suginome, N. Sato, and T. Masamune, *Bull. Chem. Soc. Japan*, **42**, 215 (1969).
1020. H. Suginome, N. Sato, and T. Masamune, *Tetrahedron Lett.*, 2671 (1969).
1021. H. Suginome, H. Ono, and T. Masamune, *Tetrahedron Lett.*, 2909 (1969).
1022. H. Suginome, I. Yamazaki, H. Ono, and T. Masamune, *Kogyo Kagaku Zasshi*, **72**, 243 (1969).
1023. D. H. R. Barton, D. Kumari, P. Welzel, L. J. Danks, and J. F. McGhie, *J. Chem. Soc., C*, 332 (1969).
1024. D. H. R. Barton, R. P. Budhiraja, and J. F. McGhie, *J. Chem. Soc., C*, 336 (1969).
1025. H. Obara and H. Kimura, *Bull. Chem. Soc. Japan*, **42**, 2705 (1969).
1026. H. Suginome, N. Sato, and T. Masamune, *Tetrahedron Lett.*, 3353 (1969).
1027. P. Scheiner, O. L. Chapman, and J. D. Lassila, *J. Org. Chem.*, **34**, 813 (1969).
1028. J. M. Surzur, M. P. Bertrand, and R. Nouguier, *Tetrahedron Lett.*, 4197 (1969).
1028a. P. D. Rieke and N. A. Moore, *Tetrahedron Lett.*, 2035 (1969).
1029. C. Chachaty and A. Forchioni, *Compt. Rend.*, **268C**, 300 (1969).
1030. A. A. Strel'tsova, L. A. Levashova, K. E. Kuznetsova, I. F. Buchnev, R. L. Megrabyan, G. A. Isayan, *Khim. Prom. (Moscow)*, **45**, 88 (1969).
1031. K. Shimizu, S. Wakamatsu, and M. Shinomiya, *Japan. Patent*, **68** 13,452.
1032. A. V. Iogansen, A. A. Strel'tsova, G. N. Semina, L. G. Zelenskaya, L. A. Levashova, and R. L. Megrabyan, *Neftekhimiya*, **8**, 447 (1968).
1033. E. Barale and A. Guillemonat, *C. R. Acad. Sci. Paris, Ser. C*, **268C**, 1201 (1969).
1034. D. H. Dvbvig, U.S. Patent 3,403,086 (1968).
1035. D. R. Eckroth and R. H. Squire, *Chem. Commun.*, 312 (1969).
1036. H. Mauser and H. Bokranz, *Z. Natur.*, **B**, 477 (1969).
1037. M. L. Scheinbaum, *Tetrahedron Lett.*, 4221 (1969).
1037a. O. Hromatka, M. Knollmuller, and D. Binder, *Monatsh. Chem.*, **100**, 872 (1969).
1038. C. Mercier and J.-P. Dubosc, *Bull. Soc. Chim. France*, 268 (1969).
1039. W. R. Dolbier, Jr., and M. Williams, *J. Amer. Chem. Soc.*, **91**, 2818 (1969).
1039a. W. R. Dolbier and W. M. Williams, 157th National Meeting of the American Chemical Society, April, 1969, Minneapolis, Minn., ORGN-143.
1040. J. Streith, A. Blind, J.-M. Cassal, and C. Sigwalt, *Bull. Soc. Chim. France*, 948 (1969).
1041. M. Ishikawa, C. Kaneko, I. Yokoe, and S. Yamada, *Tetrahedron*, **25**, 295 (1969).
1042. J. Streith and P. Martz, *Tetrahedron Lett.*, 4899 (1969).
1043. T. Tsuchiya, H. Arai, and H. Igeta, *Tetrahedron Lett.*, 2747 (1969).
1044. O. Buchardt and P. L. Kumler, *Acta Chem. Scand.*, **23**, 159 (1969).
1045. C. Kaneko, I. Yokoe, S. Yamada, and M. Ishikawa, *Chem. Pharm. Bull. Japan*, **17**, 1290 (1969).
1046. O. Buchardt, P. L. Kumler, and C. Lohse, *Acta Chem. Scand.*, **23**, 2149 (1969).

1047. M. Hamana and H. Noda, *Chem. Pharm. Bull. Japan*, **17**, 2633 (1969).
1048. C. Kaneko, S. Yamada, and M. Ishikawa, *Chem. Pharm. Bull. Japan*, **17**, 1294 (1969).
1049. H. Mantsch, V. Zanker, W. Seiffert, and G. Prell, *Justus Liebigs Ann. Chem.*, **723**, 95 (1969).
1050. J. Joussot-Dubien and J. Houdard-Pereyre, *Bull. Soc. Chim. France*, 2619 (1969).
1051. R. M. Kellogg, T. J. VanBergen, and H. Wynberg, *Tetrahedron Lett.*, 5211 (1969).
1052. J. Biedrzycki, *Zesz. Nauk., Mat., Fiz., Chem., Wyzsza. Szk. Pedagog. Gdansku*, **8**, 195 (1968); *Chem. Abstracts*, **70**, 95984f (1969).
1053. C. Azuma and A. Sugimori, *Kogyo Kagaku Zasshi*, **72**, 239 (1969).
1054. C. G. Allison, R. D. Chambers, Y. A. Cheburkov, J. A. H. MacBride, and W. K. R. Musgrave, *Chem. Commun.*, 1200 (1969).
1055. J. R. Plimmer, P. C. Kearney, and U. I. Klirgebiel, *Tetrahedron Lett.*, 3891 (1969).
1056. T. Sasaki, K. Kanematsu, and A. Kakehi, *Chem. Commun.*, 432 (1969).
1057. J. Streith and J.-M. Cassal, *Bull. Soc. Chim. France*, 2175 (1969).
1058. V. Snieckus, *Chem. Commun.*, 831 (1969).
1059. J. Biedrzycki, *Zesz. Nauk., Mat., Fiz., Chem., Wyzsza. Szk. Pedagog. Gdansku*, **8**, 195 (1968).
1060. J. C. Doty, J. L. R. Williams, and P. J. Grisdale, *Can. J. Chem.*, **47**, 2355 (1969).
1061. C. A. Taylor, M. A. El-Bayounu, and M. Kasha, *Proc. Nat. Acad. Sci.*, **63**, 253 (1969).
1062. J. P. Ferris, J. E. Kuder, and A. W. Catalan, *Science*, **166**, 765 (1969).
1063. H. Steinmaus, I. Rosenthal, and D. Elad, *J. Amer. Chem. Soc.*, **91**, 4922 (1969).
1064. E. C. Taylor, Y. Maki, and B. E. Evans, *J. Amer. Chem. Soc.*, **91**, 5181 (1969).
1065. F. R. Stermitz and C. C. Wei, *J. Amer. Chem. Soc.*, **91**, 3103 (1969).
1066. P. S. Song and W. E. Kurtin, *Mol. Photochem.*, **1**, 1 (1969).
1067. K. Nakamaru, S. Nuzuma, and M. Koizumi, *Bull. Chem. Soc. Japan*, **42**, 255 (1969).
1068. V. Zanker and G. Prell, *Ber. Bunsenges.*, **73**, 791 (1969).
1069. R. Noyori, M. Kato, M. Kawanisi, and H. Nozaki, *Tetrahedron*, **25**, 1125 (1969).
1070. T. L. Banfield and D. Husain, *Trans. Faraday Soc.*, **65**, 1985 (1969).
1071. E. J. Moriconi, R. E. Misner, and T. E. Brady, *J. Org. Chem.*, **34**, 1651 (1969).

Part 2

Photophysical Processes of Organic Compounds: Formation and Decay of Electronically Excited States

JACK E. LEONARD and
GEORGE S. HAMMOND

Division of Chemistry and Chemical Engineering
California Institute of Technology, Pasadena

INTRODUCTION

The upsurge in photochemical research in the past decade has in part been powered by the great increase in knowledge about photophysical processes that precede and are concurrent with the photochemical pathways. Understanding the nature of the physical processes of formation and decay of excited states has shed light on the chemical reactivity of these states, has given tools for measuring the rates of reaction from these states, and has lent models for predicting and characterizing organic photoreactions. Conversely, photochemical studies have provided tools for elucidating photophysical processes, such as energy-transfer processes. The fruitfulness of the interaction between the physical and chemical studies explains why a chapter such as this is included in a volume on the chemistry of excited states. The sheer quantity of material explains why it requires a separate chapter.

In order to provide some limits to the chapter, we have largely excluded organic compounds that are bound to metals, either as coordination compound ligands or by carbon-metal bonds.

The survey is divided into three major parts: the formation of electronically excited states, the properties of excited states, and the decay of excited states. This division is largely clear-cut and should provide little confusion. But one notable exception is energy transfer that involves the decay of the donor excited state and the formation of an acceptor excited state. However, we have generally excluded nonphoton production of excited states from the section on the formation of excited states, and so this process is included under decay processes.

The fare this year has been highly varied; we hope that you find our presentation provides a guide through the richness of the subject.

I. FORMATION OF ELECTRONICALLY EXCITED STATES: LIGHT ABSORPTION

Photochemistry as a theoretical science can probably be dated from the statement of the Grotthus-Draper law in the first half of the nineteenth century: only light that is absorbed by a system is capable of producing chemical change in the system. Since that time a great deal has been learned about light-absorption processes and their relation to molecular structure. At present, organic absorption spectroscopy, like melting-point measurements, is generally regarded by the bench chemist as an essential physical tool for defining a compound's characteristics, but far less diagnostic than the techniques of nuclear magnetic resonance, mass, and infrared spectroscopy.

For the physical chemist and the organic photochemist, though, absorption spectroscopy, with the inverse phenomenon—emission spectroscopy—is practically the only method for locating and characterizing electronically excited states. Coupled with a greater understanding of quantum mechanics, high-resolution electronic spectroscopy has continued to provide very valuable information on the excited states of organic molecules (see Section II and Table V for other means of characterizing excited states).

A. Absorption Spectroscopy of Molecules: Ground State to Spin-allowed Excited States

Molecular absorption spectroscopy, in both its experimental and theoretical aspects, is dominated by transitions in the visible and near-ultraviolet regions of the electromagnetic spectrum, approximately 200 to 700 nm (50,000 to 15,000 cm^{-1}). The lower limit, of course, is determined by the general lower limits on the energies of electronic transitions, but the upper limit is determined by the absorption of oxygen and by the fact that most theoretical models are not precise enough to interpret higher excitations of organic molecules.

The theoretical and synthetic interest in the nature of these high-energy electronic transitions, however, has led to commercially available spectrometers that can cover this region of the spectrum, and we may expect an increase in research in this field. Milazzo and Ceccheti [66] review the techniques required for working in this region of the energy spectrum. Aihara [213] discussed the Rydberg spectroscopy (leading to the photoionization of the molecules in the high energy limit) of aromatic compounds.

De Reilhac, Astoin, and Romand [70] report on improvements in the method of measuring the absorption coefficients of gases; they reported results on methane and carbon dioxide. Merer and Mulliken [77] discuss the theoretical and experimental results on the spectrum of ethylene. Their discussion covers both the molecular-orbital transitions, which are the kind most familiar from the near ultraviolet and visible regions, and the Rydberg, or hydrogenlike, states of the molecule. Far ultraviolet studies have been applied to a number of aliphatic systems in particular, because these compounds often have their lowest-energy electronic transitions in this region.

Concerning the usual region of electronic spectroscopy, the visible and near ultraviolet, a number of reviews appeared this past year. Parkin [3] discusses both light-absorption and photoelectron spectroscopy, with emphasis on the theoretical models involved in understanding such spectroscopy. Hida [1] and Hatano [2] give more general introductions to electronic spectroscopy. Permogorov, Serdyukova, and Frank-Kamenetskii [8] review the electronic spectroscopy of dyes. Dehler and Kresze [4] review the experimental aspects of ultraviolet spectroscopy as applied to aromatic hydrocarbons.

On experimental matters, Popova, Popov, Rautian, and Sokolovskii [15] discuss nonlinear interference effects in the absorption, emission, and lasing action of gases; they show how such effects can be separated from Doppler broadening effects in these systems. Siano and Metzler [6] examine the structureless bands in the spectra of hydroxypyridines and propose a four-parameter log-normal distribution analysis of the band shapes. They discuss the physical meaning of using such an analysis. Kajima and Toda [5] discuss the experimental use of potassium chloride pressed disks in ultraviolet spectroscopy.

The bulk of electronic spectroscopy is in the near ultraviolet and is performed on aromatic or heterocyclic molecules. These systems are also the favorite ones for theoretical calculations, because they are amenable to π-electron approximations, which are much simpler (and cheaper) than all-electron calculations. All-electron calculations were performed on a series of isoelectronic acids, amides, and acyl fluorides by Basch, Robin, and Kuebler [17] and on butadiene by Buenker and Whitten [19].

The most common procedures used are still based on molecular orbitals composed of linear combinations of atomic orbitals (LCAO). The use of computers, however, has led of late to a number of calculations using a large number of gaussian fractions, selected to approximate atomic orbitals, for the wave functions. Although gaussian functions do not closely correspond to the atomic functions, they are easily integrated by computer methods, so that their initial handicap can be overcome by evaluating a large number of them. They are treated by the self-consistent field (SCF) method developed originally for use with LCAO molecular orbitals. Gaussian treatments (GSCF) were carried out on such diverse systems and vinyl silane and propene [33], saturated and unsaturated three-membered rings [64, 65], and azulene and naphthalene [102].

The simplest LCAO-MO method is the Hückel method for extended π systems. Its great advantage, namely, that it is numerically fairly easy to handle, has largely disappeared in the computer age. It is, however, a common starting point to obtain first-approximation orbitals and for carrying out vibronic analyses [274]. Wohl [13] provides a Hückel analysis of 46 diverse aromatic compounds and uses a regression analysis to apply these results to an additional 88 compounds. He uses these results to discuss the use of the Hückel method as a routine qualitative method in spin-allowed transitions. The Pariser-Pople-Parr (PPP) method, which combines the self-consistent field approach with the mixing of excited configurations into the ground state (configuration interaction or CI) is by far the commonest calculation method in use. Recently it has been used frequently in a modified form in which the electronegativity of the molecular-framework atom is treated as a variable β and the variation principle is then applied to minimize the energy. PPP calculations have become so popular because they yield good results for both carbocyclic and heterocyclic systems and because they can allow for spin properties, which Hückel theory ignores. Modifications of the Hückel theory, such as the free-electron model and extended Hückel theory, have been developed to overcome this difficulty of simple Hückel theory. The complete neglect of differential overlap method (CNDO) has been used to calculate spectra for all kinds of π-organic systems, such as small π-bonded systems [25], benzene derivatives [84, 85], and nitrogen heterocycles [248, 254]. Favini [81a] discusses the relation between geometry and spectra in sterically hindered molecules. He considers three contributions to the energy: (1) the energy of the π system, (2) the repulsion energy between atoms, and (3) the σ-bond deformation energy.

A number of papers have appeared in the past year concerning the calculation of the vibronic structure of electronic spectra. Savin [11, 12] discusses a general equation for deriving the matrix elements for ground and excited electronic state vibrational wave functions. Lukashin, Frank-Kamenetskii, and Permogorov [10, 20] derive a set of formulas for the vibronic structure of conjugated compounds. Hochstrasser and Marzzacco [9] discuss vibronic interaction between energetically close states and apply their theoretical model to the absorption and emission spectra of pyridine and pyrazine derivatives. Sadlej [16] constructs the vibronic spectrum expected in the ethylene dimer. Lukashin, Permogorov, and Frank-Kamenetskii [20] derive general equations for the vibronic spectra of conjugated molecules. Merer and Mulliken [62] calculate the vibrational structure of the ethylene spectrum and find that both the C—C stretching and the twisting vibrations contribute to the spectrum. McDiarmid [75] also calculated the ethylene vibronic spectrum. Perrin and Gouterman [274] provide an analysis of vibronic intensity borrowing in cyclic polyenes and porphyrin.

Currie and Holmes [88] provide an interesting empirical equation for the relation between substituents and λ_{max} (wavelength of maximum absorbance) for a number of conjugated heteroenoid compounds. Hida [227] provides a

MO interpretation for an earlier such empirical equation relating λ_{max} to the number of conjugated double bonds in certain aromatic- and heteroaromatic-substituted polyenes.

One interesting experimental study by Slamnik and Sunjic [283] relates the molecular weight of imidazole derivatives to the per gram absorption coefficients of their picrate derivatives. Linear free-energy relations between λ_{max} and the Hammett substituent coefficients are noted for m-methoxylated-p-fuchsins [139], 3-substituted thianaphthenes [342], aroylphosphonic acids [179], and 2- and 4-nitrodiphenylamines [143]. Porter [174] reviews the spectra of aromatic free radicals, which are species frequently encountered in photochemical studies. Henderson [14] reviews an interesting sidelight of absorption spectroscopy in a review of asymmetric synthesis using circularly polarized light. Since the two enantiomers show different extinction coefficients for right and left circularly polarized light, it is possible to populate the transition states leading to products selectively in order to give rise to asymmetric products from racemic starting materials.

Bradley [7] provides a review of the thermal excitation of electronically excited states, as found in shock tubes. This process involves the interconversion of translational and electronic energy, a process which is usually unimportant but which can lead to interesting results when it occurs.

Theoretical and experimental studies on molecular spectroscopy are summarized in Table II. Table I lists the abbreviations used in the tables for calculation methods.

TABLE I
Abbreviations for Calculation Methods

CI	Configuration interaction
CNDO	Complete neglect of differential overlap
DECI	CI method, including doubly excited configurations
EHMO	Extended Hückel MO
FEMO	Free-electron MO
SGCF	SCF method using gaussian function for molecular orbitals
HF	Hartree-Fock-SCF
HMO	Hückel MO
LCAO	Linear combination of atomic orbitals
MM	Molecules in molecules
MO	Molecular orbital
mod	Modified
PMO	Perturbational MO
PPP	Pariser-Pople-Parr
β-PPP	PPP calculations with parametrized core potential ($=\beta$)
SCF	Self-consistent field
TECI	CI, including triply excited configurations
VE	Variable electronegativity

TABLE II
Molecular Spectroscopy:
Ground State to Spin-allowed Excited States

System	Calculation method	Reference
A. Aliphatic compounds		
I. Near ultraviolet, visible, and near infrared		
a. Theoretical models and calculations		
1. Vibronic structure of ethylene dimer spectrum		16
2. Isoelectronic amides, acids, and acyl fluorides	ab initio GSCF	17
3. Nitroethylene	PPP	18
4. Butadiene	ab initio SCF-MO-DECI	19
5. Vibronic spectra of conjugated hydrocarbons, including butadiene, benzene, and hexatriene		20
6. Ethylenic amides	Intramolecular charge transfer	21
7. Conjugated carbonyl compounds	PPP	22
8. Formaldehyde, hydrogen cyanide, and carbon dioxide	mod PPP	23
9. Croconate anion $(C_5O_5)^{2-}$	VE-SCF-MO-DECI	24
10. Small π-bonded molecules	CNDO	25
11. Acetyl and benzoyl systems	LCAO-MO	26
12. β-Diketones	LCAO-MO	27
13. Conjugated carbonyl compounds	MM	28
14. Exciton band structure in polypeptides	PPP	29
15. Thioformaldehyde	PPP	30
16. Thiocarbonic acid	PPP	31
17. Tetracyanopolymethines	VE-PPP	32
18. Vinylsilane and propene	GSCF	33
19. Urazole	GSCF, EHMO, CNDO	33a
b. Experimental results		
1. Difluoromethylene		34
2. Vibrational structure of first band of formaldehyde spectrum		35
3. Olefins		36
4. Vibrational structure of 330 nm band of cyclobutanone		37
5. Ethoxy-1, 3-dienes		38
6. Bicyclo[3.1.0]hex-3-ene-2-ones		39

TABLE II—continued

System	Calculation method	Reference
7. Lactams		40
8. Nitro compounds		41
9. Nitroso compounds		42
10. Hydrazones		43
11. Unsaturated carbonyl and β-dicarbonyl compounds		44
12. β-Diketones		45
13. Vibrational structure of electronic spectra of deuterated methylamines		46
14. Pelargonidol		47
15. Saturated bicyclic amines		48
16. Isomeric bis(trimethylsilyl)-1,3-butadienes		49
17. Silyl ketones		50
18. Permethylated polysilanes		51
19. Perethylated polygermanes and polystannanes		52
20. Trimethylgermane carboxylic acid		53
21. *trans*-1,2-Vinylene bis(di-*n*-butylphosphine) and its monoxide		54
22. Vinylarsines and vinylphosphines		55
23. Organosulfur compounds		56
24. Thioformaldehyde		57
25. Mercaptals		58
II. Far ultraviolet		
a. Theoretical models and calculations		
1. Carbon dioxide and suboxide		59
2. Nitromethane	CNDO-CI	60
3. Fluoromethane: bond exciton model		61
4. Vibrational structure of first π,π^* transition in ethylene		62
5. Acetylene, diacetylene, triacetylene, allene, and alkyl derivatives	VE-PPP-TECI	63
6. Saturated three-membered rings: cyclopropane, ethylene oxide, ethylenimine, and diaziridine	GSCF	64
7. Unsaturated three-membered rings: cyclopropene, 3,3-dimethylcyclopropene, and difluorodiazirine	GSCF	65
b. Experimental results		
1. Techniques of vacuum ultraviolet spectroscopy: a review		66
2. Carbon monoxide		67
3. Carbon suboxide		59

TABLE II—continued

System	Calculation method	Reference
4. Carbon disulfide		68
5. Cyanogen		69
6. Carbon dioxide and methane		70
7. Methyl radical		71
8. Perfluoro-*n*-paraffins		72
9. 140 nm transition in ethane		73
10. Ethylene		
(*a*) 174.4 nm Rydberg transition		74
(*b*) Vibrational intensity progression in N, V (π,π^*) transition in ethylene		75
(*c*) Matrix isolation spectrum		76
(*d*) Review of spectra of ethylene and its alkyl derivatives		77
11. Chloroacetylene		78
12. Rydberg transitions in *cis*- and *trans*-butene-2		79
13. Olefins		36
14. Aliphatic carbonyl compounds		80
15. Homopolypeptides		81
B. Aromatic hydrocarbons		
I. Near ultraviolet, visible, and near infrared		
a. Theoretical models and calculations		
1. Benzene	LCAO-MO	82
2. Fluorobenzene	β-PPP	83
3. Nitrobenzene	CNDO-CI	84
4. Aminophenols and aminoresorcinols	β-PPP	85
5. Benzyl radical		86
6. Salicylaldehyde azine	PPP	87
7. Biphenyl		
(*a*)	β-PPP	89
(*b*)	CNDO-CI	90
8. Biphenyl and biphenylene	PPP	90a
9. Diphenylmethyl cations	PPP	91
10. Monoamino and monohydroxy derivatives of benzenoid aromatics	mod PPP	92
11. Styrene	PPP	93
12. Benzaldehyde oxime		94
13. Schiff's bases derived from benzaldehyde and para-substituted phenylamines		95
14. Monosubstituted phenyl azides	HMO	96
15. Stilbene	FEMO-CI	97

TABLE II—continued

System	Calculation method	Reference
16. Stilbene and its naphthyl analogs	PPP	98
17. Comparison of t-butyl and trimethylsilyl phenyl ketones		99
18. p-Disilylbenzene	PPP	100
19. Phenylcycloheptatrienylium cation	LCAO-MO	101
20. Azulene and naphthalene	GSCF	102
21. Phenylnaphthalenes	PPP	103, 103a
22. Anthracene cation	FEMO	104
23. Aryl phenyl ketones	PPP	105
24. 2,2′-Linked biphenyls	PPP	106
25. Fluorenzylium cation and its benzo derivatives		107
26. Fluorenone	PPP	108
27. Aryl-substituted fluorenyl carpanions	PPP	109
28. Bis(biphenylene)ethylene and vinylogous compounds	mod HMO	110
29. Dehydrodianthrones	PPP	111
30. Triphenylene	MM	112
31. Annulenes	PPP	113
32. [18] Annulene	PPP	114
33. Several alternant and nonalternant hydrocarbons	MM	114a
b. Experimental results		
1. Benzene and its derivatives		
(a) Benzene and alkyl benzenes		
(1) Benzene, alkylbenzenes, and substituted alkyl benzenes		115
(2) Fluoromethyl benzenes		116
(3) Triphenylmethyl cations		117
(4) Triphenylmethyl and α-methyl benzyl cations		118
(5) Several phenylalkyl cations		119, 119a
(6) Arylmethylenes		120
(7) 2,3:6,7-Dibenzosuberan		121
(8) Carbonium ions derived from styrene		122
(9) Isomeric methyl styrenes		123
(10) Stilbene and substituted stilbenes		124
(11) Methyl substituted 1,1-diarylalkenes		125
(b) Halobenzene and alkylhalobenzenes		
(1) Dihalobenzenes		126

TABLE II—continued

System	Calculation method	Reference
(2) *m*-Difluorobenzene		127
(3) *p*-Bromostyrene		128
(4) Polyhalobiphenyls		129
(c) Phenols and anisoles		
(1) Alkylphenols		130
(2) Terpene phenols and guiacols in neutral and ionized states		131, 131a
(3) *p*-Fluorophenol		132
(4) Phenoxy radical derived from tyrosine		133
(5) Dihydroxybenzenes		134
(6) Polysubstituted phenols and anisoles		135
(7) *p*-Bromoethoxybenzene		136
(8) *o*-Fluoroanisole		137
(9) Esters and arylides of anisic acid		138
(10) *m*-Methoxylated-*p*-fuchsines (triphenylmethane dyes)		139
(d) Anilines, anilides, and nitrobenzenes		
(1) *o*- and *m*-Bromoanilines in vapor phase		140
(2) *N*-methylaniline in vapor phase		141
(3) Radicals derived from *N,N*-dimethylaniline		142
(4) 2- and 4-Nitrodiphenylamines		143
(5) Diphenylamine derivatives and corresponding diarylamine oxides		144
(6) Cation radicals derived from *p*-phenylenediamine and its *N*-methyl derivatives		145
(7) Nine Wurster salts		146
(8) Drugs derived from *p*-aminobenzoic acid		147
(9) Schiff base derivatives o acetoacetanilide		148
(10) Products of nucleophilic agents with aromatic nitro compounds		149
(11) Polarization spectra of benzoic acid, *p*-nitroaniline, *p*-nitrophenol, and *p*-aminobenzoic acid in thin films		150

TABLE II—continued

System	Calculation method	Reference
(e) Phenyl ketones, quinones, and functional group derivatives		
(1) o-Fluoroacetophenone		151
(2) Alkoxy derivatives of acetophenone		152
(3) o-Hydroxylated acetophenones and functional group derivatives		153
(4) Acetophenone N'-(N-phenylcarbonyl) hydrazones		154
(5) Benzophenone with acceptor and donor substituents		155
(6) n,π^* bands of phenyl carbonyl compounds, α-diketones, and quinones		156
(7) Substituent effects on spectra of p-benzoquinones		157
(8) Differences in spectra of o- and p-benzoquinones		158
(9) Anions of quinones in solutions and solids		159
(f) Benzaldehyde and functional group derivatives		
(1) Lowest frequency o-, m-, and p-bromobenzaldehyde in gas phase		160
(2) p-Isopropylbenzaldehyde		161
(3) o-, m-, and p-Anisaldehyde in gas phase		162
(4) Simple aromatic Schiff bases		163
(5) N-substituted benzamidines		164
(6) Bisazomethine derivatives of aromatic amines		165
(g) Benzene azo compounds		
(1) Azobenzene, azoxybenzene, and salicylanilide		166
(2) 4-Halo-4′-nitroazobenzenes		167
(3) Chloroazobenzenes		168
(4) Derivatives of bis(phenylazo)benzene		169
(5) p-Nitrophenyldiazoaminoazobenzene		170
(h) Cinnamic acid derivatives		
(1) Hydroxycinnamic acids		171, 172

TABLE II—continued

System	Calculation method	Reference
(2) N-alkyl and N,N-dialkyl derivatives of some cinnamamides		173
(i) Miscellaneous benzene derivatives		
(1) Spectra of aromatic free radicals: a review		174
(2) $3d$ orbital participation in thiophenols		175
(3) Phenylgermanes		176
(4) N=P participation in conjugation in N,P-diaryl-P, P-diethyl phosphine imides		178
(5) Linear free-energy relations in the λ_{max} of n,π^* band of esters of aroylphosphonic acids		179
(6) Triphenyl boroxin and related compounds		180
(7) Benzyl acetate in different states		181
(8) p-Toluonitrile in vapor phase		182
(9) Semicarbazones and thiosemicarbazones		183
(10) Tetracyanoquinodimethane anion radical salts		184
(11) Substituted (α-hydroxyiminoalkyl) benzenes		185, 186
2. Naphthalene and its derivatives		
(a) α-Phenylated naphthalenes		187
(b) Acetyl- and benzoyl-substituted naphthols		188
(c) α-Naphthoic acid, naphthalic anhydride, and naphthalimide		189
(d) Azonaphthalenes		190
(e) β- and β,β'-acetoxy and mesyloxy naphthalenes		191
(f) 1,4-Naphthoquinones and 9,10-anthraquinones		192
(g) Naphthazarin (5,8-dihydroxy naphthoquinone)		193
(h) Acenaphthylene photodimer		194
3. Polynuclear aromatics		
(a) Anthraquinones		
(1) 9,10-Anthraquinone and its 2-methyl derivative		195

TABLE II—continued

System	Calculation method	Reference
(2) α- and β-Haloanthraquinones		196
(3) α- and β-Aminoanthraquinones		197
(4) 1-Hydroxyanthraquinone-9-monoimine		198
(5) 1- and 2-Halo- and 1,5-dihaloanthraquinones		199
(6) 1,4,5,8-Tetrahydroxy-anthraquinones		200
(7) Naturally occurring poly-hydroxyanthraquinone		201
(8) Polarization spectra of dyes based on 1,4-diamine- and 1-amine-4-hydroxyanthraquinones		202
(b) Phenanthrene		
(1) Totally symmetric vibronic perturbations at 340 nm		203
(2) Phenanthrene cation in boric acid glass		204
(c) Phenylethynyl acenes		205
(d) Benzanthrone and its 3-bromo and 3-methoxy derivatives		206
(e) Benzo[a]anthracene photodimer		207
(f) Benzo[g,h,i]perylene		208
4. Nonalternant hydrocarbons		
(a) Rotational analysis of the 0—0 band of the 350 nm ($S_0 — S_2$) transition		209
(b) Polarization of π,π^* bands of fluorenone analogs		210
(c) Fluoranthene and its derivatives		211
(d) Carbonium ions derived from acenaphthylene and 1,1'-biace-naphthylidine (biacene)		212
C. Aromatic compounds		
I. Far ultraviolet		
a. Vacuum-ultraviolet spectroscopy and the ionization potential of aromatic hydrocarbons		213
b. Interference effects in the Rydberg spectrum of naphthalene		214
c. N-[α(and β)naphthyl]acetamide and acetyl(1-naphthylmethyl) amine		215

TABLE II—continued

System	Calculation method	Reference
D. Heterocyclic Compounds		
I. Mixed heterocycles and comparative studies		
a. Theoretical models and calculations		
1. Mixed heterocycles		
(a) Application of perturbational MO approach to color of dyes	PMO	218
(b) Anthranil, benzisoxazole, and benzoxazole	SCF-MO	219
(c) N-oxides of benzofurazan	PPP	220
(d) Phenoxazine, phenothiazine, and phenoxanthin		221
(e) Triphenedioxazines	PPP	222
2. Comparative studies		
(a) Five-membered heterocycles containing nitrogen, oxygen, and sulfur	PPP	223
(b) Five-membered heterocycles	PPP	224
(c) 1,2-Difuryl- and 1,2-dithienyl-ethylenes and benzodifurans and benzodithiophenes	HMO	225
(d) Bithiophenes and bifurans	PPP	226
(e) An MO interpretation of Hewitt's law relating λ_{max} to number of conjugated double bonds in aromatic- and heteroaromatic-substituted polyenes		227
(f) Comparison of iso-π-electronic oxygen, nitrogen, sulfur, and selenium heterocycles		228
b. Experimental results		
1. Mixed heterocycles		
(a) Oxygen-nitrogen heterocycles		
(1) Oxazole, isoxazole, and furazan		229
(2) Diphenyloxazoles		230
(3) Naphthoxazoles		231
(4) 2-Aryl derivatives of 3,4-dihydronaphtho-[1',2':5 4]-oxazoles		232
(5) Effect of intermolecular hydrogen bonding on spectra of aryl-oxadiazoles		233
(6) Phenyloxadiazoles		234
(7) Phenoxazine dyes		235

TABLE II—continued

System	Calculation method	Reference
(b) Sulfur-nitrogen heterocycles		
(1) Phenothiazine drugs		236
(2) o-Substituted azorhodanines		237
(3) 5-(2-Thiazolylazo)-2,6-diamino-pyridine		238
(4) α. ε-Bis(4-oxo-2-thioxo-3-thiazolidinyl)caproic acid and its 5-arylidene derivatives		239
(5) Some dyes containing two or more conjugated heterocyclic chromophores		240–243
(6) 5-(2-Thiazolylazo)-2,6-dihydroxy-pyridine		244
(7) 3,5-Disubstituted tetrahydro-1,3,5-thiadiazine-2-thiones		245
2. Comparative studies		
(a) Acetylenic derivatives of furan and thiophene		246
(b) Naphthostyrils and thianaphthostyrils		247
II. Nitrogen heterocycles		
a. Theoretical models and calculations		
1. N-heterocyclic benzene analogs	CNDO	248
2. Aza aromatics		249
3. Protonated N-heterocycles	β-PPP	250
4. Aza derivatives of biphenyl	PPP	251
5. Pyrrole and benzocondensed five-membered nitrogen heterocycles	SCF-CI	252
6. Imidazole	SCF-CI	253
7. Pyridine-N-oxide	PPP, CNDO	254
8. Reduced N-alkylnicotinamide	SCF	255
9. 3,5-Dicyanopyridines	HMO	256
10. trans-1,2-Dipyridyl ethylene	β-PPP	257
11. Azonaphthalene	mod SCF	258
12. Hydroquinolines	HMO	259
13. Methyl- and halo-substituted quinolines and isoquinolines	PMO	260
14. Atypical methylated purine and pyrimidine compounds	PPP	261
15. 2-, 6-, and 8-substituted purines	PPP	262
16. Homopurines		263
17. 2,6-Disubstituted purines	PPP	264
18. Optical properties for dinucleoside phosphates from time-dependent Hartree theory	FEMO	265

TABLE II—continued

System	Calculation method	Reference
19. Estimated coulomb integral for carbazole calculations	HMO	266
20. Acridan	PPP	267
21. Monoaminoacridine	β-PPP	268
22. Monohydroxyacridines	β-PPP	269
23. Acridine-N-oxide	β-PPP	270
24. Pyrido[2,3-d]pyridazine and pyrido[3,4-d]pyridazine	PPP	271
25. Pyrazolo[1,2-a]pyridazinium cation, benzo[c]pyrazolo[1,2-a]cinnolium cation, and pyrrolo[1,2-f]phenanthradine		272
26. Corrin and related compounds	PPP	273
27. Vibronic borrowing in cyclic polyenes and porphyrins		274
28. Methyl flavins	PPP	275
29. Franck-Condon principle and light absorption of merocyanines	PPP, HMO	276
30. s-Triazines		217

b. Experimental
  1. Far ultraviolet
    (a) Pyrrole in gas phase   277
    (b) L-Diketopiperazines   278
  2. Near ultraviolet, visible, and near infrared
    (a) Enol forms of 1-butyl-4-cyano- and 1-butyl-4-ethoxycarbonyl-2,3-dioxopyrrolidine   279
    (b) 4-Nitropyrrole analogs of chalcone   280
    (c) 1-(p-substituted phenyl)pyrroles   281
    (d) Rotational fine structure of 283.8 nm band of indole   282
    (e) Molecular weight of imidazole derivatives by extinction coefficients of their picrates   283
    (f) Nitroimidazoles   284
    (g) 2,5-Disubstituted 1-methyl-benzimidazoles   285
    (h) Benzimidazolethione derivatives   286
    (i) Bis(2-phenylphenanthro[9,10-d]) imidazoyl radical   287
    (j) Acetylnaphthoylenebenzimidazoles   288
    (k) Aromatic and heteroaromatic-substituted 2-pyrazolines   289
    (l) 1,3,5-Triphenyl-2-pyrazoline   290

TABLE II—continued

System	Calculation method	Reference
(m) Electron-releasing properties of pyrazole ring		291
(n) Vinylogous azolides containing pyrazole, triazole, and tetrazole group		292
(o) 4-Thiohydantoins and 2,4-dithiohydantoins		293
(p) 1,2,4-Triazole and its 4-amino derivative		294
(q) Substituent effects on pyridine chromophores		295
(r) 4-Chloropyridine		296
(s) 2-Substituted-5-nitropyridines		297
(t) 2-Substituted-3-nitropyridines		298
(u) 3,5-Dinitropyridine		299
(v) Spectra of pyridinium and quinolinium inner salts		300
(w) Pyrimidine-h_4 and -d_4 in gas phase		301
(x) Free radicals from reaction of hydroxyl free radical with pyrimidine bases		302
(y) Alkylamino-s-triazines		303
(z) s-Tetrazine		
(i) Rotational structure of 551·5 nm band		304
(ii) Vibrational structure of s-tetrazine-d_0 and -d_2		304a
(aa) 7-Azaindole		305
(bb) Purine tautomers		306
(cc) 1,2,3,4-Tetrahydromercaptoquinoline		307
(dd) Methyl derivatives of 8-mercaptoquinoline		308
(ee) 5-Sulfo-8-mercaptoquinolines		309
(ff) 5-Halo-8-mercaptoquinolines		310
(gg) Aminoacridines		311
(hh) N-Aroylcarbazoles		312
(ii) 5,10-Dihydrophenazine		313
(jj) 6-Nitro-1-methyl-2-ethyliso-β-carboline		314
(kk) n,π^* Transition of quinazoline in gas phase		315
(ll) Two n,π^* transitions in quinoxaline		316, 317
(mm) Phthalazine		318

TABLE II—continued

System	Calculation method	Reference
(*nn*) Interaction of conjugated chromophores in bishemicyanine molecules		319, 320, 321
(*oo*) Azapyridocyanines		322
(*pp*) Lumiflavine		323
(*qq*) Transition probabilities in porphyrins and metalloporphyrins		324
(*rr*) Porphyrins and porphyrin cations		325
(*ss*) Chlorophyll *a*		326, 327
(*tt*) Chlorophyll *b*		328
III. Oxygen heterocycles		
a. 1-(5-nitro-2-furyl)polyenes	FEMO	329
b. Acetals of furan series		330
c. Xanthone derivatives		331
d. Coumarin derivatives		332
e. Coumarone derivatives		333
f. 1,4-Benzodioxan derivatives		334–336
IV. Sulfur heterocycles		
a. Arenedithiocarboxylic acid esters	HMO	337
b. Mercaptotropylium ion and iso-π-electronic sulfur heterocycles	PPP	338
c. 6*a*-Thiathiophthenes	SCF-MO	339
d. 2*H*-Thiopyran-2-thiones	PPP	340
e. Naphthothiopyrans	PPP	341
f. 3-Substituted thianaphthenes		342
V. Other heteroatoms		
a. 2,4,6-Tris(ethylamino)-1,3,5-triethylborazine	mod HMO	343
b. Selenophene: review of experimental and theoretical studies		344
c. Chalcone analogs containing selenophene		345

B. Absorption Spectroscopy of Spin-forbidden Transitions

Part *A* of Table III tabulates the reports in this area during the past year. Goodman and Laurenzi [346] discuss the probability of the spin-forbidden singlet triplet absorption based on a vigorous quantum-mechanical treatment. They review the various kinds of equations for calculating such probabilities, and they discuss the various factors contributing to the intensity of S,T absorption bands.

TABLE III

Absorption Spectroscopy of Electronically Excited States and Spin-forbidden Transitions

System	Calculation method	Reference
A. Spin-forbidden transitions: singlet-triplet absorption		
I. Theoretical models and calculations		
a. Probability of S,T transitions: a general treatment		346
b. S-T separation in methylene	HF	347
c. S,T transition probabilities in linear polyacene		348
d. Pyrrole and benzocondensed five membered nitrogen heterocycles	SCF-CI	252
e. Anthranil, benzisoxazole, benezoxazole	SCF	219
II. Experimental results		
a. Effect of bromo substitution on S,T absorption of naphthalene		349
b. Pyrazine crystals		350
c. Naphthalene crystals		351
d. Dichloronaphthalenes: vapor phase and quasi-line crystalline solution spectra		352
e. Anthracene and 9,10-diphenylanthracene by oxygen perturbation		353
f. Benzophenone crystals		354
g. Quasiline S,T spectrum of dibenzofuran		355
h. Azine crystals		379
B. Absorption from electronically excited states		
I. Singlet-singlet absorption		
a. Benzene		356
b. Benzene excimer		357
c. Benzene and mesitylene		358
d. Naphthalene and biphenyl		359
e. Naphthalene excimer, 1,2-benzanthracene		360
f. Anthracene		361
g. 1,2-Benzanthracene, pyrene, 3-bromopyrene		362
h. Polynuclear aromatics		363
i. Coronene		364
II. Triplet triplet absorption		
a. Techniques		
1. Extinction coefficients of T,T absorption by esr		365, 366
2. T,T absorption spectrum by excitation spectrum of radical formation		367
3. T,T absorption spectrum by luminescence excitation		368

TABLE III—continued

System	Calculation method	Reference
4. Polarization of T,T transitions		369, 370, 371, 372, 386
5. Optical rotatory dispersion technique for excited states: T,T transitions		373, 374
b. Theoretical models and calculations		
1. Aromatic hydrocarbons	FEMO	376
2. Aromatic hydrocarbons	HMO, SCF-MO	377
3. Naphthalene, 1- and 2-naphthol	PPP	378
4. Phenylnaphthalenes	SCF-MO	366
5. Xanthene and acridine dyes	FEMO	380
6. Dehydrodianthrones	PPP	111
c. Experimental results		
1. Aliphatic hydrocarbons		
(a) Camphorquinone and biacetyl		381
(b) Lycopene		406
2. Aromatic hydrocarbons		
(a) Quasiline T,T absorption spectra: review		382
(b) Benzene		356, 383
(c) Retinol (excited by energy transfer)		384
(d) Triphenylamine		385
(e) Aniline; N,N-dimethylaniline; N,N,N,N-tetramethyl-p-phenylenediamine		386, 387
(f) Triphenylmethane derivatives		388, 389
(g) Chloranil		390
(h) Deuteronaphthalene		391
(i) Quasiline T,T absorption spectrum of		392, 405
(j) Naphthalene and monosubstituted derivative		393
(k) Naphthalene		394
(l) Anthracene		361, 370
(m) 9,10-Phenanthrenequinone		395
(n) Polynuclear aromatics		363, 369, 371, 376, 396
(o) 1,2-Benzanthracene, pyrene 3-bromopyrene		362
(p) Coronene		364
3. Heterocyclic compounds		
(a) Substituted aromatics and hetero aromatics		397
(b) Tryptophan and tyrosine		398

TABLE III—continued

System	Calculation method	Reference
(c) Uracil		399, 400
(d) 6-Uracilcarboxylic acid (orotic acid)		407
(e) Acridine		401
(f) Acridine dye adsorbates		402
(g) Cyanine dyes		403
(h) Photochromic spiropyrans		404
(i) Phenoxazine and phenothiazine		372
(j) Phenosafranine and neutral red		408
(k) Thiopyronine		409
(l) Azine crystals		379

Krishna and Salzman [348] find in a model calculation that the oscillator strength of the $A \rightarrow L_a$ (π,π^*) transition in linear polyacene should decrease with increasing size of the hydrocarbon. Their calculations show that the intensity of such a transition arises from the mixing of the (π,π) and (σ,π) states by spin-orbit interaction.

El-Sayed [349] demonstrates the intramolecular heavy atom enhancement of the intensity of S,T absorption in a series of brominated naphthalenes. Brinen and Koren [353] use oxygen-perturbation methods to obtain the spectrum of S,T absorption in anthracene and 9,10-diphenyl anthracene and find that phenyl substitution decreases the S_0-T_1 splitting. Crystal spectra were obtained and analyzed for benzophenone [354], naphthalene [351], and pyrazine [350].

C. Absorption Spectroscopy of Electronically Excited States

The spectroscopy of excited state → excited state transitions is a relatively recent but highly useful technique. Previous limitations set by the short lifetimes of these excited states have been successfully pushed back. As can be seen in Part B in Table III, this kind of spectroscopy is becoming quite common, and with the further development of flash and laser photolysis equipment, this tool will become even more important as a routine method of elucidating both excited-state structure and photolytic reaction pathways.

The microsecond and longer lifetimes of triplets have made them fairly accessible for some time. This long lifetime, combined with their paramagnetism, means that the concentration of triplets can be measured by electron paramagnetic resonance techniques and the extinction coefficients of their transitions can in some cases be measured directly [365, 366].

Besides the usual flash-photolytic method of locating T,T absorptions, Naumova and Glyadkovskii [368] use the excitation spectra of luminescence in benzene and naphthalene to locate the triplet-triplet absorptions. Zhuravleva [367] uses the formation of radicals in a matrix, monitored by epr spectroscopy, to determine the absorption in the triplet manifold. He obtains good agreement with the flash-photolytic spectra for benzene and a series of substituted acetylenes. Strong, Richtol, and Ramme [373, 374] study the optical rotatory dispersion of the T,T absorption in benzoin and in 7-keto-13-methyl-5,6,7,9,10,13-hexahydrophenanthrone. They find that the benzoin triplet, from resolved material, is optically active, but they can find no other optically active species in either system.

The short lifetime of excited singlets formerly precluded studying their spectra. The development of pulse radiolysis and Q-switched lasers means that high concentrations of singlets can be obtained in a short time, so that their spectra become measurable. As can be seen in Part B.I in Table III, a number of aromatic hydrocarbons have now been investigated by this technique.

Judeikis and Siegel [375] derive analytical expressions for the effect of singlet-singlet and triplet-triplet absorption in optically thick samples. These equations allow them to calculate the average ground-state depopulation that can be related to epr and phosphorescence intensities.

D. Polymolecular Effects in Absorption Spectroscopy

Perturbations on the spectrum of a molecule due to its environment can provide information about the molecule itself, in both its ground and excited states, and about the perturbing environment. Precisely because these perturbations carry such a wealth of information, their interpretation is often difficult if not impossible. But as studies in polymolecular effects progress, we may expect them to continue to provide a depth of understanding inaccessible by any other means. Perturbations due to solvation are sometimes among the largest observed, depending on the polarity of the solvent and the solute. Midwinter and Suppan [411] discuss the enhanced perturbation of the spectra of polar solutes in nonpolar solution when even a low concentration of polar solvent is added. This deviation from linearity of the polarity of solvent mixtures has often been noted, but Midwinter and Suppan provide a quantitative description in terms of the balance between dipole-dielectric stabilization and the entropy change due to orienting the solvent. Suppan [414] also discusses the usefulness of solvent-induced spectral shifts in calculating excited-state polarizability and dipole moments, using the Onsager dielectric model.

Bakhshiev [413] summarizes results from some of his previous work and gives some new calculations in a review of the effect of various van der Waals interactions on solvent perturbations. Included are tables for various kinds of compounds and solvents of the magnitude of solvent shifts due to orientation, induction, and dispersion-polarization of van der Waals interactions.

Irving, Byers, and Leermakers [415] point out that the solvent polarizability can also have a strong effect on the electronic spectrum, even in polar solvents. They propose a model involving enhanced stabilization of the Franck-Condon state by polarizable solvents to account for the large shift (454 to 508 nm) in the λ_{max} in the spectrum of all-*trans*-retinyl-pyrrolidinium perchlorate on going from polar, nonpolarizable solvents to polar, polarizable solvents.

Electron donor-acceptor interactions are surveyed in an article by Mulliken [432]. This article covers both absorption and emission spectra in gas, solid, and liquid phases. Lower [448] reports an interesting class of nonstoichiometric charge-transfer compounds prepared by making a solid solution of trinitrobenzene in mixed aromatic hydrocarbons. The unexcited solid solution becomes a nonstoichiometric charge-transfer complex on excitation.

Two reviews [472, 410] of the tendency of dye molecules to aggregate in solution were published. In addition Leterrier and Douzon [472a] produce a lengthy discussion of the possible biological consequences of the tendency of various dye radicals to aggregate in solution, as shown by absorption and emission spectra and electron-spin-resonance investigation. Akbarova, Levshin, and Klemenkova [475] study solvent effects on such dye aggregation. Korovina and Bakhshiev [476] demonstrates with Rhodamine 6 and malachite green that there is little or no change in the oscillator strength of the spectra in going from a film to a solution of dye molecules.

The interesting crystalline solution spectral phenomenon known as the Shpol'skii effect is discussed in Section III.A. These "quasiline spectra" are found in both absorption and emission.

Tanaka and Tanaka [508] discuss a new method for describing exciton states of molecules in a way analogous to π-electron methods in molecular quantum mechanics. They apply the method to the analysis of the spectrum of anthracene. Philpett [523] discusses dipole Davydov splitting in the exciton spectrum, allowing in his calculation for the boson character of excitons. He applied his calculations to the ^1A_g, ^1L_a transitions of naphthalene, anthracene, phenanthrene, and tetracene. He finds excellent agreement for anthracene and tetracene, and he discusses the special crystal effects producing deviations from theory in the other two systems.

O. S. Davydov [506] discusses the theory of the Urbach rule describing the line shape of the low-energy absorption in crystals. Frolova and Kurik discuss the temperature variation of the Urbach steepness parameter σ.

TABLE IV
Polymolecular Effects in Absorption Spectroscopy

System	Calculation method	Reference
A. Solvent-solute interactions		
I. Theoretical models and calculations		
a. Deviations from additivity of solvent dipole effects in mixed solvents		411
b. Validity of Doub-Vandenbelt equation in nonaqueous solvents		412
c. Effect of van der Waals solvent-solute interaction in electronic spectra of molecules		413
d. Review of applicability of Onsager dipole model in solvent-solute interactions and its use for elucidating excited-state properties (Cf. Table V, *A.* II.*a.*4 and *A.*II.*b.*)		414
e. Effect of solvent polarizability on spectrum of all-*trans*-retinyl pyrolidinium perchlorate		415
f. Effect of dispersion on measured solution spectra		416, 417
g. Electrochromism and solvatochromism		418
II. Experimental results		
a. Solvatochromic shifts and catalytic hydrogenation: empirical correlations		418a–421
b. Solvent and temperature effects on spectrum of benzene: Ham bands		422
c. Solvent effects on spectrum of 1,4-benzodioxan: intensity changes		423
d. Chloronitrobenzene complexation with tetrahydrofuran solvent		424
e. Temperature effects on hydrogen bonding to solvent: nitrophenols in aqueous solution		425
f. Solvent interaction with intermolecular hydrogen bonds		426
g. Solvent effects on n,π^* transitions		
1. *o*- and *p*-Nitrophenol		427
2. Benzophenone		428
3. *p*-Benzoquinone		429
4. Acyclic aliphatic α-enones		430
5. Hydroxyanthraquinones		431
h. Rhodamine B in organic solvents		431a
B. Charge-transfer complexes		
I. Theoretical models and calculations		
a. Survey of molecular complexes: theory and experiment		432

TABLE IV—continued

System	Calculation method	Reference
b. Hückel calculations of spectral shifts of pyridine in charge-transfer compounds		433
c. Hückel and SCF studies of interaction of nitrobenzene with aniline and phenol		434
d. CNDO-CI calculations of intramolecular charge transfer in N-alkyl-N-nitroso anilines		435
e. Hückel calculations of acceptor properties of cyanopyridinium ions		436
II. Experimental results		
a. Cyano acceptor compounds		
1. Tetracyanoethylene (TCNE) with phenylthiazole		437
2. TCNE with α,ω-diphenylpolyene ketones		438
3. 2-(Dicyanomethylene) indan-1,3-dione with π donor compounds		439
4. 2-(Dicyanomethylene)-1,1,3,3-tetracyanopropane with quinoline		440
5. α,α'-Dicyanostilbene and its derivatives		441
6. 7,7,8,8-Tetracyano-p-quinodimethane (TCNQ) complexes with p-phenylenediamine derivatives		442
7. TCNQ complexes with tetramethylbenzidine		443
b. Quinone acceptor compounds		
1. Chloranil with tetramethylbenzidine		444
2. Chloranil with hexamethylbenzene, durene, and mesitylene		445
3. Chloranil with α,ω-diphenylpolyene ketones		438
4. Chloranilic acid with amino acids		446
5. Chloranil with 9,10-dichloroanthracene		447
c. Nitrobenzene acceptor compounds		
1. Nonstoichiometric complexes of trinitrobenzene with arometic hydrocarbons		448
2. Trinitrobenzene complexes with tetraalkylammonium halides		449
3. Picric acid complexes with amine		450
d. Iodine complexes		
1. Iodine complexes with amides		451
2. Iodine complexes with disulfides		452, 453

TABLE IV—continued

System	Calculation method	Reference
3. Iodine complexes with pyridine: solvent effects on their formation		454
4. Binding energy of the hexamethyl-benzene-iodine complex		455
e. Complexes of pyridinium ions and iodide ions		456
f. Miscellaneous studies		
1. Charge-transfer donor abilities of *o,o'*-bridged biphenyls		458
2. Hexafluorobenzene and pentafluoro-benzonitrile as charge-transfer acceptor compounds		459
3. Complexes of amines with aliphatic halides		457, 460
4. Hexamethylbenzene complexes with tetrachlorophthalic anhydride and pyromellitic dianhydride		461
5. 4-Nitroquiniline-1-oxide complexes with methylbenzene and methyl anilines		462
6. Interaction of aniline with eosin		463
7. Interaction of solid layers of dyes with ammonia		464
8. Interaction of acridine orange with DNA		465
9. Charge-transfer interaction between substituted pyridinyl radicals		466
10. 4-Nitroquinoline-*N*-oxide with nicotinamide		467
11. Würster salts		146
C. Spectra of aggregates		
I. Ethylene dimer		16
II. Dimer cations		
a. Benzene		468
b. Cyclohexene, butadiene, and 1,3-pentadiene		469
c. Naphthalene and anthracene derivatives and pyrene		470
III. Clustering of terphenyl radical anions with sodium cations in tetrahydrofuran solution		471
IV. Absorption of light by dye aggregates		
a. Review		472
b. Monomer, dimer, trimer, and polymer in acriflavine and Rhodamine 3B		473
c. Azine, xanthene, and triphenylmethane dyes		474
d. Resonance interaction effects on lineshapes in spectrum of Rhodamine 6Zh		475

TABLE IV—continued

System	Calculation method	Reference
e. Differences in Einstein absorption coefficient $B(\nu)$ and absorption coefficient $K(\nu)$ in spectra of dye aggregates		476
f. Würster's cation		477
V. Anthracene aggregates in glassy solution		478, 479
VI Aggregation of chlorophyll in aqueous and nonpolar solvents		480, 481
VII. Ion-pair formation of 3,6-diaminoacridine sulfate in dimethyl formamide		482
D. Crystal spectra		
I. Mixed crystals and crystalline solutions		
a. Quasiline spectra (Shpol'skii effect)		
1. Nature of Shpol'skii effect		483
2. Nature of multiplet structure of quasiline spectrum of benzene in cyclohexane		484
3. Concentration dependence of quasiline spectra of aromatic hydrocarbons		485
4. Vibrational analysis of quasiline spectrum of dibenzofuran		355
5. Effect of crystallization rate and concentration on quasiline spectrum of dibenzofuran		486, 487
6. Vibrational structure of quasiline spectra of *p*-substituted stilbenes		488, 489 490
7. Vibrational structure of polynuclear aromatics		491
8. Isoviolanthrone		492
9. Quasiline spectrum of 1-azatriphenylene compared that calculated by PPP method		493
10. Naphthoxazoles		231
11. Porphyrins		325
12. Quasiline T,T and S,T spectra		
(*a*) Review of quasiline T,T spectra		382
(*b*) Effect of ethyl bromide on quasiline spectra of aromatic hydrocarbons		392, 405
(*c*) Quasiline S,T spectrum of dibenzofuran; analysis of vibrational structure of T_1		355

TABLE IV—continued

System	Calculation method	Reference
(d) Quasiline S,T spectra of chloronaphthalene		352
b. Mixed crystals		
1. Theoretical models and calculation		
(a) Persistence and amalgamation types in optical spectra of mixed crystals		494
(b) Retarded interactions in mixed crystals		495
(c) Electronic states of dilute mixed molecular crystals		496
2. Experimental results		
(a) Matrix and concentration effects on spectra of polymethyl benzenes in rapidly cooled n-hexane		497, 498
(b) Polarization of excitonic excitations of benzene crystals doped with heteroaromatics		499
(c) Spectrum of anthracene in rapidly cooled n-heptane		500
(d) Polarization of second singlet state of phenanthrene in polymethylbenzene crystals		501
(e) Benz(f)indan in fluorene: polarization and vibrational structure		502
(f) Polarization of acenaphylene absorption in acenaphthene and durene host crystals		503
II. Molecular crystals		
a. Theoretical models and calculations		
1. Excitons: a review		504
2. Effect of crystalline state on molecular spectra: a discussion of current theory		505
3. Theory of the Urbach rule		506
4. Use of "effective molecule" method for studying molecular crystal spectra		507
5. Calculation of excitonic states of anthracene		508
6. Calculation of exciton-phonon coupling in infinite anthracene crystal		509
7. Analysis of surface excitation absorption of polymethine dye crystal		510

TABLE IV—continued

System	Polarization study	Reference
b. Experimental results		
1. Triglycine sulfate	X	511
2. Deuterobenzenes: molecular and mixed crystals		512
3. Benzoic acid crystals		513
4. Vibrational structure of p-disubstituted benzenes	X	514
5. Würster's ion	X	515
6. Tetracyanoquinodimethan		516
7. 4,4'-Dichlorobenzophenone	X	517
8. trans-Stilbene	X	518
9. Benzil	X	519
10. 4,6,8-Trimethylazulene	X	520
11. Fluorene	X	521, 522
12. Naphthalene, anthracene, phenanthrene, tetracene: dipole Davydov splitting		523
13. Application of Urbach rule to 9,10-dichloroanthracene		524
14. Temperature effects on sintle crystals of tetracene and pentacene	X	525
15. Temperature effects on lineshapes in pentacene spectrum: nature of exciton-phonon interaction		526
16. Pentacene	X	527
17. S,T absorption in perylene crystals: Zeeman studies	X	528
18. Purine	X	529
19. 1,3,5,7-Tetranitro-1,3,5,7-tetrazacyclooctane: β-crystal form		530
20. Cyanine dyes	X	531
21. Azines		379

II. PROPERTIES OF EXCITED STATES

One goal of photochemical and photophysical studies has been to obtain information about the structure and properties of the excited states that are formed. But these studies have been long hampered by the inherently brief lifetime of these states. Some information can be obtained from the normal absorption and emission spectra. Indeed, it is from studies of emission spectra that we know that excited-state lifetimes are so brief. Thus, some information about the properties of excited states can be found in the tables and text on emission and absorption spectroscopy.

Direct evidence of excited-state properties, however, has been greatly enlarged by the development of electron-spin-resonance spectroscopy. Since triplet excited states may have lifetimes measured in milliseconds and longer, and since they are paramagnetic, they can be directly observed by esr spectroscopy. A number of such studies are listed in Table V, Part B.II.a. One interesting study is the correlation of the zero-field splitting parameter D and the energy of the triplet state above the singlet ground state. The experimental study, Riley and Rosenthal [552], on β-diketones is an extension of the experimental and theoretical results of Brinen and Orloff [552a], on condensed aromatics. Several reports have appeared on the effect of a microwave field on the phosphorescence spectra of organic compounds. Such studies were originally carried out in the esr cavity, such as the reported study of quinoxaline and naphthalene by Forman and Kwiram [552]. Two groups, however, have since extended these studies to zero-field conditions, so that zero-field splitting information can be directly obtained [568–570]. Turro [543] reviews the properties of the organic triplet state and some of the techniques used to study it.

TABLE V
Properties of Excited States

System	Calculation method	Reference
A. Singlet states		
I. Calculations of excited state geometries		
a. Ethylene	CNDO	532
b. Toluene		533
c. Azulene		534
d. Phenol and phenolate ion	LCAO-SCF-CI	535
e. N-Heterocycles	LCAO-SCF-CI	536
II. Experimental results		
a. Dipole moments of the first excited state		
1. Propynal: from electric field–induced spectrum		537
2. Phenol and aniline: from optical Stark effect		538
3. p-Fluorophenol and p-fluoroaniline: from optical Stark effect		539
4. Eight naphthalene derivatives: from shift induced by solvation		540
5. Azulene: from optical Stark effect in mixed crystals		541

TABLE V—continued

System	Calculation method	Reference
6. 1,3,5-Triphenyl-2-pyrazoline derivatives		290
7. 9-Bromoanthracene from solvent shifts		541a
8. Indole and methylindoles from solvent shifts		541b
b. Polarizability of first excited states of anthracene and naphthalene		542
B. Triplet excited states		
I. Calculation of triplet-state properties		
a. Zero-field splittings		
1. Trimethylenemethane	PPP	544
2. Aromatic hydrocarbons and nitrogen heterocycles	PPP	545
b. Differences in expected stability of triplet state of cyclic II systems containing $4n$ electrons: group theoretical considerations		546
c. Structure of two lowest acraldehyde triplet states	CNDO	547
d. Structure of triplet states of N-heterocycles		536
e. Structure of nitrogen-containing stilbene analogs in their triplet states	CNDO	548
II. Experimental results		
a. Electron-spin-resonance studies of triplet state		
1. Anisotropic saturation of esr of pyrene-d_{10}		549, 550
2. Electron-nuclear double resonance of naphthalene and deuteronaphthalenes		551
3. Correlation between zero-field splitting parameter D^* and triplet energies of β-diketonates		552
4. Benzene in benzene-d_6 crystal		553
5. Mesitylene in B-trimethylborazole		554
6. Napthalene-d_8, quinoxaline, and cyclopentanone		555
7. Phenanthrene-h_{10} and -d_{10} oriented in biphenyl crystals		556
8. Triphenylene oriented in dodecahydrotriphenylene crystals		557
9. Benzophenone single crystals		558

TABLE V—continued

System	Calculation method	Reference
10. Naphthalene carbonyl compounds		559
11. Pyrazine		560
12. Pyrazine in acid solution		561
13. Phenothiazine derivatives		562
14. Phenoxazine and phenothiazine oriented in biphenyl crystals		563
15. Coumarin and related compounds		564
16. Dicoumarol		565
17. Acridine dyes		566
18. Charge ions for complexes of aromatic hydrocarbons and 1,2,4,5-tetracyanobenzene		567
b. Zero-field microwave-optical double resonance		
1. Discussion		568
2. Tetramethylpyrazine		569
3. Quinoxaline		570
c. Optical spin polarization of spectrum of naphthalene triplet		571
d. Optical detection of magnetic resonance in naphthalene and quinoxaline (in durene and biphenyl crystals)		572
e. Zeeman and Stark effects in optical absorption of benzophenone triplet excitons: zero-field splittings		573
f. Basicity of n,π^* triplet of acridine from pH dependence of photoreduction quantum yield		574

Studies of the singlet state have increased somewhat this year. Dipole moments of the first excited state have been obtained by two basic methods. The application of a strong electric field leads to the well-known Stark splitting of states that can provide excited-state dipole information. Hochstrasser and Noe [541] use this effect in the crystal state to determine the excited-state dipole of azulene ($\Delta\mu = -0.42D$ for the first singlet and $-0.31D$ for the second). Lombardi and Huang [538, 539] use the gas-phase Stark effect to determine $|\Delta\mu|$ for p-fluoroaniline (0.82D), p-fluorophenol (0.44), phenol (0.20), and aniline (0.85), The gas-phase measurements, unfortunately, do not yield the sign of the dipole moment. Conrad and Dows [537] use an oscillating electric field to produce a special Stark effect known as the electric field-induced spectrum. By this technique they are able to determine that the dipole moment of propynal along the *a* principal rotational axis is $0.97D$. The second method involves the shift in the fluorescence and absorption maxima that occurs when a gaseous compound is dissolved. Suppan [414] discusses this effect in terms of the Onsagar dielectric model and discusses some experimental results available to him. Uzhinov, Kozachenko, and Kuz'min used this technique to measure the excited-state dipole moments of some naphthalene derivatives [540] and the polarizability of anthracene and naphthacene [542]. Kutsyna, Voevoda, Tishchenko, and Shepel [290] use this method for obtaining the dipole moments of 1,3,5-triphenyl-2-pyrazoline derivatives.

III. DECAY OF ELECTRONICALLY EXCITED STATES

Eventually all electronically excited molecules are either deactivated to ground-state molecules and heat or light or react to form photochemical products. The latter process is well reviewed in other parts of this book, and so we concern ourselves only with the loss or transfer of thermal or radiative energy by an excited molecule.

The natural classification of deactivating processes is a division of what we can see from what we cannot, that is, of radiative and nonradiative processes. In general, molecules decay from excited states by both these routes, but the radiative processes are easier to study and make up the bulk of the literature. In fact there is very little that we know about the nonradiative processes that we have not found out by studying radiative ones. Apropos to this experimental situation, we have divided this section along the same lines.

Among the nonradiative processes, we have included not only the intramolecular processes by which a molecule cascades down from one excited state to another—internal conversion and intersystem crossing—but also the processes by which it passes its excitation to another molecule, either by energy transfer or by the formation of excited complexes. Thus, under radiative processes we are concerned only with molecular luminescence and with processes that are best seen as perturbations of molecular luminescence, such as solvent effects.

A. Radiative Decay of Excited States

1. MOLECULAR LUMINESCENCE

One important discussion of molecular luminescence is a review by Drexhage [575] of the nature of molecular luminescence. In a number of recent studies he has utilized multilayered monomolecular films to provide "spacers" of appropriate thickness to study the luminescent molecule as an oscillating dipole. He has been able to show that the properties of these molecular emitters correspond very closely to the known properties of the oscillating dipoles used to emit in the radio frequencies of the electromagnetic spectrum. Thus, he is able to observe interference effects on the frequency of the emitted light and on the radiative decay rate. The latter observation is very significant, because it allows for separate observation of the radiative and nonradiative rate constants for decay. Furthermore, by studying the angular distribution of an emitter held close to a mirror, he is able to distinguish the various kinds of dipole and quadrupole emissions involved. Thus, his group has been able to show for a monomethinocyanine dye that the fluorescence, as expected, is electric dipole in character, but the phosphorescence is at least partially electric quadrupole. Although the experimental requirements of his technique excludes its usefulness for some kinds of organic molecules, this method is quite general and should prove a powerful aid to the study of both radiative and nonradiative decay. Nurmukhametov and Shigoria [578a] suggest a method of predicting the luminescence properties of molecules based on the relative placement of singlet and triplet π,π^* and n,π^* levels. Use of the method permits designing luminophors with specific properties.

Jordon, Mordecai, and Freed [576–578] discuss the theory of radiative decay and apply it to the problem of the long radiative lifetime of small molecules and to the relation between radiative and radiationless transitions.

Grabowski [579] reviewed the experimental aspects of luminescence, including the kinetics of luminescence, artifacts due to reabsorption, measurement of quantum yields, and environmental effects. Ejder [582] discusses the terminology and presentation of emission and excitation spectra. Hamilton and Naqvi [581] discuss the conditions under which the monomer and excimer fluorescence should show an isosbestic point, and show that for pyrene these conditions are met below -40 and above 50°C. Seybold, Gouterman, and Collis [583] discuss a calorimetric method of determining fluorescence yield and apply it to fluorescein and its halogenated derivatives.

One interesting aspect of emission, compared with absorption, spectroscopy is that polarization of the fluorescence can be determined even in fluid solution. Suzuki [589] reviews the principles and methods of such measurements. Several authors [590–593] discuss depolarization of emission by rotational diffusion, energy transfer, and intramolecular displacement, especially in macromolecules [592–593], where such measurements are used to study conformation and active sites.

Parker [604–605] discusses triplet-singlet luminescence in fluid solution.

Until 1968 high-yield phosphorescence had not been obtained in fluid solution. It was believed that this was due to adventitious quenching by impurities. The occurrence of such luminescence in highly purified, inert solvents seems to confirm this view.

Azumi [601–603] discusses the theory of photoselection in Zeeman split phosphorescence, deriving general equations for describing it and applying them to pyrazine.

Theories of the mechanism of phosphorescence focus on the mixing of singlet and triplet by spin-orbit coupling. Pavlopoulos [606] and Lim, Li, and Li [607] describe the vibronic spin-orbit interaction. The interaction of a heavy-atom, with its higher spin-orbit coupling constant, and an excited molecule leads to enhanced intersystem crossing that causes increased population of the triplet state and shortness of the phosphorescence lifetime. The effect is especially strong when the heavy atom is intramolecular, but it is also a significant effect in heavy-atom solvents. El-Sayed [349] discusses the effect in bromonaphthalenes; Metzger, Smith, and Meyer [610] observe the effect in naphthalene and phenanthrene embedded in rare gas and chemically inert gas matrices. In methane they find that only fluorescence occurs, in argon both fluorescence and phosphorescence, and in krypton only phosphorescence. The phosphorescence intensity in annealed samples depends on the temperature but not on the lifetime.

Henry and Siebrand [618, 626] consider the effect of the interaction of σ states, both excited and ground, in the spin-orbit coupling in naphthalene. They calculate the natural radiation lifetime of the naphthalene triplet and find that only states associated with C—H bonds are significant and that the α CH bonds make more important contributions than β CH bonds.

One of the most active areas of research has been the kinetics of luminescence decay. The observed lifetime τ_{obs} is related to the natural radiative lifetime τ_r by the equation $1/\tau_{obs} = 1/\tau_r + 1/\tau_{nr}$, where τ_{nr} is the average lifetime of the excited state undergoing only nonradiative processes. Thus, lifetime for the measurements can, under carefully specified conditions, provide a measure rate of nonradiative processes. Note, however, that some means is required to separate the radiative and nonradiative parts of the rate constant and that, as written, the equation concerns only unimolecular (or pseudounimolecular) processes. The separation of actual unimolecular processes, such as internal conversion and intersystem crossing, from pseudounimolecular processes, such as impurity quenching, is often nontrivial. Many of the studies in Part II in Table VI deal with this question of radiationless processes and, so we leave them until we come to this body of material (Section III.B in the text). Here we are concerned only with outlining the general form of the past year's radiative lifetime studies.

TABLE VI
Radiative Decay of Excited States

System	Reference
A. General radiative decay	
I. Theoretical models and calculations	
a. Long radiative lifetime of small molecules	576
b. Radiative decay of polyatomic molecules	577, 578
c. Vibrational structure of luminescence	
1. In dye molecules	8
2. In molecules with energetically close excited states	9
3. Numerical methods for calculation of vibronic states	11, 12
d. Nonlinear interference effects in emission spectra	15
e. Effect of relative positioning of π,π^* and n,π^* levels on luminescence type of molecule	578a
II. Experimental methods and applications	
a. General problems in studying luminescence: a review	579
b. Effect of nonlinear photochemical transformations on photoluminescence	580
c. Conditions for an isosbestic point in monomer-excimer emission spectra	581
d. Methods of representing emission, excitation, and photoconductivity spectra	582
e. Fluorescence yield measurements	
1. Calorimetric, photometric, and lifetime determinations of fluorescence yields	583
2. Review of fluorescence yield determination	584
3. Quinine bisulfate as a fluorescence standard	585, 586
4. Visual estimation of quantum yields in solution	587
f. Applications of luminescence	588
III. Fluorescence phenomena	
a. Polarization properties	
1. Measurement of polarization of fluorescence excitation	589
2. Effect of torsional vibrations and thermal rotation on fluorescence polarization	590

TABLE VI—continued

System	Reference
3. General expression for time dependence of depolarization	591
4. Effects of energy transfer and rotational diffusion on fluorescence polarization of macromolecules: a theoretical discussion	592
5. Effect of intramolecular displacements on fluorescence polarization in macromolecules	593
6. Measurement of emission anisotropy in solution	
(a) Aromatic hydrocarbons	594, 595
(b) Phthalimide	596–598
(c) Activation energy of depolarization of dye solutions	599
b. Relation between linewidth of vapor-phase emission and interaction of electronic states	600
IV. Phosphorescence phenomena	
a. Review	611
b. Zeeman effect: theoretical and experimental studies	601–603
c. Triplet processes in fluid solution	604, 605
d. Vibronic spin-orbit interactions	606–607a
e. Heavy-atom effects	
1. Theoretical considerations concerning intramolecular heavy-atom effect	349
2. Phosphorescence in phenyl metalloid compounds of group II, IV, and V	608
3. Intramolecular heavy-atom effect on quinoxaline phosphorescence	609
4. Heavy-atom effect due to rare gas matrices on matrix-isolated naphthalene and phenanthrene phosphorescence	610
f. Calculations on effect of CH σ-π interactions on spin-orbit coupling in naphthalene	618, 626
g. Vibronic intensity distribution in benzene- deuterated benzenes, and benzene^{13}-C_1	612
h. Unimolecular delayed fluorescence	
1. Application of delayed fluorescence	613
2. Two-photon delayed fluorescence of N,N,N',N'-tetramethyl-p-phenylenediamine	614
3. Activation-controlled delayed fluorescence of benzil and hexamethylbenzene	615

TABLE VI—continued

System	Reference
B. Lifetime of radiating molecular states (see also Table VII, Part A, for radiative lifetime studies used to determine radiationless decay rates)	
I. Experimental techniques for determining fluorescence decay rates	
a. Subnanosecond decay times	
1. Measured by cross-correlation phase fluorometer	616
2. Measured by mode-locked laser methods	617
b. Nanosecond decay times using spark-discharge excitation	619
II. Correlation of fluorescence lifetime with integrated absorption intensity (oscillator strength f)	
a. Porphyrins and metal porphyrins	324
b. Effect of totally symmetric vibronic perturbations on correlation of τ_f with f in phenanthrene	203
c. Connection of τ_f with other fluorescence characteristics	620
d. Neporent's relation and fluorescence decay of fluorescein	621

System	Decay rates studied[a]	Reference
III. Special effects on radiative decay rates		
a. Substitution effects		
1. Variation of lifetime and quantum yield with chain length in C_5 to C_{16} saturated hydrocarbons	F	622
2. Effect of fluorination and deuteration on chrysene luminescence	F,P	623
3. Phenyl and halophenyl substitution on naphthalene	P	624
4. Deuteration		
(a) Benzene and toluene in gas phase	F	625
(b) Liquid benzene	F	627
(c) Benzene and methylbenzenes	P	628, 629
(d) Dideuteronaphthalenes	P	630
(e) 2-Naphthaldehyde	P	631

[a] F, fluorescence; P, phosphorescence

TABLE VI—continued

System	Decay rates studied[a]	Reference
b. Pressure effects		
1. β-Naphthylamine vapors	F	632, 632a
2. Benzene	F	648
c. Temperature effects		
1. Benzene	P	633, 634
2. Benzene and n-allyl benzenes	P	635
3. Mesitylene	P	636
4. Polynuclear aromatics		
(a)	P	637, 638
(b)	F	644
5. Anthracene vapors	F	639
6. Pyrene	F	640, 641
7. 1,12-Benzoperylene	F	642, 643
8. Coronene and benzocoronene	F,P	645, 646
9. Fluorescein in tartaric acid	P	647
d. Wavelength effects		
1. 3,6-Bis(dimethylamino)phthalimide vapor	F	649
2. 1-Anilinonaphthalenesulfonate	F	650
3. Anti-Stokes phosphorescence in carbazole, fluorescein acridine yellow, and anthranilic acid	P	651
e. Effect of phosphorescence from different zero-field levels of triplet state		
1. Quinoxaline in durene		652
2. Pyrazine in p-dichlorobenzene		653

System	Spectral types studied[a]	Decay rates measured[a]	Reference
IV. Experimentally determined lifetimes			
a. Acetone	F	F	654
b. Thiocyanic acid, anion, and metal salts	P	P	655
c. Biphenyl and its derivatives	F	F	676
d. Benzophenone, acetophenone, and triphenylene in hydrocarbon solvents	P	P	656
e. Anthracene solutions	F	F	594
f. Triphenylamine	P	P	385
g. Photochromic aromatic and heterocyclic compounds	P	P	397

[a] F, fluorescence; P, phosphorescence

TABLE VI—continued

System	Spectral types studied[a]	Decay rates measured[a]	Reference
h. Indoles	F,P	P	657
i. Flavines	P	P	658
j. Phthalazine	F,P	P	318
k. Phthalimide derivatives	F	F	597, 659
l. Fluorescein dyes	F	F	583
m. Acenaphthylene dimers	F,P	F,P	194
n. 7-Methoxy-10b-methyl-4,5,9, 10-tetrahydro-2(10bH) pyrenone	P	P	660
o. Phenyl-substituted silanes	F,P	P	661
p. Corannulene	F,P	P	662
q. Phthalocyanines	F	F	663, 664
r. Charge-transfer complexes of hexamethylbenzene with tetrachlorophthalic anhydride and pyromellitic dianhydride	F	F	461

System	Spectral types studied[a]	Reference
C. Molecular luminescence spectra		
I. Non-S_1, S_0 fluorescence		
a. Singlet ground-state species		
1. Azulene	F,P	665
2. Azulenoid hydrocarbons: a review	F	666
3. Protonated azulenoid systems	F	667
4. β-Styrilnaphthalene	F	668
b. Doublet ground-state species		
1. Methoxyl radical	F	59
2. Benzyl radical	F	669
3. o-Xylyl radical	F	670
4. Phenoxyl radical	F	133
5. Carbonyl sulfide radical cation	F	671
6. Triplet ground states: a review of emission and absorption spectra of arylmethylenes	F	120
II. S_1, S_0 fluorescence and T_1, S_0 phosphorescence		
a. Aliphatic hydrocarbon derivatives		
1. Carbon disulfide	F	68
2. 2-Pentanone	F,P	672
3. Camphorquinone	P	673

[a] F, fluorescence; P, phosphorescence

TABLE VI—continued

System	Spectral types studied[a]	Reference
b. Aromatic hydrocarbons		
1. Benzene at low pressure	F	674, 674a
2. Benzene and its derivatives	F	115
3. Trifluorobenzenes	F	675
4. Biphenyl and biphenylene	F,P	90a
5. Benzophenones	P	677
6. m- and p-Benzaldehyde	F	678
7. p-Dimethylaminobenzal aromatic amines	F	680
9. 2,3-, 2,4-, and 2,5-Dihydroxybenzylidine-o-aminophenol	F	681
10. Hydroxycinnamic acid derivatives	F	172
11. Estrone	F	682
12. Alkanophenones and alkanonaphthones	F	683
13. Naphthalene derivatives containing carboxamide group	F	215
14. 1-Phenylnaphthalene and binaphthyl derivatives	F,P	103a
15. 1-Phenylnaphthalenes	F,P	187
16. Tetralin	F	684, 685
17. Anthracene and naphthalene	F	686
18. 9-Methylanthracene	F	687
19. Anthracene derivatives	F	688
20. Fluorenone derivatives	F	210
21. Aryl fluorenyl ethylenes	F	690
22. Anthraquinone and benzophenone	F,P	691
23. 2-Chloroanthraquinone	F	692
24. 2-Hydroxyanthraquinone	F	693
25. 1- and 2-Methoxy- and dimethoxy-anthraquinones	F,P	694
26. 1- and 2-aminoanthraquinones	F,P	197
27. 1,4,5,8-Tetrahydroxyanthraquinone	F	200
28. Anthraquinone and 2-methyl-9,10-anthraquinone	P	195
29. Anthraquinone dyes	F	694a
30. Acenaphthenequinone	P	695
31. Benzo[g,h,i]perylene	F	205
32. Polynuclear aromatics	F,P	369, 396, 696
33. Phenylethynyl-substituted polynuclear aromatics	F	205
34. Amino and hydroxy derivatives of polynuclear aromatics	P	92

[a] F, fluorescence; P, phosphorescence

TABLE VI—continued

System	Spectral types studied[a]	Reference
35. Bisazomethine derivatives of binuclear aromatic amines	F,P	165
36. 1,4-Benzodiazepine derivatives	F	689
c. Nitrogen heterocycles		
1. Aminopyridines	F,P	697
2. Azapyridocyanines	F,P	322
3. Stilbazoles and 1,2-bispyridylethylenes	F	698
4. Pyrimidine, indazole, isoquinoline, pyrazine	F,P	699
5. Acridan	F,P	267
6. Dibenzacridines	F	700
7. *N*-Arylcarbazoles	F,P	701
8. *N*-Aroylcarbazoles	F,P	312
9. Dibenzo[*f,g*]quinoxaline and dibenzo[*f,g*]quinoxaline-d_8	F	702
10. Flavines	P	703
11. Phthalocyanines	P	704
12. Chlorophyll	F	327
d. Oxygen, sulfur, and selenium heterocycles		
1. Benzofuran and coumarin derivatives	F	705
2. Naphthostyril and thionaphthostyril derivatives	F	247
3. Dibenzothiophene and dibenzoselenophene	F	706
e. Comparative and mixed heterocycles		
1. Effect of heteroatoms on electric spectra of aromatic compounds	F	216
2. Thiobenzophenone, xanthione, and *N*-methylthioacridone	F	707
3. Phenoxazine dyes	F	235
4. Phenoxazine phenoxanthin, and phenothiazine	F,P	372, 221
5. 3,4-Dihydronaphtho[1′,2′:5,4]oxazoles	F	232
6. Organic luminophors based on benzoxazones	F	708
7. Organic dyes	F	709
8. Organic dyes	P	710
f. Charge-transfer complexes of *p*-benzoquinone with aromatic hydrocarbons	P	711

[a] F, fluorescence; P, phosphorescence

TABLE VI—continued

System	Spectral types studied[a]	Reference
D. Luminescence as a structural probe		
I. Excited-state acid-base equilibria		
a. Aromatic hydrocarbons in acid media	F	714
b. Aromatic carboxylic acids	F	715
c. 1-Indianone	P	716
d. Phenols from fluorescence quenching	F	717
e. Anthracene	F,P	718
f. Aminoanthracenes	F	719
g. 9-Anthranoic acid	F	719a
h. 2- and 4-Styrylpyridines	F	720
i. Phenazine	F,P	721
j. Benzimidazoles in acid media	F	722
k. N-Heterocycles	P	723
II. Tautomeric equilibria		
a. 7-Azaindole	F	305
b. Thymine and monomethylthymine anions	F,P	724
c. Adenine	F	725
d. Monohydroxyacridines	F	269
III. Conformation equilibria		
a. Benzene	P	726
b. 2-Phenylnaphthalene derivatives	F	727
c. Phthalimide derivatives	F	728
IV. Location of unsaturated sites		
a. Isoquinoline alkaloids	F	729
b. Photochromic spiropyrans	F	730
c. Naturally occurring corrins	F	712
V. Structural and environmental effects on proteins	F	713
VI. Intramolecular hydrogen bonding in oxadiazoles	F	233
E. Environmental effects on radiative decay		
I. Solid-state effects		
a. Quasiline spectra (Shpol'skii effect)		
1. General considerations		
(a) Effect of concentration on quasiline phosphorescence: a review		382
(b) Effect of freezing rate and concentration on quasiline spectra of aromatic compounds		486, 487 731, 732

[a] F, fluorescence; P, phosphorescence

TABLE VI—continued

System	Spectral types studied[a]	Reference
(c) Nature of crystal structure giving rise to quasiline spectra		483, 484, 493
(d) Temperature dependence of line intensity in quasiline spectra		733
(e) Heavy-atom effects on quasiline spectra		392, 405
(f) Failure of oxygen to quench quasiline spectra		734
(g) Use of quasiline spectra to determine vibrational symmetries and selection rules		735
(h) Model calculations for vibrational analysis of quasiline fluorescence of stilbene		736
2. Experimental results		
(a) Benzene, toluene, and naphthalene	F,P	405, 736a
(b) Benzyl radicals and toluenes and α-deuterated derivatives	F	737
(c) Ethylbenzyl and methylphenylethyl radicals	F	750
(d) Triphenylmethyl radicals	F	738
(e) Durene	F,P	497
(f) Pentamethylbenzene	F	751
(g) Hexamethylbenzene	F	498
(h) Stilbene derivatives		
(1) p-Substituted	P	124
(2) p-Substituted	F	490
(3) trans-Stilbene	F	736
(i) Naphthalene	F,P	392
(j) Monohalo naphthalenes		
(1)	P	739, 740
(2)	F,P	352
(k) α-Naphthoic acid, naphthaldehyde and naphthalimide	F,P	189
(l) Phenanthrene	P	382
(m) 1,2-Benzanthracene	F	741
(n) Pyrene	F	742, 743
(o) Pentacene	F,P	744
(p) Polynuclear aromatics	F,P	483, 485, 734, 735, 745
(q) 1,12-Benzoperylene	F	746
(r) 3-Methylcholanthrene	F	747
(s) Quinoline and benzo[f]quinoline	P	748
(t) Naphthoxazoles	F	231

[a] F, fluorescence; P, phosphorescence

TABLE VI—continued

System	Spectral types studied[a]	Reference
(u) Azatriphenylenes	F,P	493, 749
(v) Dibenzofuran	F	486, 487
(w) Dibenzothiophene, fluorene, stilbene, diphenylbutadiene, terphenyl, and phenylnaphthoxazole	F	731
(x) Dibenzofuran, fluorene, and dibenzothiophene	F	732
(y) Pyrazine		752
b. Solid-state effects apart from quasiline behavior		
1. Crystalline and rigid glass solutions		
(a) Effect of nonequivalent sites on intensity of benzene phosphorescence and fluorescence		753, 775
(b) Temperature and solvent effects on lifetime of benzene-h_6 and -d_6 phosphorescence		754
(c) Effect of solvent structural changes rigid solutions on phosphorescence and thermally activated delayed fluorescence		755
(d) Matrix effects on phosphorescence decay of benzene and methylbenzene derivatives		629
(e) Matrix effects on phosphorescence of α-diketones in glassy solutions		756
(f) Presence of two components in phosphorescence decay of propiophenone		757
(g) Shifting of relative energy of n,π^* and π, π^* triplet states of quinone, and α-diketones in rigid solutions		758
(h) Matrix and temperature dependence of fluorescence and phosphorescence of 2- and 4-aminobenzophenone		759
(i) Phonon structure of emission of doped naphthalene crystals		579a
(j) Pressure effects on crystalline and polymeric solutions		
(1) Naphthalene-h_8 and -d_8 and bromonaphthalene phosphorescence in ethanol-isopentane-alcohol		760, 761
(2) Aromatic hydrocarbons in polystyrene		761a

[a] F, fluorescence; P, phosphorescence

TABLE VI—continued

System	Reference
(3) Coronene fluorescence in poly(methylmethacrylate)	762
(4) Triphenylene phosphorescence in various matrices	763
(5) Coronene and 1, 12-benzoperylene fluorescence in n-hexane	763a
2. Molecular crystals	
(a) Durene fluorescence and phosphorescence	764
(b) Anthracene fluorescence	765
(c) Pyrazine phosphorescence: phonon structure or structural defects?	766–769a
(d) Heavy-atom, vibronic, and host-crystal perturbations on phosphorescence of molecular crystals	770
(e) Theory of luminescence by impurity centers from nonequilibrium vibrational states	771
3. Existence of "quasiquinone" structure in crystal state luminescence of p-(halosalicylideneamino)-benzoic acid	772
4. Induction of fluorescence by adding electron donor substituents to benzene carbonyl derivatives in rigid solution	773
II. Liquid-state effects on luminescence	
a. Effect of liquid crystal transitions on decay time of pyrene in cholesterol benzoate	774
b. Solvent-induced spectral shifts	
1. Review	579
2. Failure of low-temperature solutions to achieve configuration equilibrium, and effect on fluorescence maximum ν_{max}	776
3. Correlation of ν_{max} with Kosower's solvent scale in the fluorescence of N-arylaminonaphthalenesulfonates and related molecules	777
4. Polarity and viscosity effects on ν_{max} and spectral structure in the fluorescence of 9, 9'-bianthryl	778
5. Variable solvent stabilization of ground and excited states of 1, 3-diphenyl-2-pyrazolines	779
6. Dielectric-constant effects on ν_{max} of N-arylcarbazoles	701

TABLE VI—continued

System	Reference
7. Solvent shifts of fluorescence of fat-soluble riboflavine derivatives	779a
c. Solvent effects on emission intensity	
1. Solvent isotope effects on fluorescence of 2-aminophthalimides	780
2. Temperature, excitation energy, and solvent effects on spectrum of 2, 2'-binaphthyl	781
3. Effect of pH on fluorescence of bovine bovine serum albumin and component amino acids	782
4. Effect of added ethanol or glycerol on fluorescence lifetime and quantum yield of aqueous acridine solutions	782a
d. Deviations from ideality in viscosity dependence of rotational depolarization of fluorescence	783, 784
e. Molecular aggregation in solution	
1. Anthracene fluorescence	478
2. Acriflavine and Rhodamine 3B fluorescence	473
3. Fluorescence of aggregates of radicals derived from Würster's blue and flavine mononucleotide	472a
4. Review of effects of aggregation on emission of dyes in solution	472
5. Aggregate fluorescence of chlorophyll in solution	326–328, 481
f. Solvent effects on polarization of 2-naphthaldehyde and 2-naphthyl ketone phosphorescence: second-order vibronic effects on spin-orbit coupling	784a

One major factor in the increasing number of lifetime studies has been improved equipment for carrying out such studies. Flash-photolytic apparatus, rotating-sector phosphorimeters, and nanosecond-fluorescence lifetime apparatus are commercially available. In addition, subnanosecond lifetime apparatus is being developed by several groups. Spencer and Weber [616] describe a cross-correlation phase fluorimeter and use it to study reduced nicotine adenine dinucleotide (NADH, $\tau = 0.38$ nsec). Merkelo, Hartman,

Mar, and Singhal [617] use a mode-locked laser to study fluorescence decay down to 0.08 nsec.

Because of the symmetry of the equations describing light absorption and light emission, the "natural" radiative lifetime τ_r can be described as a constant times the oscillator strength, f of the transition that is common to the two processes. In turn, f is measured as a constant times the integrated absorption of the transition. This mathematical description has several experimental consequences. The variation of τ_{obs} with such effects as temperature and pressure should be due to τ_{nr} only, since it is known experimentally that, barring special effects, f is independent of these parameters. Furthermore, the variation of τ_r with substitution of the chromophore should be easily determined from the effect of the substitution on f. Unfortunately, direct tests of this relationship are devilishly hard, since good values for the natural radiative lifetime are difficult to obtain. One study of special interest in this regard is the demonstration by Craig and Small [203] that the totally symmetric vibronic perturbations on the 340 nm vibronic band of phenanthrene destroys the absorption-fluorescence symmetry. Obviously, substitution of phenanthrene would, in some positions, remove this perturbation and lead to changes in τ_r expected from the absorption spectra. More studies of this kind are obviously in order to help provide a good empirical foundation for radiationless transition studies.

Berlman [676] uses this τ-f correlation to argue for the existence of a hidden band under the lowest absorption band of biphenyl. The observed fluorescence shows the vibrational structure expected for a $^1L_{6 \to b}$, and in addition the integrated absorption, distributed over the vibrational states, gives a reasonable estimate of ε_{max} for a $^1L_{16 \to b}$ band.

Although τ_r is probably independent of temperature and pressure effects in general and also many kinds of substitution (for example, H → D), large effects on τ_{obs} are often observed (see Table VI, Part B.III). Such effects are usually explained in terms of τ_{nr}, although Brinen, Orloff, Gallivan, and Stamm [623] find in the case of 6-fluorochrysene that τ_f, the normal fluorescence lifetime, is increased relative to chrysene, as are τ_p and the radiationless deactivation of the triplet state.

The existence of nonexponential decay, and hence of no single τ_{obs}, can arise for several reasons. One special case of particular interest in the mechanism of radiative decay is the differential phosphorescence decay rates from separate zero-field levels of the triplet state. At room temperature, or even 77°K, such decay is not normally observed, since equilibration among levels is rapid. Near absolute zero, however, the equilibration is slower than phosphorescence and can be observed. Azumi and Nakano [652] observe the decay from separate levels in quinoxaline. El-Sayed, Moore, and Tinti [653] observe it in pyrazine. In both cases the molecules were in host crystals to avoid bimolecular processes. In neither case was a magnetic field used to split the levels. Schmoerer and Sixl [571] use an esr spectrometer to split the levels and observe the phosphorescence rates; they term the process optical spin polarization (see Table IV and Part II, above.)

One of the standing curiosities of fluorescence spectroscopy is the failure of azulene and benzoazulenes to fluoresce from their first singlet state and their high-yield fluorescence from the second singlet. In general, for reasons to be discussed later, fluorescence and phosphorescence occur almost solely from the vibrationally relaxed first excited singlet or triplet, S_1 or T_1, to the ground singlet S_0. Thus the azulene anomaly is almost unique. Pook and Dhringa [666, 667] report several more examples of azulenoid S_2, S_0 fluorescence. Rentzepis [665] succeeds in obtaining S_1, S_0 fluorescence by using high-power picosecond laser pulses to saturate the S_1 state and thus overcome the low quantum yield of azulene. A new S_2, S_0 fluorescence is observed in β-styrilnaphthalene by Hammond, Shim, and Van [668]. This compound emits normal S_1, S_0 naphthalene-like fluorescence but in addition has a stryene-like fluorescence, showing that intramolecular energy transfer in this system is slow compared with emission; it is not clear why this system, unlike similar systems, has such a poorly competing energy transfer. It is possible that other systems of this kind can be located, however.

In addition to its use in elucidating electronic structure, emission spectroscopy can also be used in some cases to clarify the geometry of molecules and to elucidate excited-state equilibria. From the Franck-Condon factors, for example, the structure of the excited state can often be described in some detail, as done this past year for the triplet state of benzene [726] and the singlet states of 2-phenylnaphthalene derivatives [727] and phthalimide derivatives [728].

In addition, acid-base equilibria can be studied for both singlet and triplet states by studying the pH dependence of luminescence. Tautomeric equilibria and similar problems concerning the location of sites of unsaturation have also been studied. One main impetus for fluorescence studies has been the use of fluorescent molecules as labels for studying the conformation and reactivity of proteins. Cowgill [713] reviews structural and environmental effects on the fluorescence of tyrosine and tryptophan.

2. ENVIRONMENTAL EFFECTS ON RADIATIVE DECAY OF EXCITED MOLECULES

At this point it should be clear that this section has very indefinite limits. It is hardly a clear distinction to classify rotational depolarization rates, which depend on viscosity, and hence on solvent, in any case, as both a molecular and an environmental effect. But this we have done, for in the former case, for example, Ref. 590, it appears to be simply a viscosity effect, but in the latter, for example, Refs. 783 and 784, there are special solvent effects related to the formation of a stronger or weaker solvent "shell" that produces a larger effective molecule and hence strange solvent effects. Several of the earlier references include mention of environmental effects, but only in a secondary fashion. In this part we have tried to include papers of which one main concern is the environmental effect observed.

One very important solvent effect in luminescence work is found in crystalline solutions of aromatic and heteroaromatic compounds. The effect is characterized by a very marked sharpening of absorption or luminescence bands or both. The effect is also called the Shpol'skii effect, after the Russian scientist who first noted it in 1952. The effect has been observed only below 100°K. Most examples of it have been recorded in normal paraffins as solvents, but reported cases this past year included methylcyclopentane and aromatic hydrocarbon solvents. The sharpening is due to the isolation of molecules of the solute in the solvent crystal, so that they are vibrationally uncoupled, producing an "oriented gas" of solute molecules. Nakhimovskaya [483] and Klimova, Nersesova, Naumova, Oglobina, and Glyadkovskii [485] report observing the quasilines due to isolated molecules, or "submicrocrystals" [483], and the band spectra due to crystalline solute. Because of this necessity of isolating the molecules, both freezing rate and concentration should have a major effect on the spectrum. The expected correlations have been confirmed by several groups [382, 486, 487, 731, 732]. In addition, Teplitskaya and Personov [734] show that oxygen fails to quench such luminescence, which they attribute to the isolation of the luminescence centers. In the luminescence of some compounds a multiplet structure is observed; in their study of the quasiline electronic spectra of azatriphenylene, Ruziewicz, Olszowski, and Chojnack [493] attribute this multiplet structure to solute molecules in nonequalent sites in the n-alkane solvents.

The sharp lines observed in the Shpoliskii effect is a consequence of the lack of coupled vibrations and of rotations, so that only a few vibronically allowed vibrational states are found. The spacing of the observed lines can usually be related to the infrared or Raman spectra. Gase and Hasse [736] carry out a complete vibrational analysis of the spectrum of *trans*-stilbene in n-octane, including the Franck-Condon analysis.

The clarity of the vibrational spectrum makes quasiline spectra ideal systems for studying vibronic coupling. Mikhailenko and Teplyakov [735] study a number of aromatic hydrocarbons in order to derive selection rules for vibronic coupling.

Bolotnikova and Sichkar [392, 405] extend the experimental uses of quasiline spectra by including ethyl bromide in the crystalline solutions of aromatic hydrocarbons and hexane or cyclohexane. They observe charge-transfer bands with benzene, toluene, and naphthalene. They also observe effects on the phosphorescence and triplet-triplet absorption spectra that depend on concentration.

Bogomolov and Vedernikov [747] show that the quasiline fluorescence can be used to determine the concentration of the carcinogen 3-methylcholanthrene in the 10 to 1000 ng/ml range.

Other effects observed in solid solutions are generally due to distortion of the solute by the solvent matrix or by the existence of defect or nonequivalent sites. (Recall that special electronic effects—excimer and exciton states—present in mixed crystals are not covered in this section.)

Peuker and Trifonov [771] report a theoretical study of light emission by impurity centers in a nonequilibrated vibrational state due to vibronic absorption. Because of these effects, anti-Stokes luminescence, arising from vibrational levels of the excited state other than the lowest, should be observed even at absolute zero.

In the spectra of benzene, Martin and Kalantar [753] observe an anomaly in the lifetime of benzene phosphorescence in cyclohexane solutions that they attribute to the coexistence of cubic and monoclinic sites in the solution. Martinez and Dorignac [755] observe that crystalline phase changes account for observed temperature dependence of phosphorescence and delayed fluorescence. In studying the phosphorescence of benzene in carbon tetrachloride at 77°K, Marakami and Kanda [775] report that they have again observed a very anomalous vibrational structure in this solvent. They are unable to account for this very repeatable anomaly.

Some idea of the difficulty of interpreting crystal phenomena is provided by two separate examples of the existence radically varying interpretations of the same phenomena. In their studies on the phosphorescence of pyrazine crystals, El-Sayed, Moomaw, and associates [766, 767] suggest that the fine structure of the spectrum is due to phonon effects. After studying the effects of zone-refining pyrazine, however, Azumi and Nakano [768, 769] suggest that the effect is due to defect sites in the crystal. In a second case, Leukner [633] has published a note questioning Martin and Kalantar's interpretation of the effect of cooling rate on benzene phosphorescence in glossy solids. Where they consider this effect to be due to nonequivalent lattice sites, Leubner argues that it is solely a temperature effect.

The effects of molecular distortion by the lattice has been a continuous area of investigation of Offen's group [760-763]. By studying the pressure dependence of the emission, the activation volume can be measured, and some idea of intermolecular forces can be gained. In phosphorescence decay studies of benzene and methylbenzene derivatives, Rabalais, Maria, and McGlynn [629] find that matrix and temperature effects decrease as benzene is more heavily methylated. They tentatively propose a solvent-assisted distortion mechanism to account for their findings.

Gacoin [757] finds two separate phosphorescence decay paths in propiophenone crystalline solutions; this effect is not in the molecular crystal. He proposes isomerization, tautomerization, and excited-state isomerization as possible explanations. In his studies of aromatic and aliphatic diketones, Almgren [756] finds spectral types that he assigns to different conformations, solvated excited species, or products formed by reactions with solvent.

A phenomenon related to these is Smirnov, Kirichenko, and Gachkovskii's [772] observation of a luminescence of halo derivatives of p-(salicylideneamino)benzoates that they attribute to a "quasiquinoid" structure in the crystal. Crystalline solvent effects on o-quinones lead to a shift of the π,π^* triplet state to a higher energy than the n,π^* states [758].

An appropriate transition from solid- to liquid-state studies is provided by Tomkiewicz and Weinreb's [774] study of the decay rate of pyrene fluorescence in the liquid crystal cholesterol benzoate. As the transition point from the crystalline phase to the cholesteric phase at 145° is reached, the lifetime goes through a V-shaped minimum. That this minimum is due to the medium and not the solute is illustrated by the absence of this effect in polystyrene solution.

Solvent-effect studies on fluorescence tend to concentrate on either intensity changes or band shifts. Seliskar, Turner, Gohlke, and Brand [777] find that Kosower's empirical solvent scale parameters Z give a smooth curve when plotted against the fluorescence frequency maximum in various solvents. Fluorescence maximum shifts can occur either by solvent stabilization of the excited state or by excited-state configurational changes that alter the Franck-Condon factors in the spectrum. Grabowski [579] reviews solvent-induced fluorescence shifts. Viscous solvents often tend to be slow in reorganizing and hence often show large solvent shifts. Bakhshiev, Piterskaya, Studenov, and Altaiskaya [776] observe such a viscosity effect in the fluorescence of 4-aminophthalimide derivatives. Schneider and Lippert [778] find viscosity shifts in the fluorescence of 9,9'-bianthryl. Selective solvation by water in mixed solvents leads to deviations from inverse viscosity in the fluorescence depolarization of 1- and 2-acetylanthracene [783, 784] and fluorescin [784]. Stabilization shifts are observed in the fluorescence 1,3-diphenylpyrazoline derivatives in going from cyclohexane to isopropyl alcohol [779], accompanied by an increased radiationless deactivation.

Chakrabarti [784a] finds that hydroxylic solvents, compared with hydrocarbon solvents, alter the polarization of 2-naphthaldehyde and 2-naphthyl ketone. The effect appears to be due to second-order spin-orbit coupling due to vibronic interactions.

Solvent effects on luminescence intensity are believed to be due to the enhancement of the radiationless deactivation, although usually no actual determination is made of the effect on the separate rate constants for radiative and nonradiative relaxation. We deal with these cases further in the next section.

B. Pseudounimolecular Radiationless Deactivation

Figure 1 shows the Jablonski diagram for the formation and deactivation of molecular excited states. The two processes with which we are concerned here are internal conversion, shown, for example, as $S_2 \to S_1$, and intersystem crossing, such as $S_1 \to T_1$. These processes are usually treated as molecular processes, although they almost invariably involve collisional deactivation. These processes are very fast, ordinarily in the range of $k_{\text{unimolecular}} \geq 10^8 \text{ sec}^{-1}$ for $S_2 \to S_1$ and $S_1 \to T_1$. The intersystem crossing $T_1 \to S_0$ is slow by contrast and may have $k_{\text{unimolecular}} < 10^{-1} \text{ sec}^{-1}$. The rapidity of these processes has made their study very difficult. This is especially true of internal conversion, which often has a lifetime ($=k_{\text{unimolecular}}^{-1}$) in the picosecond range. Thus, practically all fluorescence and phosphorescence are observed from the vibrationally relaxed, lowest excited level.

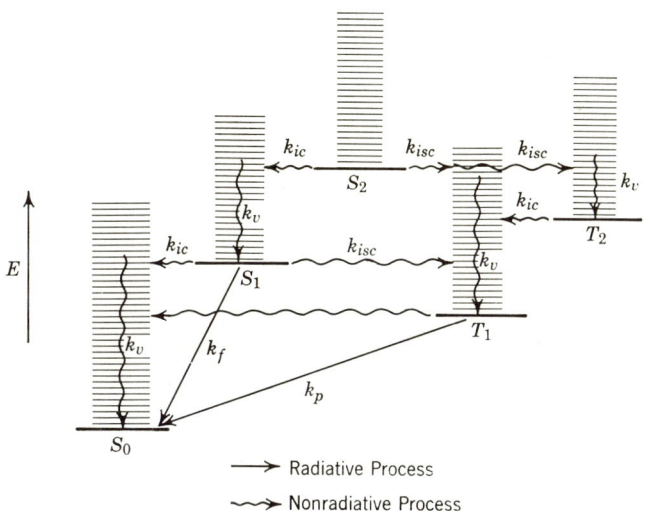

Figure 1

Because of the difficulty of observing these processes directly, theories are often difficult to formulate, and they are also hard to verify. Ting [785] proposes an alternative theory to the Robinson-Frosch stationary-perturbation approach. He treats radiationless transition as a process of virtual radiative transfer between excited states. The probability of the transition, then, is proportional to the product of the oscillator strengths connecting the initial and final states to the ground state. He is able to show that his theoretically derived relation is very close to the experimentally derived rates of a variety of radiationless processes.

Rhodes [787] examines the commonly used theoretical assumption that radiationless transitions take place between a discrete molecular state and a continuum of states. He finds a very pronounced effect of the properties of the exciting light (coherence and bandwidth) on the nonradiative-decay processes. Rhodes, Henry, and Kasha [788] further discuss the applicability of the Born-Oppenheimer approximation in radiationless-transition theory.

Englman [789] estimates the matrix elements in the Jortner theory of radiationless transition for anthracene, azulene, naphthalene, and benzene. He is unable to justify Jortner's interpretation of radiationless decay in organic molecules. Heller [790] discusses the relationship of photochemical reactivity and radiationless deactivation in terms of the participation of specific bonds in the conversion of electronic-vibrational energy to thermal or chemical energy.

The process of internal conversion is actually two processes—the loss of vibrational energy and the crossing from one electronic state to another. Schlag, Yao, and Von Weyssenhoff [786] discuss the use of the modulation excitation of specific vibronic states as a means of determining absolute internal conversion rate constants. Schlag and Von Weyssenhoff [632] apply this method to 2-aminonaphthalene vapor.

Rentzepis and Malley [791] observe vibrational relaxation of excited states directly by the use of crossed laser beams. The method is similar to Rentzepis's laser-reflection technique described in last year's "Survey." It allows the observation of processes occurring with picosecond lifetimes.

The usual method of determining the rate of vibrational processes is by studying the pressure dependence of fluorescence. At very low pressures *resonance fluorescence* is observed, in which the molecule emits light of the same frequency as that which it has absorbed. But at higher pressures collisional reactivation leads to the appearance of the more common S_1, S_0 fluorescence. Several such studies are listed in Table VII, Part *A.I.b*. Becker, Dolan, and Balke [793] use the competition between photochemistry and internal conversion to study the effect of excitation into particular vibrational states on the internal conversion to ground-state molecules.

TABLE VII

Decay of Excited States by Radiationless or Polymolecular Processes

System	Reference
A. Pseudounimolecular radiationless deactivation	
I. Internal conversion	
a. General considerations	
1. New theory of radiationless transitions	785
2. Relaxation theory for vibronic states under modulated excitation	786
3. Effects of molecular size and radiation bandwidth on radiation transitions	787, 788
4. Estimation of matrix elements in Jortner theory of radiationless transitions	789
5. Relation of radiationless relaxation and photochemical reaction	790
b. Vibrational relaxation	
1. Crossed laser beams as a probe for picosecond relaxation processes	791
2. Collisional relaxation of ν_3 vibration in CO_2	792
3. Wavelength dependence of the crosssection for vibrational relaxation of 2-pentanone	672
4. Relative efficiency of biacetyl and inert molecules in vibrational relaxation of excited biacetyl	792a
5. Vibrational relaxation of $^1B_{2u}$ state of benzene	674
6. Pressure effects on vapor-phase luminescence of pyrimidine, pyrazine, indazole, and isoquinoline	699
7. Direct timing of relaxation of selected vibronic states of β-naphthylamine by modulated luminescence	632
8. Competition between photochemistry and internal conversion	793
9. Temperature-dependent vibrational and electronic relaxation in 3- and 4-aminophthalimide observed by nanosecond-resolved emission spectroscopy	793a
c. Electronic relaxation	
1. Internal-conversion rate in cyclobutanone and cyclobutanone-2-^3H	794
2. Rate of first-singlet to ground-state deactivation in benzene vapor	648
3. Rate of internal conversion of benzene in solution at 77°K; effect of concentration and excitation wavelength	634
4. Temperature and solvent effects on internal conversion of benzene	794a

TABLE VII—continued

System	Reference
5. Comparison of internal conversion in benzene and substituted benzenes in liquid phase	115
6. Internal conversion in biphenyl derivatives in solution at 77°K	795
7. Competition between internal conversion and excimer formation in benzene, toluene, and xylene	795a
8. Direct relaxation from first singlet to ground state in azulene	665
9. Temperature dependence of internal conversion of crystalline anthracene	765
10. Activation energy for internal conversion of aromatic hydrocarbons in poly(methylmethacrylate)	644–646
11. Temperature dependence of internal conversion of 1, 12-benzoperylene	642, 643
12. Temperature dependence of internal conversion of 3, 6-bis(dimethylamine) phthalimide vapors	649
13. Rate constants for internal conversion of stilbazoles and 1, 2-bispyridylethylenes by comparison of fluorescence lifetime observed with that calculated from absorption spectrum	698
14. Effect of solvent and substituents on radiationless deactivation of 1, 3-diphenyl-2-pyrazolines	779
15. Enhancement of internal conversion rates for excited acridine in water–organic solvent mixtures	779a

II. Intersystem crossing

    a. General considerations (see also Table VI, Part *A*. IV, for similar effects on radiative decay)

1. Temperature dependence of $T_1 \to S_0$ intersystem crossing	796, 797
2. Effect of CH anharmonicity on $T_1 \to S_0$ intersystem crossing	798
3. Crossing into singlet manifold from triplet states other than T_1	799, 800
4. Vibronic effects on $S \to T$ intersystem-crossing rate of heteroaromatic and aromatic carbonyl compounds	801
5. Spectroscopic method for determining the most probable $S \to T$ intersystem-crossing route	802
6. A mechanism for intersystem crossing in aromatic ketones: spin-polarization studies	803
7. Role of intramolecular charge-transfer transitions in spin-orbit coupling of anilines	804
8. Lack of methanol solvent effect on intersystem-crossing rate of dienes	804a

    b. Triplet → singlet intersystem-crossing rates

1. Effect of loss of rotational freedom on radiationless decay of benzene triplets in cyclohexane	753

TABLE VII—continued

System	Reference
2. Environmental effects on decay of benzene at low temperatures	754
3. Thermal activation in relaxation of benzene and alkybenzene triplets	635
4. Effect of α-deuteration on radiationless deactivation of toluene triplets	628
5. Effect of external heavy atoms on intersystem crossing in benzophenone	804b
6. Pressure effects on radiationless decay of naphthalene derivatives	760, 761
7. External heavy-atom effect on vibronic structure of thermally activated delayed fluorescence of naphthalene	804c
8. Effect of deuteration on 2-naphthaldehyde triplet lifetime	631
9. Activation energy in relaxation of aromatic hydrocarbon triplets	637, 638
10. Temperature dependence of radiationless deactivation of coronene and benzocoronene	645, 646
11. Effect of deuteration and fluorination of chrysene on triplet deactivation	623
b. Singlet → triplet intersystem-crossing rates	
1. Cyclobutanone	794
2. Biacetyl at low pressure	805
3. Camphorquinone	673
4. Deuterium-isotope effect on S_1, T_1 radiationless transition in benzene	806
5. Benzene-h_6, -d_6, and -1, 4-d_2 at low pressures	807
6. Wavelength effect in monofluorobenzene	808
7. Solvent effects on biphenyl at 77°K	795
8. Benzene and derivatives in liquid state	115
9. Solvent effects in N-arylaminonaphthalenesulfonates and related compounds	777
10. Direct measurement of S,T relaxation in azulene	809
11. Anthracene	361
12. S_1, T_2 intersystem crossing in anthracene crystals	765
13. Deactivation of S_2 in anthracene vapors	639
14. Aromatic molecules in solution	810
15. Temperature effects on pyrene-h_{10} and -d_{10}	640, 641, 811
16. 1, 12-Benzoperylene	642, 643
17. Crossing into separate zero-field levels of pyrazine triplet state	812, 813
18. Azastilbenes	698
19. Solvent effects on alkyladenines	725
20. Wavelength effects on orotic acid (6-uracilcarboxylic acid)	814
21. Wavelength effects on uracil	815
22. Phthalazine	318

TABLE VII—continued

System	Reference
23. Triplet-state quenching of organic-dye lasing action	816
24. Xanthene and thiazine dyes	817
25. Deuterium-isotope effect on dibenzo[f,g]quinoxaline	702
26. Porphyrins	818
d. Induced S, T transition by heavy-atom effect	
1. Difference in susceptibility of cis → trans and trans → cis isomerization rates of m-halostilbene derivatives to halogen substituent effect	819
2. Rate constants of fluorescence decay of pyrene in heavy-atom solvents	820
3. Effect of ethylbromide and paramagnetic copper species on the intersystem-crossing rate of polynuclear hydrocarbons	821
4. Effect of heavy atoms and hydroxylic solvents on the intersystem-crossing rates of isoquinoline derivatives	822
5. Enhanced population of porphyrin triplet states by methyl iodide	823
6. Enhanced phosphorescence of flavins by potassium iodide	658
B. Relaxation of excited molecule by ground-state molecules	
I. Complex formation by excited and ground-state molecules of same chemical species: excimers and excitons	
a. Solid-state excimers and excitons	
1. Quantum theory of damping of molecular exciton states at trapping sites	824
2. Review of photoemission from aromatic organic crystals	825
3. Cooperative excitons in crystal with two molecules per unit cell	826
4. Singlet excitons	
(*a*) Isotopically mixed deuterated benzene crystals	827, 828
(*b*) *trans*-Stilbene	518
(*c*) Excimer emission of tetrachloro-*p*-xylene	840
(*d*) Naphthalene	829
(*e*) Isotopically mixed deuterated naphthalene crystals	828
(*f*) Diffusion and surface reaction of singlet excitons in anthracene	830
(*g*) Excitonic structure of anthracene fluorescence	831
(*h*) Singlet and triplet excitons in stationary photocurrent generation in anthracene crystals	868
(*i*) Tetracene	832
(*j*) Temperature, solvent, and concentration dependence of pyrene excimer fluorescence	841
(*k*) Pyrene excimer emission at high pressure	842
5. Triplet excitons	
(*a*) Factor-group splitting in benzene $^3B_{1u}$ state	833
(*b*) Biphenyl crystal phosphorescence	834

TABLE VII—continued

System	Reference
(c) Benzophenone crystal phosphorescence	354
(d) Spin relaxation and optical-spin polarization of naphthalene triplet excitons	835
(e) Zeeman populations of anthracene triplet excitons	836
(f) Anisotropy of exciton diffusion in anthracene	837
(g) Mechanism of delayed fluorescence in acene crystals	838
(h) Exciton diffusion in pyrazine and isotopically mixed deuterated pyrazine	839
(i) Excitonic structure of pyrazine phosphorescence	350
b. Liquid-state excimers	
1. General considerations and experimental techniques	
(a) Semiempirical theory of excimer luminescence	843, 844
(b) Temperature dependence of weakly coupled dimers	845
(c) Excimers: a review	846
(d) Bimolecular rate constants for excimer formation by nanosecond time-resolved spectroscopy	847
(e) Isosbestic points in excimer-monomer luminescence spectra	581
2. Experimental results	
(a) Benzene and alkylbenzene derivatives	
(1) Competition between internal conversion and excimer luminescence in liquid benzene	795a
(2) Effect of deuteration on excimer decay in liquid benzene	627
(3) Excimer emission in benzene and alkylbenzenes under electron impact	848
(4) Temperature and concentration dependence of benzene and alkylbenzene derivatives	849
(5) Review of photophysical processes in toluene	850
(6) Molecular and excimer fluorescence of toluene	851
(7) Excimer effects on o-xylene triplet states	852
(8) Excimer fluorescence on alkylbenzenes	853
(9) Effect of dilution on fluorescence lifetime and intensity of aromatic solvents	854
(b) Triplet decay of indene through excimer formation	855
(c) Naphthalene derivatives	
(1) Excimer emission of liquid naphthalene and biphenyl under pulse radiolysis	359
(2) Excimer emission of naphthalene under electron impact	848
(3) Triplet energy migration in liquid naphthalene derivatives	856
(4) Triplet exciton motion in molten naphthalene-durene mixtures	857

TABLE VII—continued

System	Reference
(5) Aggregate formation in sodium naphthionate and effect of additional solutes	858
(d) Polynuclear aromatics	
(1) Pyrene excimers	859, 860
(2) Saturation of excimer formation in pyrene	861
(3) Aggregation of 3, 4-benzopyrene in aqueous lauryl sulfonate solutions	862
(e) 4-Amino-N-phthalimide	869
(f) 2, 5-Diphenyl-1,3-oxazole (PPO)	863
(g) 2-Phenyl-5(4-biphenylyl)-1,3,4-oxadiazole (PBD)	864
(h) Nonradiative decay of Eosin Blue through excimer formation	865
(i) Nonradiative decay of phthalimide and other dyes through excimer formation	866, 869
(j) Enhanced polarization of uranin solutions by aggregation caused by potassium iodide	867
II. Complex formation by excited and ground-state molecules of different chemical species: mixed excimers and excited complexes (singlet states unless otherwise specified)	
a. Exciplexes of amines with electron acceptors	
1. Triethylamine with excited benzene, naphthalene, and p-terphenyl	871, 872
2. Triethylamine with excited fluorenone	873, 874
3. Tetraalkyldiamine diphenylmethanes with excited benzanthracene	875
4. Dimethylaniline with excited dimethylanilinium salts	876
5. Dimethylaniline with excited halobenzenes	877
6. Aromatic amines with excited bromobenzene	878
7. Aromatic amines with excited benzanthracene hydrocarbons	879
8. Aromatic amines with aromatic electron acceptors	880
9. Dimethylaniline with excited pyrene	881
10. Aminopyrine with excited aminobenzoic acids	882
11. Alkaloids with excited heterocycles	883
12. Excited tetramethyl-p-phenylenediamine with organic and inorganic electron acceptors	884
13. Excited aryl amines with free radicals	885
b. Excited fluorenone with dimethyl-N-(cyclohexyl)ketenimine	886, 887
c. Acetonitrile with excited aromatics and aza aromatics	888
d. Biacetyl	
1. With amines and phenols (singlet and triplet)	889, 890
2. With triphenylphosphine and triphenylamine	891
e. Olefins	
1. Excited iodine with olefins	892
2. Internal exciplex formation in 6-phenyl-2-hexene	893

TABLE VII—continued

System	Reference
3. Fluorescence quenching by quadricyclene	894
4. Mechanism of quenching of aromatic hydrocarbon singlets by conjugated dienes	895
5. Excited acylnitriles by piperylene	896
6. Excited 2,3-diazabicyclo[2. 2. 2]oct-2-ene	897
f. Carbon tetrachloride and related compounds	
1. Carbon tetrachloride with excited naphthalene derivatives	898
2. Excited 1-methylnaphthalene and 2-phenyl-5-(α-naphthyl) oxazole with carbon tetrachloride	899
3. Carbon tetrachloride with excited pyrene	900
4. Alkylhalides with polycyclic aromatics	901
g. Mixed excimers and excitons	
1. Mixed excimers of anthracene derivatives	902
2. Perylene-benzo[*ghi*]perylene mixed crystals	903
3. Interaction of naphthalene singlet and triplet excitons with impurity molecules	904
h. Quenching of quinine sulfate and rhodamine dyes by halide ion complexation	905
i. Excited-state behavior of ground-state charge-transfer complexes	
1. Tetracyanobenzene and tetrachlorophthalic anhydride complexes with aromatic hydrocarbons	906
2. Solvent effects on tetrachlorophthalic anhydride complex with hexamethylbenzene	907
3. Fluorescence and phosphorescence of complexes of aromatic hydrocarbons with *N*-methyl-9-chloroacridinium and 2,4,6-trimethylpyridinium cations	908
4. Vacuum ultraviolet-induced emission of various kinds of charge-transfer complexes	909
j. Aliphatic sulfides as quenching agents	910
k. H atom and proton-transfer-quenching processes	
1. Effect of inter- and intramolecular hydrogen bonding on the position of n,π^* and π,π^* levels	911
2. Fluorescence self-quenching through hydrogen bonding through 2-phenylindole	912
3. Indirect evidence of the existence of excited 2-*H*-indole	913
4. Intramolecular proton-transfer quenching of 3-hydroxy-2-napthoic acid	914
5. Proton transfer in oxalic acid solutions of 3,6-diaminoacridine	915
l. Effect of excited complex formation on triplet state	
1. Classification and formation of charge-transfer complexes in the excited states	916
2. Guest-host interactions in benzil-benzoin mixed crystals	917

TABLE VII—continued

System	Reference
3. Nonexponentiality of phosphorescence of mixed crystalline solution	918
4. Internal electron transfer in triplet state of N,N-dibenzylphenacylamine	919
5. Excited biacetyl with triethylborane	920
6. Long-lived luminescence of rigid solutions of charge-transfer complexes	921
7. Triplet-triplet absorption spectrum and triplet-lifetime studies of naphthalene and phenanthrene with tetrachlorophthalic anhydride and pyronellitic dianhydride	922
8. Pyromellitic dianhydride-mesitylene complex	923
9. Tetracyanobenzene with benzene and methylbenzenes	924
III. Oxygen quenching	
a. Fluorescence of 2-phenyl-5-(α-naphthyl)oxazole	899
b. Role of exciplex in photooxidation of toluene	925
c. Fluorescence quenching of aromatic hydrocarbons by oxygen	926
d. Detection of naphthalene-photosensitized generation of singlet oxygen by electron paramagnetic resonance	927
e. Reaction by singlet oxygen and by way of transient eosine-oxygen exciplex	928
f. Energy transfer from singlet oxygen to methylene blue	929
g. Energy transfer and complex formation between oxygen and triplet sensitizers	929a
h. Formation of singlet oxygen by energy transfer from benzaldehyde	929b
IV. "Quenching"	
a. Mechanisms of concentration quenching of aromatic hydrocarbons	930, 931
b. Concentration effects in photochemical cis-trans isomerization of difurylethylene and dithienylethylene	932
c. Analytical method for calculating constants of temperature quenching of fluorescence	933
d. "Local temperature" effects in impurity quenching	934
e. Fluorescence and phosphorescence quenching in tryptophan and tyrosine-free acids and in proteins	
1. Thiols and sulfides as quenchers of fluorescence of tyrosyl and tryptophanyl residues	935
2. Viscosity effects in fluorescence of tryptophan derivatives	936
3. Temperature, isotope, and solvent effects on fluorescence of tryptophan derivatives	937, 938
4. Environmental effects on fluorescence of tyrosine and its homologs	939
5. Phosphorescence from tryptophan and tyrosine in different microenvironments	940

TABLE VII—continued

System	Reference
V. Electronic energy transfer	
a. Solid-state transfer (triplet energy unless otherwise specified)	
1. Resonance transfer in crystal	942
2. Effect of relative orientation on energy transfer	943
3. Donor-acceptor distance relations by sensitized fluorescence measurements	943a
4. Theory of singlet energy transfer in polymeric matrices by virtual excitons	944
5. Donor fluorescence as probe for energy transfer from several donors in polymeric matrices	944a
6. Resonance transfer from triplets to singlets in rigid solutions	945
7. Excitation diffusion in poly(vinylcarbazole) and N-isopropylcarbazole	946, 947
8. From triplets higher than T_1 in rigid solutions	948
9. Temperature and concentration effects in two-component mixed crystal systems	949
10. Temperature and concentration effects in three-component mixed crystal systems	950
11. Solvent-sensitized phosphorescence in toluene	994, 995
12. Exciton diffusion and trap-trap singlet energy transfer in anthracene-doped naphthalene	951
13. Singlet exciton diffusion in phenanthrene-doped naphthalene	952
14. Resonance singlet transfer from pyrene to perylene in polymeric matrices	953
15. Triplet transfer from polystyrene and poly(methyl-methacrylate) to solute molecules	996
16. Exciton diffusion in benzophenone crystals	954
17. In phenanthrene-doped napthhalene	955, 956
18. In solid solutions of aromatic compounds	957
19. Photosensitization of poly(vinyl cinnamate)	958
b. Liquid-phase energy transfer	
1. Singlet energy transfer	
(a) Effect of energy transfer on fluorescence polarization	592, 595
(b) Concentration quenching of fluorescein fluorescence by energy transfer	959, 1006
(c) Resonance transfer in dye solutions	960, 961
(d) Phase and modulation technique for studying resonance transfer in viscous solutions	962
(e) Dependence of resonance-transfer kinetics on spectral overlap	963
(f) Diffusion and energy transfer as mechanisms of energy migration in solution	964, 965
(g) Energy transfer in luminescent mixed solutions	966

System	Reference
(h) Energy-transfer quenching of fluorescence of aromatic compounds by rare-earth ions	967
(i) Excitation intensity and concentration effects in energy transfer from pyrene to perylene	968, 1005
(j) Aminopyridines as singlet sensitizers	969
(k) From aromatic hydrocarbons to alkyl azides	970
(l) From azulene to stilbene	971
(m) In chemiluminescence of hydrazides	972
(n) In scintillator solutions	9973,74
2. Triplet energy transfer	
(a) Review of triplet energy transfer	975, 976
(b) Energy transfer in radiation	977, 978
(c) Proton-magnetic-resonance study of triplet energy transfer in pyrene	979
(d) Conditions for diffusion-controlled energy transfer	1004
(e) Asymmetric induction through energy transfer from optically active naphthyl amines	980
(f) Secondary isotope effects in the cis-trans isomerization of β-methylstyrene-β-d	981
(g) Role of second triplet state of anthracene in energy transfer and intersystem crossing	982, 983
(h) From 7-oxo-13-methyl-5,6,7,9,19,13-hexahydrophenanthrone to naphthalene	374
(i) From various sensitizers to bicyclo[3. 3. 0]octanones	984
(j) From polynuclear aromatic to lycopene	406
(k) From various sensitizers to all-*trans*-retinal	985
(l) From acetophenone (formed from autoxidation of ethylbenzene) to various acceptors	986
(m) Reversible energy transfer between benzophenone and biphenyl, fluorene, and triphenylene	987
(n) Concentration effects on triplet energy transfer efficiency of aromatic ketones	988
(o) From dibenzofuran isologs to phenyl-3-thienyl ketone	989
(p) Benzene photosensitized cis-trans isomerization of 2-butene	990
(q) Photosensitized isomerization of 2-butene by various triplet energy donors	991
(r) Relative rates of benzene energy transfer to various olefins	992
(s) Competition between exchange energy transfer and intermediate formation in photosensitized isomerization of 2-pentene	993
(t) Nonvertical energy transfer in organic azides	997
(u) From various donors to thionine	998
(v) From chlorophyll *a* to carotenoid molecules	999

TABLE VII—continued

System	Reference
(w) From anthracene to β-carotene	1000
(x) From aromatic ketones to rare-earth ions	1001
(y) From acridine to paramagnetic metal ions	1002
(z) In electron transfer luminescence	1003
c. Vapor-phase energy transfer (triplet energy unless otherwise specified)	
1. Xenon-sensitized vacuum ultraviolet photolysis of ethane	1007
2. From cadmium to benzene to 2-butenes	1008, 1009
3. From mercury to naphthalene	1010
4. From benzene and toluene to π-bonded molecules	1011
5. Pressure effects on fluorescence lifetime of phthalimide-quenching by singlet energy transfer	1012
6. From 1,3,5- and 1,2,4-trifluorobenzene to biacetyl	675
7. From benzene to naphthalene	1013
d. Intramolecular energy transfer	
1. Review	1014
2. Monitoring of intramolecular energy transfer by photochemical reaction: Diels-Alder reaction of 1,4-naphthoquinone with cyclopentadiene	1015
3. Energy transfer in chemiluminescence of phthalic hydrazide joined to acridone moiety	1016
4. Induction of fluorescence polarization in intramolecular energy transfer	1017
5. In a,ω-diphenyl alkanes	1018
6. In 1,1-dimethyl-1-benzoyl-2,4-pentadiene	1019
7. In some noncoplanar bis(2,2′biphenylylene)spirans	1020
8. In1,4-dimethoxy-5,8-methano-6,7-exo[fluorene-9′-spiro-1′-cyclopropane]naphthalene	1021
9. In rigid spiro compounds	1022, 1023
10. In some indole alkaloids	1024
11. In some N-acyl carbazoles	312, 701, 1025
12. In estrone	
13. Between naphthalene chromophore and benzophenone chromophore separated by rigid steroid bridge	1026
C. Relaxation processes involving two excited molecules	
I. Triplet-triplet annihilation	
a. In vapor phase of naphthalene, anthracene, and phenanthrene	1027
b. Liquid phase	
1. Solute reencounter probabilities from delayed excimer fluorescence of polynuclear aromatics	1028
2. Flash-spectroscopic studies on annihilation in tris(p-carbomethoxyphenyl)amine	1029
3. Delayed fluorescence in N, N, N', N'-tetramethyl-p-phenylenediamine solutions	1030

TABLE VII—continued

System	Reference
4. Population of S_2 by T, T annihilation in porphyrins	1031
5. Mixed T, T annihilation between eosine and anthracene	1032
c. Solid solutions and mixed crystals	
1. Resonance energy-transfer mechanism for triplet-triplet annihilation between carbazole and naphthalene	1033
2. Mechanisms of T, T annihilation in mixed crystals	1034
3. Interaction of triplet excitons with excited molecules of naphthalene in benzophenone	1035
4. Mixed T, T annihilation between benz[f]indan guest molecules and biphenyl or fluorene host crystals	1036
5. Effect of concentration on decay time of naphthalene and phenanthrene phosphorescence	1037
6. Annihilation of naphthalene triplets in ether-ethanol glass and of 2-bromonaphthalene triplets in naphthalene crystals	1038
7. Stimulated recombination luminescence in rigid, amorphous solutions of N, N, N', N'-tetramethyl-p-phenylenediamine, mesitylene, p-xylene, and durene	1039
8. Amine acid and proteins	1040
9. DNA-acridine dye complexes in frozen aqueous solution	1041
d. Molecular crystals	
1. Review of radiationless loss mechanisms and intermolecular interactions	1042
2. Theory of triplet exciton annihilation in polyacene crystals	1043
3. Nonlinear quenching of luminescence by annihilation of excited species	1044
4. Effect of electronic field on annihilation in anthracene	1045
5. Failure of square dependence of intensity of exciting light on delayed fluorescence of anthracene	1046, 1047
6. Phenanthrene single crystals	1048
II. Reverse triplet-triplet annihilation: fission of tetracene singlet excitons into two triplet excitons	1049–1052

By using time-resolved nanosecond emission spectroscopy Ware, Chow, and Lee [793a] observe fluorescence from both the Franck-Condon state and the vibrationally relaxed excited state of phthalimide and can also observe the effect of solvent relaxation on the fluorescence lifetime.

Internal conversion between excited states is seldom observed directly. The internal conversion of the first singlet to the ground state, however, competes, along with S_1, T_1 intersystem crossing, with radiative decay, and rate constants for this process can be determined.

Rentzepis [665] determines the rate constant for the S_1, S_0 internal conversion of azulene by observing, for the first time, the S_1, S_0 fluorescence. He uses high-power densities to populate the first singlet enough to overcome the very low transition probability from this state. The unusually high S_1,S_0 radiationless decay rate in azulene accounts for the low fluorescence quantum yield.

A number of studies given in Table VII, Part *A. I. c*, show that the S_1,S_0 internal conversion rate often depends on temperature, presumably because the vibrationally relaxed excited state has a poor correspondence to isoenergetic vibrational states of the electronic ground state. Wavelength dependence is also observed, especially when both n,π^* and π,π^* excited states are present.

Intersystem crossing involves both $S \rightarrow T$ pressures and $T \rightarrow S$ processes. In internal-conversion processes we noted that there was a large rate difference between $S_n \rightarrow S_1$ and $S_1 \rightarrow S_0$ processes. Similarly, in intersystem crossing $S_n \rightarrow T_m$ processes are faster in general than $T_1 \rightarrow S_0$. The former process competes with fluorescence, a spin-allowed radiative process; the latter process occurs on approximately the same time scale as the spin-forbidden phosphorescent radiation.

In the gas phase and solid state, where phosphorescence and fluorescence decay can both be observed routinely, the relative quantum yields of the two processes can give some measure of the amount of $S_n \rightarrow T_m$ intersystem crossing. Parmenter and Poland [805] found that in biacetyl vapor intersystem crossing can occur even without collisional deactivation of the singlet states. Triplet emission is observed even at 40 μatm, where the collision with the walls becomes the main triplet-destroying process.

Measuring solution intersystem crossing directly has been achieved by Drent, Van der Deijl, and Zandstra [809], using the laser technique of Rentzepis described above for measuring picosecond processes. They use enhancement of the S, T process by heavy-atom solvent to observe the competition between internal conversion and fluorescence.

Studying the excitation spectrum of camphorquinone phosphorescence, Tsai and Charney [673] find that crossing from upper singlet states to upper triplet states appears to be efficient; they interpret this to mean that the internal conversion between triplet states is as rapid as internal conversion between singlet states. In their studies on anthracene vapors Borisevich and Tolstorozhev [639] find significant intersystem crossing from the second singlet level.

In addition to the participation of excited singlets in intersystem crossing the second triplet state of anthracene participates in an S_1,T_2 intersystem-crossing process both in the solid state [765] and in solution [982, 983]. The same S_1,T_2 process appears to be important in other polynuclear hydrocarbons [821], including some cases in which this process requires thermal activation [640, 641, 811].

In addition to the deuterium isotope effect on the T_1,S_0 crossing discussed below, there is evidence from the vapor-phase study of benzene and benzene-d_6 that there is also an enhancement of the S_1,T_1 crossing [806, 807].

As in all other processes involving triplet states, perturbation by external and internal heavy atoms enhances both $S \to T$ and $T \to S$ processes [805b, 819–823, 658].

One of the important recent kinds of studies on both $S \to T$ and $T \to S$ processes is spin polarization at very low temperature. Near absolute zero the transitions between zero-field levels of the triplet state becomes very slow, and phosphorescence and electron-spin-resonance studies can elucidate the rates for crossing into and out of the separate levels. Such studies have been carried out on pyrazine [802, 812, 813] and aromatic ketones [803], both of which show important n, π^*; π,π^* interactions in their intersystem-crossing mechanisms. Lim [702] observes the effect of vibrational mixing of n,π^* and π,π^* states in quinoxaline and quinoxaline-d_8 due to out-of-plane CH and CD wagging modes.

Crossing from T_1 to S_0 is usually observed as a competitive process with phosphorescence. In order to determine the T_1,S_0 radiationless decay rate accurately, the population of the triplet state and the quantum yield and lifetime of phosphorescence all have to be accurately known. Brinen, Orloff, Gallivan, Stamm, and Roberts [623] conduct such studies on chrysene, chrysene-d_{12}, and 6-fluorochrysene and find that deuteration in this case does not significantly alter S,T intersystem crossing or radiative processes but does effect the T,S crossing rate. Monofluorination, on the other hand, affects both radiationless and radiative T,S rates and the fluorescence rate. Clearly measurement of the phosphorescence lifetime alone would not discriminate these differences, since it is a composite of both radiative and nonradiative processes.

Within the limits of deuteration of aromatic hydrocarbons, there have been carefully studied systems, such as chrysene, where it appears that only the T, S radiationless transition is affected. As we noted, however, there is evidence in gas-phase benzene that the population of the triplet state is also affected [806]. Whether or not the latter result will be found to be more general is not clear. For now, the general assumption is that it is not and that changes in phosphorescence lifetime are due to changes in the radiationless decay rate.

Operating on the latter assumption, several studies have been made and are included in the studies in Part A. II. b in Table VII.

Keller [799, 800] uses the monitoring of fluorescence following T,T absorption to gain valuable information about T,S crossing in the excited-state manifold. In general he finds that internal conversion between triplet states is much faster than T,S intersystem crossing.

C. Deactivation of Excited Molecules by Ground-state Molecules

In contrast to the relative passivity of molecules that carry off vibrational energy in the pseudounimolecular process described above, there are a number of specific interactions between excited molecules and ground-state molecules that deactivate excited molecules. In general these processes are complex formation of various kinds and electronic energy transfer. The distinction is not sharp. Excitation in crystals is transferred through a special kind of complex, namely, excitons, and there is evidence of the complex formation in fluid-state energy transfer. But, although the distinction is soft, it is nevertheless useful, particularly in an area where experimental techniques for accurate classification are only now being developed. Stephenson and Hammond [941] review the various modes by which quenching by ground-state molecules can occur, emphasizing the experimental techniques and models. Complex formation is summarized in Part B.I. and II in Table VII and energy transfer in Part B.V.

Perhaps the most thoroughly studied case of interaction of ground- and excited-state molecules is exciton formation. Strictly speaking, it is not simply an interaction between an identifiable excited-state molecule with ground-state molecules but a cooperative phenomenon that is most widely viewed as a bound state of a positive hole and an electron in the crystal. Excitation into exciton states has been discussed in Section I.D. and Table IV. Schmidt [824] discusses the quantum theory of the damping of molecular exciton states in a crystal and finds analogies to the radiationless transition mechanism described by Robinson and Frosch. Kocki [825] reviews photoemission from organic crystals. The cooperative nature of excitons in excitation processes is underscored by a study on excitation of cooperative excitons in a crystal with two molecules per unit cell [826]. Burland and Castro [833] report on the location of all four possible factor-group levels which are found in benzene phosphorescence at low temperature. Haaner and Wolf [835] apply optical spin polarization techniques (see Section II above) to triplet excitons in naphthalene. Soos [836] studies the Zeeman population of anthracene triplet excitons.

Defect and impurity sites play a very important role in the deactivation of exciton states. Examples of this are seen in the singlet excitons of anthracene [831, 868] and *trans*-stilbene and in the triplet exciton of several doped acene crystals [838], pyrazine [350], and biphenyl [834].

Solid-state excimer formation is also found. Excimers, or excited dimers, are formed by a reaction of the type $A^* + A \to (AA)^*$ (* indicates an electronically excited species). Excimers are obviously related to photochemical dimerization intermediates but are distinguished from the latter species in that they decay to two ground-state molecules, namely, $(AA)^* \to 2A +$ heat. Excimer emission has been measured from tetrachloro-p-xylene [840] and pyrene [841] solid solutions. In addition, Kim, Beardslee, Phillips, and Often [842] report on pressure effects on the pyrene excimer and monomer and find that both kinds of emission are remarkably insensitive to pressures even to 30 bars.

Excimers are more commonly studied in the liquid state, where exciton formation does not occur. The pyrene excimer mentioned above has been extensively studied [859–861], because of the ease with which it forms. But as can be seen in Part B. I. b in Table VII, there are many other systems that have been studied this past year. Foerster [846] gives an extensive review of excimers. Chandra and Lim [843, 844] present a semiempirical theory for the luminescence from excimers. Sudlej [845] discusses the temperature dependence of weakly coupled dimers. In experimental studies, Hamilton and Naqvi [581] discuss the conditions for an isosbestic point in the emission spectrum of a solution showing both monomer and excimer.

Speed and Sellinger [847] apply time-resolved nanosecond spectroscopy to the pyrene excimer. They find that in solvents of low viscosity excimer formation does not proceed at the diffusion-controlled rate, although it does in solvents of higher viscosity.

Lawson, Hirayama, and Lipsky [795a] find that in liquid benzene the direct formation of excimer from the S_3 state is slower than internal conversion from the same state. The excimer formation occurs predominantly after relaxation to S_1.

Studies by Helman [627] and Hirayama and Lipsky [849] show that excimer formation in benzene proceeds essentially without activation energy. Deuteration slows the deactivation of the excimer more than of the monomer.

Goldschmidt, Tomkiewicz, and Berlman [861] find that under high-intensity excitation, the formation of pyrene excimers decreases with intensity, apparently because of depletion of the ground-state molecules.

The study of excimers has led to a fuller understanding of the nature of interaction of excited and ground-state molecules. In complexes where the ground and excited molecules are of the same chemical species, net charge transfer, as opposed to charge resonance, interaction cannot be a major stabilizing force. When the donor and acceptor species are not of the same chemical species, however, we should expect it to become an important contribution to the total stabilization. Weller, whose group has been active in studying the deactivation of excited aromatics with amines, provides a review of his researches in this field [870]. He is able to spectroscopically distinguish solvated ion pairs $(A_S^- \ldots D_S^+)$ and excited molecule complexes $^1(A^-D^+)$. The latter species, which we shall refer to as exciplexes, are characterized by their broadened long wavelength emission, relative to the molecular fluorescence, and by their high dipole moment. Table VII, Part *B*. II. *a*, lists a number of studies of such amine-hydrocarbon exciplexes. Exciplex formation has been found by exciting either the acceptor and the amine under appropriate conditions.

Although not yet so well-studied as the amine-acceptor exciplexes, a number of other kinds of exciplexes have been found involving other nitrogen compounds, olefins, and carbon tetrachloride with various excited molecules.

Day and Wright [897] find that steric effects seem to be operative in the exciplex quenching of 2,3-diazabicyclo[2.2.2]oct-2-ene by dienes. It appears that the diene must be able to achieve planarity in order for efficient quenching to occur.

Just as we noted that excimers are clearly similar to the intermediates in photodimerization, so we should expect exciplexes to resemble intermediates in photoaddition reactions. Singer, Davis, and Muralidhavan [887] propose the reversible formation of an exciplex between fluorenone and dimethyl-*N*-(cyclohexyl)ketenimine as an intermediate step in the photoaddition reaction of the two reagents. Morrison and Ferree [893] propose an intramolecular exciplex intermediate in the photocycloaddition in 6-phenyl-2-hexene.

Mixed excimers, which are distinguished from the exciplexes above because the donor and acceptor molecules in excimers tend to be very similar, are found in polynuclear aromatic molecules [902, 903].

Not only do the charge-transfer exciplexes resemble photoaddition intermediates, but as electron-transfer species they are formally similar to proton-transfer species [911–915], which can also lead to the deactivation of excited states.

A few triplet exciplexes [916–924] have been studied, but such species do not seem so important as singlet exciplexes. That many of these triplet exciplexes are also related to stable ground-state complexes also distinguishes them from the singlet species.

Oxygen quenching is included as a special topic in Table VII, because oxygen, being a ground-state triplet, can cause deactivation by spin-orbit perturbation, energy transfer, and complex formation. The last process is frequently overlooked, but it seems to be involved in certain cases. Wei and Adelman [925] invoke such an exciplex in the photooxidation of toluene. Koizumi and Usui [928] suggest an exciplex in the interaction of triplet eosine and oxygen. Canva, Baling, Douzon, and Bourdon [929a] find an addition compound formed between triplet sensitizers and oxygen that are stable at low temperatures (200°K) but decompose on warming.

Part *B*.IV in Table VII is labeled "Quenching," because the topics are a potpourri of studies that derive unity only from the fact that they involve some emission-quenching process that is not easily classifiable under the topics above.

Foerster [930] reviews the mechanisms involved in the concentration quenching of the fluorescence of aromatic compounds. Type I quenching involves the formation of excimers that do not involve changes in the absorption spectrum. Type II quenching involves the formation of excimers and aggregates that do involve changes in the absorption spectrum as well. The final, type III, mode of quenching involves excitation transfer, which increases the probability of radiationless transfer. Rohatgi and Singhal [931] argue that fluorescein at pH 12 has concentration quenching by mechanisms corresponding to Foerster's types II and III. They find that type II quenching predominates at moderate concentrations and that the energy quenching type III mode is the most important at high concentrations.

Turoverov [933] proposes an equation for determining the activation energy for fluorescence quenching by studying the temperature dependence and assuming a quantum yield for fluorescence with the form $q = (1 + K + ke^{-\Delta E/kT})^{-1}$.

A number of studies [935–940] deal with the quenching of fluorescence and phosphorescence of tryptophane and tyrosines, both as free amino acids and bound in polypeptides and proteins. The use of the natural fluorescence of proteins as a structural probe (see Table VI, Part E, V above) depends on the ability to understand the various possible effects involved in quenching these residues. Studies reported this year focus on quenching by thiols and sulfides, such as are present in cysteine and cystine residues, by viscosity effects, temperature effects, isotope effects, and solvent effects.

The second major bimolecular process by which ground-state molecules deactivate excited-state molecules is electronic energy transfer. Electronic energy can be transferred either with or without intermolecular contact. In the former class are "trivial" or absorption-reemission energy transfer and resonance energy transfer. In the latter class are exciton migration, excimer and exciplex decay, and exchange energy transfer.

The trivial-transfer process involves the absorption of emitted light by other molecules in the environment. Although this process is obviously what finally occurs to most emitted light—at the walls of the laboratory if not within the test tube—it is not a very important event in most photophysical experiments and has not been studied to any great extent. Dombi [966] gives a careful treatment of the effect of both trivial and resonance transfer on the observed radiation of mixed solutions. His equations give good agreement with experimental results.

By contrast, resonance transfer is very important. It is also called Förster transfer, after Theodor Förster, who gave a time-dependent pertubation-theoretical treatment of the phenomenon. Resonance transfer corresponds to energy transfer between two coupled oscillators; it falls off as r^6_{AB}, where r_{AB} is the separation between the donor and acceptor molecules. Its efficiency depends on the overlap of the donor's emission spectrum and the acceptor absorption spectrum, and its allowedness is proportional to the probability of donor emission and acceptor absorption. Möbius [943a] tests the distance requirement for resonance transfer, using the monolayer techniques of Drexhage discussed in radiative processes. He is able to determine the critical distance for such transfer in his system. Levshin and Grineva [960–961] find that resonance transfer between properly selected dye molecules is very efficient; at 55 Å half of the excitation is transferred. They find that mixed solutions of the dye molecules can have a higher emission yield than the sum of the two separate solutions.

The product of probabilities involved in its allowedness means that resonance transfer involving spin-multiplicity charges on the part of the acceptor molecules should be of very low probability. Triplet donor–singlet acceptor combinations, however, can have high probability. Martinez [945] observes such transfer between benzene triplets and azulene singlets in rigid glass at low temperature.

Georghion [962] gives expressions for the experimental parameters for studying resonance transfer by phase and modulation techniques. He also performs calculations on a phenanthrene-acridine solution in cyclohexanol. Dexter, Förster, and Knox [942] answer Davydov's objections to the possibility of resonance energy transfer in mixed crystals.

The dominant mode of energy transfer in crystals seems to be exciton migration. In their study of solid solutions of anthracene in naphthalene, Mansour and Weinreb [951] find that both free and localized excitons are involved in the decay of excitations. Hunter [952] finds that in phenanthrene-doped naphthalene, the excitation is transferred both by naphthalene exciton migration and by resonance transfer from naphthalene defect traps and phenanthrene molecules. In solid solutions of perylene in N-isopropylcarbazole, Kloepffer [947] estimates that, by the hopping model of exciton migration, the singlet exciton takes 40,000 steps and has an average displacement of 1350 Å from the initial site of excitation. Adam and O'Dwyer [949–950] study the kinetics of triplet exciton migration in both two- and three-component mixed crystal systems. Nakahara, Koyanagi, and Kanda [954] study the rate of triplet exciton migration in benzophenone crystals. Hunter, McAlpine, and Hochstrasser [956] find in naphthalene-doped benzophenone that separate mechanisms are required to account for exciton migration in crystalline and glassy solutions. Terskoi and Brudz [944] give a theory for energy transfer in transparent polymers involving virtual excitons.

Energy transfer involving excimer formation is found by Goldschmidt, Tomkiewicz, and Weinreb [1005] in excitation transfer from perylene to pyrene. They find that the rate of energy transfer is equal to that of pyrene excimer formation, indicating that the pyrene monomer is a fairly inefficient energy-transfer agent. By contrast, Chapman and Wampfler [988] find that at high concentrations of triplet π,π^* sensitizers, excimer formation between sensitizer molecules competes with photosensitization of triplet energy acceptors.

The most important energy transfer in organic photochemistry is the transfer of triplet energy in order to produce high triplet concentrations selectively. Koizumi [976] reviews the application of triplet energy transfer to photochemical reactions. Parker [975] reviews the scope and mechanism of triplet energy transfer.

In most cases studied triplet transfer seems to occur by an exchange mechanism involving actual contact, or very close approach, of the donor and acceptor. If the separation of the donor and acceptor levels is sufficiently exoergic, the transfer is generally found to go as close to the diffusion-controlled rate. Wagner [987] finds that some endothermic energy transfer can occur, so that biphenyl, fluorene, and triphenylene all quench the photoreduction of benzophenone, although all three have higher triplet energies than benzophenone. Nordin and Strong [1004] find by computational analysis that, even with the reaction endothermic by more than 3 kcal mole^{-1}, reverse transfer can occur.

Eisenthal [943] finds that planar and nonplanar donor-acceptor pairs (benzophenone-phenanthrene-d_{10} and anthrone-phenanthrene-d_{10}) in rigid glass show no orientation effects, as measured by polarization, on the energy-transfer efficiency. That steric effects are involved to some extent is found by Cole [980], who finds that using an asymmetric sensitizer on a racemic acceptor gives some optical activity in the product.

Caldwell and Sovocool [981] find in the photosensitized isomerization of trans-β-methylstyrene-β-d that the initially formed product is higher in cis than in the protonated compound.

Cocivera [979] finds that in the nuclear-magnetic-resonance spectrum of pyrene under ultraviolet irradiation, a 6.6 Hz line broadening occurs which he argues is due to triplet excitation transfer between pyrene molecules. He estimates an exchange rate of 4×10^7 l mole^{-1} sec^{-1}.

Liu and Kellog [982] and Liu and Edman [983] find spectroscopic and chemical evidence that the second triplet state of anthracenes is involved in energy transfer. They estimate a T_2, T_1 internal conversion rate in excess of 4×10^9 sec^{-1}. Alfimov, Batekha, and Smirnov [948] find in rigid solution transfer from higher triplet states of aromatic hydrocarbons to various acceptors.

Although triplet energy exchange transfer is usually considered to occur by an exchange mechanism, Saltiel, Neuberger, and Wrighton [993] argue that both exciplex (Schenck intermediates) and exchange mechanisms are operative in the cis-trans isomerization of 2-pentenes sensitized by acetone and acetophenone.

In considering energy transfer in solution, it is worth recalling, as has been shown [964, 965, 974], that diffusion can lead to anomalously high energy transfer in solutions of low viscosity, even in singlet energy transfer.

Part B.V.d in Table VII lists a number of systems for which intramolecular energy transfer occurs between separate chromophones in the molecule. Strictly speaking, of course, this is an internal conversion process, although one which is restricted to only a certain class of molecules. Such transfer has been used to study various aspects of energy-transfer phenomena. White, Roberts, and Roswell [1016] use a special combination of a phthalic hydrazide and an acridone moiety to give a chemiluminescent species. Rauh, Evans, and Leermakers [1017] observe induction of fluorescence polarization in intramolecular polarization in reserpine and 17-(1-napthyl)yohimbol.

Although most intramolecular energy-transfer studies, in contrast to intermolecular-transfer studies, are carried out on molecular luminescence properties, Filipescu and Menter [1915] observe chemical reaction following internal excitation transfer in the Diels-Alder adduct of cyclopentadiene and 1,4-naphthaquinone. Leermakers, Montillier, and Rauh [1019] observe cis-trans isomerization of the diene part of 1,1-dimethyl-1-benzoyl-2,4-pentadiene following excitation of the benzoyl moiety. Filipescu [1014] reviews intramolecular energy-transfer processes.

D. Deactivation of Excited Molecules by Processes Involving Two Excited Molecules: Annihilation Processes

Annihilation involves energy transfer from one excited molecule to another, apparently through resonance, exciplex, or exchange mechanisms. Since the kinetics of annihilation processes are bimolecular in excited molecules, three factors favor its occurrence: high light intensity, long-lived excited states, and efficient pathways for excitation to migrate in the medium. Thus, triplet-triplet annihilation in condensed phases is the most commonly observed bimolecular process of excited states. With the high power densities found in laser beams, however, the observation of the less probable singlet-singlet processes, even in the gas phase, will become experimentally accessible. The triplet-triplet process is an important deactivation route for triplets, because the reaction $^3A^* + {}^3B^* \rightarrow {}^1A^* + {}^1B$ is spin-allowed, in contrast to all the other pathways we have discussed other than the energy transfer to a ground-state molecule: $^3A^* + {}^1B \rightarrow {}^1A + {}^3B^*$.

Annihilation processes are studied almost exclusively by delayed fluorescence, that is, fluorescence that has the lifetime properties of phosphorescence. Such fluorescence can arise either from thermally activated intersystem crossing ($^3A^* + \text{heat} \rightarrow {}^1A^* \rightarrow A + h\nu$) or from triplet-triplet annihilation. The two processes can be distinguished by the square dependence on light intensity for annihilation.

Finger and Zahlan [1027] study the vapor-phase annihilation processes in naphthalene, anthracene, and phenanthrene. From the temperature dependence of the process they propose that there is an excimer step in the process, that is, $^3A^* + {}^3B^* \rightarrow {}^1(AB)^* \rightarrow {}^1A^* + B$. Stevens [1028] finds that in fluid solution of polynuclear aromatics the sequence of events seems to be $^3A^* + {}^3B^* \rightarrow {}^1A^* + B \rightarrow {}^1(AB)^*$. Thus, excimer formation requires reencounter of the excited and ground-state molecules formed in the annihilation step. He calculates temperature and concentration dependence of the ratio of excimer and molecular fluorescence by a one-dimensional random-walk model.

Swenberg [1043] gives an improved theoretical formulation of bimolecular triplet exciton annihilation kinetics in polyacene crystals. His calculations show that the inverse triplet-triplet annihilation (singlet exciton fission) should have a nonnegligible probability for pentacene and tetracene. Pope, Geacintov, and Vogel [1051, 1052] and Merrifield, Avakian, and Groff [1049, 1050] find that exciton fission does occur in tetracene crystals, each singlet exciton undergoing fission giving rise to two triplet excitons. By using an external magnetic field, the probability of this process can be reduced, giving rise to an enhanced quantum yield for exciton fluorescence.

Martinez [1033] finds a process formally corresponding to the process $^3A^* + {}^3B^* \rightarrow {}^1A + {}^3B^{*\prime}$. The process is a resonance, or Förster, transfer involving overlap of the donor (carbazole) phosphorescence and the acceptor (biphenyl) T, T absorption spectrum.

In crystals containing impurities, either deliberate or accidental, the main deexcitation processes can occur between these impurity traps. When the concentration of such traps is low, saturation of them can occur at low light intensity, as has been observed in anthracene [1046, 1047] and phenanthrene [1048] crystals. Adventitious impurity traps can interfere with deliberate impurity traps, giving rise to skewed results. Such guest-impurity annihilation has been found, by a study of the temperature dependence of delayed fluorescence in benz[f]indan in biphenyl and fluorene host crystals [1036].

Kikuchi, Kokubun, and Koizumi [1932] use acetophenone as a sensitizer to produce significant populations of both anthracene and eosine triplets in ethanol solution so that the delayed fluorescence of both compounds can be observed. They find that mixed triplet-triplet annihilation significantly favored eosine as the energy acceptor. Eosine does not efficiently undergo T, T self-annihilation, although anthracene does.

REFERENCES

This chapter covers the literature abstracted in *Chemical Abstracts* from volume 70, number 5 (1969), to volume 72, number 4 (1970). This completes the 1968 literature survey that could not be included in last year's volume because of labor disputes at Chemical Abstracts Services.

1. Mitsuhiko Hida, *Senryo to Yakuhin*, **13**, 436–443 (1968).
2. Masahiro Hatano, *Kobunshi*, **17**, 1033–1040 (1968).
3. James E. Parkin, *Ann. Rep. Prog. Chem.*, **64**, 205-215 (1967; published 1968).
4. Juergen Dehler and Guenter Kresze, *Chem. Unserer Zeit*, **2**, 123–126 (1968).
5. Yukio Kojima and Shozo Toda, *Kanagawa-ken Kogyo Shikensho Kenkyo Hokoku*, **1968**, 81–85.
6. Donald B. Siano and David E. Metzler, *J. Chem. Phys.*, **51**, 1856–1861 (1969).
7. John N. Bradley, *Transfer Stor. Energy Mol.*, **1**, 64–93 (1969).
8. V. I. Permogorov, L. A. Serdyukova, and M. D. Frank-Kamenetskii, *Izv. Akad. Nauk SSSR, Ser. Fiz.*, **32**, 1560–1563 (1968).
9. Robin M. Hochstrasser and Charles Marzzacco, *Mol. Lum., Int. Conf*, **1968**, ed. E. C. Lim (New York: Benjamin, 1969).
10. A. V. Lukashin and M. D. Frank-Kamenetskii, *Dokl. Akad. Nauk SSSR*, **188**, 391–394 (1969).
11. F. A. Savin, *Opt. Spektrosk.*, **25**, 836–842 (1968).
12. F. A. Savin, *Opt. Spektrosk.*, **26**, 114–116 (1969).
13. Arnold J. Wohl, *Tetrahedron*, **24**, 6889–6896 (1968).
14. Giles Lee Henderson, *Trans. Ill. State Acad. Sci.*, **61**, 360–366 (1968).
15. T. Ya. Popova, A. K. Popov, S. G. Rautian, and R. I. Sokolovskii, *Zh. Eksp. Teor. Fiz.*, **57**, 850–863 (1969).
16. Andrzej J. Sadlej, *Int. J. Quantum Chem.*, **3**, 569–580 (1969).
17. Harold Basch, Melvin B. Robin, and Norman A. Kuebler, *J. Chem. Phys.*, **49**, 5007–5018 (1968).
18. Karl R. Loos, Urs P. Wild, and Hans H. Guenthard, *Spectrochem. Acta*, **25A**, 275–281 (1969).
19. Robert J. Buenker and Jerry L. Whitten, *J. Chem. Phys.*, **49**, 5381–5387 (1968).
20. A. V. Lukashin, V. I. Permogorov, and M. D. Frank-Kamenetskii, *Dokl. Akad. Nauk SSSR*, **183**, 874–877 (1968).
21. Paul M. Vay, *J. Chim. Phys. Physicochim. Biol.*, **65**, 2043–2049 (1968).
22. T. G. Edwards and Roger Grinter, *Mol. Phys.*, **15**, 357–365 (1968).
23. Ilyas Absar, Thesis, University of Sasketchewan, 1968; *Diss. Abstr.*, **29B**, 4607 (1969).
24. Kazuyoshi Sakamoto and Yasumasa Ihaya, *Theor. Chim. Acta*, **13**, 220–229 (1969).
25. Janet Del Bene and Hans H. Jaffe, *J. Chem. Phys.*, **50**, 1126–1129 (1969).
26. Keith Yates, Sandra L. Klemenko, and Imre G. Csizmadia, *Spectrochim. Acta*, **25A**, 765–778 (1969).
27. M. P. Noskova and N. N. Kazanova, *Zh. Strukt. Khim.*, **10**, 718–721 (1969).
28. T. G. Edwards and Roger Grinter, *Theor. Chim. Acta*, **12**, 387–396 (1968).
29. Sigeo Yomosa, Mitsuo Honda, and Hideao Suzuki, *J. Phys. Soc. Jap.*, **26**, 1485–1194 (1969).
30. Juergen Fabian and Achim Mehlhorn, *Z. Chem.* **9**, 271–272 (1969).
31. Juergen Fabian, *Theor. Chim. Acta*, **12**, 200–205 (1968).
32. Mario Bossa, G. Ciullo, and Antonio Symaelloti, *Tetrahedron*, **25**, 1991–1996 (1969).
33. E. Zeeck, *Theor. Chim. Acta*, **15**, 86–88 (1969).
33a. Robert W. Kramling, thesis, Washington State University, 1968; *Dissertation Abstr.*, **29B**, 3277–3278 (1969).
34. Anthony P. Modica, *J. Phys. Chem.*, **72**, 4594–4598 (1968).
35. V. A. Job, V. Sethuraman, and K. K. Innes, *J. Mol. Spectrosc.*, **30**, 365–426 (1969).

36. Alfred Tze-Hua Lee, thesis, University of California, Los Angeles, 1968; *Diss. Abstr.*, **29B**, 3263–3264 (1969).
37. D. C. Moule, *Can. J. Phys.*, **47**, 1235–1236 (1969).
38. Gerard Jean Martin, Marie C. Jouet, Jean P. Dorie, Jean P. Gouesnard, Simone Odiot, and Maryvonne L. Martin, *Bull. Soc. Chim. Fr.*, **1969**, 2508–2510.
39. Renee Fraisse Jullien and Claudine Frejaville, *Bull. Soc. Chim. Fr.*, **1969**, 2095–2100.
40. Catherine Y. S. Chen, and Charles A. Swenson, *J. Phys. Chem.*, **73**, 1642–1647 (1969).
41. C. N. Ramachandra Rao, *Chem. Nitro Nitroso Groups*, ed. by Henry Feuer (New York: Wiley-Interscience Publisher, 1969), 79–135.
42. C. N. Ramachandra Rao and K. R. Bhaskar, *ibid.*, 137–163.
43. Don C. Iffland, Michael P. McAneny, and Dennis Joseph Weber, *J. Chem. Soc.*, **1969C**, 1703–1706.
44. W. W. Haskell and I. A. Read, *Appl. Spectrosc.*, **23**, 532–535 (1969).
45. Liang-Tsai Cheng, thesis, Louisiana State University, 1968; *Diss. Abstr.*, **29B**, 3275 (1969).
46. Masamichi Tsuboi, Akiko Y. Hirakawa, and Kiroyuki Kawashima, *J. Mol. Spectrosc.*, **29**, 216–229 (1969).
47. Sloboder Pistic, Jelisaveta M. Baranac, and Aurora A. Muk, *Glas. Hem. Drus.*, *Beograd*, **32**, 153–165 (1967).
48. Ted M. McKinney, *Spectrochim. Acta*, **25A**, 501–506 (1969).
49. H. Bock and H. Seidl, *J. Amer. Chem. Soc.*, **90**, 5694–5700 (1968).
50. Hans Bock, Hartmuth Alt, and H. Seidl, *J. Amer. Chem. Soc.*, **91**, 355–361 (1969).
51. Colin G. Pitt, *J. Amer. Chem. Soc.*, **91**, 6613–6622 (1969).
52. Wiendelt Drenth, J. G. Noltes, E. J. Bulten, and H. M. J. C. Creemers, *J. Organomet. Chem.*, **17**, 173–174 (1969).
53. Omar W. Steward and Joseph E. Dziedsic, *J. Organomet. Chem.*, **16**, P5–P7 (1969).
54. Michael A. Weiner and George Pasternack, *J. Org. Chem.*, **34**, 1130–1133 (1969).
55. George Pasternack, thesis, City Univ. of New York, 1969; *Diss. Abstr.*, **29B**, 4108–4109 (1969).
56. R. D. Obolentsev and N. S. Lyubopytova, *Khim. Seraorg. Soedin., Soderzh. Neftyakh Nefteprod.*, **8**, 232–236 (1968).
57. Anthony B. Callear, J. Connor, and D. R. Dickson, *Nature*, (*London*), **221**, 1238 (1969).
58. Jack Barrett and M. J. Hitch, *Spectrochim. Acta*, **25A**, 407–415 (1969).
59. John L. Roebber, *U.S. Govt. Res. Develop. Rep.*, **69**, 80 (1969).
60. Bernard Tinland, *Spectrosc. Lett.*, **1**, 407–411 (1968).
61. Lawrence Edwards and John W. Raymonda, *J. Amer. Chem. Soc.*, **91**, 5937–5939 (1969).
62. A. J. Merer and Robert S. Mulliken, *J. Chem. Phys.*, **50**, 1026–1027 (1969).
63. Julia C. Tai and Norman L. Allinger, *Theor. Chim. Acta*, **12**, 261–266 (1969).
64. Harold Basch, Melvin B. Robin, Norman A. Kuebler, Clive Baker, and David Warren Turner, *J. Chem. Phys.*, **51**, 52–66 (1969).
65. Melvin B. Robin, Harold Basch, Norman A. Kuebler, K. B. Wiberg, and G. B. Ellison, *J. Chem. Phys.*, **51**, 45–52 (1969).
66. Giulio Milazzo and Gaetano Ceccheti, *Appl. Spectrosc.*, **23**, 197–203 (1969).
67. Joe D. Simmons, Arnold M. Bass, and Shelby G. Tilford, *Astrophys. J.*, **155**, 345–358 (1969).
68. Gilbert Richard Cook and Masaru Ogawa, *J. Chem. Phys.*, **51**, 2419–2424 (1969).

69. Stephen Bell, G. J. Cartwright, G. B. Fish, D. O. O'Hare, Robert K. Ritchie, Arthur D. Walsh, and P. A. Warsop, *J. Mol. Spectrosc.*, **30**, 162–163 (1969).
70. L. De Reilhac, Nicole Damany Astoin, and Jacques Romand, *Spectrochim. Acta*, **25A**, 19–30 (1969).
71. Jose M. Riveros, *J. Chem. Phys.*, **51**, 1269–1270 (1969).
72. G. Belanger, P. Sauvageau, and Camille Sandorfy, *Chem. Phys. Lett.*, **3**, 649–651 (1969).
73. Edwin F. Pearson and K. Keith Innes, *J. Mol. Spectrosc.*, **30**, 232–240 (1969).
74. A. J. Merer and L. Schoonveld, *Can. J. Phys.*, **47**, 1731–1743 (1969).
75. Ruth McDiarmid, *J. Chem. Phys.*, **50**, 1794–1800 (1969).
76. Benjamin Katz and Joshua Jortner, *Chem. Phys. Lett.*, **2**, 4379 (1968).
77. A. J. Merer and Robert S. Mulliken, *Chem. Rev.*, **69**, 639–656 (1969).
78. R. Thomson and P. A. Warsop, *Trans. Faraday Soc.*, **65**, 2806–2814 (1969).
79. Ruth McDiarmid, *J. Chem. Phys.*, **50**, 2328–2336 (1969).
80. A. E. Lyuts, A. E. Cherkashin, and Yu. A. Kushnikov, *Izv. Akad. Nauk. Kaz. SSR, Ser. Khim.*, **18**, 55–64 (1968).
81. Seinosuke Onari, *J. Phys. Soc. Jap.*, **27**, 269 (1969).
81a. Giorgio Favini, *Corsi Semin. Chim.*, **14**, 56–57 (1968).
82. M. M. Mestechkin and L. S. Gutyra, *Opt. Spekt.*, **26**, 159–167 (1969).
83. Claude Leibovici, *Compt. Rend.*, **267C**, 1121–1122 (1968).
84. Bernard Tinland, *Theor. Chim. Acta*, **13**, 171–174 (1969).
85. Bernard Tinland, *Tetrahedron*, **24**, 6833–6838 (1968).
86. Lydie Watmann-Grajcar, *J. Chim. Phys. Physicochim. Biol.*, **66**, 1018–1022 (1969).
87. Mario Bossa, Antonio Sgamelloti, and F. A. Gianturco, *J. Chem. Soc.*, **1969B**, 742–745.
88. D. J. Currie and Henry L. Holmes, *Can. J. Chem.*, **47**, 863–864 (1969).
89. Bernard Tinland, *Acta Phys. (Budapest)*, **25**, 111–114 (1968).
90. Bernard Tinland, *Spectrosc. Lett.*, **2**, 25–30 (1969).
90a. Robert D. McAlpine, thesis, University of Pennsylvania, 1968; *Dissertation Abstr.*, **29B**, 3870–3871 (1969).
91. W. Th. A. M. Van der Lugt, Henricus M. Buch, and Luitzen J. Oosteroff, *Ind. Chim. Belge*, **32**, 161–164 (1967).
92. Milos Tichy and Rudolf Zahradnik, *J. Phys. Chem.*, **73**, 534–547 (1969).
93. Claude Leibovici, *Compt. Rend.*, **268C**, 596–597 (1969).
94. Claude Leibovici, *An. Quim.*, **55**, 119–26 (1969).
95. S. A. Houlden and Imre G. Csizmadia, *Tetrahedron*, **25**, 1137–1153 (1969).
96. Takahiro Tsunoda, Tsuguo Yamaoka, and Kunihiro Ikari, *Kogyo Kagaku Zasshi*, **72**, 156–162 (1969).
97. Rolf Gase, *Z. Phys. Chem. (Leipzig)*, **239**, 307–320 (1968).
98. Gunnar Wettermark, Lars Tegner, and Olle Martensson, *Ark. Kemi*, **30**, 185–212 (1968).
99. Hans Bock, H. Alt, and Helmut Seidl, *Angew. Chem. Int. Ed. Engl.*, **7**, 885–886 (1968).
100. Peter G. Perkins, *Theor. Chim. Acta*, **12**, 427–430 (1968).
101. Peter Schuster, Dagmar Vedrilla, and Oskar E. Polansky, *Monatsh. Chem.*, **100**, 1–27 (1969).
102. Robert J. Buenker and Sigrid D. Peyerimhoff, *Chem. Phys. Lett.*, **3**, 37–42 (1969).
103. Homer E. Holloway, Robert V. Nauman, and James H. Wharton, *J. Phys. Chem.*, **72**, 4474–4482 (1968).

103a. Jimmie R. McDonald, thesis, Louisiana State University, 1968; *Dissertation Abstr.*, **29B**, 1628 (1969).
104. Z. H. Zaidi and Baij N. Khanna, *J. Chem. Phys.*, **50**, 3291–3296 (1969).
105. Cyril Parkanyi, E. J. Baum, J. Wyatt, and J. N. Pitts, Jr., *J. Phys. Chem.*, **73**, 1132–1138 (1969).
106. Georg Hohlneicher, Friedrich Doerr, Norbert Mika, and Siegfried Schneider, *Ber. Bunsenges. Phys. Chem.*, **72**, 1144–1155 (1968).
107. Rudolf Zahradnik, J. Pancir, and A. Krohn, *Coll. Czech. Chem. Commun.*, **34**, 2831–2832 (1969).
108. Patricia A. Mullen and Malcolm K. Orloff, *J. Mol. Spectrosc.*, **30**, 140–143 (1969).
109. Guenter Haefelinger, Andrew Streitweiser, Jr., and James S. Wright, *Ber. Bunsenges. Phys. Chem.*, **73**, 456–465 (1969).
110. Horst Stegemeyer, *Ber. Bunsenges. Phys. Chem.*, **73**, 612–619 (1969).
111. Urs P. Wild, *Chimia*, **22**, 473–477 (1968).
112. Aldo Gamba, *Rend. Ist. Lombardo Sci., Lett.*, **102A**, 538–546 (1968).
113. N. Trinajstic and Roger J. Wratten, *J. Mol. Struct.*, **3**, 395–402 (1969).
114. Frederic A. Van-Catledge and Norman L. Allinger, *J. Amer. Chem. Soc.*, **91**, 2582–2589 (1969).
114a. Giorgio Favini, Aldo Gamba, and Massimo Simonetta, *Theor. Chim. Acta*, **13**, 175–194 (1969).
115. V. M. Berenfel'd and V. A. Krongauz, *Izv. Akad. Nauk SSSR. Ser. Fiz.*, **32**, 1575–1579 (1968).
116. G. V. Klimusheva, L. M. Yagupol'skii, and R. V. Yaremko, *Teor. Eksp. Khim.*, **5**, 392–396 (1969).
117. Edward O. Holmes, Jr., *J. Phys. Chem.*, **73**, 273–274 (1969).
118. Grady L. Roberts, *Appl. Spectrosc.*, **23**, 165–168 (1969).
119. V. Bertoli and Peter H. Plesch, *Spectrochim. Acta*, **25A**, 447–465 (1969).
119a. Robert Vinson Morris, Thesis, University of Iowa, 1968; *Dissertation Abstr.*, **29B**, 549 (1969).
120. Anthony M. Trozzolo, *Acc. Chem. Res.*, **1**, 329–335 (1968).
121. Ladislav Hruban, A. Klasek, and Frantisek Santavy, *Acta Univ. Palacki. Olumuc., Fac. Med.*, **48**, 9–19 (1968).
122. V. Bertoli and Peter H. Plesch, *J. Chem. Soc.*, **1968B**, 1500–1516.
123. B. J. Ansari and D. Sharma, *Ind. J. Pure Appl. Phys.*, **6**, 614–616 (1968).
124. T. N. Bolotnikova, L. Ya. Malkes, A. I. Nazarenko, and V. N. Yakovenko, *Opt. Spektrosk.*, **25**, 621–623 (1968).
125. Robert Van der Linde, Jauke U. Veenland, and Thymen J. De Boer, *Spectrochim. Acta*, **25A**, 487–499 (1969).
126. K. G. Srinivasacharya, *Ind. J. Pure Appl. Phys.*, **7**, 62–63 (1969).
127. P. D. Singh and A. N. Pathak, *Ind. J. Pure Appl. Phys.*, **7**, 39–46 (1969).
128. Kamalesh Singh and V. B. Singh, *Ind. J. Phys.*, **42**, 668–671 (1968).
129. David E. Fenton, *Chem. Ind. (London)*, **1969**, 695–696.
130. T. N. Pliev, *Dokl. Akad. Nauk SSSR*, **184**, 1113–1116 (1969).
131. T. N. Pliev, V. N. Poletova, I. S. Aul'chenko, and L. A. Kheifits, *Dokl. Akad. Nauk SSSR*, **184**, 867–870 (1969).
131a. T. N. Pliev, V. N. Poletova, I. S. Aul'chenko, and L. A. Kheifitz, *Zh. Prikl. Spektrosk.*, **10**, 855–859 (1969).
132. S. N. Thakur and S. K. Tiwari, *Ind. J. Pure Appl. Phys.*, **7**, 570–572 (1969).
133. E. E. Fesenko and D. I. Roshchupkin, *Zh. Prikl. Spektrosk.*, **9**, 461–465 (1968).
134. J. Sunkel and Herbert Staude, *Ber. Bunsenges. Phys. Chem.*, **73**, 203–209 (1969).

135. Hermann Kaemmerer and W. Lewenz, *Spectrochim. Acta*, **24A**, 2059–2069 (1968).
136. Y. P. Srivastava, Devendra Sharma, and Nitish K. Sanyal, *Ind. J. Pure Appl. Phys.*, **6**, 617–620 (1968).
137. Laxmi N. Tripathi, *Ind. J. Pure Appl. Phys.*, **7**, 357–359 (1969).
138. M. K. Verzilina and A. V. Belotsvetov, *Zh. Obshch. Khim.*, **39**, 660–664 (1969).
139. Waldemar Broser and Wolfgang Harrer, *Z. Naturforsch.*, **24B**, 542–547 (1969).
140. C. G. Rama Rao, *Ind. J. Phys.*, **42**, 354–361 (1968).
141. C. P. D. Dwivedi, *Ind. J. Pure Appl. Phys.*, **7**, 186–189 (1969).
142. Satoshi Arimitsu, Katsumi Kimura, and Hiroshi Tsubomura, *Bull. Chem. Soc. Jap.*, **42**, 1858–1861 (1969).
143. M. Day and Arnold T. Peters, *J. Soc. Dyers Colour.*, **85**, 8–13 (1969).
144. A. A. Efimov and L. A. Skripko, *Zh. Prikl. Spectrosk.*, **9**, 650–653 (1968).
145. Yoichi Iida and Yoshio Matsunaga, *Bull. Chem. Soc. Jap.*, **41**, 2535–2536 (1968).
146. Yoichi Iida and Yoshio Matsunaga, *Bull. Chem. Soc. Jap.*, **41**, 2619 (1968).
147. Ahmed Mustafa, R. Abu-Eittah, and S. Elgendi, *Appl. Spectrosc.*, **23**, 254–256 (1969).
148. Jan Maszew and Stanislaw Kusmierczyk, *Zesz. Nauk. Uniw. Jagiellon., Pr. Chem.*, **14**, 59–66 (1969).
149. A. Ya. Kaminskii, V. Sh. Golubchik, and S. S. Gitis, *Zh. Prikl. Spektrosk.*, **10**, 109–114 (1969).
150. K. R. Popov, *Opt. Spektrosk.*, **25**, 843–848 (1968).
151. G. N. R. Tripathi, *Ind. J. Pure Appl. Phys.*, **7**, 582–583 (1969).
152. Monica M. Byrne and N. H. P. Smith, *Spectrochim. Acta*, **25A**, 313–316 (1969).
153. Panos Grammaticakis, *Compt. Rend.*, **268C**, 730–733 (1969).
154. Panos Grammaticakis, *Compt. Rend.*, **267C**, 976–979 (1968).
155. R. S. Tshekhanskii and I. K. Serebryakova, *Zh. Prikl. Spektrosk.*, **9**, 248–254 (1968).
156. Akira Kuboyama, Ryuichi Yamazaki, Sanae Yabe, and Yoshio Uehara, *Bull. Chem. Soc. Jap.*, **42**, 10–15 (1969).
157. Gerhard Wegner, Thomas F. Keyes III, Nobuo Nakabayashi, and Harold G. Cassidy, *J. Org. Chem.*, **34**, 2822–2826 (1969).
158. Anton Rieker, Wolfgang Rundel, and Horst Kessler, *Z. Naturforsch*, **24B**, 547–562 (1969).
159. A. Fulton, *Aust. J. Chem.*, **21**, 2847–2852 (1968).
160. R. M. P. Jaiswal, *Ind. J. Pure Appl. Phys.*, **7**, 47–51 (1969).
161. K. G. Srinivasacharya, *Ind. J. Pure Appl. Phys.*, **7**, 583–584 (1969).
162. C. P. D. Dwivedi, *Ind. J. Pure Appl. Phys.*, **7**, 410–412 (1969).
163. Jan Moszew and Zenon Klapyta, *Zesz. Nauk. Uniw. Jagiellon., Pr. Chem.*, **13**, 67–72 (1968).
164. Jiri Sevcik, *Monatsh. Chem.*, **100**, 1307–1309 (1969).
165. B. M. Krasovitskii, N. I. Mal'tseva, A. I. Nazarenko, and V. B. Smelyakova, *Proc. Int. Conf. Lumin., 1966*, ed. by Szigeti (Budapest, Hung.: Akad. Kiado, 1968), **1**, 618–625.
166. Camillo Tosi, *Corsi Semin. Chim.*, **14**, 50–51 (1968).
166a. Rosenda A. Yunes and Alberto J. Terenzani, *Rev. Fac. Ing. Quim., Univ. Nac. Litoral*, **36**, 83–93 (1967).
167. Harald A. B. Linke and D. Pramer, *Z. Naturforsch*, **24B**, 997–999 (1969).
168. A. I. Kiprianov and V. Yu. Buryak, *Ukr. Khim. Zh.*, **34**, 1016–1020 (1968).
169. V. P. Razumova, *Zh. Obshch. Khim.*, **39**, 1418–1419 (1969).
170. V. P. Razumova, *Zh. Obshch. Khim.*, **39**, 1418–1419 (1969).
171. L. I. Dranik and T. A. Shubert, *Dokl. Akad. Nauk SSSR*, **185**, 705–706 (1969).
172. Frederick T. Wolf, *Advan. Front. Plant Sci.*, **21**, 169–172 (1968).

173. A. D. Delaney, D. J. Currie, and Henry L. Holmes, *Can. J. Chem.*, **47**, 3278–3280 (1969).
174. George G. Porter, *Proc. Fourth Mol. Spectrosc. Conf.*, **1968**, 305–311.
175. G. Di Lonardo and C. Zauli, *J. Chem. Soc.*, **1969A**, 1305–1306.
176. James Melvin Meyer, thesis, Northwestern University, 1968; *Diss. Abstr.*, **29B**, 2340 (1969).
178. B. I. Stepanov and T. G. Edel'mann, *Zh. Obshch. Khim.*, **39**, 1549–1551 (1969).
179. Kenichi Terauchi and Hiroshi Sakurai, *Bull. Chem. Soc. Jap.*, **42**, 821–823 (1969).
180. Ludovico Santucci and Carlo Triboulet, *J. Chem. Soc.*, **1969A**, 392–396.
181. S. C. Bag, *Ind. J. Phys.*, **42**, 235–239 (1968).
182. Ram Nath Singh, *J. Sci. Res. Banaras Hindu Univ.*, **16**, 129–138 (1966).
183. E. V. Vladzimirskaya and Yu. M. Pashkevich, *Farm. Zh. (Kiev)*, **24**, 23–26 (1969).
184. Yoichi Iida, *Bull. Chem. Soc. Jap.*, **42**, 71–75 (1969).
185. Ludovic Fey and N. Bodor, *Rev. Roum. Chim.*, **14**, 225–234 (1969).
186. Ludovic Fey and N. Bodor, *Rev. Roum. Chim.*, **14**, 481–493 (1969).
187. Arnold Zweig and James B. Gallivan, *J. Amer. Chem. Soc.*, **91**, 260–264 (1969).
188. Lech Skulski, *Bull. Acad. Pol. Sci., Ser. Sci. Chim.*, **17**, 253–258 (1969).
189. I. L. Belaits, R. N. Nurmukhametov, and D. N. Shogorin, *Zh. Fiz. Khim.*, **43**, 869–874 (1969).
190. M. A. Chekalin and N. A. Zhukova, *Zh. Prikl. Spektrosk.*, **10**, 1008–1010 (1969).
191. Lidia Prajer-Janczewska, Anna Postawka, Krystyna Kocma, and Krystyna Rudolph, *Rocz. Chem.*, **42**, 1437–1451 (1969).
192. Roger R. Hill and G. H. Mitchell, *J. Chem. Soc.*, **1969B**, 61–64.
193. G. D. Baruah, S. Nath Singh, and Rama S. Singh, *Ind. J. Pure Appl. Phys.*, **7**, 283–285 (1969).
194. Noboru Mataga, Yoshikazu Torihashi, and Yoshifumi Furutani, *Kogyo Kagaku Zasshi*, **72**, 120–122 (1969).
195. Jerry D. Scott and William Harold Watson, *J. Chem. Phys.*, **49**, 4246–4247 (1969).
196. L. Sh. Tushishvii, N. A. Sheheglova, D. N. Shigorin, and N. S. Dokunikhin, *Zh. Fiz. Khim.*, **43**, 981–984 (1969).
197. N. A. Shcheglova, D. N. Shigorin, and N. S. Dokunikhin, *Zh. Fiz. Khim.*, **42**, 2724–2734 (1968).
198. Achim Mueller, Karoly Kormendy, Ferenc Ruff, and Miklos Vajda, *Acta Chim. (Budapest)*, **59**, 109–118 (1969).
199. V. Ya. Fain, *Reakts. Sposobnost. Org. Soedin*, **5**, 735–749 (1968).
200. G. D. Baruah, S. Nath Singh, and Rama S. Singh, *Ind. J. Pure Appl. Phys.*, **7**, 281–282 (1969).
201. A. F. Kovalev and V. I. Litvinenko, *Khim. Farm. Zh.*, **3**, 22–25 (1969).
202. N. V. Platanova, K. R. Popova, and L. V. Smirnov, *Opt. Spektrosk.*, **26**, 357–363 (1969).
203. David P. Craig and G. J. Small, *J. Chem. Phys.*, **50**, 3827–3834 (1969).
204. Z. H. Zaidi and B. N. Khanna, *Curr. Sci. (India)*, **37**, 580–582 (1968).
205. D. R. Maulding and B. G. Roberts, *J. Org. Chem.*, **34**, 1734–1736 (1969).
206. N. S. Proskuryakova and R. N. Nurmukhametov, *Opt. Spektrosk.*, **27**, 224–227 (1969).
207. Shobha Singh and Camille Sandorfy, *Chem. Phys. Lett.*, **2**, 507–508 (1968).
208. Takashi Kajiwara, Ichimin Shirotani, Hiroo Inokuchi, and Satoshi Iwashima, *J. Mol. Spectrosc.*, **29**, 454–60 (1969).
209. A. J. Mellugh, Donald A. Ramsay, and Ian Gordon Ross, *Aust. J. Chem.*, **21**, 2835–2845 (1968).

210. Juergen Dehler and Klaus Fritz, *Tetrahedron Lett.*, **1969**, 2157–2160.
211. Josef Michl, *J. Mol. Spectrosc.*, **30**, 66–76 (1969).
212. R. N. Young, *J. Chem. Soc.*, **1969B**, 896–900.
213. Junichi Aihara, *Kagaku No Ryoiki*, **23**, 35–40 (1969).
214. Joshua Jortner and Graeme C. Morris, *J. Chem. Phys.*, **51**, 3689–3691 (1969).
215. N. Ya. Dodonova and I. P. Vinogradov, *Opt. Spektrosk.*, **27**, 179–181 (1969).
216. Stephen Finney Mason, *Photoelec. Spectrom. Group, Bull.*, **18**, 547–553 (1968).
217. Giorgio Favini, *Corsi Semin. Chim.*, **14**, 54–55 (1968).
218. G. Hallas, *J. Soc. Dyers Colour.*, **84**, 510–516 (1968).
219. Susan D. Carson and Herbert M. Rosenberg, *J. Mol. Spectrosc.*, **32**, 242–246 (1969).
220. R. Sosa and L. Paoloni, *Tetrahedron*, **25**, 4197–4205 (1969).
221. Horst Mantsch and J. Dehler, *Can. J. Chem.*, **47**, 3173–3178 (1969).
222. Bernard Tinland, *Theor. Chim. Acta*, **12**, 85–86 (1968).
223. F. P. Billingsley II and J. E. Bloor, *Theor. Chim. Acta.* **11**, 325–343 (1968).
224. F. Momicchioli and G. Del Re, *J. Chem. Soc.*, **1969B**, 674–679.
225. Hans Guesten, Leo Klasic, and Otto Volkert, *Z. Naturforsch*, **24B**, 12–15 (1969).
226. N. Trinajstic and A. Hinchliffe, *Croat. Chem. Acta*, **40**, 163–169 (1968).
227. Mitushiko Hida, *Senryo To Yakuhin*, **14**, 88–94 (1969).
228. Juergen Fabian, Achim Mehlhorn, and Rudolf Zahradnik, *Theor. Chim. Acta*, **12**, 247–255 (1968).
229. Enzo Borello, *Corsi Semin. Chim.*, **14**, 72 (1968).
230. Desmond J. Brown and P. B. Ghosh, *J. Chem. Soc.*, **1969B**, 270–276.
231. G. V. Gobov, R. N. Nurmukhametov, and L. N. Pushkina, *Zh. Fiz. Khim.*, **43**, 57–63 (1969).
232. O. A. Gunder, I. F. Mikhailova, and V. F. Poduzhailo, *Monokrist., Stsintill. Org. Lyuminofory*, **2**, 73–79 (1967); from a Russian abstract; *Chem. Abstr.*, **70**, Abstr. no. 55255h (1968).
233. Isadore B. Berlman, *Chem. Phys. Lett.*, **3**, 61–63 (1969).
234. Roy E. Ballard and C. H. Park, *Spectrochim. Acta*, **24A**, 1975–1980 (1968).
235. Vaclav Stuzka, A. P. Golovina, and I. P. Alimarin, *Collect. Czech. Chem. Commun.*, **34**, 221–228 (1969).
236. Ahmed Mustafa, R. Abu-Eittah, and S. Elgendi, *Appl. Spectrosc.*, **23**, 249–253 (1969).
237. Pernendu B. Talukdar and S. K. Segupta, *Ind. J. Chem.*, **7**, 129–131 (1969).
238. A. I. Busev, V. M. Ivanov, and Usama El Dbik, *Vestn. Mosk. Univ. Khim.*, **23**, 66–69 (1968).
239. Yu. D. Kovaliv, *Farm. Zh. (Kiev)*, **24**, 19–22 (1969).
240. A. I. Kiprainov and G. G. Dyadyusha, *Ukr. Khim. Zh.*, **35**, 608–615 (1969).
241. S. G. Fridman and A. I. Kiprianov, *Zh. Org. Khim.*, **5**, 373–376 (1969).
242. A. I. Kiprianov and I. L. Mushkalo, *Zh. Org. Khim.*, **4**, 2222–2225 (1968).
243. A. I. Kiprianov and T. M. Verbovskaya, *Zh. Org. Khim*, **4**, 1991–1995 (1968).
244. V. M. Ivanov, A. I. Busev, and Usama El Dbik, *Vestn. Mosk. Univ., Khim.*, **23**, 64–68 (1968).
245. Pavol Kristian and J. Bernat, *Coll. Czech. Chem. Commun.*, **34**, 2952–2958 (1969).
246. J. Eiduss, S. P. Korshunov, L. I. Vereshchagin, K. Venters, and S. Hillers, *Khim. Atsetilena*, ed. by M. Shostakovskii, (Moscow: Izd. "Nauka", 1968), 411–414.
247. I. L. Belaits, R. N. Nurmukhametov, D. N. Shigorin, and G. I. Bystritskii, *Zh. Fiz. Khim.*, **43**, 1673–1678 (1969).
248. Jay Jea-Yong Rhee, Thesis, University of New Mexico, 1968; *Diss. Abstr.*, **29B**, 4620 (1969).
249. Bimal Kumar Das, *J. Ind. Chem. Soc.*, **45**, 1075–1078 (1968).

250. Kichisuke Nishimoto, Kimiko Nakatsukasa, Ryoichi Fujishiro, and Shunji Kato, *Theor. Chim. Acta*, **14**, 80–90 (1969).
251. Bernard Tinland, *J. Mol. Struct.*, **3**, 516–519 (1969).
252. Nikolai Tyutyulkov and F. Dietz, *Izv. Otd. Khim. Nauki. Bulg. Akad. Nauk*, **2**, 55–64 (1969).
253. Henryk Chojnacki, *Theor. Chim. Acta*, **12**, 373–378 (1968).
254. K. Seibold, Georges Wagniere, and Heinrich Labhart, *Helv. Chim. Acta*, **52**, 789–796 (1969).
255. Gerald Maggiora, H. Johansen, and Lloyd L. Ingraham, *Arch. Biochem. Biophys.*, **131**, 352–358 (1969).
256. Joseph Kuthan and J. Prochazkova, *Coll. Czech. Chem. Commun.*, **34**, 1190–1203 (1969).
257. Jaroslav Leska, *Monatsh. Chem.* **100**, 545–552 (1969).
258. William R. Carper and J. Stengl, *Mol. Phys.*, **16**, 627–631 (1969).
259. R. P. Vorob'eva, M. V. Kopytina, and N. G. Roazhkova, *Zh. Prikl. Spektrosk.*, **9**, 837–841 (1968).
260. Bhupeshchandra Das, *J. Ind. Chem. Soc.*, **46**, 479–481 (1969).
261. Armelle Denis and Helene Berthod, *J. Chim. Phys. Physicochim. Biol.*, **65**, 1815–1822 (1968).
262. J. S. Kwiatkowski, *Acta Phys. Pol.*, **34**, 365–385 (1968).
263. Bernard Pullman and Helene Berthod, *Compt. Rend.*, **268D**, 980–983 (1969).
264. J. S. Kwiatkowski, *Theor. Chim. Acta*, **13**, 149–154 (1969).
265. Allan S. Schneider and Robert Arthur Harris, *J. Chem. Phys.*, **50**, 5204–5215 (1969).
266. Masatane Kuroki, *Nippon Kagaku Zasshi*, **90**, 421–423 (1969).
267. Valentin Zanker, B. Schneider, and W. Seiffert, *Tetrahedron Lett.*, **1969**, 1497–1500.
268. W. Seiffert, V. Zanker, Horst Mantsch, and B. Schneider, *Tetrahedron Lett.*, **1968**, 5655–5660.
269. W. Seiffert, V. Zanker, and Horst Mantsch, *Tetrahedron*, **25**, 1001–1012 (1969).
270. Horst Mantsch, W. Seiffert, and V. Zanker, *Rev. Roum. Chim.*, **14**, 125–133 (1969).
271. Giorgio Favini and G. Buemi, *Theor. Chim. Acta.* **13**, 79–80 (1969).
272. Vinicio Galasso and Giancarlo De Alti, *Tetrahedron*, **25**, 2259–2264 (1969).
273. Helge Johansen and Lloyd L. Ingraham, *J. Theor. Biol.*, **23**, 191–204 (1969).
274. Marilyn H. Perrin, Martin Gouterman, and Charles L. Perrin, *J. Chem. Phys.*, **50**, 4137–4150 (1969).
275. Pill-Soon Song, *Int. J. Quantum Chem.*, **3**, 303–316 (1969).
276. Guenter Scheibe, E. Daltrozzo, O. Woerz, and J. Heiss, *Z. Phys. Chem. (Frankfurt am Main)*, **64**, 97–114 (1969).
277. Patricia A. Mullen and Malcolm K. Orloff, *J. Chem. Phys.*, **51**, 2276–2278 (1969).
278. Maria D'Alagni, Basilio Pispisa, and Franco Quadrifoglio, *Ric. Sci.*, **38**, 910–913 (1968).
279. B. A. Arbuzov, O. A. Erastov, A. B. Remizov, and S. N. Ignat'eva, *Izv. Akad. Nauk SSSR, Ser. Khim.*, **1969**, 763–767.
280. S. V. Tsukerman, V. P. Izvekov, V. F. Lavrushin, and Yu. S. Rozum, *Khim. Geterotsikl. Soedin.*, **1969**, 513–520.
281. Taku Matsuo, Hideto Shosenji, and Reiji Miyamoto, *Bull. Chem. Soc. Jap.*, **41**, 2849–2852 (1968).
282. A. Mani and John R. Lombardi, *J. Mol. Spectrosc.*, **31**, 308–317 (1969).
283. M. Slamnik and V. Sunjik, *Bull. Sci., Cons. Acad. Sci. Arts RSF Yougoslavie*, **14A**, 76–77 (1969).

284. L. V. Epishina, V. I. Slovetskii, O. V. Lebedev, L. I. Filippova, L. I. Khmel'nitskii, V. V. Sevost'yanova, T. S. Novikova, *Khim. Geterotsikl. Soedin.*, Sb. 1: *Azotso-derzhashchie Geterotsikly*, 1967, ed. by S. Hillers (Riga, USSR: Izd. "Zinatne," 1967), 102-108.
285. Adolf Jurasek, Rudolf Kada, and T. Sticzay, *Coll. Czech. Chem. Commun.*, **34**, 572-581 (1969).
286. Peter Nuhn, Guenther Wagner, and Siegfried Leistner, *Z. Chem.*, **9**, 152-153 (1969).
287. Yoichiro Nagai and Yoshiko Sakaino, *Nippon Kagaku Zasshi*, **90**, 309-313 (1969).
288. B. M. Krasovitskii and E. A. Shevchenko, *Khim. Geterotsikl. Soedin.*, **4**, 756-758 (1968).
289. S. V. Tsukerman, E. G. Buryakovskaya, and V. F. Lavrushin, *Opt. Spektrosk.*, **26**, 541-546 (1969).
290. L. M. Kutsyna, L. V. Voevoda, V. G. Tishchenko, and A. V. Shepel, *Opt. Speltrosk.*, **26**, 168-172 (1969).
291. C. Wijnberger and Clarisse L. Habraken, *J. Heterocycl. Chem.*, **6**, 545-551 (1969).
292. Siegfried Hoffman, *Z. Chem.*, **9**, 25 (1969).
293. John T. Edward and K. Liu, *Can. J. Chem.*, **47**, 1117-1122 (1969).
294. Giancarlo De Alti, Vinicio Galasso, and Adriano Bigotto, *Corsi Semin. Chim.*, **14**, 52-53 (1968).
295. Josef Kracmar, J. Kracmarova, and Jaroslav Zyka, *Pharmazie*, **23**, 567-573 (1968).
296. B. R. Pandey, *Ind. J. Phys.*, **42**, 269-270 (1968).
297. Piotr Tomasik, *Rocz. Chem.*, **42**, 2037-2042 (1968).
298. Piotr Tomasik and Zofia Skrowaczewska, *Rocz. Chem.*, **42**, 1427-1434 (1968).
299. Piotr Tomasik and Zofia Skrowaczewska, *Rocz. Chem.*, **42**, 1583-1589 (1968).
300. Swami P. Tandon, M. P. Bhutra, P. C. Mehta, Jagdeesh P. Saxena, and Kamala Tandon, *Ind. J. Pure Appl. Phys.*, **6**, 694-697 (1968).
301. K. Keith Innes, H. D. McSwinney, Jr., J. D. Simmonds, and S. G. Tilford, *J. Mol. Spectrosc.*, **31**, 76-94 (1969).
302. Lawrence S. Myers, Jr., Mary L. Hollis, L. M. Theard, and F. C. Peterson, *Nucl. Sci. Abstr.*, **23**, 33207 (1969).
303. Thomas M. Ward and Jerome Bernard Weber, *Spectrochim. Acta*, **25A**, 1167-1176 (1969).
304. John Michael Brown, *Can. J. Phys.*, **47**, 233-236 (1969).
304a. Larry Allen Franks, thesis, Vanderbilt University, 1968; *Dissertation Abstr.*, **29B**, 2571 (1969).
305. Carl Al Taylor, M. Ashraf El-Bayoumi, and Michael Kasha, *Proc. Nat. Acad. Sci. U.S.*, **63**, 253-260 (1969).
306. Bernard Pullman, Helene Berthod, Ernst D. Bergmann, Felix Bergmann, Zohar Nieman, and Hannah Weller-Feilchenfeld, *Compt. Rend*, **267C**, 1461-1463 (1968).
307. J. Bankovskis and A. Sturis, *Latv. Psr. Zinat. Akad. Vestis, Kim. Ser.*, **1968**, 671-682.
308. J. Bankovskis, A. Sturis, A. Ievins, and I. Labrence, *Latv. Psr. Zinat. Akad. Vestis, Kim. Ser.*, **1968**, 39-51.
309. J. Bankovskis, J. Asaks, M. Abolina, and A. Ievins, *Latv. Psr. Zinat. Akad. Vestis, Kim. Ser.*, **1968**, 52-59.
310. J. Bankovskis et al., *Latv. PSR Zinat. Akad. Vestis, Kim. Ser.*, **1968**, 60-71.
311. A. V. Shablya and G. I. Lashkov, *Izv. Akad. Nauk SSSR, Ser. Fiz.*, **32**, 1529-1533 (1968).
312. Maximilian Zander, *Ber. Bunsenges. Phys. Chem.*, **72**, 1161-1166 (1968).
313. David N. Bailey, David K. Roe, and David M. Hercules, *Appl. Spectrosc.*, **22**, 785-786 (1968).

314. A. N. Lomakin and L. A. Moskaleva, *Zh. Obshch. Khim.*, **39**, 941 (1969).
315. Yoshinori Hasegawa, Yoshito Amako, and Hiroshi Azumi, *Bull. Chem. Soc. Jap.*, **41**, 2608–2611 (1968).
316. Yoshinori Hasegawa, Yoshito Amako, Hiroshi Azumi, and Mitsuo Ito, *Bull. Chem. Soc. Jap.*, **42**, 840 (1969).
317. Yoko Kaizu and Mitsuo Ito, *J. Mol. Spectrosc.*, **30**, 149–153 (1969).
318. Hiroaki Baba, Iwao Yamazaki, and Takeshi Takemura, *Bull. Chem. Soc. Jap.*, **42**, 276 (1969).
319. A. I. Kiprianov and V. Yu. Buryak, *Zh. Org. Khim.*, **5**, 368–372 (1969).
320. A. I. Kiprianov and V. Yu. Buryak, *Zh. Org. Khim.*, **4**, 2030–2033 (1968).
321. A. I. Kiprianov and V. Yu. Buryak, *Ukr. Khim. Zh.*, **35**, 179–183 (1969).
322. Ingo Leubner, Juetgen Dehler, and Guenter Scheibe, *Ber. Bunsenges. Phys. Chem.*, **72**, 1133–1140 (1968).
323. Jacek Koziol, *Photochem. Photobiol.*, **9**, 45–53 (1969).
324. V. A. Mashenkov, K. N. Solov'ev, and S. F. Shkirman, *Dokl. Akad. Nauk. Beloruss. SSR*, **13**, 507–510 (1969).
325. A. N. Sevchenko, *Proc. Int. Conf. Lumin., 1966*; ed. by G. Szigeti (Budapest: Akad. Kiado, 1968), **1**, 183–193.
326. Janos Hevesi and G. S. Singhal, *Spectrochim. Acta*, **25A**, 1751–1758 (1969).
327. G. S. Singhal, W. P. Williams, and Eugene Rabinowitch, *J. Phys. Chem.*, **72**, 3941–3951 (1968).
328. Danuta Frackowiak, H. Manikowski, and Z. Salamon, *Acta Phys. Pol.* **34**, 669–674 (1968).
329. L. A. Stolyarova and J. Eiduss, *Latv. PSR Zinat. Akad. Vestis, Kim. Ser.*, **1969**, 434–438.
330. Z. I. Zelikman, A. I. Suprunova, and V. G. Kul'nevich, *Izv. Vyssh. Ucheb. Zaved., Pishch. Tekhnol.*, **1968**, 27–31.
331. A. Nagana Goud and Srinivasa Rajagopal, *Monatsh. Chem.*, **100**, 1368–1369 (1969).
332. J. Mendez and M. I. Loho, *Microchem. J.*, **13**, 506–512 (1968).
333. Boris Arventiev, Mihai Delibas, A. Galearschi, and V. Delibas, *An. Stiint. Univ. "Al. I. Cuza" IASI*, **14, 1c**, 163–167 (1968).
334. V. Urba and V. Salna, *Liet. Fiz. Rinkinys*, **8**, 693–706 (1968).
335. V. Urba, V. Salna, and H. Johaitis, *Nauch. Konf. Molodykh Uch. Litov. SSR, Rab. Obl. Fiz., Mat. Kibern*; ed. by P. Brazdziunas 193–195. (Vilnius, USSR: Akad. Nauk Litov. SSR, 1967).
336. H. Jonaitis, V. Urba, and V. Salna, *Liet. Fiz. Rinkinys*, **7**, 861–876 (1968).
337. Juergen Fabian, Steffen Scheithauer, and Roland Mayer, *J. Prakt. Chem.*, **311**, 45–60 (1969).
338. Juergen Fabian, K. Fabian, and Horst Hartmann, *Theor. Chim. Acta*, **12**, 319–324 (1969).
339. Robert A. W. Johnstone, and S. D. Ward, *Theor. Chim. Acta*, **14**, 420–425 (1969).
340. Juergen Fabian and Gunter Laban, *Tetrahedron*, **25**, 1441–1447 (1969).
341. Juergen Fabian, *Z. Chem.*, **9**, 272–273 (1969).
342. Norbert Kucharczyk, Bohumil Kakac, and Vaclav Horak, *Coll. Czech. Chem. Commun.*, **34**, 2959–2970 (1969).
343. Marcel Graffeuil and Jean F. Labarre, *J. Chim. Phys. Physicochim. Biol.*, **66**, 177–179 (1969).
344. Giulio Milazzo, *Gazz. Chim. Ital.*, **98**, 1511–17 (1968).

345. S. V. Tsukerman, V. D. Orlov, and V. F. Lavrushin, *Khim. Geterotsikl. Soedin.*, **1969**, 67–69.
346. Lionel Goodman and Bernard J. Laurenzi, *Adv. Quantum Chem.*, **4**, 153–169 (1968).
347. Marvin C. Goldberg and John R. Riter, Jr., *J. Chem. Phys.*, **50**, 547–549 (1969).
348. Vanulapalli G. Krishna and W. R. Salzman, *J. Chem. Phys.*, **50**, 3875–3879 (1969).
349. Mostafa A. El-Sayed, *Proc. Int. Conf. Lumin.*, *1966*, ed. by G. Szigeti (Budapest: Akad. Kiado, 1968), **1**, 373–379.
350. Paul Pesteil, Louise Pesteil, and Jean P. Danoy, *Compt. Rend.*, **268B**, 1774–1777 (1969).
351. George Castro and G. Wilse Robinson, *J. Chem. Phys.*, **50**, 1159–1164 (1969).
352. M. M. Val'dman, N. N. Malykhina, and L. P. Khudyakova, *Zh. Prikl. Spektrosk.*, **9**, 470–474 (1968).
353. Jacob S. Brinen and John G. Koren, *Chem. Phys. Lett.*, **2**, 671–672 (1968).
354. Yasuo Udagawa, Tohru Azumi, Mitsuo Ito, and Saburo Nagakura, *J. Chem. Phys.*, **49**, 3764–3771 (1968).
355. G. V. Gobov and L. A. Nakhimovskaya, *Vop. Opt. Mol. Spektrosk.*, **1968**, 39–45 (from a Russian abstract; see *Chem. Abstr.*, **71**, Abstr. no. 43960w).
356. John B. Birks, *Chem. Phys. Lett.*, **3**, 567–568 (1969).
357. John Kerry Thomas and I. Mani, *J. Chem. Phys.*, **51**, 1834–1838 (1969).
358. Roland Bonneau, Jacques Joussot-Dubien, and Rene Bensasson, *Chem. Phys. Lett.*, **3**, 353–356 (1969).
359. L. M. Theard, F. C. Peterson, and R. A. Holroyd, *J. Chem. Phys.*, **51**, 4126–4132 (1969).
360. John Kerry Thomas, *J. Chem. Phys.*, **51**, 770–778 (1969).
361. D. S. Kliger and A. C. Albrecht, *J. Chem. Phys.*, **50**, 4109–4111 (1969).
362. Alexander Mueller and Udo Sommer, *Ber. Bunsenges. Phys. Chem.*, **73**, 819–826 (1969).
363. George Porter and Michael R. Topp, *Nature (London)*, **220**, 1228–1229 (1968).
364. Maurice W. Windsor and J. R. Novak, *U.S. Govt. Res. Develop. Rep.*, **68**, 70 (1968).
365. Jacob Solomon Brinen, *Mol. Lumin., Int. Conf.*, 1968, ed. by E. C. Lim (New York: W. A. Benjamin, 1969), 333–349.
366. Jacob Solomon Brinen and Malcolm K. Orloff, *J. Chem. Phys.*, **51**, 527–531 (1969).
367. T. S. Zhuravleva, *Opt. Spektrosk.*, **25**, 696–700 (1969).
368. T. M. Naumova and V. I. Glyadkovskii, *Opt. Spektrosk.*, **27**, 228–231 (1969).
369. J. B. Gallivan and Jacob S. Brinen, *J. Chem. Phys.*, **50**, 1590–1595 (1969).
370. Yves Meyer and Raymond Astier, *J. Phys. (Paris)*, **29**, 1075–1080 (1968).
371. A. Lavalette, *Chem. Phys. Lett.*, **3**, 67–70 (1969).
372. Jean M. Lhoste and Jean B. Merceille, *J. Chim. Phys. Physicochim. Biol.*, **65**, 1889–1901 (1968).
373. Robert L. Strong and Herbert H. Richtol, *Fast React. Prim. Proc. Chem. Kinet., Proc. Fifth Nobel Symp.*, ed. by Stig. Claesson, (Stockholm: Almqvist and Wiksell, 1967), 71–79.
374. Goran Ramme, Robert L. Strong, and Herbert H. Richtol, *J. Amer. Chem. Soc.*, **91**, 5711–5714 (1969).
375. Henry S. Judeikis and Seymour Siegel, *U.S. Govt. Res. Develop. Rep.*, **69**, 71 (1969).
376. Guy Nouchi, *J. Chim. Phys. Physicochim. Biol.*, **66**, 554–565 (1969).
377. Daniel Lavalette, *Chem. Phys. Lett.*, **3**, 264–266 (1969).
378. Leo Klasinc and U. Sommer, *Chem. Phys. Lett.*, **3**, 107–109 (1969).
379. Charles J. Marzzacco, Thesis, University of Pennsylvania, 1968; *Dissertation Abstr.*, **29B**, 3703 (1969).

380. Guy Nouchi, *J. Chim. Phys. Physicochim. Biol.*, **66**, 548–553 (1969).
381. Ajit Singh, Arthur Reginald Scott, and F. Sopchyshyn, *J. Phys. Chem.*, **73**, 2633–2643 (1969).
382. T. N. Bolotnikova, T. M. Naumova, and L. K. Artemova, *Izv. Akad. Nauk SSSR, Ser. Fiz.*, **32**, 1475–1479 (1968).
383. Raymond Astier and Yves H. Meyer, *Chem. Phys. Lett.*, **3**, 339–401 (1969).
384. A. Sykes and T. G. Truscott, *J. Chem. Soc.*, **1969D**, 929–930.
385. G. C. Terry, V. E. Uffindell, and F. W. Willets, *Nature (London)*, **223**, 1050–1051 (1969).
386. Kevin D. Cadogan and Andreas C. Albrecht, *J. Phys. Chem.*, **73**, 1868–1877 (1969).
387. Hiroshi Tsubomura and Naoto Yamamoto, *Kogyo Kagaku Zasshi*, **72**, 6–11 (1969).
388. M. A. Pak, D. N. Shigorin, and G. A. Ozerova, *Dokl. Akad. Nauk SSSR*, **186**, 369–372 (1969).
389. Guy Nouchi and Claude Silvie, *Compt. Rend.*, **268B**, 546–548 (1969).
390. Donald R. Kemp and George Porter, *J. Chem. Soc.*, **1969D**, 1029–1030.
391. A. S. Faevskii, P. N. Logvinenko, and A. N. Faidysh, *Izv. Vyssh. Ucheb. Zaved., Fiz.*, **12**, 157–159 (1969).
392. T. N. Bolotnikova and O. N. Sichkar, *Zh. Prikl. Spektrosk.*, **9**, 500–503 (1968).
393. W. H. Melhuish, *J. Chem. Phys.*, **50**, 2779 (1969).
394. Edward J. Land and Albert J. Swallow, *J. Chem. Phys.*, **49**, 5552–5553 (1968).
395. Patricia A. Carapelluci, Harold P. Wolf, and Karl Weiss, *J. Amer. Chem. Soc.*, **91**, 4635–4639 (1969).
396. Ian A. Ramsay and Ian H. Munro, *Proc. Int. Conf. Lumin*, **1966**, ed. by G. Szigeti (Budapest: Akad. Kiado, 1968), **1**, 333–338.
397. John L. Kropp and Maurice W. Windsor, *U.S. Govt. Res. Develop. Rep.*, **68**, 56 (1968).
398. E. E. Fesenko and D. I. Roshchupkin, *Zh. Prikl. Spektrosk.*, **8**, 834–839 (1968).
399. D. W. Whillans, M. A. Herbert, John W. Hunt, and Harold E. Johns, *Biochem. Biophys. Res. Commun.*, **36**, 912–918 (1969).
400. Elie Hayon, *J. Amer. Chem. Soc.*, **91**, 5397–5398 (1969).
401. Valentin Zanker and Dirk Benicke, *Z. Phys. Chem. (Frankfurt am Main)* **66**, 34–47 (1969).
402. O. D. Dmitrievskii, *Opt. Spektrosk.*, **25**, 883–888 (1968).
403. Richard A. Pierce and Robert A. Berg, *J. Chem. Phys.*, **51**, 1267 (1969).
404. T. A. Shakhverdov, *Izv. Akad. Nauk SSSR, Ser. Fiz.*, **32**, 1564–1568 (1968).
405. T. N. Bolotnikova and O. N. Sichkar, *Opt. Spektrosk.*, **26**, 752–757 (1969).
406. A. Sykes and T. G. Truscott, *J. Chem. Soc.*, **1969D**, 274–276.
407. M. A. Herbert, John Wilfred Hunt, and Harold E. Johns, *Biochem. Biophys. Res. Commun.*, **33**, 643–648 (1968).
408. A. K. Chibisov, B. V. Skvortsov, A. V. Karyakin, and L. N. Rygalov, *Khim. Vys. Ener.*, **3**, 210–216 (1969).
409. Masaharu Morita and Shunji Kato, *Bull. Chem. Soc. Jap.*, **42**, 25–35 (1969).
410. Stephen F. Mason, *J. Soc. Dyers Colour.*, **84**, 604–612 (1968).
411. J. E. Midwinter and P. Suppan, *Spectrochim. Acta*, **25A**, 953–958 (1969).
412. Guenter P. Schiemenz, *Spectrochim. Acta*, **25A**, 439–445 (1969).
413. N. G. Bakhshiev, *Opt. Spektrosk.*, **27**, 42–49 (1969).
414. Paul Suppan, *J. Chem. Soc.*, **1968A**, 3125–3133.
415. Charles S. Irving, Gary W. Byers, and Peter A. Leermakers, *J. Amer. Chem. Soc.*, **91**, 2141–2143 (1969).
416. V. M. Korovina and L. I. Al'perovich, *Opt. Spektrosk.*, **25**, 506–508 (1968).

417. H. Lang and G. Loeber, *Ber. Bunsenges. Phys. Chem.*, **73**, 710–716 (1969).
418. Wolfgang Liptay, *Angew. Chem. Int. Ed. Engl.*, **8**, 177–188 (1969).
418a. A. V. Finkel'shtein and G. A. Reutov, *Reakts. Sposobnost Org. Soedin.*, **5**, 341–349 (1968).
419. A. V. Finkel'shtein and G. S. Krasnoshchekova, *Zh. Obshch. Khim.*, **39**, 310–312 (1969).
420. A. V. Finkel'shtein and V. V. Ivanov, *Kinet. Katal.*, **10**, 921–923 (1969).
421. A. V. Finkel'shtein and Z. M. Kuz'mina, *Zh. Fiz. Khim.*, **43**, 333–338 (1969).
422. Agostino Trombetti and Carlo Zauli, *J. Chem. Soc.*, **1968B**, 1574–1577.
423. V. Salna, V. Urba, and H. Jonaitis, *Nauch. Konf. Molodykh Uch. Litov SSR, Rab. Obl. Fiz., Mat. Kibern* (Vilnius, USSR: Akad. Nauk Litov, SSR, 1967), 196–198.
424. D. J. Cowley and L. H. Sutcliffe, *Spectrochim. Acta*, **25A**, 989–997 (1969).
425. E. E. Yudovich and V. V. Pal'chevskii, *Zh. Obshch. Khim.*, **39**, 62–70 (1969).
426. S. H. Dandegaonker, *J. Ind. Chem. Soc.*, **46**, 148–152 (1969).
427. A. E. Lutskii and Z. A. Tret'yak, *Ukr. Khim. Zh.*, **34**, 1095–1099 (1968).
428. Robert Glenn Lewis and J. J. Freeman, *J. Mol. Spectrosc.*, **32**, 24–38 (1969).
429. Michel Lamotte and Jacques Joussot-Dubien, *J. Chim. Phys. Physicochem. Biol.*, **66**, 161–165 (1969).
430. Alain Bienvenue and Jacques E. Dubois, *Bull. Soc. Chim. Fr.*, **1969**, 391–396.
431. Ivan Petrov and Bojan Soptrajanov, *Glas. Hem. Drus., Beograd*, **32**, 389–399 (1967).
431a. Hiromu Imai, *Tech. Rep. Kansai Univ.*, **9**, 49–57 (1967).
432. Robert S. Mulliken, *Proc. Robert A. Welch Found. Conf. Chem. Res.*, **1967**, **11**, 109–150 (1968).
433. Karl Henson and Walter Sarholz, *Theor. Chim. Acta*, **12**, 206–213 (1968).
434. V. I. Danilova and S. Ya. Belomyttsev, *Izv. Vyssh. Ucheb. Zaved., Fiz.*, **11**, 126–133 (1968).
435. Joan T. D'Agostino and Hans H. Jaffe, *J. Amer. Chem. Soc.*, **91**, 3383–3384 (1969).
436. J. W. Verhoeven, I. P. Dirkx, and Thymen J. De Boer, *Tetrahedron*, **25**, 4037–4055 (1969).
437. Roger Arnaud and Jane M. Bonnier, *J. Chim. Phys. Physicochim. Biol.*, **66**, 954–959 (1969).
438. E. Paspaleev, *Dokl. Bulg. Akad. Nauk*, **22**, 313–316 (1969).
439. Suprabhat Chatterjee, *J. Chem. Soc.*, **1969B**, 725–729.
440. Seiki Sekanoue, Toshiaki Tamamura, Shigekazu Kusabayashi, Hiroshi Mikawa, Nobutami Kasai, Masao Kakudo, and Haruo Kuroda, *Bull. Chem. Soc. Jap.*, **42**, 2407 (1969).
441. O. B. Nagy and A. Bruylants, *Bull. Cl. Sci., Acad. Roy. Belg.*, **53**, 1159–1176 (1967).
442. Haruo Kuroda, Shoji Hiroma, and Hideo Akamatur, *Bull. Chem. Soc. Jap.*, **41**, 2855–2858 (1968).
443. Takako Amano, Haruo Kuroda, and Hideo Akamatu, *Bull. Chem. Soc. Jap.*, **42**, 671–676 (1969).
445. Thomas F. Hunter and D. H. Norfolk, *Spectrochim. Acta*, **25A**, 193–197 (1969).
446. M. A. Slifkin, B. M. Smith, and R. H. Walmsley, *Spectrochim. Acta* **25A**, 1479–1489 (1969).
447. M. V. Kurik, A. A. Motuz, and E. K. Frolova, *Ukr. Fiz. Zh.* (*Russ. Ed.*), **14**, 679–681 (1969).
448. Stephen K. Lower, *Mol. Cryst. Liquid Cryst.*, **5**, 363–368 (1969).
449. George H. Schenk, Doina Triff, *Anal. Lett.*, **2**, 61–69 (1969).
450. A. E. Lutskii, I. I. Men'shova, L. A. Fedotova, and M. G. Voronkov, *Zh. Obshch. Khim.*, **39**, 879–886 (1969).

451. C. A. Indira Chary and K. Venkata Ramiah, *Curr. Sci. (India)* **37**, 580 (1968).
452. Bengt Nelander, *Acta Chem. Scand.*, **23**, 2127–2135 (1969).
453. Bengt Nelander, *Acta Chem. Scand.*, **23**, 2136–2148 (1969).
454. William J. McKinney and Alexander I. Popov, *J. Amer. Chem. Soc.*, **91**, 5215–5218 (1969).
455. Tatsuaki Kuroi, *Oita Daigaku Kyoikugakubu Kenyu Kiyo, Shizenkagaku*, **3**, 31–36 (1968).
456. J. W. Verhoeven, I. P. Dirkx, and Thymen J. DeBoer, *Tetrahedron*, **25**, 3395–3405 (1969).
457. You Sun Kim and Jung Hee Oh, *Daehan Hwahak Hwoejee*, **11**, 126–131 (1967).
458. Tapan, K. Mukherjee, *J. Phys. Chem.*, **73**, 3442–3445 (1969).
459. Gerald A. Corker and Melvin Calvin, *J. Chem. Phys.*, **49**, 5547–5548 (1968).
460. You Sun Kim and Jung Hee Oh, *Daehan Hwahak Hwoejee*, **11**, 121-125 (1967).
461. J. Prochorow and R. Siegoczynski, *Chem. Phys. Lett.*, **3**, 635–639 (1969).
462. Teisuke Okano, Tadayoshi Miura, Matsuko Yoshida, and Kaneto Uekama, *Yakugaku Zasshi*, **89**, 1379–1385 (1969).
463. John Crysochoos and Cheng-Schen Huang, *Spectrosc. Lett.*, **1**, 367–371 (1968).
464. G. N. Meshkova and A. T. Vartanyan, *Zh. Fiz. Khim.*, **43**, 664–670 (1969).
465. Vladimir Kleinwachter and Jaromir Koudelka, *Biophysik.*, **5**, 119–125 (1968).
466. M. Itoh, *Chem. Phys. Lett.*, **2**, 371–3 (1968).
467. Teisuke Okano and Kazuyuki Tsuji, *Yakugaku Zasshi*, **89**, 297–301 (1969).
468. B. Badger and Brian Brocklehurst, *Trans. Faraday Soc.*, **65**, 2582–2587 (1969).
469. B. Badger and Brian Brocklehurst, *Trans. Faraday Soc.*, **65**, 2576–2581 (1969).
470. B. Badger and Brian Brocklehurst, *Trans. Faraday Soc.*, **65**, 2588–2594 (1969).
471. P. Biloen, T. Fransen, A. Tulp, and G. J. Hoytink, *J. Phys. Chem.*, **73**, 1581–1583 (1969).
472. Eric Coates, *J. Soc. Dyers Colour.*, **85**, 355–368 (1969).
472a. Francois Leterrier and Pierre Douzon, *Photochem. Photobiol.*, **8**, 369–381 (1968).
473. D. D. Pant, H. C. Pant, and K. C. Pant, *Ind. J. Pure Appl. Phys.*, **7**, 182–185 (1969).
474. D. M. Akbarova and L. V. Levshin, *Zh. Prikl. Spektrosk.*, **10**, 269–275 (1969).
475. D. M. Akbarova, L. V. Levshin, and Z. S. Klemenkova *Zh. Prikl. Spektrosk.* **11**, 148–152 (1969).
476. V. M. Korovina and N. G. Bakhshiev, *Opt. Spektrosk.*, **26**, 100–102 (1969).
477. T. G. Beaumont and Keith M. C. Davis, *J. Chem. Soc.*, **1969B**, 575–578.
478. L. A. Klimova, G. N. Nersesova, and V. I. Glyadkovskii, *Opt. Spektrosk.*, **25**, 290–292 (1968).
479. Guy Goumet, Michel Martinaud, and Guy Nouchi, *Compt. Rend.*, **268B**, 1572–1574 (1969).
480. Seymour S. Brody and Suse B. Broyde, *Biophys. J.*, **8**, 1511–1533 (1968).
481. Claude Balny, Seymour S. Brody, and Gaston Hui Bon Hoa, *Photochem. Photobiol.*, **9**, 445–454 (1969).
482. Hartmut Lang and G. Loeber, *Z. Phys. Chem. (Frankfurt am Main)*, **66**, 69–72 (1969).
483. L. A. Nakhimovskaya, *Izv. Akad. Nauk SSSR, Ser. Fiz.*, **32**, 1521–1524 (1968).
484. G. S. Kembrovskii, V. P. Bobrovich, and A. N. Sevchenko, *Zh. Prikl. Spektrosk.*, **10**, 829–833 (1969).
485. L. A. Klimova, G. N. Nersesova, T. M. Naumova, A. I. Ogloblina, and V. I. Glyadkovskii, *Izv. Akad. Nauk SSR, Ser. Fiz.*, **32**, 1471–1474 (1968).
486. L. N. Ustyugova and L. A. Nakhimovskaya, *Zh. Prikl. Spektrosk.*, **9**, 1053–1056 (1968).
487. L. A. Nakhimovskaya, *Vop. Opt. Mol. Spektrosk.* **1968**, 46–50 (from a Russian abstract; *Chem. Abstr.*, **71**, Abstr. no. 44026h).

488. T. N. Bolotnikova, L. Ya. Malkes, A. I. Nazarenko, and V. N. Yakovenko, *Opt. Spektrosk.*, **26**, 132–134 (1969).
489. T. N. Bolotnikova, L. Ya. Malkes, A. I. Nazarenko, and V. N. Yakovenko, *Zh. Prikl. Spektrosk.*, **9**, 680–687 (1968).
490. T. N. Bolotnikova, L. Ya. Malkes, A. I. Nazarenko, and V. N. Yakovenko, *Zh. Prikl. Spektrosk.*, **9**, 858–863 (1968).
491. L. P. Utkina, *Zh. Prikl. Spektrosk.*, **9**, 466–469 (1968).
492. V. A. Butlar and N. A. Kovrizhnykh, *Vop. Radiofiz. Spektrosk.*, **3**, 11–16 (1967) (from a Russian abstract; *Chem. Abstr.*, **71**, Abstr. no. 44059w).
493. Zdzislaw Ruziewicz, Andrzej Olszowski, and Henryk Chojnacki, *Acta Phys. Pol.*, **34**, 801–819 (1968).
494. Hirotsugu Matsuda and Takashi Miyata, *Progr. Theor. Phys.*, *Suppl.*, **1968**, 450–463.
495. David P. Craig and P. D. Dacre, *Proc. Roy. Soc.*, **310A**, 297–312 (1969).
496. Bat-Sheva Sommer and Joshua Jortner, *J. Chem. Phys.*, **50**, 187–203 (1969).
497. Gilles Durocher and Sydney Leach, *J. Chim. Phys. Physicochim. Biol.*, **66**, 628–636 (1969).
498. Gilles Durocher, *J. Chim. Phys. Physicochim. Biol.*, **66**, 637–641 (1969).
499. N. I. Ostapenko and M. T. Shpak, *Phys. Status Solidi*, **31**, 531–534 (1969).
500. E. G. Moisya, *Opt. Spektrosk.*, **27**, 177–178 (1969).
501. Gad Fischer, *Mol. Cryst. Liquid Cryst.*, **6**, 105–123 (1969).
502. Alan Bree and R. Zwarich, *Spectrochim. Acta*, **25A**, 713–722 (1969).
503. Alan Bree, C. Y. Fan, and R. A. Kydd, *Spectrochim. Acta*, **25A**, 1375–1380 (1969).
504. Martin Pope and Nicholas E. Geacintov, *Ind. Res.*, **11**, 68–70 (1969).
505. A. L. Butucelea, *Sutd. Cercet. Chim.*, **17**, 199–216 (1969).
506. O. S. Davidov, *Ukr. Fiz. Zh.* (*Ukr. Ed.*), **13**, 1326–1331 (1968).
507. C. Radvilavicius and A. Bolotin, *Leit. Fiz. Rinkinys*, **7**, 723–728 (1968).
508. Masashi Tanaka and Jiro Tanaka, *Mol. Phys.*, **16**, 1–15 (1969).
509. M. R. Philpott, *J. Chem. Phys.*, **50**, 3925–3929 (1969).
510. Bruno M. Fanconi, George A. Gerhold, and William T. Simpson, *Mol. Cryst. Liquid Cryst.*, **6**, 41–81 (1969).
511. V. N. Vishnevskii, I. F. Viblyi, L. N. Kulik, and N. A. Romanyuk, *Opt. Spektrosk.* **26**, 748–751 (1969).
512. V. L. Broude, *Fiz. Tverd. Tela*, **11**, 1159–1167 (1969).
513. Bruno Wyncke, Armand Hadni, H. Wendling, Xavier Gerbaux, Pierre Strimer, and Guy Morlot, *J. Phys.* (*Paris*), **29**, 851–856 (1968).
514. I. A. Zhigunova, G. V. Klimusheva, and L. P. Yatsenko, *Ukr. Fiz. Zh.* (*Russ. Ed.*), **14**, 66–71 (1969).
515. Jiro Tanaka and Masagi Mizuno, *Bull. Chem. Soc. Jap.*, **42**, 1841–1852 (1969).
516. T. Mookherji and G. M. Arnett, *J. Chem. Phys.*, **50**, 4090–4092 (1969).
517. Martin Vala, Jr., and Jiro Tanaka, *J. Chem. Phys.*, **49**, 5222–5234 (1968).
518. M. K. Chauduri and S. C. Ganguly, *Proc. Phys. Soc., London* (*Solid State Physics*), **2**, 1560–1565 (1969).
519. Subhas Ch. Bera, Ranajit K. Mukherjee, and Mihir Chowdhury, *J. Chem. Phys.*, **51**, 754–761 (1969).
520. A. F. Prikhot'ko and A. F. Skorobogat'ko, *Opt. Spektrosk.*, **26**, 375–378 (1969).
521. Alan Bree and R. Zwarich, *J. Chem. Phys.*, **51**, 903–912 (1969).
522. S. C. Chakrovorty and S. C. Ganguly, *Proc. Phys. Soc., London* (*at Mol. Phys.*), **2**, 1235–1239 (1969).
523. M. R. Philpott, *J. Chem. Phys.*, **50**, 5117–5128 (1969).
524. E. K. Frolova and M. V. Kurik, *Phys. Status Solidi*, **33**, K99–K102 (1969).

525. Valentin Zanker and Joerg Preuss, *Z. Angew. Phys.*, **27**, 363–365 (1969).
526. M. V. Kurik and L. I. Tsikora, *Fiz. Tverd. Tela*, **11**, 2624–2626 (1969).
527. A. F. Prikhot'ko, A. F. Skorobogat'ko, and L. I. Tsikora, *Opt. Spektrosk.*, **26**, 966–971 (1969).
528. Richard H. Clark and Robin M. Hochstrasser, *J. Mol. Spectrosc.*, **32**, 309–319 (1969).
529. Holly H. Chen and Leigh B. Clark, *J. Chem. Phys.*, **51**, 1862–1870 (1969).
530. John Norman Maycock, V. R. Pai Verneker, and W. Lochte, *Phys. Status Solidi*, **35**, 849–860 (1969).
531. Hiroshi Hada, Hideo Matsusaki, and Mikio Tamura, *Nippon Kagaku Zasshi*, **90**, 115–116 (1969).
532. G. H. Kirby and K. Miller, *Chem. Phys. Lett.*, **3**, 643–645 (1969).
533. I. A. Zhigunova, *Opt. Spektrosk.*, **26**, 173–176 (1969).
534. A. J. McHugh and Ian Gordon Ross, *Aust. J. Chem.*, **21**, 3055–3057 (1968).
535. Nikolai Tyutyulkov, G. Hibaum, and A. Gochev, *Dokl. Bolg. Akad. Nauk.*, **21**, 1081–1084 (1968).
536. Nikolai Tyutyulkov and G. Hiebaum, *Theor. Chim. Acta*, **14**, 39–47 (1969).
537. Roddy Merl Conrad and David A. Dows, *J. Mol. Spectrosc.*, **32**, 276–286 (1969).
538. John R. Lombardi, *J. Chem. Phys.*, **50**, 3780–3783 (1969).
539. Kwo-Tsair Huang and John R. Lombardi, *J. Chem. Phys.*, **51**, 1228–1230 (1969).
540. B. M. Uzhinov, A. I. Kozachenko, and M. G. Kuz'min, *Zh. Prikl. Spektrosk.*, **9**, 1041–1046 (1968).
541. Robin M. Hochstrasser and Lewis J. Noe, *J. Chem. Phys.*, **50**, 1684–1688 (1969).
541a. Barbara Pasztor-Bartoszewicz, *Zesz. Nauk, Mat., Fiz., Chem., Wyzsza Szk. Pedagog. Gdansku*, **8**, 55–64 (1968).
541b. Emile Vander Donckt, *Bull. Soc. Chim. Belg.*, **78**, 69–75 (1969).
542. B. M. Uzhinov, A. I. Kozachenko, and M. G. Kuz'min, *Zh. Prikl. Spektrosk.*, **9**, 525–527 (1968).
543. Nicholas J. Turro, *J. Chem. Ed.*, **46**, 2–6 (1969).
544. Yasuhiko Gondo and August H. Maki, *J. Chem. Phys.*, **50**, 3638–3639 (1969).
545. Yasuhiko Gondo and August H. Maki, *J. Chem. Phys.*, **50**, 3270–3279 (1969).
546. Weston T. Borden, *J. Chem. Soc.*, **1969D**, 881–882.
547. John J. McCullough, Helen Ohorodnyk, and D. P. Santry, *J. Chem. Soc.*, **1969D**, 570–571.
548. Lee G. Pedersen, David G. Whitten, and M. T. McCall, *Chem. Phys. Lett.*, **3**, 569–572 (1969).
549. Peter Hans Hermann Fischer and A. B. Denison, *Mol. Phys.*, **17**, 297–304 (1969).
550. Arthur B. Dension and Peter H. H. Fischer, *Proc. Colloq. Ampere (at Mol. Etud. Radio Elec.), 1968*, ed. by P. Averbuch (Amsterdam: North-Holland Publishing, 1969), **15**, 455–459.
551. P. Ehret and Hans Christolph Wolf, *Z. Naturforsh.*, **23A**, 1740–1746 (1968).
552. Reed F. Riley and Jack Rosenthal, *J. Chem. Phys.*, **50**, 1034–1035 (1969).
552a. J. S. Brinen and M. K. Orloff, *Chem. Phys. Lett.*, **1**, 276–278 (1967).
553. Mart S. De Groot, I. A. M. Hesselmann, and J. H. Van der Waals, *Mol. Phys.*, **16**, 45–60 (1969).
554. Mart S. De Groot, I. A. M. Hesselmann, and J. H. Van der Waals, *Mol. Phys.*, **16**, 61–68 (1969).
555. Mark Sharnoff, *Mol. Cryst. Liquid Cryst.*, **9**, 265–283 (1969).
556. Roger E. Gerkin and Arthur M. Winer, *J. Chem. Phys.*, **50**, 3114–3115 (1969).
557. Roger E. Gerkin and Peter Szerenyi, *J. Chem. Phys.*, **50**, 4095–4106 (1969).

558. Mark Sharnoff, *J. Chem. Phys.*, **51**, 451–452 (1969).
559. C. H. J. Wells, A. Horsfield, and J. Paxton, *J. Chem. Soc.*, **1969D**, 393–394.
560. Mark Sharnoff, *Chem. Phys. Lett.*, **2**, 498–500 (1968).
561. Jean P. Grivet, *Compt. Rend.*, **268B**, 1186–1189 (1969).
562. Corrine Thiery, Jean Capette, Jean Meunier, and Francois Leterrier, *J. Chim. Phys. Physicochim.*, **66**, 134–139 (1969).
563. Jean M. Lhoste, Marius Ptak, and Doris Lexa, *J. Chim. Phys. Physicochim. Biol.*, **65**, 1876–1888 (1968).
564. David R. Graber, Michael W. Grimes, and Alfred Haug, *J. Chem. Phys.*, **50**, 1623–1626 (1969).
565. Susan J. Gull, D. R. Graber, and A. Haug, *Photochem. Photobiol.*, **10**, 139–140 (1969).
566. Yukio Kubota and Masaji Miura, *Bull. Chem. Soc. Jap.*, **42**, 2763–2767 (1969).
567. Hisaharu Hayashi, Suchiro Iwata, and Saburo Nakagura, *J. Chem. Phys.*, **50**, 993–1000 (1969).
568. Dino S. Tinti, Mostafa A. El-Sayed, August H. Maki, and Charles B. Harris, *Chem. Phys. Lett.*, **3**, 343–346 (1969).
569. Jan Schmidt and J. H. Van der Waals, *Chem. Phys. Lett.*, **2**, 640–642 (1968).
570. Jan Schmidt and H. Van der Waals, *Chem. Phys. Lett.*, **3**, 546–549 (1969).
571. M. Schwoerer and H. Sixl, *Z. Naturforsch.*, **24A**, 952–967 (1969).
572. Arthur Forman and Alvin L. Kwiram, *J. Chem. Phys.* **49**, 4714–4715 (1968).
573. Robin M. Hochstrasser and Tien-Sung Lin, *J. Chem. Phys.*, **49**, 4929–4945 (1968).
574. Katsumi Nakamura, Shigeya Niizuma, and Masao Koizumi, *Bull. Chem. Soc. Jap.*, **42**, 255–257 (1969).
575. Karl A. Drexhage, *Sci. Am., 212* (March), 108 (1970).
576. Mordechai Bixon and Joshua Jortner, *J. Chem. Phys.*, **50**, 3284–3290 (1969).
577. Mordechai Bixon and Joshua Jortner, *J. Chem. Phys.*, **50**, 4061–4070 (1969).
578. Karl F. Freed and Joshua Jortner, *J. Chem. Phys.*, **50**, 2916–2927 (1969).
578a. R. N. Nurmukhametov and D. N. Shigorin, *Proc. Int. Conf. Lumin.*, **1966**, ed. by G. Szigeti (Budapest: Akad. Kiado, 1968), **1**, 353–356.
579. Zbigniew R. Grabowski, *ibid.*, 315–326.
580. Yu. V. Naboikin, L. A. Ogurtsova, A. P. Podgornyi, and F. S. Pokrovskaya, *Opt. Spektrosk.*, **27**, 307–309 (1969).
581. T. D. S. Hamilton and K. Razi Naqvi, *Chem. Phys. Lett.*, **2**, 374–378 (1968).
582. E. Ejder, *J. Opt. Soc. Amer.*, **59**, 223–224 (1969).
583. Paul G. Seybold, Martin Gouterman, and James Callis, *Photochem. Photobiol.*, **9**, 229–242 (1969).
584. Anthony C. Testa, *Fluorescence News*, **4**, 1–3 (1969).
585. Aaron N. Fletcher, *Photochem. Photobiol.*, **9**, 439–444 (1969).
586. James E. Gill, *ibid.*, 313–322 (1969).
587. A. K. Babko, S. P. Baranov, and L. V. Kalabina, *Zh. Anal. Khim.*, **24**, 485–489 (1969).
588. Alfred Bril, *Proc. Int. Conf. Lumin.*, **1966**, ed. by G. Szigeti (Budapest: Akad. Kiado, 1968), **2**, 2112–2120.
589. Satoshi Suzuki, *Oyo Denki Kenyusho Hokoku*, **19**, 20–27 (1967).
590. A. Jablonski, *Bull. Acad. Pol. Sci., Ser. Sci., Math., Astron., Phys.*, **16**, 601–604 (1968).
591. Terence Tao, *Mol. Lumin., Int., Conf.*, **1968**, ed. by E. C. Lim (New York: W. A. Benjamin, 1969), 851–861.
592. Gregoric Weber and Sonia R. Anderson, *Biochemistry*, **8**, 361–371 (1969).

593. James L. Speyer and Marvin H. Winkler, *Biochim. Biophys. Acta*, **188**, 345–347 (1969).
594. Richard Karol Bauer, *Acta Phys. Pol.*, **35**, 101–116 (1969).
595. Richard Karol Bauer, *Acta Phys. Pol.*, **35**, 975–987 (1969).
596. K. I. Rudik, *Bull. Acad. Pol. Sci., Ser. Sci., Math., Astron. Phys.*, **16**, 911–915 (1968).
597. Richard K. Bauer and K. I. Rudik, *Acta Phys. Pol.*, **35**, 259–270 (1969).
598. R. K. Bauer and K. I. Rudik., *Bull. Acad. Sci., Ser. Sci., Math., Astron. Phys.*, **16**, 543–549 (1968).
599. Erzsebet Tombacz, L. Vize, and Laszlo Szalay, *Proc. Int. Conf. Lumin.*, *1966*, ed. by G. Szigeti (Budapest: Akad. Kiado, 1968), **1**, 285–291.
600. Lloyd M. Logan, J. P. Byrne, and Ian G. Ross, *ibid.*, 194–199.
601. Tohru Azumi and Saburo Nagakura, *Bull. Chem. Soc. Jap.*, **42**, 2203–2208 (1969).
602. Tohru Azumi, Mitsuo Ito, and Saburo Nagakura, *Bull. Chem. Soc. Jap.*, **42**, 685–689 (1969).
603. Tohru Azumi, *Mol. Lumin., Int. Conf.*, **1968**, ed. by E. C. Lim (New York: W. A. Benjamin, 1969), 79–92.
604. Cecil Allen Parker, *Ber. Bunsenges. Phys. Chem.*, **73**, 764–772 (1969).
605. Cecil Allen Parker and Thelma A. Joyce, *Trans. Faraday Soc.*, **65**, 2823–2829 (1969).
606. T. G. Pavlopoulos, *J. Chem. Phys.*, **51**, 2936–2940 (1969).
607. Edward C. Lim, R. Li, and Y. H. Li, *J. Chem. Phys.*, **50**, 4925–4933 (1969).
607a. Edward C. Lim and Jack M. H. Yu, *J. Chem. Phys.*, **49**, 3878–3884 (1968).
608. M. S. Fadeeva, V. A. Kuznetsov, and Yu. A. Kaplin, *Zh. Prikl. Spektrosk.*, **9**, 489–491 (1968).
609. Yoshio Murakami, Ryoichi Shimada, and Yoshiya Kanada, *Mol. Lumin., Int. Conf.*, *1968*, ed. by E. C. Lim (New York: W. A. Benjamin, 1969), 119–134.
610. J. L. Metzger, Bryan Edward Smith, and Beat Meyer, *Spectrochim. Acta*, **25A**, 1177–1188 (1969).
611. Mostafa A. El-Sayed, *Acta Phys. Pol.*, **34**, 649–668 (1968).
612. George C. Nieman, *J. Chem. Phys.*, **50**, 1660–1673 (1969).
613. C. A. Parker, *Fast React. Primary Processes Chem. Kinet., Proc. Nobel Symp.*, *5th*, ed. by Stig Claesson (Stockholm: Almqvist and Wiksell, 1967), 317–324.
614. Marc Ewald and Bernard Muel, *Compt. Rend.*, **268B**, 973–976 (1969).
615. Cecil A. Parker and Thelma A. Joyce, *Chem. Commun.*, **1968**, 1421–1422.
616. Richard D. Spencer and Gregorio Weber, *Ann. N.Y. Acad. Sci.*, **158**, 361–376 (1969).
617. Henry Merkelo, Steward R. Hartman, T. Mar, G. S. Singhal, and Govindjee, *Science*, **164**, 301–302 (1969).
618. Bryan R. Henry and Willem Siebrand, *Chem. Phys. Lett.*, **3**, 90–92 (1969).
619. J. R. Greenleaf and Terence A. King, *Proc. Int. Conf. Lumin.*, *1966*, ed. by G. Szigeti (Budapest: Akad. Kiado, 1968), **1**, 212–220.
620. L. Gati, *Acta Phys. Chem.*, **15**, 5–17 (1969).
621. I. Ketskemety and Laszlo Kozma, *Acta Phys. Chem.*, **14**, 75–76 (1968).
622. Fumio Hirayama and Sanford Lipsky, *J. Chem. Phys.*, **51**, 3616–3617 (1969).
623. Jacob Solomon Brienen, Malcolm K. Orloff, J. B. Gallivan, R. F. Stamm, and Bernard George Roberts, *J. Mol. Spectrosc.*, **32**, 368–374 (1969).
624. James B. Gallivan, *J. Phys. Chem.*, **73**, 3070–3075 (1969).
625. George M. Breuer and Edward K. C. Lee, *J. Chem. Phys.*, **51**, 3130–3132 (1969).
626. Bryan R. Henry and Willem Siebrand, *Chem. Phys. Lett.*, **3**, 327–328 (1969).
627. W. P. Helman, *J. Chem. Phys.*, **51**, 354–357 (1969).
628. J. W. Rabalais, H. J. Maria, and Sean P. McGlynn, *Chem. Phys. Lett.*, **3**, 59–60 (1969).

629. J. W. Rabalais, H. J. Maria, and Sean P. McGlynn, *J. Chem. Phys.*, **51**, 2259–2273 (1969).
630. Timothy D. Gierke, Richard J. Watts, and S. J. Strickler, *J. Chem. Phys.*, **50**, 5425–5426 (1969).
631. Nien-Chu Yang, Steven L. Murov, and Tsu-Chia Shieh, *Chem. Phys. Lett.*, **3**, 6–8 (1969).
632. Edward W. Schlag and Hanns Von Weyssenhoff, *J. Chem. Phys.*, **51**, 2508–2514 (1969).
632a. Michael E. Starzak, thesis, Northwestern Univ., 1968; *Diss. Abstr.*, **29B**, 2835–2836 (1969).
633. Ingo H. Leubner, *J. Phys. Chem.*, **73**, 2088–2090 (1969).
634. M. D. Lumb, C. Lloyd Braga, and L. C. Pereira, *Trans. Faraday Soc.*, **65**, 1992–1999 (1969).
635. Ingo H. Leubner and Joe E. Hodgkins, *J. Phys. Chem.*, **73**, 2545–2550 (1969).
636. M. D. Erlitz and George C. Nieman, *J. Chem. Phys.*, **50**, 1479–1480 (1969).
637. Peter Frank Jones and Seymour Siegel, *J. Chem. Phys.*, **50**, 1134–1140 (1969).
638. Peter Frank Jones, *U.S. Govt. Res. Develop. Rep.*, **68**, 56 (1968).
639. N. A. Borisevich and G. B. Tolstorozhev, *Dokl. Akad. Nauk SSSR*, **188**, 308–310 (1969).
640. John L. Kropp, William R. Dawson, and M. W. Windsor, *U.S. Govt. Res. Develop. Rep.*, **69**, 76 (1969).
641. John L. Kropp, William R. Dawson, and Maurice W. Windsor, *J. Phys. Chem.*, **73**, 1747–1752 (1969).
642. William R. Dawson and John L. Kropp, *U.S. Govt. Res. Develop. Rep.*, **69**, 76 (1969).
643. William R. Dawson and John L. Kropp, *J. Phys. Chem.*, **73**, 1752–1758 (1969).
644. John L. Kropp and William R. Dawson, *U.S. Govt. Res. Develop. Rept*, **68**, 70 (1968).
645. William R. Dawson, and John L. Kropp, *U.S. Govt. Res. Develop. Rep.*, **68**, 64 (1968).
646. William R. Dawson and John L. Kropp, *J. Phys. Chem.*, **73**, 693–699 (1969).
647. Henryk Wardzinski, *Zesz. Nauk., Mat., Fiz., Chem., Wyzsza Szk. Pedagog. Gdansku*, **8**, 65–71 (1968).
648. Teh-Hsuan Chen and Edward W. Schlag, *Mol. Lumin., Int. Conf., 1968*, ed. by E. C. Lim (New York: W. A. Benjamin, 1969), 381–392.
649. L. G. Pikulik, V. A. Yakovenko, and M. Ya. Kostko, *Izv. Akad. Nauk SSSR, Ser. Fizt*, **32**, 1496–1499 (1968).
650. Richard D. Spencer, William M. Vaughan, and Gregario Weber, *Mol. Lumin., Int. Conf., 1968*, ed. by E. C. Lim (New York: W. A. Benjamin, 1969), 607–269.
651. G. M. Kislyak, G. M. Lysenko, and V. I. Ponochovnyi, *Ukr. Fiz. Zh. (Russ. Ed.)*, **14**, 338–340 (1969).
652. Tohru Azumi and Yasuko Nakano, *J. Chem. Phys.*, **50**, 539–541 (1969).
653. Mostafa A. El-Sayed, William R. Moomaw, and Dino S. Tinti, *J. Chem. Phys.*, **50**, 1888–1889 (1969).
654. Joseph E. Knoll, thesis, Polytechnic Institute of Brooklyn, 1968; *Diss. Abstr.*, **29B**, 4559–4560 (1969).
655. Jimmie R. McDonald, V. M. Scherr, and Sean P. McGlynn, *J. Chem. Phys.*, **51**, 1723–1731 (1969).
656. W. D. K. Clark, A. D. Litt, and C. Steel, *J. Amer. Chem. Soc.*, **91**, 5413–5415 (1969).
657. Pill-Soon Song and William E. Kurtin, *J. Amer. Chem. Soc.*, **91**, 4892–4906 (1969).
658. William E. Kurtin and Pill-Soon Song, *Photochem. Photobiol.*, **9**, 127–142 (1969).
659. L. G. Pikulik, M. Ya. Kostko, and V. A. Yakovenko, *Vestsi. Akad. Navuk. Belarus. SSR, Ser. Fiz.-Mat. Navuk*, **1969**, 89–95.

660. Howard Elliot Zimmerman and Guilford Jones, II, *J. Amer. Chem. Soc.*, **91**, 5678–5679 (1969).
661. E. S. Moyer and W. J. McCarthy, *Anal. Chim. Acta.* **45**, 13–19 (1969).
662. James F. Verdieck and W. A. Jankowski, *Mol. Lumin., Int. Conf., 1968*, ed. by E. C. Lim (New York: W. A. Benjamin, 1969). 829–836.
663. Sol E. Harrison and Walter F. Kosonocky, *Proc. Int. Conf. Lumin., 1966*, ed. by G. Szigeti (Budapest: Akad. Kiado, 1968), **1**, 327–332.
664. Paul G. Seybold and Martin Gouterman, *J. Mol. Spectrosc.*, **31**, 1–13 (1969).
665. Peter M. Rentzepis, *Chem. Phys. Lett.*, **3**, 717–720 (1969).
666. J. A. Poole and R. C. Dhingra, *Mol. Lumin., Int. Conf., 1968*, ed. by E. C. Lim (New York: W. A. Benjamin, 1969), 813–827.
667. R. C. Dhingra and John A. Poole, *J. Phys. Chem.*, **72**, 4577–4580 (1968).
668. George S. Hammond, Sang C. Shim, and Shui P. Van, *Mol. Photochem.*, **1**, 89–106 (1969).
669. A. N. Singh and Indra S. Singh, *Ind. J. Pure Appl. Phys.*, **7**, 349–351 (1969).
670. A. N. Singh and Indra S. Singh, *Ind. J. Pure Appl. Phys.*, **7**, 714–715 (1969).
671. D. L. Judge and Masaru Ogawa, *J. Chem. Phys.*, **51**, 2035–2036 (1969).
672. F. Sheldon Wettack, *J. Phys. Chem.*, **73**, 1167–1169 (1969).
673. Lin Tsai and Elliot Charney, *J. Phys. Chem.*, **73**, 2462–2463 (1969).
674. Charles S. Parmenter and Anne H. White, *J. Chem. Phys.*, **50**, 1631–1643 (1969).
674a. Lloyd M. Logan, Ilze Buduls, and Ian Gordon Ross, *Mol. Lumin., Int. Conf., 1968*, ed. by E. C. Lim (New York: W. A. Benjamin, 1969), 53–62.
675. George P. Semeluk, R. D. S. Stevens, and I. Unger, *Can. J. Chem.*, **47**, 597–603 (1969).
676. Isadore B. Berlman, *Proc. Int. Conf. Lumin., 1966*, ed. by G. Szigeti (Budapest: Akad. Kiado, 1968), **1**, 514–519.
677. Nien-Chu Yang and Ruth L. Dusenbery, *Mol. Photochem.*, **1**, 159–171 (1969).
678. G. D. Baruah, O. N. Singh, and Rama S. Singh, *Ind. J. Pure Appl. Phys.*, **7**, 352–353 (1969).
679. M. G. Jayswal and R. S. Singh, *Proc. Int. Conf. Lumin., 1966*, ed. by G. Szigeti (Budapest: Akad. Kiado, 1968), **1**, 357–362.
680. Ai Nakanishi, *Eisei Kagaku*, **14**, 198–202 (1968).
681. Lubomir Kabrt and Zavis Holzbecher, *Coll. Czech. Chem. Commun.*, **34**, 796–812 (1969).
682. Arye Weinreb and A. Werner, *Chem. Phys. Lett.*, **3**, 231–232 (1969).
683. B. N. Tripathi and C. L. Garg, *Ind. J. Chem.*, **7**, 778–779 (1969).
684. M. Grossman, G. P. Semeluk, and I. Unger, *J. Phys. Chem.*, **73**, 1149–1150 (1969).
685. M. Grossman, George P. Semeluk, and I. Unger, *J. Phys. Chem.*, **73**, 3175–3176 (1969).
686. T. S. Jaseja, Ved Parkash, and M. K. Dheer, *J. Appl. Phys.*, **40**, 1882–1883 (1969).
687. V. P. Klochkov and S. M. Korotkov, *Opt. Spektrosk.*, **25**, 970–972 (1968).
688. I. N. Kozlov, A. M. Sarzhevskii, and M. I. Khomich, *Zh. Prikl. Spektrosk.*, **9**, 666–669 (1968).
689. Julian Braun, G. Caille, and Etienne Adonai Martin, *Can. J. Pharm. Sci.*, **3**, 65–68 (1968).
690. B. M. Krasovitskii and A. I. Nazarenko, *Monokrist., Stsintill. Org. Lyuminofory*, **2**, 115–118 (1967). (Translation of a Russian abstract from *Chem. Abstr.*, **70**, abstract no. 55260f.)
691. N. A. Borisevich and V. V. Guzinskii, *Proc. Int. Conf. Lumin., 1966*, ed. by G. Szigeti (Budapest: Akad. Kiado, 1968), **1**, 200–204.

692. S. Nath Singh, G. D. Baruah, and K. P. R. Nair, *Ind. J. Pure Appl. Phys.*, **7**, 133–135 (1969).
693. G. D. Baruah, K. P. R. Nair, and B. B. Lal, *Ind. J. Pure Appl. Phys.*, **7**, 713–714 (1969).
694. N. A. Shchegloa, D. N. Shigorin, and N. S. Dokunikhin, *Zh. Fiz. Chim.*, **43**, 1700–1708 (1969).
694a. S. M. Eremenko and V. A. Kalibabchuk, *Teor. Eksp. Khim.*, **4**, 847–849 (1968).
695. S. Nath Singh, M. G. Jayswal, and Rama Shankar Singh, *Bull. Chem. Soc. Jap.*, **42**, 2048–2050 (1969).
696. D. N. Shigorin, V. M. Voznyak, G. A. Ozerova, R. N. Nurmukhametov, and A. K. Piskunov, *Proc. Int. Conf. Lumln., 1966,* ed. by G. Szigeti (Budapest: Akad. Kiado, 1968), **1**, 540–545.
697. M. R. Padhye and V. V. Bhujle, *Curr. Sci. (India)*, **38**, 215 (1969).
698. David G. Whitten and M. T. McCall, *J. Amer. Chem. Soc.*, **91**, 5097–5103 (1969).
699. Lloyd M. Logan and Ian Gordan Ross, *Acta Phys. Pol.*, **34**, 721–732 (1968).
700. Heinz H. Perkampus, A. Knop, and J. V. Knop, *Spectrochim. Acta*, **25A**, 1589–1602 (1969).
701. Maximilian Zander, *Z. Naturforsch.*, **24A**, 254–256 (1969).
702. Edward C. Lim and Y. H. Li, *Chem. Phys. Lett.*, **4**, 25–26 (1969).
703. Pill-Soon Song and William E. Kurtin, *Photochem. Photobiol.*, **10**, 211–214 (1969).
704. Klaus E. Rieckhoff and E. M. Voigt, *Mol. Lumin., Int. Conf., 1968*, ed. by E. C. Lim (New York: W. A. Benjamin, 1969), 295–308.
705. Kazuyoshi Oka, Tadao Hinohara, and Kohji Matsui, *Kogyo Kagaku Zasshi*, **71**, 1010–1015 (1968).
706. Jane M. Bonnier, Pierre Jardon, and Jean P. Blanchi, *Bull. Soc. Chim. Fr., 1968*, 4787–4790.
707. R. N. Nurmukhametov, L. A. Mileshina, D. N. Shigorin, and G. T. Khachaturova, *Zh. Fiz. Khim.*, **43**, 51–56 (1969).
708. B. M. Bolotin, D. A. Drapkina, and V. G. Brudz, *Proc. Int. Conf. Lumin., 1966*, ed. by G. Szigeti (Budapest: Akad. Kiado, 1968), **1**, 626–631.
709. William J. McCarthy and D. A. Wiegand, *Spectrosc. Lett.*, **1**, 349–353 (1968).
710. Richard W. Chambers and David R. Kearns, *Photochem. Photobiol.*, **10**, 215–219 (1969).
711. Guenther Briegleb, Walter Herre, and D. Wolf, *Spectrochim. Acta*, **25A**, 39–46 (1969).
712. A. J. Thomson, *J. Amer. Chem. Soc.*, **91**, 2780–2785 (1969).
713. Robert Cowgill, *Mol. Lumin., Int. Conf., 1968*, ed. by E. C. Lim (New York: W. A. Benjamin, 1969), 589–605.
714. Stephen F. Mason and Barry Edward Smith, *J. Chem. Soc.*, **1969A**, 325–328.
715. Emile Vander Donckt and George Porter, *Trans. Faraday Soc.*, **64**, 3215–3217 (1968).
716. Yoshiya Kanda, J. Stanislaus, and Edward C. Lim, *J. Amer. Chem. Soc.*, **91**, 5085–5090 (1969).
717. I. Avigzl, Jehuda Feitelson, and Michael Ottolenghi, *J. Chem. Phys.*, **50**, 2614–2617 (1969).
718. Emile Vander Donckt and George Porter, *Trans. Faraday Soc.*, **64**, 3218–3223 (1968).
719. Zbigniew Grabowski, K. Rotkiewicz, and A. J. Saklej, *Proc. Int. Conf. Lumin., 1966*, ed. by G. Szigeti (Budapest: Akad. Kiado, 1968), **1**, 310–314.
719a. T. C. Werner and David M. Hercules, *J. Phys. Chem.*, **73**, 2005–2011 (1969).
720. J. C. Doty, Jack L. R. Williams, and Patrick J. Grisdale, *Can. J. Chem.*, **47**, 2355–2359 (1969).
721. Anna Grabowska and Barbara Pakula, *Photochem. Photobiol.*, **9**, 339–350 (1969).

722. Michio Kondo and Harumitsu Kuwano, *Bull. Chem. Soc. Jap.*, **42**, 1433–1435 (1969).
723. Anna Crabowska and Barbara Pakula, *Proc. Int. Conf. Lumin.*, *1966*, ed. by G. Szigeti (Budapest: Akad. Kiado, 1968), **1**, 368–372.
724. K. Berens and K. L. Wierzchowski, *Photochem. Photobiol.*, **9**, 433–438 (1969).
725. John W. Eastman, *Ber. Bunsenges. Phys. Chem.*, **73**, 407–412 (1969).
726. George C. Nieman, *J. Chem. Phys.*, **50**, 1674–1683 (1969).
727. Homer E. Holloway, Robert V. Nauman, and James H. Wharton, *J. Phys. Chem.*, **72**, 4468–4473 (1968).
728. T. V. Veselova, I. I. Reznikova, A. S. Cherkasov, and V. I. Shirkov, *Opt. Spektrosk.*, **26**, 972–976 (1969).
729. Hanjiro Ito, Tomoko Kita, and Toshiaki Tomimatsu, *Tokushima Daigaku Yakugaku Kenkyu Nempto*, **16**, 5–9 (1967).
730. Claude Balny, Pierre Douzou, Tuvio Bercovici, and Ernst Fischer, *Mol. Photochem.*, **1**, 225–233 (1969).
731. G. V. Gobov and V. S. Tambovtsev, *Zh. Prikl. Spektrosk.*, **9**, 1014–1018 (1968).
732. G. V. Gobov, L. A. Nakhimovskaya, N. S. Proskuryakova, V. S. Tambovtsev, and L. N. Ustyugova, *Izv. Akad. Nauk SSSR, Ser. Fiz.*, **32**, 1542–1545 (1968).
733. R. I. Personov and V. V. Solodunov, *Fiz. Tverd. Tela*, **11**, 2890–2893 (1969).
734. T. A. Teplitskaya and R. I. Personov, *Zh. Fiz. Khim.*, **43**, 1679–1681 (1969).
735. V. I. Mikhailenko and P. A. Teplaykov, *Izv. Akad. Nauk SSSR, Ser. Fiz.*, **32**, 1591–1595 (1968).
736. Rolf Gase and G. Hasse, *Z. Phys. Chem.* (*Leipzig*), **242**, 42–48 (1969).
736a. Sydney Leach, A. Lopez-Campillo, R. Lopez-Delgado, and M. C. Tomas-Magos, *Proc. Int. Conf. Lumin.*, *1966*, ed. by G. Szigeti (Budapest: Akad. Kiado, 1968), **1**, 270–274.
737. Lydie Watmann-Grajcar, *J. Chim. Phys. Physichochim. Biol.*, **66**, 1023–1040 (1969).
738. Annick Pellois, Jean C. Navatte, and Jean Ripoche, *Compt. Rend.*, **268B**, 1183–1185 (1969).
739. Kh. I. Mamedov and I. K. Nasibov, *Izv. Akad. Nauk SSSR, Ser. Fiz.*, **32**, 1508–1510 (1968).
740. Kh. I. Mamedov, I. K. Nasibov, and E. N. Niyakov, *Opt. Spektrosk.*, **25**, 689–695 (1968).
741. N. A. Fenina, *Vop. Radiofiz. Spektrosk.*, **3**, 3–10 (1967). (Translation of a Russian abstract in *Chem. Abstr.*, **70**, abstract no. 44079c.) (1969).
742. Annick Pellois and Jean Ripoche, *Chem. Phys. Lett.*, **3**, 280–282 (1969).
743. Jean Ripoche and Annick Pellois, *Compt. Rend.*, **268B**, 237–240 (1969).
744. A. F. Prikhot'ko, A. F. Skorobogat'ko, and L. I. Tsikora, *Opt. Spektrosk.*, **26**, 214–219 (1969).
745. G. E. Fedoseeva and A. Ya. Khesina, *Zh. Prikl. Spektrosk.*, **9**, 282–288 (1968).
746. R. I. Personov and V. V. Solodonov, *Fiz. Tverd. Tela*, **11**, 2890–2893 (1969).
747. S. G. Bogomolov and G. S. Vedernikov, *Zh. Prikl. Spektrosk.*, **9**, 480–484 (1968).
748. R. Z. Laipanov and Kh. I. Mamedov, *Zh. Prikl. Spektrosk.*, **9**, 475–479 (1968).
749. Zdzislaw Ruziewicz, *Proc. Int. Conf. Lumin.*, *1966*, ed. by G. Szigeti (Budapest: Akad. Kiado, 1968), **1**, 363–367.
750. Annick Pellois, Jean C. Navatte, Jean Ripoche, *Compt. Rend.*, **268B**, 1134–1137 (1969).
751. Gilles Durocher, *J. Chim. Phys. Physicochim. Biol.*, **66**, 985–987 (1969).
752. Edward F. Zalewski and Donald S. McClure, *Mol. Lumin., Int. Conf.*, *1968*, ed. by E. C. Lim (New York: W. A. Benjamin, 1969), 739–749.
753. T. E. Martin and Alfred H. Kalantar, *J. Chem. Phys.*, **50**, 1486–1487 (1969).

754. G. F. Hatch, M. D. Erlitz, and G. C. Nieman, *Mol. Lumin., Int. Conf., 1968*, ed. by E. C. Lim (New York: W. A. Benjamin, 1969), 21–38.
755. Andre Martinez and Dominique Dorignac, *J. Chim. Phys. Physicochim. Biol.*, **66**, 817–824 (1969).
756. Mats Almgren, *Photochem. Photobiol.*, **9**, 1–6 (1969).
757. Philippe Gacoin, *Compt. Rend.*, **269B**, 86–89 (1969).
758. Akira Kuboyama and Sanae Masuzaki, *Tokyo Kogyo Shikensho Hokoku*, **64**, 105–110 (1969).
759. E. J. O'Connell, Jr., *J. Chem. Soc.*, **1969D**, 571–572.
759a. R. Ostertag and Hans C. Wolf, *Phys. Status Solidi*, **31**, 139–146 (1969).
760. Henry W. Offen and David E. Hein, *Mol. Lumin., Int. Conf.*, *1968*, ed. by E. C. Lim (New York: W. A. Benjamin, 1969), 1–14.
761. Henry W. Offen and David E. Hein, *U.S. Govt. Res. Develop. Rep.*, **68**, 70 (1968).
761a. Ichimin Shirotani, Takashi Kajiwara, and Hiroo Inokuchi, *Bull. Chem. Soc. Jap.*, **42**, 2387–2389 (1969).
762. Henry W. Offen, S. A. Balbo, and R. L. Tanquary, *Spectrochim. Acta*, **25A**, 1023–1025 (1969).
763. Henry W. Offen and David E. Hein, *J. Chem. Phys.*, **50**, 5274–5278 (1969).
763a. David E. Hein and Henry W. Offen, *Mol. Cryst.*, **5**, 217–227 (1969).
764. Michel Bouyer, J. Seite, and J. Martelli, *Proc. Int. Conf. Lumin., 1966*, ed. by G. Szigeti (Budapest: Akad, Kiado, 1968), **1**, 533–535.
765. Lloyd M. Logan, I. H. Munro, Digby F. Williams, and Frederick R. Lipsett, *Mol. Lumin., Int. Conf., 1968*, ed. by E. C. Lim (New York: W. A. Benjamin, 1969), 773–785.
766. Mostafa A. El-Sayed and R. Moomaw, *Excitons, Magnons Phonons Mol. Cryst., Proc. Int. Symp., 1968*, ed. by A. B. Zahlan (Cambridge, England: Cambridge Univ. Press, 1968), 103–123.
767. Mostafa A. El-Sayed, L. Hall, A. Armstrong, and W. R. Moomaw, *ibid.*, 125–133.
768. Tohru Azumi, Yasuko Nakano, and Mitsuo Ito, *Bull. Chem. Soc. Jap.*, **41**, 2551 (1968).
769. Tohru Azumi and Yasuko Nakano, *J. Chem. Phys.*, **51**, 2515–2528 (1969).
769a. Dino S. Tint, W. R. Moomaw, and Mostofa A. El-Sayed, *J. Chem. Phys.*, **50**, 1035–1036 (1969).
770. Stanley M. Ziegler, thesis, Univ. of California, Los Angeles, 1968; *Diss. Abstr.*, **29B**, 3284 (1969).
771. K. Peuker and E. D. Trifonov, *Phys. Status Solidi*, **30**, 479–484 (1968).
772. E. A. Smirnov, L. N. Kirchenko, and V. F. Gachkovskii, *Zh. Fiz. Khim.*, **42**, 2658–2660 (1968).
773. V. G. Trishchenko and L. M. Egupova, *Monokrist., Stsintill. Org. Lyuminofory*, **2**, 129–131 (1967). (Translation of a Russian abstract in *Chem. Abstr.*, **70**, abstract no. 44075y (1969).
774. Y. Tomkiewicz and Ayre Weinreb, *Chem. Phys. Lett.*, **3**, 229–230 (1969).
775. Yoshio Marakami and Yoshiya Kanda, *Bull. Chem. Soc. Jap.*, **41**, 2599–2603 (1968).
776. N. G. Bakhshiev, I. V. Piterskaya, V. I. Studenov, and A. V. Altaiskaya, *Opt. Spektrosk.*, **27**, 349–351 (1969).
777. Carl J. Seliskar, David Charles Turner, James R. Gohlke, and Ludwig Brand, *Mol. Lumin., Int. Conf., 1968*, ed. by E. C. Lim (New York: W. A. Benjamin, 1969), 677–696.
778. Fritz Schneider and Ernst Lippert, *Ber. Bunsenges. Phys. Chem.*, **72**, 1155–1160 (1968).
779. Zbigniew Raciszewski and J. F. Stephen, *J. Amer. Chem. Soc.*, **91**, 4338–4341 (1969).

779a. Kunio Yagi, Nobuko Ohishi, Makoto Naoi, and Akira Kotaki, *Arch. Biochem. Biophys.*, **134**, 500–505 (1969).
780. Theodor Förster and K. Rokos, *Z. Phys. Chem. (Frankfurt am Main)*, **63**, 208–211 (1969).
781. Donald L. Horrocks, *Mol. Lumin., Int. Conf., 1968*, ed. by E. C. Lim (New York: W. A. Benjamin, 1969), 63–78.
782. L. F. Gladchenko, M. Ya. Kostko, and L. G. Pikulik, *Izv. Akad. Nauk SSSR, Ser. Fiz.*, **32**, 1584–1587 (1968).
782a. Hiroshi Kokubun, *Bull. Chem. Soc. Jap.*, **42**, 919–922 (1969).
783. A. M. Sarzhevskii and M. I. Khomich, *Zh. Prikl. Spektrosk.*, **9**, 1005–1007 (1968).
784. A. M. Sarzhevskii and M. I. Khomich, *Izv. Akad. Nauk SSSR, Ser. Fiz.*, **32**, 1517–1520 (1968).
784a. S. K. Chakrabarti, *Mol. Phys.*, **16**, 467–479 (1969).
785. Chen-Hanson Ting, *Photochem. Photobiol.*, **9**, 17–31 (1969).
786. Edward W. Schlag, Shang Jeong Yao, and Hanns Von Weyssenhoff, *J. Chem. Phys.*, **50**, 732–736 (1969).
787. William Rhodes, *J. Chem. Phys.*, **50**, 2885–2895 (1969).
788. William Rhodes, Bryan R. Henry, and Michael Kasha, *Proc. Nat. Acad. Sci. U.S.*, **63**, 31–35 (1969).
789. Robert Englman, *Isr. J. Chem.*, **7**, 221–225 (1969).
790. Adam Heller, *Mol. Photochem.*, **1**, 257–269 (1969).
791. M. M. Malley and P. M. Rentzepis, *Chem. Phys. Lett.*, **3**, 534–536 (1969).
792. J. A. F. Alexander, John T. Houghton, W. B. McKnight, *Proc. Phys. Soc., London (At. Mol. Phys.)*, **1**, 1225–1226 (1968).
792a. E. Caro, J. Grotewold, and E. A. Lissi, *J. Chem. Soc.*, **1969D**, 318–319.
793. Ralph S. Becker, Edward Dolan, and David E. Balke, *J. Chem. Phys.*, **50**, 239–245 (1969).
793a. W. R. Ware, P. Chow, and S. K. Lee, *Chem. Phys. Lett.*, **2**, 356–358 (1968).
794. Norman E. Lee and Edward K. C. Lee, *J. Chem. Phys.*, **50**, 2094–2107 (1969).
794a. John W. Eastman, *J. Chem. Phys.*, **49**, 4617–4621 (1968).
795. A. V. Aristov and E. N. Viktorova, *Opt. Spektrosk.*, **25**, 509–515 (1968).
795a. Craig Lawson, Fumio Hirayama, and Sanford Lipsky, *Mol. Lumin., Int. Conf., 1968*, ed. by E. C. Lim (New York: W. A. Benjamin, 1969), 837–849.
796. Bernard A. Baldwin, *J. Chem. Phys.*, **50**, 1039–1040 (1969).
797. Willem Siebrand, *J. Chem. Phys.*, **50**, 1040–1041 (1969).
798. Terence Edward Martin and Alfred H. Kalantar, *Mol. Lumin., Int. Conf., 1968*, ed. by E. C. Lim (New York: W. A. Benjamin, 1969), 437–451.
799. Richard A. Keller, *ibid.*, 453–468.
800. Richard A. Keller, *Chem. Phys. Lett.*, **3**, 27–29 (1969).
801. Edward C. Lim. *Mol. Lumin., Int. Conf., 1968*, ed. by E. C. Lim (New York: W. A. Benjamin, 1969), 469–478.
802. Mostafa A. El-Sayed, Dino S. Tinti, and D. V. Owens, *Chem. Phys. Lett.*, **3**, 339–342 (1969).
803. S. Dym and Robio M. Hochstrasser, *J. Chem. Phys.*, **51**, 2458–2468 (1969).
804. E. C. Lim, *Int. Conf. Photochem. (Prepr.)* (Muelheim/Ruhr, Germany: Max-Planck Inst. Kohlenforsch., 1967), **2**, 507.
804a. Jack Saltiel and Oliver C. Zafiriou, *Mol. Photochem.*, **1**, 319–324 (1969).
804b. Henry Van Zwet, *Rec. Trav. Chim. Pays-Bas*, **87**, 1201–1210 (1968).
804c. A. Baczynski, *Bull. Acad. Pol. Sci., Ser. Sci., Math., Astron., Phys.*, **16**, 609–614 (1968).

805. Charles S. Parmenter and H. M. Poland, *J. Chem. Phys.*, **51**, 1551–1558 (1969).
806. George M. Breuer and Edward K. C. Lee, *J. Chem. Phys.*, **51**, 3615–3616 (1969).
807. E. M. Anderson and George B. Kistiakowsky, *J. Chem. Phys.*, **51**, 182–188 (1969).
808. M. E. McBeath, G. P. Semeluk, and I. Unger, *J. Phys. Chem.*, **73**, 995–1000 (1969).
809. E. Drent, G. Makkes Van der Deijl, and P. J. Zandstra, *Chem. Phys. Lett.*, **2**, 526–528 (1968).
810. V. L. Ermolaev and E. B. Sveshnikova, *Acta Phys. Pol.*, **34**, 771–790 (1968).
811. Peter Frank Jones and Seymour Siegel, *Chem. Phys. Lett.*, **2**, 486–488 (1968).
812. Mostafa A. El-Sayed and Lawrence H. Hall, Jr., *J. Chem. Phys.*, **50**, 3113–3114 (1969).
813. Mostafa A. El-Sayed, *U.S. Govt. Res. Develop. Rep.*, **69**, 77 (1969).
814. D. W. Williams and Harold E. Johns, *Photochem. Photobiol.*, **9**, 323–330 (1969).
815. I. H. Brown and Harold E. Johns, *Photochem. Photobiol.*, **8**, 273–286 (1968).
816. A. V. Buettner, Benjamin B. Snavely, and O. G. Peterson, *Mol. Lumin., Int. Conf., 1968*, ed. by E. C. Lim (New York: W. A. Benjamin, 1969), 403–422.
817. Masae Nemoto, Hiroshi Kokubun, and Masao Koizumi, *Bull. Chem. Soc. Jap.*, **42**, 1223–1230 (1969).
818. G. P. Gurinovich, E. K. Kruglik, and A. I. Patsko, *Izv. Akad. Nauk SSSR, Ser. Fiz.*, **32**, 1450–1455 (1968).
819. Klaus Krueger and Ernst Lippert, *Z. Phys. Chem. (Frankfurt am Main)*, **66**, 293–297 (1969).
820. M. F. Thomaz and Brian Stevens, *Mol. Lumin., Int. Conf., 1968*, ed. by E. C. Lim (New York: W. A. Benjamin, 1969), 153–165.
821. Attila Yildiz and Charles Reilley, *Spectrosc. Lett.*, **1**, 335–343 (1968).
822. Teisuke Okano and Hitoshi Matsumoto, *Yakugaku Zasshi*, **89**, 510–516 (1969).
823. A. T. Gradyushko, V. A. Mashenkov, K. N. Solov'ev, and M. P. Tsvirko, *Zh. Prikl. Spektrosk.*, **9**, 514–518 (1968).
824. P. P. Schmidt, *Mol. Cryst.*, **5**, 185–210 (1968).
825. Masakatsu Kocki, *Kagaku no Ryoiki*, **23**, 30–34 (1969).
826. David P. Chock and Stuart A. Rice, *J. Chem. Phys.*, **49**, 4345–4355 (1968).
827. V. K. Dolganov and E. F. Sheka, *Fiz. Tverd. Tela*, **11**, 2427–2434 (1969).
828. Bat-Sheva Sommer and Joshua Jortner, *J. Chem. Phys.*, **50**, 822–838 (1969).
829. T. B. El-Kareh, *Proc. Int. Conf. Lumin., 1966*, ed. by G. Szigeti (Budapest: Akad. Kiado, 1968), **1**, 380–386.
830. B. J. Mulder, *Philips Res. Rep., Suppl.*, **4**, 128 pp. (1968).
831. E. Glockner and Hans C. Wolf, *Z. Naturforsch.*, **24A**, 943–951 (1969).
832. Nicholas Geacintov, Martin Pope, and Hartmut Kallman, *Proc. Int. Conf. Lumin., 1966*, ed. by G. Szigeti (Budapest: Akad. Kiado, 1968), **1**, 260–264.
833. D. M. Burland and G. Castro, *J. Chem. Phys.*, **50**, 4107–4108 (1969).
834. Alan Bree and R. Zwarich, *Mol. Cryst. Liquid Cryst.*, **5**, 369–379 (1969).
835. D. Haarer and Hans C. Wolf, *Phys. Status Solidi*, **33**, K117–120 (1969).
836. Zoltan G. Soos, *J. Chem. Phys.*, **51**, 2107–2112 (1969).
837. Vladimiro Ern, *Phys. Rev. Lett.*, **22**, 343–345 (1969).
838. Yvan Rousset, *Compt. Rend.*, **268B**, 1644–1646 (1969).
839. Motohiko Koyanagi, Tadayoshi Shigeoka, and Yoshiya Kanda, *Mol. Lumin., Int. Conf., 1968*, ed. by E. C. Lim (New York: W. A. Benjamin, 1969), 765–772.
840. S. K. Chakrabarti, *Mol. Phys.*, **16**, 417–419 (1969).
841. E. Loewenthal, Y. Tomkiewicz, and A. Weinreb, *Spectrochim. Acta*, **25A**, 1501–1513 (1969).
842. J. J. Kim, R. A. Beardslee, David T. Phillips, and H. W. Offen, *U.S. Govt. Res. Develop. Rep.*, **69**, 74 (1969).

843. Asish K. Chandra and Edward C. Lim., *Mol. Lumin., Int. Conf., 1968*, ed. by E. C. Lim (New York: W. A. Benjamin, 1969), 249–266.
844. Asish K. Chandra and Edward C. Lim, *J. Chem. Phys.*, **49**, 5066–5072 (1968).
845. Andrzej J. Sadlej, *Chem. Phys. Lett.*, **2**, 451–453 (1968).
846. Theodor Förster, *Angew. Chem., Int. Ed. Engl.*, **8**, 333–343 (1969).
847. R. Speed and B. K. Selinger, *Aust. J. Chem.*, **22**, 9–17 (1969).
848. L. G. Christophorou, M. E. M. Abu-Zeid, and James G. Carter, *J. Chem. Phys.*, **49**, 3775–3782 (1968).
849. Fumio Hirayama and Sanford Lipsky, *J. Chem. Phys.*, **51**, 1939–1951 (1969).
850. John B. Birks, *Mol. Lumin., Int. Conf., 1968*, ed. by E. C. Lim (New York: W. A. Benjamin, 1969), 219–236.
851. Jeffrey R. Greenleaf, Michael D. Lumb, and John B. Birks, *Proc. Phys. Soc., London (At. Mol. Phys.)*, **1**, 1157–1159 (1968).
852. Robert B. Cundall and A. J. R. Voss, *Chem. Commun.*, **1969**, 116.
853. Fumio Hirayama and Sanford Lipsky, *Mol. Lumin., Int. Conf., 1968*, ed. by E. C. Lim (New York: W. A. Benjamin, 1969), 237–247.
854. Jean Klein, Francine Heisel, H. Lami, and Gilbert Laustriat, *Proc. Int. Conf. Lumin., 1966*, ed. by G. Szigeti (Budapest: Akad. Kiado, 1968), **1**, 205–211.
855. Charles DeBoer, *J. Amer. Chem. Soc.*, **91**, 1855–1856 (1969).
856. P. Holzman and Richard C. Jarnagin, *J. Chem. Phys.*, **51**, 2251–2253 (1969).
857. Heinz Baessler, *J. Chem. Phys.*, **49**, 5198–5199 (1968).
858. Emanoil Lucatu and P. Suciu, *Proc. Int. Conf. Lumin., 1966*, ed. by G. Szigeti (Budapest: Akad. Kiado, 1968), **1**, 503–506.
859. John B. Birks, *Acta Phys. Pol.*, **34**, 603–617 (1968).
860. C. R. Goldschmidt, Y. Tomkiewicz, and Isadore B. Berlman, *Chem. Phys. Lett.*, **2**, 520–522 (1968).
861. C. R. Goldschmidt, Y. Tomkiewicz, and Isadore B. Berlman, *Chem. Phys. Lett.*, **2**, 536–538 (1968).
862. Guenter Reske, *Z. Naturforsch.*, **23A**, 2137–2141 (1968).
863. Donald L. Horrocks, *Fluorescence News*, **3**, 1–4 (1968).
864. Donald L. Horrocks, *J. Chem. Phys.*, **50**, 4962–4966 (1969).
865. S. S. Rathi, K. Gopalakrishnan, and Jugal Kishore, *Acta Phys. (Budapest)*, **25**, 245–250 (1968).
866. L. V. Levshin and N. Nizamov, *Vestn. Mosk. Univ., Fiz., Astron.*, **24**, 42–48 (1969).
867. Emanoil Lucatu, *Compt. Rend.*, **267B**, 1146–1148 (1968).
868. Herbert Bauser and H. H. Ruf, *Phys. Status Solidi*, **32**, 135–149 (1969).
869. V. V. Zelinskii and L. G. Pikulik, *Zh. Prikl. Spektrosk.*, **10**, 684–686 (1969).
870. Albert Weller, *Fast React. Primary processes Chem. Kinet., Proc. Nobel Symp., 5th*, ed. by Stig Claesson (Stockholm: Almqvist and Wiksell, 1967), 413–428.
871. M. G. Kuz'min and L. N. Guseva, *Chem. Phys. Lett.*, **3**, 71–72 (1969).
872. L. N. Guseva, N. A. Sadovskii, and M. G. Kuz'min, *Khim. Vys. Energ.*, **3**, 44–47 (1969).
873. Joseph B. Guttenplan and Saul G. Cohen, *Tetrahedron Lett.*, **1969**, 2125–2128.
874. Richard A. Caldwell, *Tetrahedron Lett.*, **1969**, 2121–2124.
875. I. L. Edilashvili and A. S. Cherkasov, *Zh. Fiz. Khim.*, **42**, 2462–2466 (1968).
876. Chyongjin Pac and Hiroshi Sakurai, *Kogyo Kagaku Zasshi*, **72**, 230–235 (1969).
877. Takafumi Tosa, Chyongjin Pac, and Hiroshi Sakurai, *Tetrahedron Lett.*, **1969**, 3635–3638.
878. Tadeusz Latowski, *Zesz. Nauk., Mat., Fiz., Chem., Wyzsza Szk. Pedagog. Gdansku*, **8**, 189–193 (1968).

879. I. L. Edilashvili and A. S. Cherkasov, *Izv. Akad. Nauk SSSR, Ser. Fiz.*, **32**, 1538–1541 (1968).
880. H. Beens and Albert Weller, *Acta Phys. Pol.*, **34**, 593–602 (1968).
881. Tadashi Okada, H. Matsui, H. Oohari, H. Matsumoto, and Noboru Mataga, *J. Chem. Phys.*, **49**, 4717–4718 (1968).
882. Teisuke Okano, Yoshihiro Yamazaki, and Kaneto Uekama, *Yakugaku Zasshi*, **89**, 44–50 (1969).
883. Estera Kunec-Vajic and Karlo Weber, *Crcat. Chem. Acta*, **40**, 205–212 (1968).
884. D. Greatorex, Terence J. Kemp, and J. P. Roberts, *J. Phys. Chem.*, **73**, 1616–1617 (1969).
885. Yu. I. Kiryukhin and Kh. S. Bagdasar'yan, *Khim. Vys. Energ.*, **3**, 179–80 (1969).
886. Lawrence A. Singer, *Tetrahedron Lett.*, **1969**, 923–926.
887. Lawrence A. Singer, Gene A. Davis, and V. P. Muralidharan, *J. Amer. Chem. Soc.*, **91**, 897–902 (1969).
888. Dieter Rehm and Albert Weller, *Ber. Bunsenges. Phys. Chem.*, **73**, 834–839 (1969).
889. Nicholas J. Turro and Robert Engel, *Mol. Photochem.*, **1**, 143–146 (1969).
890. Nicholas J. Turro and Robert Engel, *Mol. Photochem.*, **1**, 235–238 (1969).
891. Robert Stephen Davidson and Paul F. Lambeth, *J. Chem. Soc.*, **1969D**, 1098–1099.
892. Robert J. Donovan, David Husain, and C. D. Stevenson, *Trans. Faraday Soc.*, **65**, 2941–2947 (1969).
893. Harry A. Morrison and William I. Ferree, Jr., *Chem. Commun.*, **1969**, 268–269.
894. Barton S. Solomon, Colin Steel, and Albert Weller, *J. Chem. Soc.*, **1969D**, 927–928.
895. Leonard M. Stephenson, Jr., thesis, California Institute of Technology, 1968; *Diss. Abstr.*, **29B**, 1311 (1969).
896. Ted R. Evans and Peter A. Leermakers, *J. Amer. Chem. Soc.*, **91**, 5898–5900 (1969).
897. A. Colin Day and T. R. Wright, *Tetrahedron Lett.*, **1969**, 1067–1070.
898. D. Schulte-Frohlinde and R. Pfefferkorn, *Int. Conf. Photochem.* (*Prepr.*) (Muelheim/Ruhr, Germany: Max-Planck Inst. Kohlenforsch., 1967), **2**, 672–685.
899. Charles Tanielian, *Compt. Rend.*, **268C**, 1031–1034 (1969).
900. M. D. Lumb and D. A. Weyl, *Proc. Int. Conf. Lumin., 1966*, ed. by G. Szigeti (Budapest: Akad. Kiado, 1968), **1**, 477–485.
901. M. S. S. C. Leite and K. Razi Naqvi, *Chem. Phys. Lett.*, **4**, 35–38 (1969).
902. A. S. Cherkasov and I. E. Obyknovennaya, *Proc. Int. Conf. Lumin., 1966*, ed. by G. Sziegeti (Budapest: Akad. Kiado, 1968), **1**, 496–502.
903. Takashi Kajiwara, Ichimin Shirotani, and Hiroo Inokuchi, *J. Mol. Spectrosc.*, **32**, 1–12 (1969).
904. V. R. Priimachek and A. N. Faidysh, *Opt. Spektrosk.*, **27**, 431–438 (1969).
905. A. K. Babko and A. P. Kostyshina, *Ukr. Khim. Zh.*, **35**, 544–551 (1969).
906. Noboru Mataga and Yoshimitsu Murata, *J. Amer. Chem. Soc.*, **91**, 3144–3152 (1969).
907. Keith Michael Charles Davis, *Nature (London)*, **223**, 728 (1969).
908. Guenther Briegleb, G. Betz, and W. Herre, *Z. Phys. Chem. (Frankfurt am Main)*, **64**, 85–96 (1969).
909. Tomohiko Hirooka, Masakatsu Kochi, Junichi Aihara, Hiroc Inokuchi, and Yoshiya Harada, *Bull. Chem. Soc. Jap.*, **42**, 1481–1486 (1969).
910. Joseph Guttenplan and Saul G. Cohen, *Chem. Commun.*, **1969**, 247–248.
911. D. N. Shigorin, N. A. Sheglova, and Yu. I. Kozlov, *Proc. Int. Conf. Lumin., 1966*, ed. by G. Sziegeti (Budapest: Akad. Kiado, 1968), **1**, 528–532.
912. Donald L. Horrocks, *J. Chem. Phys.*, **50**, 4151–4156 (1969).
913. C. M. Chopin and J. H. Wharton, *Chem. Phys. Lett.*, **3**, 552–555 (1969).
914. William R. Ware, P. R. Shukla, P. J. Sullivan, and R. V. Bremplis, *U.S. Govt. Res. Develop. Rep.*, **69**, 87 (1969).

915. A. N. Terenin, A. V. Shablya, G. J. Lashkov, and K. B. Demidov, *Proc. Int. Conf. Lumin.*, *1966*, ed. by G. Szigeti (Budapest: Akad. Kiado, 1968), **1**, 137–145.
916. Henk Beens and Albert Weller, *Mol. Lumin.*, *Int. Conf.*, *1968*, ed. by E. C. Lim (New York: W. A. Benjamin, 1969), 203–217.
917. J. C. Mackie, *Aust. J. Chem.*, **22**, 255–258 (1969).
918. G. M. Lysenko, G. M. Kislyak, and V. I. Ponochovnyi, *Ukr. Fiz. Zh.* (*Ukr. Ed.*), **13**, 2074–2076 (1968).
919. Albert Padwa, William Eisenhardt, Robert Gruber, and Deran Pashayan, *J. Amer. Chem. Soc.*, **91**, 1857–1859 (1969).
920. J. Grotewold, E. A. Lissi, *Chem. Commun.*, **1968**, 1367–1368.
921. Glyn D. Short, *Chem. Commun.*, **1968**, 1500–1501.
922. Guenther Briegleb and H. Schuster, *Angew. Chem., Int. Ed. Engl.*, **8**, 771 (1969).
923. R. Potashnik, C. R. Goldschmidt, and M. Ottolenghi, *J. Phys. Chem.*, **73**, 3170–3171 (1969).
924. Hisaharu Hayashi, Saburo Nagakura, and Suchiro Iwata, *Mol. Lumin.*, *Int. Conf.*, *1968*, ed. by E. C. Lim (New York: W. A. Benjamin, 1969), 351–363.
925. Kei Sin Wei and Albert H. Adelman, *Tetrahedron Lett.*, **1969**, 3297–3300.
926. Charles S. Parmenter and J. D. Rau, *J. Chem. Phys.*, **51**, 2242–2246 (1969).
927. David R. Kearns, Ahsan U. Khan, Christopher K. Duncan, and August H. Maki, *J. Amer. Chem. Soc.*, **91**, 1039–1040 (1969).
928. Masao Koizumi and Yoshiharu Usui, *Tetrahedron Lett.*, **1968**, 6011–6014.
929. Joachim Stauff and H. Fuhr, *Ber. Bunsenges. Phys. Chem.*, **73**, 245–251 (1969).
929a. Jennifer Canva, Claude Balny, Pierre Douzou, and Jean Bourdon, *Compt. Rend.*, **268C**, 1027–1030 (1969).
929b. R. H. Kummler and M. H. Bortner, *Environ. Sci. Technol.*, **3**, 944–946 (1969).
930. Theodor Förster, *Proc. Int. Conf. Lumin.*, **1966**, ed. by G. Szigeti (Budapest: Akad. Kiado, 1968), **1**, 160–165.
931. Krishna K. Rohatgi and G. S. Singhal, *Ind. J. Chem.*, **7**, 1020–1024 (1969).
932. Abraham A. Zimmerman, Charles M. Orlando, Jr., Michael H. Gianni, and Karl Weiss, *J. Org. Chem.*, **34**, 73–77 (1969).
933. K. K. Turoverov, *Opt. Spektrosk.*, **26**, 564–570 (1969).
934. E. Balint and Janos Hevesi, *Acta Phys. Chem.*, **14**, 77–84 (1968).
935. G. M. Bhatnagar, L. C. Gruen, and John A. Maclaren, *Aust. J. Chem.*, **21**, 3005–3013 (1968).
936. Ira Weinryb, *Biochem. Biophys. Res. Commun.* **34**, 865–868 (1969).
937. Joseph Eisinger and Gil Navon, *J. Chem. Phys.*, **50**, 2069–2077 (1969).
938. Joseph Eisinger, *Mol. Lumin.*, *Int. Conf.*, *1968*, ed. by E. C. Lim (W. A. Benjamin, 1969), 185–201.
939. Jehuda Feitelson, *Photochem. Photobiol.*, **9**, 401–410 (1969).
940. Eloise Kuntz, Robert Canada, Richard Wagner, and Leroy Augenstein, *Mol. Lumin., Int. Conf.*, *1968*, ed. by E. C. Lim (New York: W. A. Benjamin, 1969), 551–567.
941. Leonard M. Stephenson and George S. Hammond, *Angew. Chem., Int. Ed. Engl.*, **8**, 261–270 (1969).
942. David L. Dexter, Theodor Förster, and Robert S. Knox, *Phys. Status Solidi*, **34**, K159–162 (1969).
943. Kenneth B. Eisenthal, *J. Chem. Phys.*, **50**, 3120–3122 (1969).
943a. Dietmar Möbius, *Z. Naturforsch.*, **24A**, 251–253 (1969).
944. Ya. A. Terskoi and V. G. Brudz, *Opt. Spektrosk.*, **25**, 877–882 (1968).
944a. Robert M. Pearlstein, *Photochem. Photobiol.*, **8**, 341–347 (1968).
945. Andre Martinez, *J. Chim. Phys. Physicochim. Biol.*, **65**, 1663–1664 (1968).

946. Walter Kloepffer, *Ber. Bunsenges. Phys. Chem.*, **73**, 864–867 (1969).
947. Walter Kloepffer, *J. Chem. Phys.*, **50**, 1689–1694 (1969).
948. M. V. Alfimov, I. G. Batekha, V. A. Smirnov, *Dokl. Akad. Nauk SSSR*, **185**, 626–628 (1969).
949. K. R. Adam and M. F. O'Dwyer, *Aust. J. Chem.*, **22**, 2061–2084 (1969).
950. K. R. Adam and M. F. O'Dwyer, *Aust. J. Chem.*, **22**, 2085–2090 (1969).
951. S. Mansour and Arye Weinreb, *Chem. Phys. Lett.*, **2**, 653–656 (1968).
952. Thomas F. Hunter, *Photochem. Photobiol.*, **9**, 377–383 (1969).
953. Noboru Mataga, Harumichi Obashi, and Tadashi Okada, *J. Phys. Chem.*, **73**, 370–374 (1969).
954. Akahiko Nakahara, Motohiko Koyanagi, and Yoshiya Kanda, *J. Chem. Phys.*, **50**, 552–554 (1969).
955. Tokihisa Nakamura and Sohachiro Hayakawa, *Jap. J. Appl. Phys.*, **8**, 85–90 (1969).
956. Thomas F. Hunter, Robert D. McAlpine, and Robin M. Hochstrasser, *J. Chem. Phys.*, **50**, 1140–1141 (1969).
957. Keith G. Stolzle, thesis, Louisiana State University, Baton Rouge, 1968; *Diss. Abstr.*, **29B**, 3649–3659 (1969).
958. Minoru Tsuda, *Bull. Chem. Soc. Jap.*, **42**, 905–908 (1969).
959. L. T. Kantardzhyan, S. S. Chirkinyan, and M. B. Chryan, *Izv. Akad. Nauk SSSR, Ser. Fiz.*, **32**, 1534–1537 (1968).
960. V. L. Levshin and Yu. I. Grineva, *Zh. Prikl. Spektrosk.*, **9**, 630–636 (1968).
961. V. L. Levshin and Yu. I. Grineva, *Acta Phys. Pol.*, **34**, 791–800 (1968).
962. S. Georghiou, *Proc. Phys. Soc., London (At. Mol. Phys.)*, **2**, 1084–1089 (1969).
963. Richard P. Haugland, Juan Yguerabide, and Lubert Stryer, *Proc. Nat. Acad. Sci. U.S.*, **63**, 23–30 (1969).
964. Janos Hevesi, *Mol. Lumin., Int. Conf., 1968*, ed. by E. C. Lim (New York: W. A. Benjamin, 1969), 167–183.
965. Z. Varkonyi, *Acta Phys. Chem.*, **15**, 19–26 (1969).
966. J. Dombi, *Acta Phys. (Budapest)*, **25**, 287–305 (1968).
967. V. L. Ermolaev and T. A. Shakhverdov, *Opt. Spektrosk.*, **26**, 845–847 (1969).
968. Isadore B. Berlman, C. R. Goldschmidt, Y. Tomkiewicz, and Arye Weinreb, *Chem. Phys. Lett.*, **2**, 657–658 (1968).
969. Anthony C. Testa, A. Weisstuch, and J. Hennessy, *Mol. Lumin., Int. Conf., 1968*, ed. by E. C. Lim (New York: W. A. Benjamin, 1969), 863–877.
970. Frederick Dunbar Lewis and J. Christopher Dalton, *J. Amer. Chem. Soc.*, **91**, 5260–5263 (1969).
971. Jack Saltiel and Eldon D. Megarity, *J. Amer. Chem. Soc.*, **91**, 1265–1267 (1969).
972. David F. Roswell, thesis, Johns Hopkins Univ., 1968; *Diss. Abstr.*, **29B**, 3270 (1969).
973. Joao C. Conte, *Trans. Faraday Soc.*, **65**, 2382–2388 (1969).
974. Charles Tanielian, *Compt. Rend.*, **267C**, 1532–1534 (1968).
975. C. A Parker, *Fast React. Primary Processes. Chem. Kinet., Proc. Nobel Symp., 5th*, ed. by Stig Claesson (Stockholm: Almqvist and Wiksell, 1967), 325–331.
976. Masao Koizumi, *Kogyo Kagaku Zasshi*, **72**, 1–6 (1969).
977. Robert B. Cundall, *Energ. Mech. Radiat. Biol., Proc. Nato Advan. Study Inst., 1967*, ed. by Glyn O. Phillips (London: Academy Press, 1968), 227–242.
978. Milton Burton, Koichi Funabashi, Robert R. Hentz, Peter K. Ludwig, John L. Magee, and Asokendu Mozumder, *Transfer Stor. Energy Mol.*, ed. by George M. Burnett (London: Wiley-Interscience, 1969), **1**, 161–218.
979. Michael Cocivera, *Chem. Phys. Lett.*, **2**, 529–532 (1968).

980. Ronald S. Cole, thesis, California Institute of Technology, 1968; *Diss. Abstr.*, **29B**, 933 (1969).
981. Richard A. Caldwell and G. Wayne Sovocool, *J. Amer. Chem. Soc.*, **90**, 7138–7139 (1968).
982. Robert S. H. Liu and Reid E. Kellogg, *J. Amer. Chem. Soc.*, **91**, 250–252 (1969).
983. Robert S. H. Liu and James R. Edman, *J. Amer. Chem. Soc.*, **91**, 1492–1497 (1969).
984. Robert L. Cargill, A. C. Miller, Dwid M. Pond, Paul De Mayo, M. F. Tchir, K. R. Neuberger, and J. Saltiel, *Mol. Photochem.*, **1**, 301–317 (1969).
985. Anthony V. Guzzo and Gary L. Pool, *J. Phys. Chem.*, **73**, 2512–2515 (1969).
986. Manfred Hoefert, *Photochem. Photobiol.*, **9**, 427–432 (1969).
987. Peter John Wagner, *Mol. Photochem.*, **1**, 71–87 (1969).
988. Orville L. Chapman and Gene Wampfler, *J. Amer. Chem. Soc.*, **91**, 5390–5392 (1969).
989. Jane M. Bonnier, Pierre Jardon, and Jean P. Blanchi, *Bull. Soc. Chim. Fr.*, **1968**, 4787–4790.
990. Shin Sato, Hajime Kobayashi, and Kazuyuki Fukano, *Kogyo Kagaku Zasshi*, **72**, 209–212 (1969).
991. Kazuyuki Fukano and Shin Sato, *Kogyo Kagaku Zasshi*, **72**, 213–215 (1969).
992. G. A. Hanniger, Jr., and Edward K. C. Lee, *J. Phys. Chem.*, **73**, 1815–1822 (1969).
993. Jack Saltiel, Kenneth R. Neuberger, and Mark Wrighton, *J. Amer. Chem. Soc.*, **91**, 3658–3659 (1969).
994. M. V. Alfimov, I. G. Batekha, and Yu. B. Shekk, *Izv. Akad. Nauk SSSR, Ser. Fiz.*, **32**, 1488–1491 (1968).
995. I. G. Batekha, M. V. Alfimov, and Yu. B. Shekk, *Khim. Vys. Energ.*, **3**, 48–53 (1969).
996. S. K. Ho and Seymour Siegel, *J. Chem. Phys.*, **50**, 1142–1152 (1969).
997. Frederick D. Lewis and William H. Saunders, Jr., *J. Amer. Chem. Soc.*, **90**, 7033–7038 (1968).
998. Horst E. A. Kramer, Martin Hafner, and Markus Zuegel, *Z. Phys. Chem. (Frankfurt am Main)*, **65**, 276–289 (1969).
999. Paul Mathis, *Photochem. Photobiol.*, **91**, 55–63 (1969).
1000. Takuji Miwa and Tomoko Miwa, *Sci. Pap. Inst. Phys. Chem. Res.*, **62**, 141–144 (1968).
1001. M. E. Movsesyan, V. A. Gevorkyan, and D. Kh. Grigoryan, *Zh. Prikl. Spektrosk.*, **10**, 458–461 (1969).
1002. T. L. Banfield and D. Husain, *Trans. Faraday Soc.*, **65**, 1985–1991 (1969).
1003. Robert Signore, *Compt. Rend.*, **268B**, 763–765 (1969).
1004. S. Nordin and Robert L. Strong, *Chem. Phys. Lett.*, **2**, 429–432 (1968).
1005. C. R. Goldschmidt, Y. Tomkiewicz, and A. Weinreb, *Spectrochim. Acta*, **25A**, 1471–1477 (1969).
1006. K. K. Rohatigi and G. S. Singhal, *Proc. Int. Conf. Lumin.*, *1966*, ed. by G. Szigeti (Budapest: Akad. Kiado, 1968), **1**, 454–461.
1007. Guenther Von Buenau, Klaus Nieswandt, Dieter Henneberg, and Gerhard Schomburg, *Ber. Bunsenges. Phys. Chem.*, **73**, 891–897 (1969).
1008. Shigeru Tsunashima, Shin Satoh, and Shin Sato, *Bull. Chem. Soc. Jap.*, **42**, 1531–1533 (1969).
1009. Heinrich E. Hunziker, *J. Chem. Phys.*, **50**, 1294–1298 (1969).
1010. Heinrich E. Hunziker, *Chem. Phys. Lett.*, **3**, 504–507 (1969).
1011. Edward K. C. Lee, Manfred W. Schmidt, Robert G. Shortridge, Jr., and G. A. Haninger, Jr., *J. Phys. Chem.*, **73**, 1805–1815 (1969).
1012. V. A. Yakovenko, L. G. Pikulik, and M. Ya. Kosto, *Zh. Prikl. Spektrosk.*, **10**, 933–939 (1969).

1013. Masaru Nishikawa and Myran C. Sauer, Jr., *J. Chem. Phys.*, **51**, 1–9 (1969).
1014. Nicolae Filipescu, *Mol. Lumin., Int. Conf., 1968*, ed. by E. C. Lim (New York: W. A. Benjamin, 1969), 697–714.
1015. Nicolae Filipescu and Julian M. Menter, *J. Chem. Soc.*, **1969B**, 616–620.
1016. Emil Henry White, David R. Roberts, and David F. Roswell, *Mol. Lumin., Int. Conf., 1968*, ed. by E. C. Lim (New York: W. A. Benjamin, 1969), 479–492.
1017. R. David Rauh, Ted R. Evans, and Peter A. Leermakers, *J. Amer. Chem. Soc.*, **91**, 1868–1870 (1969).
1018. Inna Kules and M. Ero-Gecs, *Acta Chim. (Budapest)*, **58**, 389–397 (1968).
1019. Peter A. Leermakers, Jean P. Montillier, and R. David Rauh, *Mol. Photochem.*, **1**, 57–69 (1969).
1020. Albert L. Shain, J. P. Ackerman, and M. Warfield Teague, *Chem. Phys. Lett.*, **3**, 550–551 (1969).
1021. Angelo A. Lamola, *J. Amer. Chem. Soc.*, **91**, 4786–4790 (1969).
1022. Joseph A. Hudson and Richard M. Hedges, *Mol. Lumin., Int. Conf., 1968*, ed. by E. C. Lim (New York: W. A. Benjamin, 1969), 667–676.
1023. Joseph A. Hudson, thesis, Texas A & M Univ., 1968; *Diss. Abstr.*, **29B**, 4589 (1969).
1024. R. David Rauh, Ted R. Evans, and Peter A. Leermakers, *J. Amer. Chem. Soc.*, **90**, 6897–6904 (1968).
1025. Maximilian Zander, *Z. Naturforsch.*, **24A**, 1387–1390 (1969).
1026. Richard A. Keller and Lloyd J. Dolby, *J. Amer. Chem. Soc.*, **91**, 1293–1299 (1969).
1027. Gerhard Finger and A. B. Zahlan, *J. Chem. Phys.*, **50**, 25–30 (1969).
1028. B. Stevens, *Chem. Phys. Lett.*, **3**, 233–236 (1969).
1029. K. H. Grellman, *Ber. Bunsenges. Phys. Chem.*, **73**, 827–833 (1969).
1030. R. Potashnik, M. Ottolenghi, and R. Bensasson, *J. Phys. Chem.*, **73**, 1912–1918 (1969).
1031. G. I. Kobyshev and A. N. Terenin, *Proc. Int. Conf. Lumin., 1966*, ed. by G. Szigeti (Budapest: Akad. Kiado, 1968), **1**, 520–527.
1032. Koichi Kikuchi, Hiroshi Kokubun, and Masao Koizumi, *Z. Phys. Chem. (Frankfurt am Main)*, **62**, 79–82 (1968).
1033. Andre Martinez, *Compt. Rend.*, **268B**, 41–44 (1969).
1034. Sean P. McGlynn, Minoru Kinoshita, Michael McCarville, B. N. Srinivasan, and John W. Rabalais, *Photochem. Photobiol.*, **8**, 349–359 (1968).
1035. A. S. Gaevskii, V. R. Priimachek, and A. N. Faidysh, *Izv. Vyssh. Ucheb. Zaved., Fiz.*, **12**, 12–16 (1969).
1036. Tiguna N. Misra, *J. Chem. Phys.*, **51**, 2386–2395 (1969).
1037. T. N. Bolotnikova, T. M. Naumova, F. I. Gurov, and V. G. Kazachkov, *Opt. Spektrosk.*, **25**, 523–525 (1968).
1038. Thomas F. Hunter, *Photochem. Photobiol.*, **10**, 147–152 (1969).
1039. Andreas C. Albrecht, Philip M. Johnson, and W. M. McClain, *Proc. Int. Conf. Lumin., 1966*, ed. by G. Szigeti (Budapest: Akad. Kiado, 1968), **1**, 405–411.
1040. Michael Edward McCarville and S. P. McGlynn, *Photochem. Photobiol.*, **10**, 171–181 (1969).
1041. Yukio Kubota, Yasuo Fujisaki, and Masaji Miura, *Bull. Chem. Soc. Jap.*, **42**, 853 (1969).
1042. Mostafa A. El-Sayed, *U.S. Govt. Res. Develop. Rep.*, **69**, 77–78 (1969).
1043. Charles E. Swenberg, *J. Chem. Phys.*, **51**, 1753–1764 (1969).
1044. N. A. Tolstoi and A. P. Abramov, *Proc. Int. Conf. Lumin., 1966*, ed. by G. Szigeti (Budapest: Akad. Kiado, 1968), **2**, 1403–1407.
1045. B. M. Rumyantsev and E. L. Frankevich, *Opt. Spektrosk.*, **25**, 938–942 (1968).

1046. David H. Goode and R. F. Lipsett, *J. Chem. Phys.*, **51**, 122–127 (1969).
1047. D. H. Goode, *Mol. Lumin., Int. Conf., 1968*, ed. by E. C. Lim (New York: W. A. Benjamin, 1969), 751–763.
1048. W. B. Whitten, R. A. Arndt, and A. C. Damask, *Mol. Cryst. Liquid Cryst.*, **9**, 239–248 (1969).
1049. Richard E. Merrifield, Peter Avakian, and R. P. Groff, *Chem. Phys. Lett.*, **3**, 386–388 (1969).
1050. Richard E. Merrifield, Peter Avakian, and R. P. Groff, *Chem. Phys. Lett.*, **3**, 155–157 (1969).
1051. Nicholas Geacintov, Martin E. Pope, and F. Vogel, *Phys. Rev. Lett.*, **22**, 593–596 (1969).
1052. Martin Pope, Nicholas Geacintov, and Frank Vogel, *Mol. Cryst. Liquid Cryst.*, **6**, 83–104 (1969).

Part 3

Emission Spectroscopy and Photochemistry of Coordination Complexes

JOHN F. ENDICOTT and TIMM KELLY

Department of Chemistry Wayne State University, Detroit

INTRODUCTION

In 1969 there was an improved ratio of original research reports to review papers dealing with the photochemistry of coordination complexes, the single relevant review appearing in Volume 1 of this series [1]. In this part we survey the literature within the limits of our resources [2] and interests. A few articles listed in indices but not available to us for review are included. The material selected for review includes the photochemistry and aspects of the emission spectroscopy of coordination complexes of the transition metals. No attempt is made to survey the related literature on metal carbonyl, organometallic, or rare earth complexes or on spectroscopy or theoretical studies of bonding and electronic structure.

The development of this discipline has suffered because of the untimely death of H. L. Schlaefer (August, 1969). Professor Schlaefer's rational and elegant contributions will be greatly missed.

Judging from the contributed papers presented at national and international meetings, the photochemistry of coordination complexes is growing in a healthy manner. Relevant papers have been contributed during the past year at the National Meetings of the American Chemical Society, the International Conference on Coordination Chemistry, and the Farkas Symposium on Photochemistry in Aqueous and Polar Systems.

A number of reviews of related subjects have appeared this year. Most notable are Forster's [3] fine review of the spectroscopy, including emission, of chromium(III) and Perumareddi's very detailed review and discussion [4] of the spectroscopy of quadrate complexes of chromium(III).

An intriguing discussion of the chemiluminescence arising from electron-transfer reactions has appeared [5]. From the vast literature of the reactions of radicals, two reviews that we find particularly interesting are Weiss' [6] discussion of free radicals involved in oxidation-reduction reactions in solution and Hart's [7] review of the variety of reactions of the hydrated electron.

CHAPTER ONE

I. EMISSION SPECTROSCOPY, ENERGY TRANSFER, AND SENSITIZED REACTIONS

Camassei and Forster [8] determined quantum yields of luminescence and ^2E_g lifetimes as a function of temperature for Cr^{3+} in the crystalline hosts, $K_3[Co(CN)_6]$, $NaMg[Al(C_2O_4)_3] \cdot 9H_2O$, $C(NH_2)_3Al(SO_4)_2 \cdot 6H_2O$ (GASH), $C(NH_2)_3Al(SO_4)_2 \cdot 6D_2O$ (GASD), $AlCl_3 \cdot 6H_2O$, and $AlCl_3 \cdot 6D_2O$. In the $Cr^{3+}:K_3[Co(CN)_6]$ system the ^2E_g lifetimes are independent of Cr^{3+} concentration in dilute crystals (0.05 to 10%) Cr^{3+}. The emission from pure $K_3[Cr(CN)_6]$ is weaker and has a much shorter lifetime than is observed for the dilute mixed crystals. The decay is exponential in the dilute mixed crystals and nonexponential for pure $K_3[Cr(CN)_6]$. The emission spectra of pure and dilute crystals are identical except that the 0-0 line in the pure crystals is more intense relative to the vibronic structure. The observations that Kirk, Ludi, and Schlaefer [9] made in their study of the series of mixed crystals of $K_3[Cr(CN)_6]$ and $K_3[Co(CN)_6]$ support the findings of Camassei and Forster. A maximum is observed for the quantum yield of phosphorescence at about 15% Cr^{3+}. Thus the important quenching mechanism for phosphorescence in $K_3[Cr(CN)_6]$ involves short-range Cr^{3+}—Cr^{3+} interactions. Possible quenching mechanisms considered include radiationless conversion of the ^2E_g state to the ground state by interaction with an adjacent Cr^{3+} ion or a process involving the transfer of excitation energy, either resonant or phonon-assisted transfer, producing a vibrationally excited $^4A_{2g}$ on an adjacent ion or energy transfer to an adjacent Cr^{3+} ion that may either internally convert, producing phonons, or further luminesce in the infrared.

For all the systems studied by Camassei and Forster [8], phosphorescence was observed only at low temperatures, but, on warming, the aquo complexes fluoresced. In the $Cr^{3+}:NaMg[Al(C_2O_4)_3] \cdot 9H_2O$ system no fluorescence was detected at any temperature. The replacement of H_2O by D_2O in the $Cr^{3+}:AlCl_3 \cdot 6H_2O$ system resulted in an increase in the luminescence intensity, indicating that $^2E_g \rightsquigarrow {}^4A_{2g}$ is reduced by the "deuterium" effect. The authors interpreted the change from phosphorescence at low temperature to fluorescence as the temperature increases as establishing the involvement of $^2E_g \rightsquigarrow {}^4T_{2g} \rightarrow {}^4A_{2g}$. Similarly with Cr^{3+}:GASH and Cr^{3+}:GASD the luminscence intensity of the deuterium-substituted system was greater, and the same kind of temperature dependence of the luminescence was observed. The ^2E_g decay was found to be nonexponential. This behavior was explained in terms of two different Cr^{3+} sites in GASH, C_3 and C_{3v}, each with a slightly different decay lifetime. The persistence of two strong R lines, separated by 17 cm^{-1} even at 4°K, was thought to be strong support for this explanation [10]. A general parallelism between the Arrhenius activation energy of phosphorescence lifetimes and the difference in zero-point energies of ^2E_g and $^4T_{2g}$ was interpreted as implying that the dominant mode of degradation of ^2E_g involves return to the ground state by way of $^4T_{2g}$.

The emission bands of hexaureachromium(III) were studied in the temperature range of 1.5 to 300°K by Dingle [11]. The broad emission at 12,300 cm^{-1} was assigned to the $^4T_{2g} \rightarrow {}^4A_{2g}$ transition and the (0,0) $^4T_{2g}$ (O$_h$) transition was considered to occur at 14,222 cm^{-1}. The emissions at 14,196 ± 4 cm^{-1} and 14,300 cm^{-1} were assigned to transitions originating from the ^2E_g state and either a different doublet state or a vibronic level of ^2E_g, respectively.

Crystal-field considerations were applied to intersystem crossover in Cr(III) compounds by Hempel and Matsen [12]. They define a parameter $R \equiv v(^4T_{ag})/v(^2E_g)$, where $v(^4T_{2g})$ is the frequency of the $^4A_{2g} \rightarrow {}^4T_{2g}$ transition and $v(^2E_g)$ is the frequency of the $^4A_{2g} \rightarrow {}^2E_g$ transition from absorption spectra. For compounds with R less than 1, the authors predict that only fluorescence will be observed, for $1 < R > 2$ there is a possibility of observing both fluorescence and phosphorescence, and for $R > 2$ only phosphorescence will be observed.

Several studies of emission spectra among transition-metal complexes were concerned with d^6 systems. Lytle and Hercules [5, 13] established that the luminescence of [Ru(2,2'-bipyridine)$_3$]Cl$_2$ is due to a charge-transfer spin-forbidden $\pi^* \rightarrow d$ transition. The temperature effects on the charge-transfer luminescence intensity of metal chelates of this kind were examined by Fink and Ohnesorge [14]. These authors postulate a reduction of ligand-field splittings in the excited state of the complexes due to an increase in metal-ligand distances caused by a redistribution of electron density from an orbital centered on the central metal ion to one centered on the ligand. If the ligand-field splitting were lowered sufficiently, an equilibrium could be established in the excited state between the diamagnetic (t_{2g}^5,π^*) and the paramagnetic $(t_{2g}^3 e_g^2,\pi^*)$ configurations. Luminescence intensity should decrease approximately in proportion to the extent of crossover to the high-spin state. The extent of crossover should depend highly on temperature. Experimental observations that seem to substantiate these predictions are cited.

Rossiello [15] reported that both Pt(o-phen)Cl_4 and Pt(py)$_2$$Cl_4$ emit a single apparently structureless band peaked at 14,100 cm^{-1}. The intensity of the emission was found to decrease slightly as the temperature increased from $-190°C$. In Pt(bipy)Cl_4 the emission was centered at 15,800 cm^{-1} and resembled that of other transition-metal 2,2'-bipyridine complexes in the shape of its vibrational structure; at $-190°$ C its intensity was at least an order of magnitude greater than the emission of the other two compounds, but the intensity fell off very rapidly with increasing temperature. The similarity of the emission spectrum of Pt(py)$_2$$Cl_4$ and Pt(o-phen)Cl_4 to that of $[PtCl_6]^{2-}$ was attributed to a similar mechanism of luminescence, namely, radiative decay from a low-lying triplet ligand-field excited state. The emission band of Pt(bipy)Cl_4 was different in position and apparent activation energy; these features were attributed to the decay of a charge-transfer excited state.

The complex compounds of the general formula $K_2[PtCl_{6-n}Br_n]$, $n = 0, 1, 2, \ldots, 6$, were studied by Lipnitskii and Umreiko [16]. These compounds luminesce at liquid-nitrogen temperature, both in the solid state and in glassy aqueous solutions at 77°K. The luminescence spectrum of each compound consisted of a symmetric band with a maximum at 635.3 nm for K_2PtCl_6 and at 692.7 nm for K_2PtBr_6.

Nugent et al [17] have examined a series of β-diketone complexes of +3-actinides in attempts to detect ultraviolet-excited sharp-line–sensitized luminescence. Highly efficient sensitized luminescence was detected only in the case of the curium(III) complexes.

Schlaefer, Gausmann, and Moebius [18] undertook the luminescence studies of a series of complex double salts, $[CrA_6][Cr(CN)_6] \cdot XH_2O$ [A = antipyrine (atp), imidazolone (imid), or ammonia] and $[Cr(A-A)_3][Cr(CN)_6] \cdot XH_2O$ [A-A = ethylenediamine (en), propylenediamine (pn), or trimethylenediamine (tn)], to clarify the luminescence behavior of these complex double salts. All double complex salts exhibited luminescence behavior parallel to that of $[Cr(urea)_6][Cr(CN)_6]$ [19].

When the double salt $[Cr(atp)_6][Cr(CN)_6]$ was irradiated with 580 nm light, the phosphorescence ^2E_g (ct) → $^4A_{2g}$ (ct) and also the fluorescence $^4T_{2g}$ (ct) → $^4A_{2g}$ (ct) of the cation were totally quenched but a strong signal was seen for the phosphorescence, ^2E_g(an) → $^4A_{2g}$(an), of the anion. Since irradiation with 580 nm light excited the compound to $^4T_{2g}$ (ct) localized within the cation (donor), the authors [18] conclude that energy transfer to ^2E_g (an) localized within the ligand must take place, followed by phosphorescence ^2E_g(an) → $^4A_{2g}$(an). The authors inferred that an excitation energy transfer by way of ^2E_g (ct), for example, doublet-doublet transfer, was operating. For D and A donor and acceptor:

D^* (doublet) + A(quartet) → D(quartet + A^*(doublet)

For the complexes containing oxygen donor ligands, the phosphorescence ^2E_g (an) → $^4A_{2g}$ (an) seemed amplified by a factor of 10 to 40. The magnitude of the amplification increased as the Cr—Cr distance decreased. In the case of nitrogen coordination to Cr^{3+} such a relationship was not found, and there was little increase in the intensity of anion phosphorescence ocmpared with $K_3[Cr(CN)_6]$. Phosphorescence lifetimes ranged from 2.1 μsec for $[Cr(NH_3)_6][Cr(CN)_6]$ at 24° to 2.2 msec for $[Cr(atp)_6][Cr(CN)_6]$ at −193°.

Recent studies [20–23] have extended the technique of photosensitized reactions to the study of the photochemical behavior of inorganic complexes. Porter [20] studied the biacetyl-sensitized aquation of $Co(CN)_6^{3-}$ in order to assess the role of excited electronic states in the photoaquation of this anion. The addition of $Co(CN)_6^{3-}$ to solutions of biacetyl was found to quench the biacetyl phosphorescence without affecting the fluorescence. The biacetyl-$Co(CN)_6^{3-}$ energy-transfer rate constant was calculated to be 2.6×10^7 M^{-1} sec^{-1}. The irradiation of biacetyl-$K_3[Co(CN)_6]$ solutions at 436 nm results in the production of $Co(CN)_5H_2O^{2-}$. The quantum yield for photosensitized aquation is 0.23 ± 0.04, based on the number of $Co(CN)_6^{3-}$ ions excited. No sensitized decomposition was observed for areated solutions where oxygen completely quenches the biacetyl phosphorescence. It was postulated that in this system energy transfer occurs from the triplet state of biacetyl to a triplet state of the $Co(CN)_6^{3-}$, most probably $^3T_{1g}$. The observed photosensitized aquation was attributed to the formation of this excited state. A similar photoaquation occurs in the direct photolysis [24], and a common mechanism was suggested [20] for the two processes. It was further proposed that in the direct photolysis the $^1T_{1g}$ state, populated by absorption, of $Co(CN)_6^{3-}$ converts to the $^3T_{1g}$ state by means of an intersystem crossing process with a yield near unity. A fraction, 0.3, of these triplet species aquate, and the rest are degraded to the ground state by some path that is nearly independent of temperature.

Sastri and Langford [21] studied the photosensitized reaction in the biacetyl-$PtCl_4^{2-}$ system. The phosphorescence of biacetyl was found to be completely quenched by 10^{-3} M $PtCl_4^{2-}$, with a quenching rate constant of approximately 3×10^7 M^{-1} sec^{-1}. For irradiation of a biacetyl-$PtCl_4^{2-}$ solution at 408 nm and 0.034 M $PtCl_4^{2-}$, the quantum yield of aquation, based on light absorbed by biacetyl, was 0.26 ± 0.02. Direct photolysis [25] of $PtCl_4^{2-}$ in 0.001 M $HClO_4$ at 313, 404, and 472 nm produced aquation with a quantum yield of 0.14. The wavelength independence of the quantum yield was attributed to the complete deactivation of the higher-energy excited states to the lowest-energy triplet excited state, which could be the common precursor for photoaquation. The following scheme was proposed for the sensitized reaction:

$(\text{biacetyl})_{S_0} \xrightarrow{h\nu} (\text{biacetyl})_{S_1}$

$(\text{biacetyl})_{S_1} \rightarrow (\text{biacetyl})_T$
$(\text{biacetyl})_T \rightarrow (\text{biacetyl})_{S_0} + h\nu'$ (phosphorescence)
$(\text{biacetyl})_T + (\text{complex})_{S_0} \rightarrow (\text{complex})_T \, (^1A_{1g} \rightarrow {}^3E_g)$
$(\text{complex})_T \rightarrow \text{products}$

From the higher yield of the sensitized reaction the authors inferred that the direct photoreaction does not proceed by complete conversion of all higher excited states to the lowest triplet (compare discussion in Section III.F below).

Adamson et al [22] reported the photosensitized aquation of chromium(III) complexes. Ammonia aquation resulted from the quenching of biacetyl phosphorescence by $Cr(NH_3)_5NCS^{2+}$. Irradiation at 410 nm resulted in ammonia aquation, with a quantum yield based on light absorbed by biacetyl of 0.21. It was suggested that the complex intercepts a triplet state of biacetyl. A comparison of the ratio of ammonia to thiocyanate aquations, greater than 100 : 1 for the sensitized reactions versus 22 : 1 for direct irradiation of the first ligand-field band and 8 : 1 for irradiation in the region of the doublet band, was used to infer that the photoactive state is the first quartet excited state rather than the lowest doublet excited state.

Measurements were made [23] of the eosin photosensitized rate of oxidation of iron[(II) by molecular oxygen in aqueous perchlorate medium in the pH range of 4.5 to 6.0. The authors argued that absorption of light within the strong 516 nm band of eosin produced an excited state that transferred energy to the $^3\Sigma_g^-$ ground state of oxygen. Either the $^1\Delta_g$ or $^1\Sigma_g^+$ state of oxygen may be formed. Since the $^1\Sigma_g^+$ state is known to be strongly quenched by water, probably to $^1\Delta_g$, the latter was assumed to be the active state in the oxidation of iron(II).

II. PHOTOCHEMISTRY OF TRANSITION-METAL COMPLEXES

A. d² Systems

During this year there were some relatively definitive reports of the photochemical reactions of $Mo(CN)_8^{4-}$ and $W(CN)_8^{4-}$. A careful determination of the changes in [OH$^-$] in phosphate buffered solutions has established [26] that irradiation (365 nm) of the high-energy d-d bands results in photoaquation ($\phi = 0.8$) per reaction (1):

$$OH_2 + Mo(CN)_8^{4-} + h\nu \rightarrow Mo(CN)_7OH_2^{3-} + CN^- \tag{1}$$

Many previous studies have been complicated by the back reaction (2)

$$M(CN)_7OH_2^{3-} + CN^- \rightarrow M(CN)_8^{4-} + H_2O \tag{2}$$

and equilibrium changes in the aquo-hydroxy ratio in unbuffered solutions —reaction (3):

$$M(CN)_7OH_2^{3-} \rightleftarrows M(CN)_7OH^{4-} + H^+ \tag{3}$$

The identity of the red aquo product has been definitively extablished by isolation of the silver salt, $Ag_3[Mo(CN)_7OH_2]$, following irradiation of $Mo(CN)_8^{4-}$ in strongly acidic solution [27].

That the ultraviolet irradiation of the charge-transfer bands of $Mo(CN)_8^{4-}$ and $W(CN)_8^{4-}$ produces solvated electrons has been established in flash photolysis and N_2O scavenging studies [28]. Quantum yields of photoelectron production (from N_2 yields) were found to be 0.27 and 0.34 respectively.

B. Chromium(III)

The photochemistry of chromium(III) complexes continues to afford opportunity for well-designed experiments and controversial interpretations of results. Interestingly one study in 1969 found that the photosolvolysis of $Cr(SCN)_6^{3-}$ with sunlight in mixed solvent media resulted in products that were independent of the solvent shell composition; this was interpreted as implying the formation of a long-lived photoexcited state [30].

Geis and Schlaefer [31] have reported that irradiation of $Cr(en)_3^{3+}$ in the wavelength region of 450 to 250 nm, which includes both d-d and charge-transfer absorption bands, produces as a primary product the one-ended detachment of ethylenediamine—reaction (4)

$$Cr(en)_3^{3+} + h\nu \rightarrow Cr(en)_2(H_2NCH_2CH_2NH_3)OH_2^{4+} \tag{4}$$

with a quantum yield of 0.37 (10°C) independent of the irradiating wavelength. Continued irradiation produced *cis*-diaquo-bisethylenediamine-chromium(III).

Zinato, Lindholm, and Adamson [32] have investigated the photochemistry of $Cr(NH_3)_5NCS^{2+}$. The photolabilization of this complex was found to produce $Cr(NH_3)_4(OH_2)NCS^{2+}$ and $Cr(NH_3)_5OH_2^{3+}$—reactions (5) and (6) respectively:

$$Cr(NH_3)_5NCS^{2+} + h\nu + H_2O \begin{cases} \xrightarrow{H^+} Cr(NH_3)_4(OH_2)NCS^{2+} + NH_4^+ & (5) \\ \rightarrow Cr(NH_3)_5OH_2^{3+} + NCS^- & (6) \end{cases}$$

for irradiation at 373, 492, and 652 nm, but the relative yields seem to depend on wavelength. The authors identify the aquothiocyanato product with the trans isomer. The predominant process was found to be reaction (5), in accordance with Adamson's rules [34]. These results are to be contrasted with the irradiation of $Cr(NH_3)_5Cl^{2+}$, which is reported [35] to produce cis-$Cr(NH_3)_4OH_2Cl^{2+}$, contrary to the prediction of Adamson's rules. The predominant process in the photosensitized aquation of $Cr(NH_3)_5NCS^{2+}$ is also NH_3 aquation [21].

Wasgestian [33] has investigated the photoaquation of $Cr(CN)_6^{3-}$ at low temperatures ($\geq 7°$) where the back reactions are not important. The reaction has an apparent activation energy of ~ 3 kcal/mole and $\phi = 0.12$ at $20°$.

C. d^4 Systems

Hartmann and Filss [37] have reported that the ultraviolet irradiation of deareated chloride solutions of chromium(II) produce H_2.

Two papers on the photochemistry of pentacyanonitrosyl manganate(III) were listed in indices [38, 39] but were not available to us for review.

D. d^5 Systems

The quantum yields for photoreduction of $K_3[Fe(C_2O_4)_3] \cdot 3H_2O$ in the solid state are low (0·07 to 0.17) and depend on wavelength and intensity [40], in contrast to the solution behavior of this complex [41].

Traverso et al [42] have examined the aqueous photochemistry of dibenzenechromium(I) with 254, 313, 334, 365, and 404 nm radiation. The net photoreaction was found to be dissociation to form chromium(I) and benzene. Product yields were highest in the shortest-wavelength region.

Van-den-Berg [43] has used ^{14}C-labeled oxalate to investigate the photochemical decomposition of $Fe(C_2O_4)_3^{3-}$.

E. Cobalt(III) and Other d^6 Systems

The photochemistry of coordination complexes of cobalt(III) continues to be the focus of much work. It often seems that the known experimental "facts" as well as any mechanistic patterns evolve, hopefully converging, as successive approximations. For example, it is now clear that the photoreduction of $Co(NH_3)_5Cl^{2+}$ at 254 nm leads to Co^{2+}, Cl^- and the oxidation of coordinated ammonia [44]. Furthermore, a chemical-scavenging study of this system has provided evidence for two distinguishable intermediates, one which is and one which is not scavengeable [44b]. Since the band irradiated is reasonably assigned as spin-allowed, $Cl^- \to Co(III)$, the observations are evidence for facile internal conversion (yield ~ 0.5) of the initial singlet excited state to a chemically scavengeable metastable state. All in all the simple "radical pair" mechanism of Adamson [34, 45], with a primary radical-pair yield of unity, cannot be a general mechanistic description of the photoreduction of cobalt(III) complexes. The photochemical intermediates are so short-lived and elusive, however, that noncircumstantial evidence concerning them has been extremely difficult to obtain.

The photoreduction of $Co(NH_3)_5(C_2O_4)_2^-$, $Co(NH_3)_4C_2O_4^+$, and $Co(NH_3)_5C_2O_4H^{2+}$ has been investigated by Way and Filipescu [46]. These authors have rationalized their observations in terms of a multistep mechanism involving the intermediacy of an excited state that decomposes into Co^{2+} and $C_2O_4^-$ radicals followed by radical reduction of the cobalt(III) substrate. This mechanism has a long and honorable tradition—for example, see Refs. 41, 47, and 48—that to some extent obscures its deficiencies in this application. The oxalate radical anions, like most radicals, should exhibit a radical-radical combination type of decay mode—for example, $2C_2O_4^- \to C_2O_4^= + 2CO_2$. When the oxalato substrate is a relatively poor oxidant, as are the higher ammine complexes, radical-radical coupling should compete with radical reduction of the substrate. In the case of the tetrammine complex this competition is presumably the reason for the sensitivity of ϕ_{Co}^{2+} to dissolved O_2 and for the decrease of ϕ_{Co}^{2+} with I_a [49]. Also, the exact identity of the reactive radical—for example, whether $C_2O_4^-$ or CO_2^-—is not known.

Adamson and coworkers have continued to report interestingly designed experiments. The irradiation of $Co(NH_3)_5(TSC)^{2+}$ (for TSC = trans-4-stilbenecarboxylate) at 365 nm apparently an absorption of the stilbene ligand, results in photoredox decomposition of the complex [50]. The authors have attributed this to "intramolecular energy transfer" from the ligand-centered excited state to a CTTM excited state. Comparisons are made with similar photochemistry of $Co(NH_3)_5O_2CCH_3^{2+}$ and $Co(NH_3)_5O_2CC_6H_5^{2+}$. In all these systems both Co^{2+} and "free acid" were found to be present after irradiation; these products were accounted for in terms of a mechanism involving the attack of a carboxalate radical on the coordinated NH_3 of the $Co(NH_3)_5^{2+}$ fragment trapped on the solvent cage [50]. No gas yields were reported. It is to be noted that the 254 nm irradiation of $Co(NH_3)_5O_2CCH_3^{2+}$ produces Co^{2+}, CO_2, CH_4, and C_2H_6 with $\phi_{Co}^{2+} \simeq \phi_{CO_2} \simeq \phi_{CH_4} + 2\phi_{C_2H_6}$ at high intensity ($I_a \sim 6 \times 10^{-3}$ein l^{-1} min^{-1}) [51]. There is no evidence for the generation of N_2 at 254 or 360 nm [51b]. The chemical scavenging, with alcohols, of intermediates in the latter work has shown the following:

(1) the acetoxy radical is so short-lived that it cannot be scavenged even at high alcohol concentrations (up to 13 M); (2) $\phi_{CO_2}/\phi_{Co}^{2+}$ decreases as $\phi_{CH_4}/\phi_{C_2H_6}$ increases; and (3) absorption of 254 nm radiation produces scavengable oxidants that are not precursors to the formation of the gaseous products. Absorption of 360 nm radiation produces a variety of products, apparently including $Co(NH_3)_5OH_2^{3+}$ and $Co(NH_3)_4(OH_2)O_2CCH_3^{2+}$ (these complexes were separated by ion-exchange techniques and identified spectroscopically [51b]). The benzoxy and stilbenecarboxy radicals would be expected to be longer-lived, and less reactive, than acetoxy, and the peculiar product yields, $\phi_{Co}^{2+} = 2\phi_{\text{"free acid"}}$ observed by Adamson et al. [50] may result from radical-radical decay modes.

In a different paper Adamson et al. [52] have reported that the 340 to 380 nm irradiation of $Co(CN)_5X^{3-}$ (X = CN and I) results only in the formation of $Co(CN)_5OH_2^{2-}$ and X^- and that irradiation of $Co(CN)_5OH_2^{2-}$ in the presence of anions (N_3^-, SCN^-, and I^-) results in the formation of the respective $Co(CN)_5X^{3-}$ complexes. The iodo complex was also irradiated in the 500 nm region, with lower yields of the same products. The authors have proposed that their observations are best accounted for in terms of a five-coordinate $Co(CN)_5^{2-}$ species generated following absorption of radiation. The authors suggest that their "chemical" mechanism is to be contrasted with the excited-state mechanisms such as that proposed by Porter [20] to account for the direct and sensitized photoaquation of $Co(CN)_6^{3-}$. It would be surprising if the 380 nm irradiation of $Co(CN)_5I^{3-}$ did not produce some photoreduction, since the absorption is presumed to be significantly CTTM in character; on the other hand the authors [52] did not report any attempts to detect radical species, for example, $Co(CN)_5^{3-}$ and I_2^-, that might have recombined to give the final product yields.

Hughes et al. have pointed out that the irradiation of charge-transfer band combined with a standard method for analyzing Co^{2+} affords a clean method for determining the cobalt content in simple coordination complexes of cobalt(III) [53].

Filipescu and Way [54] have photolyzed tris(acetylacetonato)cobalt(III), tris(benzoylacetonato)cobalt(III), and tris(dibenzoylmethanato)cobalt(III) in different organic solvents. The photochemical reaction produced the bis-β-ketoenolate of cobalt(II) and oxidative fragmentation of the remaining ligand. Quantum yields varied from 10^{-3} for irradiation of d-d bands to as high as 0.6 for ultraviolet irradiation.

The quantum yields for the photoreduction of $K_3[Co(C_2O_4)_3]\cdot 3H_2O$ in the solid state were reported to be dependent on wavelength but independent of absorbed intensity [40].

Busch and coworkers [55] have irradiated a series of related cobalt(III) alkyl complexes using unfiltered light of an "ordinary tungsten lamp." The authors have used a crude relationship to approximate the relative intensity of the light absorbed; from a correlation between these relative intensities and the pseudo first-order rates of photodecomposition of the complexes the authors have suggested that the photoactive bands are in the ultraviolet region. The products of the photoreaction are apparently cobalt(II) complexes and alkyl radicals. By way of contrast the irradiation (with a 750 W tungsten lamp; unfiltered) of dichloro-[^{14}C]methylcobalamin in a polystyrene matrix has been reported [56] to give B_{12r} and a ClCH diradical trapped in the organic matrix.

Information concerning the photochemistry of other nd^6 complexes has continued to be tantalizingly sparce. Bauer and Basolo [57] have reported that irradiation (unfiltered; medium-pressure mercury arc) of $trans$-Ir(en)$_2$X$_2^+$ yields $trans$-Ir(en)$_2$X(OH$_2$)$^{2+}$ and that irradiation of the cis complexes results in aquation of one end of ethylenediamine (for X = Cl, Br, and I). The authors discuss relative rates and suggest that their observations indicate that the rules that Adamson suggested [34] to systematize the photochemistry of chromium(III) complexes are applicable to the photochemistry of iridium (III) (it should be noted that Adamson's rules are not always observed in the photochemistry of chromium(III); see Section II.B, above). Since this study does not report quantum yields or even consistently irradiate the same absorption band of the various complexes, mechanistic generalizations seem inappropriate. It should also be recalled that irradiation of d-d absorption bands of Rh(NH$_3$)$_5$Cl^{2+} apparently leads only to Cl$^-$ aquation [58], contrary to the predictions of Adamson's rules.

Two puzzling short reports of the photochemistry of ruthenium(II) ammine complexes have appeared this year. Sigwart and Spence [59] reported that ultraviolet irradiation of Ru(NH$_3$)$_5$N$_2^{2+}$ and [Ru(NH$_3$)$_5$]$_2$N$_2^{4+}$ produced some ruthenium(III) ammine and N$_2$ but not H$_2$. Ford and coworkers [60] have reported that, although irradiation of metal-to-ligand charge-transfer bands (in the 360 to 460 nm region) of Ru(NH$_3$)$_5$A^{2+} (A = pyridine, acetonitrile) produces some Ru(NH$_3$)$_5$OH$_2^{2+}$, irradiation at 254 nm again produces some ruthenium(III) complexes. The irradiation of Ru(NH$_5$)$_6^{2+}$ and Ru(NH$_3$)$_5$OH$_2^{2+}$ at these shorter wavelengths also produces some " photo-oxidation " of the metal center and H$_2$ [60].

The photoelectron yields from the ultraviolet irradiation (254 nm) of Fe(CN)$_6^{4-}$ and Ru(CN)$_6^{4-}$ have been reported to be 0.66 and 0.36 respectively by Waltz and Adamson [28].

F. d^8, d^9, and d^{10} Systems

Blake and Nyman [61] reported the synthesis of platinum (0) and palladium (0) complexes through the ultraviolet irradiation of ML$_2$C$_2$O$_4$ complexes, where M = Pt or Pd and L = triphenylphosphine, triphenylarsine, or L_2 = 1,2-bis(diphenylphosphinato)ethane. Two moles of carbon dioxide were evolved per mole of complex. In the case of [Pt(PPh$_3$)$_2$C$_2$O$_4$], the reaction gave a colorless product, which, when recrystallized from benzene, was formulated as [Pt(PPh$_3$)(PPh$_2$)]$_2$ · 1/2C$_6$H$_6$ on the basis of elemental analysis and molecular weight determination.

Carassiti and coworkers [62] have carefully reinvestigated the photoaquation of PtCl$_4^{2-}$. The observed photochemical reaction was aquation of a single chloride regardless of the absorption band irradiated (254, 313, 404, and 472 nm irradiations were used); however, the product yield was much higher (0.9 compared with about 0.2) for irradiation at the shortest wavelength in a charge-transfer absorption band of the complex. The authors have interpreted their results in terms of the population of different excited states following irradiation of charge-transfer and d-d absorption bands. The yield for long-wavelength irradiation reported in this study is higher than those reported by Sastri and Langford [21] and is sufficiently close to the photosensitized yields reported in the latter study to raise the possibility that there may be no difference in yields between the direct photolysis ($\lambda \geq 313$ nm) and the photosensitized reaction of PtCl$_4^{2-}$.

Bartocci et al. [63] have found that the ultraviolet irradiation of Pt(diene)Br^{2+} in water produces no net photochemical reaction, but when the irradiation was performed in the presence of NO$_2^-$, Pt(diene)NO$_2^+$ was formed as a product. Product yields in the presence of various ratios of [NO$_2^-$] to [Br$^-$] suggested a competition of NO$_2^-$ and Br$^-$ for some reactive intermediate.

Loelieger [64] observed that the irradiation of lead(IV) trifluoroacetate in C$_6$F$_6$ solution with ultraviolet light at 77°K resulted in the appearance of an epr signal that was attributed to ·CF$_3$ radicals. The photolysis of dioxala to zinc(II)-oxalate solutions was reinvestigated [65] by van Eldik and van den Berg [66]. They concluded that the dioxalato zinc(II) ion was photochemically inert. The observed changes in the irradiated solutions could be ascribed to the photochemical decomposition of the oxalate medium in which the complex was dissolved.

Two papers on the photochemistry of copper(II) complexes were not available for review [66, 68].

REFERENCES

1. D. A. Valentine, Jr., *Annual Surveys of Photochem.*, **1**, 459 (1969).
2. Papers published during 1969 in the following journals have been covered relatively thoroughly: *J. Am. Chem. Soc., J. Chem. Soc. A., J. Phys. Chem., Inorg. Chem., J. Chem. Phys., Chem. Commun., Chem. Titles, Can. J. Chem.* It has not been possible to examine other journals quite so systematically.

3. L. S. Forster, *Transition Metal Chemistry*, **5**, 1 (1969).
4. J. R. Perumareddi, *Coord. Chem. Rev.*, **4**, 73 (1969).
5. D. M. Hercules, *Accounts Chem. Research*, **2**, 301 (1969).
6. J. J. Weiss, *Ber. Bunsengesellschaft physik. Chem.*, **73**, 131 (1969).
7. E. J. Hart, *Accounts Chem. Research*, **2**, 161 (1969).
8. F. D. Camassei and L. S. Forster, *J. Chem. Phys.*, **50**, 2603 (1969).
9. A. D. Kirk, A. Ludi, and H. L. Schlaefer, *Ber. Bunsengesellschaft physik. Chem.*, **73**, 669 (1969).
10. F. D. Camassei and L. S. Forster, *J. Mol. Spectrosc.*, **31**, 129 (1969).
11. R. Dingle, *J. Chem. Phys.*, **50**, 1952 (1969).
12. J. C. Hempel and F. A. Matsen, *J. Phys. Chem.*, **73**, 2502 (1969).
13. F. E. Lytle and D. M. Hercules, *J. Am. Chem. Soc.*, **91**, 253 (1969).
14. D. W. Fink and W. E. Ohnesorge, *ibid.*, **91**, 4995 (1969).
15. L. A. Rosiello, *J. Chem. Phys.*, **51**, 5191 (1969).
16. I. V. Lipnitskii and D. S. Umreiko, *Zh. Prikl. Spektrosk*, **11**, 670 (1969); *Chem. Absts.* **72**, 7668U (1970).
17. L. J. Nugent, J. L. Burnett, R. D. Baybarz, G. K. Werner, S. P. Tanner, J. R. Tarrant, and O. L. Keller, Jr., *J. Phys. Chem.*, **73**, 1540 (1969).
18. H. L. Schlaefer, H. Gausmann, and C. Moebius, *Inorg. Chem.*, **8**, 1137 (1969).
19. H. L. Schlaefer, H. Gausmann, and H. Witzke, *J. Chem. Phys.*, **46**, 1423 (1967).
20. G. B. Porter, *J. Am. Chem. Soc.*, **91**, 3980 (1969).
21. V. S. Sastri and C. H. Langford, *ibid.*, **91**, 7533 (1969).
22. A. W. Adamson, J. E. Martin, and F. D. Camessei, *ibid.*, **91**, 7530 (1969).
23. S. Bagger and J. Ulstrup, *Chem. Commun.*, 1190 (1969).
24. L. Moggi, F. Bolletta, V. Balzani, and F. Scandola, *J. Inorg. Nucl. Chem.*, **20**, 2589 (1966).
25. V. Balzani and V. Carassiti, *J. Phys. Chem.*, **72**, 383 (1969).
26. V. Balzani, M. F. Manfrin, and L. Moggi, *Inorg. Chem.*, **8**, 47 (1969).
27. R. P. Mitra, B. K. Sharma, and H. Mohan, *Can. J. Chem.*, **47**, 2317 (1969).
28. W. L. Waltz and A. W. Adamson, *J. Phys. Chem.*, **73**, 4250 (1969).
29. A. A. Nemodruk and E. V. Bezrogova, *Zhur. Anal. Khim.*, **24**, 408 (1969).
30. S. Behrendt, C. H. Langford, and L. S. Frankel, *J. Am. Chem. Soc.*, **91**, 2236 (1969).
31. W. Geis and H. L. Schlaefer, *Z. Physik. Chem. (Frankfurt)*, **65**, 107 (1969).
32. E. Zinato, R. D. Lindholm, and A. W. Adamson, *J. Am. Chem. Soc.*, **91**, 1076 (1969).
33. H. F. Wasgestian, *Z. physik. Chem.*, **67**, 39 (1969).
34. A. W. Adamson, W. L. Waltz, E. Zinato, D. W. Watts, P. D. Fleischauer, and R. D. Lindholm, *Chem. Rev.*, **68**, 541 (1968).
35. H. F. Wasgestian and H. L. Schlaefer, *Z. physik. Chem.*, **62**, 127 (1968).
36. K. L. Stevenson and J. F. Verdieck, *Mol. Photochem.*, **1**, 271 (1969).
37. (a) H. Hartmann and P. Filss, *Z. Phys. Chem. (Frankfurt)*, **66**, 48 (1969); (b) *ibid.*, **66**, 60 (1969).
38. W. Jakob, T. Senkowski, J. Czaja, and D. Rudowska, *Rocz. Chem.*, **43**, 253 (1969).
39. T. Senkowski, *Rocz. Chem.*, **42**, 2007 (1969).
40. H. E. Spencer, *J. Phys. Chem.*, **73**, 23 (1969).
41. C. G. Hatchard and C. A. Parker, *Proc. Roy. Soc.*, *A*235, 518 (1954).
42. O. Traverso, F. Scandola, V. Balzani, and S. Valcher, *Mol. Photochem.*, **1**, 289 (1969).
43. J. A. van den Berg, *J. S. Afr. Chem. Inst.*, **22**, 12 (1969).
44. (a) R. G. Hughes, J. F. Endicott, and M. Z. Hoffman, *Chem. Commun.*, 195 (1969); (b) R. G. Hughes, G. Caspari, J. F. Endicott, and M. Z. Hoffman, *J. Am. Chem. Soc.*, submitted for publication.

45. A. W. Adamson, *Discussions Faraday Soc.*, **29**, 163 (1960).
46. H. Way and N. Filipescu, *Inorg. Chem.*, **8**, 1609 (1969).
47. T. Copestake and N. Uri, *Proc. Roy Soc.*, *A228*, 252 (1955).
48. J. G. Calvert and J. N. Pitts, Jr., "Photochemistry," Wiley, New York, 1967.
49. E. Papaconstantinou, J. F. Endicott, M. Z. Hoffman, and B. M. Mollicone, unpublished observations.
50. A. W. Adamson, A. Vogler, and I. Lantzke, *J. Phys. Chem.*, **73**, 4183 (1969).
51. (a) E. R. Kantrowitz, J. F. Endicott, and M. Z. Hoffman, *J. Am. Chem. Soc.*, **92**, in press (1970); (b) E. R. Kantrowitz, unpublished observations.
52. A. W. Adamson, A. Chiang, and E. Zinato, *J. Am. Chem. Soc.*, **91**, 5467 (1969).
53. R. G. Hughes, J. F. Endicott, M. Z. Hoffman, and D. A. House, *J. Chem. Ed.*, **46**, 440 (1969).
54. N. Filipescu and H. Way, *Inorg. Chem.*, **8**, 1863 (1969).
55. E. Ochiai, K. M. Long, C. R. Sperati, and D. H. Busch, *J. Am. Chem. Soc.*, **91**, 3201 (1969).
56. F. S. Kennedy, T. Buckman, and J. M. Wood, *Biochem. Biophis. Acta.* **177**, 661 (1969).
57. A. Bauer and F. Basolo, *Inorg. Chem.*, **8**, 223 (1969).
58. L. Moggi, *Gazz. Ital.*, **97**, 1089 (1967).
59. C. Sigwart and J. Spence, *J. Am. Chem. Soc.*, **91**, 3992 (1969).
60. P. C. Ford, D. H. Stuermer, and D. P. McDonald, *J. Am. Chem. Soc.*, **91**, 6209 (1969).
61. D. M. Blake and C. J. Nyman, *Chem. Commun.*, 482 (1969).
62. F. Scandola, O. Traverso, and V. Carassiti, *Mol. Photochem.*, **1**, 11 (1969).
63. C. Bartocci, F. Scandola, and V. Balzani, *J. Am. Chem. Soc.*, **91**, 6948 (1969).
64. H. Loeliger, *Helv. Chim. Acta.*, **52**, 1516 (1969).
65. K. V. Krishnamurty and G. N. Harris, *Chem. Rev.*, **61**, 213 (1961).
66. R. Van. Eldik and J. A. van der Berg, *J. S. Afr. Chem. Inst.*, **22**, 24 (1969).
67. L. A. Il'iukevic and G. A. Shagisultanova, *Khim. Vys. Energy*, **3**, 207 (1969).
68. A. L. Poznyak, *ibid.*, 380 (1969).

AUTHOR INDEX

Numbers in parentheses are reference numbers and show that an author's work is referred to although his name is not mentioned in the text. Numbers in *italics* indicate the pages on which the full references appear.

Aaronson, A. M., 172(916), *207*
Abolina, M., 231(309,310), *299*
Abrahamson, E. W., 77(402), *194*
Abraitys, V. Y., 146(705), *202*
Abram, I. I., 168(883), *207*
Abramov, A. P., 280(1044), *321*
Abramovitch, R. A., 50(236), 51(236), *190,* 164(838), 172(838), *206*
Absar, I., 220(23), *291*
Abu-Eittah, R., 224(147), 229(236), *295, 297*
Abu-Zeid, M. E. M., 273(848), *316*
Acheson, R. M., 154(754), *204*
Ackerman, J. P., 279(1020), *321*
Adam, G., 10(18), *185*
Adam, K. R., 277(949,950), 287(949,950), *319*
Adam, W., 116(563), 119(563), *199*
Adamson, A. W., 330(22), 331-336, *338, 339*
Adamson, J., 175(941), *208*
Adelman, A. H., 276(925), 285, *318*
Adomeit, M., 178(1000), *209*
Agata, I., 139(682,683), 173(922,925), *202, 207, 208*
Agosta, W. C., 69(369), 132(659), 134(668), *193, 201*
Ah, H., 221(50), *292*
Aihara, J., 216, 227(213), 275(909), *297, 317*
Aimar, N., 10(12), *185*
Akamatur, H., 239(442,443), *303*
Akbarova, D. M., 237, 240(474,475), *304*
Akermark, B., 160(792), *204*
Aktipis, S., 71(380), *194*
Albrecht, A. C., 233(361), 234(361,386), 271(361), 280(1039), *301, 302, 321*
Alexander, E., 102(483), *197*
Alexander, J. A. F., 269(792), *314*

Alexander, J. E., 160(793), *204*
Alfimov, M. V., 277(948,994,995), 288, *319, 320*
Alimarin, I. P., 228(235), 256(235), *297*
Allinger, N. L., 221(63), 223(114), *292, 293*
Allison, C. G., 183(1054), *211*
Allred, E. L., 168(885,889), *207*
Almgren, M., 259(756), 266, *313*
Al'perovich, L. I., 238(416), *302*
Alt, H., 223(99), *293*
Altaiskaya, A. V., 260(776), 266, *313*
Altman, J., 35(134,135), 36(134), *188*
Amako, Y., 231(315,316), *300*
Amano, T., 239(443), *303*
Anand, N., 92(443), *195*
Anastassiou, A. G., 40(158,159,161,162), 41(166), 42(172), 43(161), *188, 189*
Anderson, C. M., 85(427), 152(427), 171(906), *195, 207*
Anderson, D. J., 121(588), 165(588), *199*
Anderson, E. M., 271(807), 282(807), *315*
Anderson, S. R., 248(592), 251(592), 277(592), *307*
Ando, W., 176(963-966), *208, 209*
Andre, J.-C., 158(785), *204*
Andrews, S. D., 170(901), *207*
Anet, F. A. L., 151(742), *203*
Angier, R. A., 173(924), *208*
Anisuzzaman, A. K. M., 172(911), *207*
Ansari, B. J., 223(123), *294*
Anselme, J.-P., 172(910), *207*
Ao, M. S., 175(940), *208*
Aoyama, H., 9(9), *185*
Aponte, G. S., 116(563), 119(563), *199*
Appleyard, G. D., 90(442a), *195*
Arai, H., 181(1043), *210*
Arbuzov, B. A., 230(279), *298*
Archer, R. A., 10(14), *185*

Ardemagni, L., 121(589,590), *199*
Arimitsu, S., 224(142), *295*
Aristov, A. V., 270(795), 271(795), *314*
Armbruster, D. C., 131(639), *201*
Armstrong, A., 260(767), 265(767), *313*
Arnaud, R., 239(437), *303,* 175(944), 176 (944), *208*
Arndt, R. A., 280(1048), 290(1048), *322*
Arnett, G. M., 243(516), *305*
Arnold, D. R., 4, 105(493), 117(567), 168 (891), *197, 199, 207*
Artemova, L. K., 234(382), 241(382), 257 (382), 258(382), 264(382), *302*
Arventiev, B., 232(333), *300*
Asaks, J., 231(309,310), *299*
Astier, R., 234(370,383), *301, 302*
Astoin, N. D., 216, 222(70), *293*
Auge, W., 153(749), *203*
Augenstein, L., 276(940), 286, *318*
Aul'chenko, 224(131,131a), *294*
Austel, V., 49(231), *190*
Avakian, P., 280(1049,1050), 290, *322*
Avaro, M., 176(956), *208*
Avigal, I., 49(223), *190,* 257(717), *311*
Azuma, C., 114(549), 125(549), 183 (1053), *198, 211*
Azumi, H., 231(315,316), *300*
Azumi, T., 233(354), 235(354), 249, 251 (601,602,603), 253(652), 260(768,769), 262, 265, 273(354), *301, 308, 309, 313*

Baba, H., 231(318), 254(318), 271(318), *300*
Babad, E., 35(134,135), 36(134), 130(634), *188, 201*
Babko, A. K., 250(587), 275(905), *307, 317*
Bach, F. L., 175(938), *208*
Baczynski, A., 271(804c), *314*
Badcock, C. C., 85(429), *195*
Badger, B., 240(468-470), *304*
Badger, R. A., 143(694), *202*
Baessler, H., 273(857), *316*
Bag, S. C., 226(181), *296*
Bagdasaryan, K. S., 48(214), *190,* 274 (885), *317*
Bagger, S., 330(23), 331(23), *338*
Baggiolini, E., 68(368), 77(398), 78(368), 91(398), *193, 194*
Bagli, J. F., 131(654), *201*

Bailey, D. N., 231(313), *299*
Baker, C., 221(64), *292*
Baker, P. M., 40(155), *188*
Bakhshiev, N. G., 237, 241(476), 260(776), 266, *302, 304, 313*
Balbo, S. A., 260(762), 265(762), *313*
Baldwin, B. A., 270(796), *314*
Baldwin, C. M., 49, *190*
Baldwin, J. E., 33(129), 90(440), 107(516), 131(643), 157(773), *188, 195, 197, 201, 204*
Balint, E., 276(934), *318*
Balke, D. E., 52(250), *191,* 268, 269(793), *314*
Ballard, R. E., 228(234), *297*
Balny, C., 52(253), *191,* 241(481), 257 (730), 261(481), 276(929a), 285, *304, 312, 318*
Balogh, G., 164(833), *205*
Balzani, V., 330(24,25), 331(26), 333(42), 337, *338, 339*
Banfield, T. L., 185(1070), *211,* 279(1002), *320*
Bankovskis, J., 231(307-310), *299*
Bansal, R. C., 152(747), *203*
Barale, E., 180(1033), *210*
Baranac, J. M., 221(47), *292*
Baranov, S. P., 250(587), *307*
Barbarella, G., 178(985), *209*
Barber, L. L., 80(414), 142(414), 151(737), *195, 203*
Barbieri, W., 113(538), *198*
Barker, W. D., 176(959), *208*
Barlow, M. G., 49(232), 50(232), 173(929), *190, 208*
Barltrop, J. A., 86(430), 151(740), *195, 203*
Barrett, J., 221(58a), *292*
Barringer, W. C., 149(720), *203*
Bartocci, C., 337(63), *339*
Barton, D. H. R., 179(1023,1024), *210*
Barton, T. J., 53(267,269), 164(835), *191, 206*
Baruah, G. D., 226(193), 227(200), 255 (200,678,692,693), *296, 310, 311*
Basch, H., 217, 220(17), 221(64,65), *291, 292*
Basolo, F., 336(57), *339*
Bass, A. M., 221(67), *292*
Basselier, J. J., 9(8), 55(275), *185, 191*

AUTHOR INDEX

Batekha, I. G., 277(948,994,995), 288, *319, 320*
Bates, R. B., 26(103), *187*
Battisti, A., 92(452), *196*
Bauer, A., 336(57), *339*
Bauer, R. K., 251(594,595,597,598), 253 (594), 254(597), 277(595), *308*
Bauer, W., 165(856), 172(856), *206*
Baum, A. A., 104(487), *197*
Baum, E., 64(344), 96(344), *193*
Baum, E. J., 79(410), 223(105), *194, 294*
Bauser, H., 272(868), 283(868), *316*
Baybarz, R. D., 329(17), *338*
Baylor, C., 168(872), *206*
Beak, P., 5, 166(859), *206*
Beardslee, R. A., 272(842), 284, *315*
Beaumont, T. G., 241(477), *304*
Becker, H.-D., 106(497), 108(497), 148 (712), *197, 203*
Becker, R. S., 52(250,255), 268, 269(793), *191, 314*
Bedford, C. T., 150(732), *203*
Beens, H., 274(880), 275(916), 285(916), *317, 318*
Behrendt, S., 332(30), *338*
Belaits, I. L., 226(189), 229(247), 256 (247), 258(189), *296, 297*
Belanger, G., 222(72), *293*
Bell, S., 222(69), *293*
Bellamy, A. J., 131(647), *201*
Bellus, D., 4, 85(428), 96(471), 105(471), 106(471), 140(428,686), *195, 196, 202*
Belomyttsev, S. Ya., 239(434), *303*
Belotsvetov, A. V., 224(138), *295*
Belyakov, V. A., 12(35), *186*
Benati, L., 53(263), *191*
Bender, C. O., 24(96), 61(96), *187*
Bengelmans, R., 113(541), *198*
Benham, J., 132(659), *201*
Ben-Hur, E., 162(814), *205*
Benicke, D., 48(212), *190,* 235(401), *302*
Ben-Ishai, R., 162(814), *205*
Bensasson, R., 233(358), 279(1030), *301, 321*
Bentrude, W. G., 77(405), 79(411), 158 (405), *194*
Bera, S. Ch., 243(519), *305*
Berchtold, G. A., 92(447,448), 138(447), *196*
Bercovici, T., 52(251-253), 257, *191, 312*

Berems, K., 257(724), *312*
Berenfel'd, V. M., 223(115), 255(115), 270(115), 271(115), *294*
Berg, J. A., van der, 333, 337, *338, 339*
Berg, R. A., 235(403), *302*
Bergmann, E. D., 62(324), 231(306), *192, 299*
Bergmann, F., 231(306), *299*
Bergmark, W., 164(832), *205*
Berkovici, T., 111(528), *198*
Berlman, I. B., 228(233), 253(676), 257 (233), 262, 274(860,861), 278(968), 284, *297, 310, 316, 319*
Bernardi, L., 113(538), *198*
Bernat, J., 229(245), *297*
Berthod, H., 229(261,263), 231(306), *298, 299*
Bertoli, V., 223(119,122), *294*
Bertrand, M. P., 180(1028), *210*
Betz, G., 275(908), *317*
Beugelmans, R., 11(23), 21(83), 164(848), 165(848), *185, 187, 206*
Beutel, J., 156(764), *204*
Beverung, W., 177(970), *209*
Bevilacqua, R., 162(818), *205*
Bezrogova, E. V., *338*
Bhacca, N. S., 12(26), *185*
Bhandari, K. S., 177(975), *209*
Bhardwaj, I., 155(760), *204*
Bhaskar, K. R., 221(42), *292*
Bhatnagar, G. M., 276(935), 286(935), *318*
Bhujle, V. V., 256(697), *311*
Bhutra, M. P., 231(300), *299*
Bibart, C. H., 73(392,392a), *194*
Bidzilya, V. A., 49(224), *190*
Biedrzycki, J., 183(1052), 184(1059), *211*
Bienick, D., 18(70), *186*
Bienvenue, A., 137(676), *202,* 238(430), *303*
Biersmith, E. L., 86(432), *195*
Bigotto, A., 231(294), *299*
Billingsley, F. P., II, 228(223), *297*
Biloen, P., 240(471), *304*
Binder, D., 180(1037a), *210*
Bindra, A. P., 42(171), *189*
Binkley, R. W., 7(3), 23(88), 167(868-870), *185, 187, 206*
Binkley, W. W., 7(3), *185*
Birks, J. B., 233(356), 234(356), 273(850, 851), 274(859), 284(859), *301, 316*

Bishop, R. P., 159(790,791), *204*
Bixon, M., 248, 250(576,577), *307*
Black, E. D., 178(1006), *209*
Blackburn, E. V., 4, 57(293), *192*
Blake, D. M., 336, *339*
Blanchi, J. P., 106(512), 108(512), 256 (706), 278(989), *197, 311, 320*
Blind, A., 181(1040), 183(1040), *210*
Block, E., 10(16), 40(16), 55(274), *185, 191*
Blomquist, A. T., 165(852,854), *206*
Bloomfield, J. J., 35(133), *188*
Bloor, J. E., 179(1010), 228(223), *210, 297*
Bluhn, A. L., 178(993), *209*
Blum, J., 62(324), *192*
Bobrovich, V. P., 241(484), 258(484), *304*
Bock, H., 221(49,50), 223(99), *292, 293*
Bodor, N., 226(185,186), *296*
Bogomolov, S. G., 258(747), 265(747), *312*
Bogri, T., 131(654), *201*
Boikess, R. S., 40(154), *188*
Boire, B. A., 131(655), 143(655), *201*
Bokranz, H., 180(1036), *210*
Boldt, P., 12(27), *185*
Boleij, J., 111(527), *198*
Bolesov, I. G., 44(174), *189*
Bolletta, F., 51(239), 330(24), *191, 338*
Bolotin, A., 242(507), *305*
Bolotin, B. M., 256(708), *311*
Bolotnikova, T. N., 223(124), 234(382,392, 405), 241(382,392,405,488,490), 257 (382), 258(124,382,392,405,490), 264 (382,392), 280(1037), *294, 302, 305, 321*
Bond, F. T., 15(56), *186*
Bonneau, R., 233(358), *301*
Bonnier, J. M., 239(437), 256(706), 278 (989), *303, 311, 320*
Boocock, D. G. B., 178(988b), *209*
Borden, W. T., 245(546), *306*
Bordin, F., 162(818), *205*
Borello, E., 228(229), *297*
Borisevich, N. A., 253(639), 255(691), 271 (639), 281(639), *309, 310*
Borkman, R. F., 72(389), *194*
Bortner, M. H., 276(929b), *318*
Bortolus, P., 57(298,299), *192*
Bos, H. J. T., 111(527), *198*
Bosco, S. R., 120(579), *199*
Bosisio, G., 113(538), *198*

Bossa, M., 220(32), 222(87), *291, 293*
Bossenbroek, B., 63(337), *193*
Bouas-Laurent, H., 64(345-347), *193*
Bouchy, M., 158(785), *204*
Bourdon, J., 276(929a), 285, *318*
Bouyer, M., 260(764), *313*
Bowman, R. M., 51(247), *191*
Boyer, J. H., 172(912), 173(934,934a), 177 (970), *207, 208, 209*
Boyle, P. H., 126(616), *200*
Bozzato, G., 141(691), *202*
Bradley, J. N., 219, *291*
Bradshaw, J. S., 119(578), 125(605), *199, 200*
Brady, T. E., 185(1071), *211*
Braga, C. L., 253(634), 269(634), *309*
Brand, L., 260(777), 266, 271(777), *313*
Braun, A. M., 27(107), *187*
Braun, J., 256(689), *310*
Bree, A., 242(502,503), 243(521), 272 (834), 283(834), *305, 315*
Bremmer, J. B., 85(427), 152(427), *4, 195*
Bremplis, R. V., 275(914), 285(914), *317*
Brennan, J. F., 156(764), *204*
Breslow, R., 110(526), *198*
Breuer, G. M., 252(625), 271(806), 282 (806), *308, 315*
Briegleb, G., 256(711), 275(908), 276(922), 285(922), *311, 317, 318*
Brienen, J. S., 252(623), 262, 271(623), 282, *308*
Bril, A., 250(588), *307*
Brinen, J. S., 233(353,365,366), 234(365, 366,369), 235(365,366), 244, 245(552a), 255(369), *301, 306*
Brinkley, R. W., 164(839,840), *206*
Brizzolara, D. F., 142(692), *202*
Brocklehurst, B., 240(468,470), *304*
Brody, S. S., 241(480,481), *304*
Bromberg, A., 57(296), *192*
Brook, A. G., 77(398a), *194*
Brook, P. R., 176(957), *208*
Brophy, B., 176(957), *208*
Broser, W., 219(139), 224(139), *295*
Broude, V. L., 243(512), *305*
Brown, D. J., 228(230), *297*
Brown, I., 173(923), *207*
Brown, I. H., 271(815), *315*
Brown, J. E., 157(773), *204*
Brown, J. M., 231(304), *299*

AUTHOR INDEX

Brown, P. R., 10(15), *185*
Broyde, S. B., 241(480), *304*
Bruce, J. M., 156(767), *4, 204*
Brudz, V. G., 256(708), 277(944), 287, *311, 318*
Brunn, E., 122(596), *199*
Bruylants, A., 239(441), *303*
Bryce-Smith, D., 44(176), 46(191), 155(759), *189, 204*
Buch, H. M., 222(91), 293
Buchachenko, A. L., 178(988), *209*
Buchardt, O., 181(1044,1046), 182(1044), *210*
Bucheck, D. J., 131(648,656,657), 132(648,656), 162(657), *201*
Buchecker, C., 169(894), 170(902), 175(902), *207*
Buchi, G., 129(626), *200*
Buckman, T., 336(56), *339*
Buchnev, I. F., 180(1030), *210*
Budhiraja, R. P., 179(1024), *210*
Buduls, I., 255(674a), *310*
Buemi, G., 230(271), *298*
Buenker, R. J., 217(102), 220(19), 223(102), *291, 293*
Buettner, A. V., 272(816), *315*
Bull, J. G., *4*
Bulten, E. J., 221(52), *292*
Bunau, G., van, 106(506), *197*
Bunbury, D. L., 160(794), *204*
Bunce, N. J., 12(33), *186*
Bunting, J. R., 51(241), *191*
Burger, M., 150(733), *203*
Burgess, E. M., 175(940), *208*
Burka, L. T., 42(167), 94(167), *189*
Burland, D. M., 272(833), 283, *315*
Burnett, J. L., 278(978), 329(17), *319, 338*
Burton, M., 278(978), *319*
Buryak, V. Yu., 225(168), 232(319-321), *295, 300*
Buryakovskaya, E. G., 230(289), *299*
Busch, D. H., 335, *339*
Buschhoff, M., 175(942,943), *208*
Busev, A. I., 229(238,244), *297*
Butlar, V. A., 241(492), *305*
Butler, J. M., 8(6), *185*
Butucelea, A. L., 242(505), *305*
Byers, G. W., 237, *302*
Bylina, A., 57(285), *192.*
Byrne, J. P., 251(600), *308*

Byrne, M. M., 225(152), *295*
Bystritskii, G. I., 229(247), 256(247), *297*

Cadogan, K. D., 234(386), *302*
Caille, G., 256(689), *310*
Cain, E. N., 116(561), *199*
Caird Ramsay, G., 108(517), *197*
Caldwell, R. A., 12(47), 34(132), 97(476), 98(476), 107(47), 111(532), 274(874), 278(981), 288, *186, 188, 196, 198, 316, 320*
Callear, A. B., 221(58), *292*
Callis, J., 248, 250(583), 254(583), *307*
Calvert, J. G., 334(48), *3, 339*
Calvin, M., 240(459), *304*
Camaggi, G., 46(190), *189*
Camassei, F. D., 327, 328(10), 330(22), 331(22), *338*
Camerman, N., 162(813,815a), *205*
Campbell, T. C., 46(185), 61(185), *189*
Canada, R., 276(940), 286(940), *318*
Canonica, L., 151(735,738), *203*
Cantrell, T. S., 135(669), *201*
Canva, J., 276(929a), 285, *318*
Capette, J., 246(562), *307*
Capon, B., *4*
Carapelluci, P. A., 111(531), 160(797), 234(395), *198, 205, 302*
Carassiti, V., 330(25), 337, *338, 339*
Cargill, R. L., 131(651), 132(651,658), 278(984), *201, 320*
Carlson, R. G., 86(432), 93(455), *195, 196*
Carlsson, D. J., 9(10), 75(397), 182(10), *185, 194*
Carnduff, J., 149(726), *203*
Caro, E., 269(792a), *314*
Carper, W. R., 229(258), *298*
Carruthers, R. A., 13(51), *186*
Carson, M. S., 88(434), *195*
Carson, S. D., 228(219), 233(219), *297*
Carter, J. G., 273(848), *316*
Cartwright, G. J., 222(69), *293*
Carty, D. T., 8(4), *185*
Caserio, M. C., 104(492), *197*
Casida, J. E., 116(559a), *198*
Caspari, G., 334(44b), *338*
Cassal, J.-M., 181(1040), 183(1040,1057), *210, 211*
Cassidy, H. G., 27(107), 225(157), *187, 295*
Castanzo, L. L., 164(841), *206*

Castellan, A., 64(345,347), *193*
Castro, G., 233(351), 235(351), 272(833), 283, *301, 315*
Catalan, A. W., 184(1062), *211*
Cauzzo, G., 57(298,299), *192*
Cavalieri, E., 131(636), 142(636), *201*
Cecchet, G., 216, 221(66), *292*
Cellura, R. P., 40(162), 42(172), *189*
Cerutti, P. A., 163(826), *205*
Chabaud, J. B., 38(147), 141(688), *188, 202*
Chachaty, C., 180(1029), *210*
Chadha, V. K., 164(847), *206*
Chakrabarti, S. K., 105(493a), 261(784a), 266, 272(840), 284, *197, 314, 315*
Chakrovorty, S. C., 243(522), *305*
Challard, B. D., 96(464), 134(666), *196, 201*
Chambers, R. D., 183(1054), *211*
Chambers, R. W., 256(710), *311*
Chandra, A. K., 273(843,844), 284, *316*
Chapman, O. L., 64(341), 80(414), 88 (436), 96(463), 108(463), 117(566), 135 (436), 142(414), 144(698,699), 151(737), 180(1027), 278(988), *195, 196, 199, 202, 203, 210, 320*
Charlier, M., 162(816), *205*
Charney, E., 158(780a), 254(673), 271 (673), 281, *204, 310*
Chary, C. A. I., 239(451), *304*
Chatterjee, S., 239(439), *303*
Chauduri, M. K., 243(518), 272(518), *305*
Chauncy, B., 57(300), *192*
Cheburkov, Y. A., 183(1054), *211*
Chekalin, M. A., 226(190), *296*
Chelnokova, Z. B., 110(525), *198*
Chen, C. Y. S., 221(40), *292*
Chen, H. H., 243(529), *306*
Chen, T.-H., 253(648), 269(648), *309*
Chen, Y. L., 116(559a), *198*
Cheng, K. F., 148(719), *203*
Cheng, L.-T., 221(45), *292*
Cherkashin, A. E., 222(80), *293*
Cherkasov, A. S., 257(728), 263(728), 274 (875,879), 275(902), 285(902), *312, 316, 317*
Cherton, J.-C., 55(275), *191*
Chiang, A., 355(52), *339*
Chibisov, A. K., 235(408), *302*
Chirkinyan, S. S., 277(959), *319*

Chock, D. P., 272(826), 283, *315*
Choi, S. K., 15(60), *186*
Chojnacki, H., 229(253), 241(493), 258 (493), 259(493), 264, *298, 305*
Chopin, C. M., 53(262), 275(913), 285 (913), *191, 317*
Chow, L. C., 104(492), *197*
Chow, P., 269(793a), 280, *314*
Chow, Y. L., 178(989,991), *209*
Chowdhury, M., 243(519), *305*
Christie, J., 60(318), *192*
Christophorou, L. G., 273(848), *316*
Chryan, M. B., 277(959), *319*
Chuang, T. T., 160(794), *204*
Ciabattoni, J., 131(642), *201*
Cicolella, A., 158(785), *204*
Cirillo, A., 120(579), *199*
Ciullo, G., 220(32), *291*
Clark, L. B., 243(529), *306*
Clark, R. H., 243(528), *306*
Clark, W. D. K., 96(460,461), 106(460, 461), *196*
Closs, G. L., 106(502), 108(502), 168(877), 171(877,903,908), 177(908), *196, 206, 207*
Closs, L. E., 106(502), 108(502), 171(908), 177(908), *197, 207*
Coates, E., 237(472), 240(472), 261(472), *304*
Cocivera, M., 278(979), 288, *319*
Cocker, W., 88(434), *195*
Cohen, E., 149(724), *203*
Cohen, J. I., 106(508), *197*
Cohen, M. D., 64(340), *193*
Cohen, S. G., 71(380), 96(466), 106(501, 508,513,514), 108(501,513,514), 111 (533), 274(873), 275(910), *194, 196, 197, 198, 316, 317*
Cole, R. S., 278(980), 288, *320*
Cole, T., 80(415), *195*
Collier, J. R., 104(488), *197*
Collin, P. J., 53(265), *191*
Collins, P. M., 92(445a,446), *195*
Compaignon, H., de Marcheville, 11(23), *185*
Condorelli, G., 164(841), *206*
Conia, J. M., 127(620), *200*
Connor, J., 221(58), *292*
Conrad, R. M., 244(537), 246, *306*
Conrow, K., 94(458a), *196*

AUTHOR INDEX

Conte, J. C., 278(973), *319*
Converse, C. A., 176(951), *208*
Cook, G. R., 222(68), 254(68), *292*
Cookson, R. C., 51(242), 52(258), *191*
Cooper, J. L., 167(863), *206*
Cope, A. C., 26(102), *187*
Copen, G., 97(475), 99(475), 100(475), *196*
Copestake, T., 334(47), *339*
Corey, E. J., 10(16), 40(16), 55(274), *185, 191*
Corker, G. A., 240(459), *304*
Cornelisse, J., 179(1011), *210*
Cortes, L., 16(64), *186*
Costa, S. M., deB, 52(258), *191*
Cotruvo, J. A., 177(976), *209*
Cotter, R. J., 68(368a), *193*
Cottrell, C., 126(617), 128(617), *200*
Courtney, J. L., 31(122), *188*
Courtot, P., 55(276), *191*
Couture, A., 42(169), *189*
Cowan, D. O., 104(487), *197*
Cowgill, R., 257(713), 263, *311*
Cowley, D. J., 238(424), *303*
Cox, O., 93(456), *196*
Coyle, J. D., 86(430), *195*
Cozzens, R. F., 63(338), *193*
Crabbe, P., 126(616), *200*
Craig, D. P., 227(203), 242(495), 252(203), 262, *296, 305*
Cram, D. J., 45(182), *189*
Crandall, J. K., 14(55), 78(406), *186, 194*
Creagh, L., 177(978), *209*
Creed, D., 156(767), *204*
Creemers, H. M. J. C., 221(52), *292*
Crellin, R. A., 13(51), *186*
Cremer, S. E., 69(370), *193*
Cridland, J. S., 178(994), *209*
Criegee, R., 8(5), *185*
Cristol, S. J., 14(53,54), 25(99), 45(99), *186, 187*
Crosby, D. G., 96(459b), 123(599a), *196, 200*
Crozet, M. P., 10(11,12), *185*
Crumrine, D. S., 146(706), *202*
Cruz, A., 126(616), *200*
Crysochoos, J., 240(463), *304*
Csizmadia, I. G., 220(26), 222(95), *291, 293*

Cundall, R. B., 12(43), 158(781), 273(852), 278(977), *4, 186, 204, 316, 319*
Cuong, N. Y., 8(7), *185*
Curran, W. V., 138(679), 149(679), 173(924), *202, 208*
Currie, D. J., 218, 226(173), *293, 296*
Curry, H. M., 63(337), *193*
Curtis, H. C., 160(796), 161(796), *205*
Cuetanovic, R. J., 13(50), *186*
Czaja, J., 333(38), *338*
Cziesla, M., 106(506), *197*

Dacre, P. D., 242(495), *305*
D'Agostino, J. T., 239(435), *303*
D'Alagni, M., 230(278), *298*
Dall'Acqua, F., 162(817), *205*
Dalton, D. R., 177(972), *209*
Dalton, J. C., 77(403), 172(915), 278(970), *5, 194, 207, 319*
Daltrozzo, E., 230(276), *298*
Damask, A. C., 280(1048), 290, *322*
Dandegaonker, S. H., 238(426), *303*
Daniel, B., 113(543), 120(543), 151(735, 738), *198, 203*
Danilova, V. I., 239(434), *303*
Danks, L. J., 179(1023), *210*
Danoy, J. P., 233(350), 235(350), 273(350), 283(350), *301*
Darnall, K. R., 77(405), 158(405,788), 161(788), *194, 204*
Das, B., 229(260), *298*
Das, B. K., 229(249), *297*
Das, B. P., 179(1009), *210*
Datta, P., 40(154), *188*
Dauben, W. G., 29(114), 69(373), 78(407, 408), 88(408), 137(675), *187, 194, 202*
David, C., 107(515), *197*
Davidov, O. S., 237, 242(506), *305*
Davidson, R. S., 48(207), 62(333), 64(333), 106(498,500,509), 107(207), 108(522), 111(509,522), 168(875), 274(891), *190, 193, 197,* 206, 317
Davis, A. J., *4*
Davis, C. B., 131(640), *201*
Davis, G. A., 108(519), 111(519,531), 274(887), 285, *197, 198, 317*
Davis, K. E., 78(409), *194*
Davis, K. M. C., 241(477), 275(907), *304* 317
Dawes, K., 156(767), *204*

Dawson, W. R., 66(360,361), 253(640-646), 270(642-646), 271(640-643,645, 646), 281(640,641), *193, 309*
Day, A. C., 168(892), 169(892), 170(900, 901), 275(897), 285, *4, 5, 207, 317*
Day, J. L., 47(193), *4, 189*
Day, M., 219(143), 224(143), *295*
De Alti, G., 230(272), 231(294), *298, 299*
Deblecker, M., 64(351), *193*
DeBoer, C., 51(245), 273(855), *191, 316*
DeBoer, T. J., 178(992), 223(125), 239(436), 240(456), *209, 294, 303, 304*
DeBoer, Th. J., 177(977), 178(984), *209*
Deglise, X., 158(785), *204*
De Groot, M. S., 245(553,554), *306*
DeGunst, G. P., 57(295), 58(295), *192*
Dehler, J., 217, 227(210), 228(221), 232(322), 255(210), 256(221,322), *291, 297, 300*
Dekker, E. E. J., 178(992), *209*
Delaney, A. D., 226(173), *296*
Del Bene, J., 218(25), 220(25), *291*
Delibas, V., 232(333), *300*
Delibas, M., 232(333), *300*
Del Re, G., 228(224), *297*
Demarco, P. V., 10(14), *185*
Demarteau, W., 107(515), *197*
De Mayo, P., 278(984), *320*
Demeyer, D. E., 57(287), *192*
Demidov, K. B., 275(915), 285(915), *318*
Denis, A., 229(261), *298*
Denison, A. B., 245(549,550), *306*
DeSchryver, F. C., 155(760), *204*
Detsina, A. M., 156(767a), 157(774), *204*
Devaquet, A., 141(689), *202*
Deviny, E. J., 78(408), 88(408), *194*
DeVrieze, J. D., 37(145), *188*
Dexter, D. L., 277(942), 287, *318*
DeZwaan, J., 53(266), *191*
Dheer, M. K., 255(686), *310*
Dhingra, R. C., 254(666,667), 263, *310*
Dickson, D. R., 221(58), *292*
Diebert, C., 81(417), *195*
Dietz, F., 229(252), 233(252), *4, 298*
Dilling, W., 12(40), 29(40), *4, 186*
Dilling, W. L., 19(77), *187*
DiLonardo, G., 226(175), *296*
DiMarco, P. G., 51(239), *191*
Dingle, R., 328, *338*
Dirkx, I. P., 239(436), 240(456), *303, 304*

Dmitrievskii, O. D., 235(402), *302*
Dodonova, N. Ya., 227(215), 255(215), *297*
Doerr, F., 223(106), *294*
Dogliotti, L., 178(1006), *209*
Dokunikhin, N. S., 227(196,197), 255(197, 694), *296, 311*
Dolan, E., 52(250), 268, 269(793), *191, 314*
Dolbier, W. R., Jr., 181(1039,1039a), *210*
Dolby, L. J., 279(1026), *321*
Dolganov, V. K., 272(827), *315*
Domb, S., 141(691), *202*
Dombi, J., 277(966), 286, *319*
Dombrowski, L. J., 111(529,530), *198*
DoMinh, T., 115(555), *198*
Donckt, E. V., 245(541b), 257(715,718), *306, 311*
Donovan, R. J., 274(892), *317*
Dopp, D., 146(706), *202*
Dopper, J. H., 53(264,266), *191*
Doppler, T., 174(937), *208*
Dorer, F. H., 168(893a), *207*
Dorie, J. P., 220(38), *292*
Dorignac, D., 259(755), 265, *313*
Dornand, J., 126(608), *200*
Dotty, J. C., 58(314), *192*
Doty, J. C., 184(1060), 257(720), *211, 311*
Dougherty, R. C., 126(617), 128(617), *200*
Dourlent, M., 162(816), *205*
Douzon, P., 52(253), *191*
Douzou, P., 237, 240(472a), 257(730), 261(472a), 276(929a), 285, *304, 312, 318*
Dows, D. A., 244(537), 246, *306*
Dranik, L. I., 225(171), *295*
Drapkina, D. A., 256(708), *311*
Drent, E., 271(809), 281, *315*
Drenth, W., 221(52), *292*
Drexhage, K. A., 247, *307*
Drisko, R. L. E., 62(330), 72(330), *193*
Droste, W., 51(243), 130(243), *191*
Dubois, J.-E., 137(676), 238(430), *202, 303*
DuBois, J. P., 164(843), *206*
Dubosc, J.-P., 180(1038), *210*
Dubose, J.-P., 178(1004), *209*
Duff, J. M., 77(398a), *194*
Dumenil, G., 52(249), *191*
Duncan, C. K., 276(927), *318*
DuPont, R., 8(7), *185*
Dupuy, A. E., Jr., 120(583,584), 121(583), *199*

AUTHOR INDEX

Dupuy, C., 10(12), *185*
Durocher, G., 242(497,498), 258(497,498, 751), *305, 312*
Durr, H., 57(305), 133(661), 175(948-950), *192, 201, 208*
Dusenbery, R. L., 96(467), 106(467), 255 (677), *196, 310*
Dvbvig, D. H., 180(1034), *210*
Dwivedi, C. P. D., 224(141), 225(162), *295*
Dyadyusha, G. G., 229(240), *297*
Dym, S., 270(803), 282(803), *314*
Dziedsic, J. E., 221(53), *292*

Eastman, D., 102(484), *197*
Eastman, J. W., 257(725), 269(794a), 271 (725), *312, 314*
Eaton, P. E., *4*
Eberbach, W., 152(744), *203*
Eckhard, I. F., 23(92), *187*
Eckroth, D. R., 180(1035), *210*
Edel'mann, T. G., 226(178), *296*
Edilashvili, I. L., 274(875,879), *316, 317*
Edman, J. R., 17(69), 23(90), 45(69), 64 (90), 278(983), 281(983), 288, *186, 187, 320*
Edward, J. T., 231(293), *299*
Edwards, J. A., 126(616), *200*
Edwards, J. O., 10(15), *185*
Edwards, L., 221(61), *292*
Edwards, O. E., 173(923), *207*
Edwards, T. G., 220(22), *291*
Efimov, A. A., 224(144), *295*
Ege, S. N., 122(592), 156(592), *199*
Eguchi, S., 148(718), *203*
Egupova, L. M., 260(773), *313*
Ehret, P., 245(551), *306*
Ehrig, B., 176(953a), *208*
Eichen, K. R., 121(587), 160(798), *199, 205*
Eiduss, J., 229(246), 232(329), *297, 300*
Eilerman, R. G., 132(659), *201*
Eisenhardt, W., 96(470), 100(470), *196*, 276(919), 285(919), *318*
Eisenthal, K. B., 277(943), 288, *318*
Eisinger, J., 162(812), *205*, 276(937,938), 286(937,938), *318*
Eizember, R. F., 26(104), 37(104), 93 (453), 94(456a), 142(453), *187, 196*
Ejder, E., 248, 250(582), *307*
Elad, D., 4, 5, 71(385,386), 163(827,828), 184(827,828,1063), *194, 205, 211*
El-Bayoumi, M. A., 231(305), 257(305), *299*
El-Bayounu, M. A., 184(1061), *211*
El Dbik, U., 229(238,244), *297*
Elgendi, S., 224(147), 229(236), *295, 297*
Elix, J. A., 42(171), *189*
El-Kareh, T. B., 272(829), *315*
Elliott, S. P., 73(393,394), 74(394), *194*
Ellison, G. B., 221(65), *292*
El-Sayed, M. A., 233(349), 235, 246(568), 249, 251(349), 251(611), 253(653), 260 (766,767,769a), 262, 265, 270(802), 271 (812,813), 280(1042), 282(802,812,813), *301, 307-309, 313, 314, 315, 321*
Elser, W. R., 137(673), *202*
Enderer, K., 170(898), *207*
Endicott, J. F., 333(44a), 334(44b,49), 335 (53), *338, 339*
Engbuts, J. B. F. N., 178(992), *209*
Engel, P. S., 84(422), 168(890), 169(890), *195, 207*
Engel, R., 158(782-784), *204*, 274(889, 890), *317*
Englman, R., 268, 269(789), *314*
Enisor, E. T., 46(188), *189*
Epishina, L. V., 230(284), *299*
Erastov, O. A., 230(279), *298*
Erb, R., 179(1018), *210*
Eremenko, S. M., 255(694a), *311*
Erlitz, M. D., 253(636), 259(754), 271 (754), *309, 313*
Erman, W. F., 135(671), 136(671), *201*
Ermolaev, V. L., 271(810), 278(967), *315, 319*
Ern, V., 273(837), *315*
Ero-Gecs, M., 279(1018), *321*
Etzemuller, J., 12(27), *185*
Evani, S., 170(895), *207*
Evans, B. E., 184(1064), *211*
Evans, G. B., 158(781), *204*
Evans, S. M., 88(434), *195*
Evans, T. R., 116(559c), *199*, 275(896), 279(1017,1024), 289, *317, 321*
Evnin, A. B., 168(891), *207*
Ewald, M., 251(614), *308*

Fabian, J., 220(30,31), 228(228), 232(337, 338,340,341), *291, 297, 300*
Fabian, K., 232(338), *300*

Fadeeva, M. S., 251(608), *308*
Faevskii, A. S., 234(391), *302*
Fahr, E., 4, 162(808), *205*
Faidysh, A. N., 234(391), 275(904), 280 (1035), *302, 317, 321*
Fain, V. Ya., 227(199), *296*
Fallis, A. G., 83(421), *195*
Fan, C. Y., 242(503), *305*
Fanconi, B. M., 242(510), *305*
Farid, A. M., 172(914), *207*
Farid, S., 57(297), 105(495), 160(495,800), *192, 197, 205*
Favini, G., 218, 222(81a), 223(114a), 230 (217,271), *293, 294, 297, 298*
Fayadh, J. M., 155(761), *204*
Feairheller, S. H., 129(626), *200*
Fedoseeva, G. E., 258(745), *312*
Fedotova, L. A., 239(450), *303*
Feitelson, J., 49(223), *190,* 257(717), 276 (939), 286(939), *311, 318*
Feler, G., 29(112), *187*
Fenina, N. A., 258(741), *312*
Fenselau, C., 137(675), *202*
Fenton, D. E., 224(129), *294*
Ferree, W. I., 44(177), *189*
Ferree, W. I., Jr., 274(893), 285, *317*
Ferris, J. P., 156(763), 184(1062), *204, 211*
Fesenko, E. E., 224(133), 234(398), 254 (133), *294, 302*
Fey, L., 226(185,186), *296*
Filipescu, N., 51(241), 103(486), 106(510), *191, 197,* 279(1014,1015), 289, 334, 335, *321, 339*
Filippova, L. I., 230(284), *299*
Filss, P., 333, *338*
Findlay, D. M., 173(932), *208*
Finger, G., 279(1027), 290, *321*
Fink, D. W., 328, *338*
Fink, P. M., 97(476), 98(476), *196*
Finke, M., 46(189a), *189*
Finkel'shtein, A. V., 238(418a-421), *303*
Finucane, B. W., 11(24), 141(687), *185, 202*
Fisch, M., 141(690), 148(714), *202, 203*
Fisch, R. H., 104(492), *197*
Fischer, E., 4, 52(251-253), 57(291), 111 (528), *191, 192, 198, 257, 312*
Fischer, G., 57(291), *192,* 242(501), *305*
Fischer, M., 113(542), 121(542,591, 595, 595a), 122(542,591,595,595a), 174 (595a), *198, 199*
Fischer, P. H. H., 245(549,550), *306*
Fischler, H. M., 18(72), 19(72), *186*
Fish, G. B., 222(69), *293*
Fisher, G. S., 36(140), *188*
Fishman, M., 143(695), *202*
Fitzek, A., 134(662a), *201*
Fitzgerald, E. A., 46(189), *189*
Fleischaver, P. D., 333(34), 334(34), 336 (34), *338*
Fleming, J. C., 177(971), *209*
Fletcher, A. N., 250(585), *307*
Flowerday, P., 173(935), *208*
Floyd, J. C., 146(707), *202*
Floyd, T. C., 131(645), *201*
Fokin, E. P., 156(767a), 157(774), *204*
Fonken, G. F., 4
Fonken, G. J., 32(124), 36(139), *188*
Forbes, E. J., 5
Forchioni, A., 180(1029), *210*
Ford, P. C., 154(757), *204,* 336(60), *339*
Forman, A., 244, 246(572), *307*
Forshult, S., 178(980-982), *209*
Forster, D. L., 175(941), *208*
Forster, L. S., 325, 327, 328(10), *338*
Forster, T., 66(357), *193*
Förster, T., 261(780), 273(846), 276(930), 277(942), 284(286, 287, *314, 316, 318*
Forsythe, G. D., 26(103), *187*
Forward, G. C., 126(619), 127(619), *200*
Foster, J. P., 179(1014), *210*
Fox. B. L., 164(845), 165(845), *206*
Fox, R. B., 63(338), *193*
Frackowiak, D., 232(328), 261(328), *300*
Fraenkel, G., 126(617), 128(617), 168 (876), *200, 206*
Franck-Neuman, M., 168(886), 169(894), 170(902), 175(902), *207*
Francois, H., 15(58), *186*
Franke, W. H., 48(217), *190*
Frankel, L. S., 332(30), *338*
Frankevich, E. L., 280(1045), *321*
Frank-Kamenetskii, M. D., 217, 218, 220 (20), 250(8), *291*
Franks, L. A., 231(304a), *299*
Fransen, T., 240(471), *304*
Frasca, A. R., 166(861), *206*
Fraser, R. R., 175(939a), *208*
Fraser-Reid, B., 164(834), *205*

AUTHOR INDEX

Freed, K. F., 248, 250(578), *307*
Freeman, J. J., 238(428), *303*
Freeman, P. K., 165(853), 167(867), *206*
Frejaville, C., 220(39), *292*
Fridman, S. G., 229(241), *297*
Fried, J. H., 126(616), *200*
Friedman, I., 46(184), *189*
Friedrich, L. E., 126(615), *200*
Friedricksen, W., 160(801), *205*
Friend, E. W., Jr., 56(283), *191*
Fritschi, G., 118(573), *199*
Fritz, K., 227(210), 255(210), *297*
Frolova, E. K., 237, 239(447), 243(524), *303, 305*
Frunze, N. K., 122(593), *199*
Fuchs, B., 85(425,426), *195*
Fuhr, H., 276(929), *318*
Fujisaki, Y., 280(1041), *321*
Fujishiro, R., 229(250), *298*
Fujita, S., 52(248), *191*
Fukano, K., 12(44,45), 44(175), *186, 189,* 278(990,991), *320*
Fukui, K., 30(120), 113(548), *188, 198*
Fukunaga, J. Y., 32(126), *188*
Fulton, A., 225(159), *295*
Funabashi, K., 278(978), *319*
Furutachi, N., 139(684,685), *202*
Furutani, Y., 226(194), 254(194), *296*

Gachkovskii, V. F., 260(772), 266, *313*
Gacoin, P., 259(757), 266, *313*
Gaevskii, A. S., 280(1035), *321*
Galasso, V., 230(272), 231(294), *298, 299*
Gale, D. M., 29(115), *187*
Galearschi, A., 232(333), *300*
Galiazzo, G., 53(270), 57(299), *191, 192*
Gallivan, J. B., 226(187), 234(369), 252 (623,624), 255(187,369), 262, 271(623), 282, *296, 301, 308*
Gamba, A., 223(112,114a), *294*
Gandhi, R. P., 164(847), *206*
Ganguly, S. C., 243(518,522), 272(518), *305*
Gano, J. E., 113(539), *198*
Ganter, C., 92(449-451), *196*
Gardner, P. D., 28(110), *187*
Gare, C. L., 109(523), *197*
Garg, C. L., 255(683), *310*
Garin, D. L., 116(559), *198*
Garrett, J. M., 32(124), *188*

Gase, R., 222(97), 258(736), 264, *293, 312*
Gati, L., 252(620), *308*
Gauri, K. K., 163(831), *205*
Gausmann, H., 329, *338*
Geacintov, N., 272(832), 280(1051,1052), 290, *315, 322*
Geacintov, N. E., 242(504), *305*
Gebrian, J. H., 40(158,159,161), 41(166), 43(161), *188, 189*
Geis, W., 332, *338*
Gelbart, W. M., 57(289), *192*
Gellert, E., 57(300), *192*
Gennari, G., 53(270), *191*
Georghiou, S., 277(962), 287, *319*
Gerbaux, X., 243(513), *305*
Gerhold, G. A., 242(510), *305*
Gerkin, R. E., 245(556,557), *306*
Gerlock, J. L., 178(997), *209*
Getoff, N., 7(2), *185*
Geushens, G., 107(515), *197*
Gevorkyan, V. A., 279(1001), *320*
Ghosh, N. R., 176(952,961), *208*
Ghosh, P. B., 228(230), *297*
Ghoshal, C. R., 176(952,961), *208*
Gianni, M. H., 276(932), *318*
Gianturco, F. A., 222(87), *293*
Gibson, T. W., 135(671), 136(671), *201*
Gierke, T. D., 252(630), *309*
Gilbert, A., 46(191), *189*
Gilbert, R., 176(959), *208*
Gilchrist, T. L., 121(588), 165(588), 175 (941), *199, 208*
Giles, D., 151(740), *203*
Giles, W. B., 160(799), *205*
Gill, J. E., 250(586), *307*
Gillan, T., 168(878), *206*
Gillespie, J. P., 153(751,753), *203, 204*
Gillis, L. B., 105(493), *197*
Ginistry, J. C., 163(830), *205*
Ginsburg, D., 35(134,135), 36(134), *188*
Gitis, S. S., 224(149), *295*
Givens, R. S., 23(88), 40(153), 117(570), *187, 188, 199*
Gladchenko, L. F., 261(782), *314*
Glemser, O., 12(29), *186*
Glockner, E., 272(831), 283(831), *315*
Gloria, H. R., 12(32), *186*
Glyadkouskii, V. I., 233(368), 236, 241 (478,485), 258(485), 261(478), 264,

301, 304
Gmiro, V. E., 448(221), *190*
Gobov, G. V., 228(231), 233(355), 241 (231,355), 257(731,732), 258(231), 259 (731,732), 264(731,732), *297, 301, 312*
Gochev, A., 244(535), *306*
Godard, A., 151(741), *203*
Goe, G. L., 26(102), *187*
Goedicke, C., 62(329), *193*
Gohlke, J. R., 260(777), 266(777), 271 (777), *313*
Goldberg, M. C., 233(347), *301*
Golde, G., 70(379), *194*
Goldfarb, T. D., 40(154), *188*
Goldfinger, P., 34(130), *188*
Goldschmidt, C. R., 274(860,861), 276 (923), 278(968,1005), 284, 285(923), 287, *316, 318-320*
Goldschmidt, Z., 144(696,697), *202*
Golovina, A. P., 228(235), 256(235), *297*
Golson, T. H., 120(583), 121(583), *199*
Golub, M. A., 12(30,49), 72(388), *186, 194*
Golubchik, V. Sh., 224(149), *295*
Gondo, Y., 245(544,545), *306*
Goode, D. H., 280(1046,1047), 290(1046, 1047), *322*
Goodman, L., 232(346), 233(346), *301*
Gopalakrishnan, K., 274(865), *316*
Gorman, A. A., 16(63), 152(63), *186*
Goto, T., 4, 162(809), *205*
Goud, A. N., 232(331), *300*
Gouesnard, J. P., 220(38), *292*
Goumet, G., 241(479), *304*
Gouterman, M., 218, 230(274), 248, 250 (583), 254(583,664), *298, 307, 310*
Govindjee, 252(617), 262(617), *308*
Gozzo, F., 46(190), *189*
Grabbe, R. R., 28(110), *187*
Grabner, D. R., 92(444), *195,* 246(564, 565), *307*
Grabowska, A., 257(721), *311*
Grabowski, Z., 257(719), *311*
Grabowski, Z. R., 57(285), *192,* 248, 250 (579), 259(579), 260(579), *307*
Gradyushko, A. T., 272(823), 282(823), *315*
Graf, W., 86(431), *195*
Graffeuil, M., 232(343), *300*
Grammaticakis, P., 225(153,154), *295*
Grauer, F., 62(324), *192*

Gray, R. T., 137(677), *202*
Gream, G. E., 160(795), 161(795,803), *205*
Greatorex, D., 274(884), *317*
Green, B., 96(466), *196*
Green, J. A., 177(969), *209*
Greene, A. E., 20(80), *187*
Greenleaf, J. R., 252(619), 273(851), *308, 316*
Grellman, K. H., 44(178), *189,* 279(1029), *321*
Gresser, J. D., 111(531), *198*
Grevels, F. W., 38(146), *188*
Griffin, C. E., 79(411), *194*
Griffin, G. W., 48(208-211), 119(575), *190, 199*
Grigoryan, D. Kh., 279(1001), *320,* 92(444), *195*
Grimes, M. W., 92(444), *195,* 246(564), *307*
Grineva, Yu. I., 277(960,961), *319*
Grinter, R., 220(22,28), *291*
Grisdale, P. J., 58(314), 184(1060), *192, 211,* 257(720), *311*
Grishina, G. I., 57(294), *192*
Grivet, J. P., 246(561), *307*
Groen, M. B., 42(168), *189*
Groff, R. P., 280(1049,1050), 290, *322*
Groncki, C. L., 111(529), *198*
deGroot, A., 160(802), *205*
Grossman, M., 158(786), *204,* 255(684, 685), *310*
Grotewold, J., 269(792a), 276(920), 285 (920), *314, 318*
Grovenstein, E., 46(185), 61(185), *189*
Grover, P. K., 92(443), *195*
Gruber, G. W., 52(259), *191*
Gruber, R., 96(470), 100(470), *196,* 276 (919), 285(919), *318*
Gruen, L. C., 276(935), 286(935), *318*
Grunewald, G. L., 23(88), *187*
Grunwell, J. R., 125(603), *200*
Guenthard, H. H., 220(18), *291*
Guenzi, W. D., 54(271), *191*
Guesten, H., 228(225), *297*
Guglielmetti, R., 52(249), *191*
Guillemonat, A., 180(1033), *210*
Gull, S. J., 246(565), *307*
Gunder, O. A., 228(232), 256(232), *297*
Gunning, H. E., 115(555), *198*
Gupta, D. N., 31(121), *188*
Gupta, P., 92(445a,446), *195*

AUTHOR INDEX

Gurinovich, G. P., 272(818), *315*
Gurov, F. I., 280(1037), *321*
Gurudata, 175(939a), *208*
Gurvich, Y. A., 110(525), *198*
Guseva, L. N., 274(871,872), *316*
Gusten, H., 133(662), *201*
Gustorf, E. K. von, 38(146), *188*
Guttenplan, J., 106(501,514), 108(501, 514), 111(533), *197, 198,* 274(873), 275 (910), *316, 317*
Gutyra, L. S., 222(82), *293*
Guzinskii, V. V., 255(691), *310*
Guzzo, A. V., 278(985), *320*

Haaf, A., 148(715), *203*
Haarer, D., 273(835), 283, *315*
Haas, A., 113(545a), *198*
Habermehl, G., 148(715), *203*
Habraken, C. L., 231(291), *299*
Hada, H., 243(531), *306*
Hadni, A., 243(513), *305*
Haefelinger, G., 223(109), *294*
Hafner, K., 165(856), 172(856), *206*
Hafner, M., 278(998), *320*
Hagan, G., 83(420), *195*
Hageman, H. J., 115(554), 157(768), *198, 204*
Hah, S., 176(952), *208*
Hahn, R. C., 25(97,98), 45(97), 137(674), *187, 202*
Hale, R. L., 173(933), *208*
Hall, L., 260(767), 265(767), *313*
Hall, L. H., 271(812), 282(812), *315*
Hallas, G., 228(218), *297*
Haller, W. S., 135(669), *201*
Hamada, A., 165(855), *206*
Hamana, M., 182(1047), *211*
Hamer, J., 4
Hamer, N. K., 159(790,791), *204*
Hamilton, J. B., 46(184), *189*
Hamilton, T. D. S., 248, 250(581), 273 (581), 284, *307*
Hamlow, H. P., 68(368), 78(368), *193*
Hammond, G. S., 3, 4, 29(117), 32(123), 62(325), 67(364), *188, 192, 193,* 168 (879), *206,* 254(668), 263, 283, *310, 318*
Hammond, H. A., 57(287), *192*
Hammond, W. B., 27(107), *187*
Hamner, A., 69(370a), *194*
Hand, E. S., 149(721), *203*

Hanifin, J. W., 149(724), *203*
Haninger, G. A., Jr., 279(1011), 278(992), *320*
Hann, R. A., 70(378), *194*
Hansen, H.-J., 174(937), *208*
Haque, K. E., 175(939a), *208*
Harada, Y., 275(909), *317*
Harrer, W., 219(139), 224(139), *295*
Harris, C. B., 246(568), *307*
Harris, G. N., 337(65), *339*
Harris, R. A., 229(265), *298*
Harrison, A. M., 175(947), 176(947), *208*
Harrison, S. E., 254(663), *310*
Hart, D. J., 149(728), *203*
Hart, E. J., 325, *338*
Hart, H., 23(93,94), 37(144), 45(94), 90 (442,442a), 145(701), 149(728), 178 (1002), *187, 188, 195, 202, 203, 209*
Hartman, S. R., 252(617), 261, *308*
Hartmann, H., 232(338), *300,* 333, *338*
Hartmann, W., 154(758), *204*
Hartung, L. D., 177(973), *209*
Hasegawa, M., 58(306,307), 146(306), *192*
Hasegawa, Y., 231(315,316), *300*
Hashimoto, S., 178(996,1001), 179(1008), *209, 210*
Haskell, W. W., 221(44), *292*
Hasse, G., 258(736), 264, *312*
Haszeldine, R. N., 49(232), 50(232), 173 (929), *190, 208*
Hata, K., 45(180), 51(238), *189, 190*
Hata, Y., 170(896,897), 176(954), *207, 208*
Hatano, M., 217, *291*
Hatch, G. F., 259(754), 271(754), *313*
Hatchard, C. G., 333(41), 334(41), *338*
Haug, A., 92(444), *195,* 246(564,565), *307*
Haugland, R. P., 277(963), *319*
Hautala, R. R., 179(1013), *210*
Havinga, E., 62(328), 179(1011), *193, 210*
Hayakaway, S., 277(955), *319*
Hayashi, H., 246(567), 276(924), 285(924), *307, 318*
Hayashi, J., 139(685), *202*
Hayon, E., 162(821), *205,* 235(400), *302*
Heaney, H., 23(92), *187*
Heathcock, C. H., 143(694), *202*
Hedaya, E., 153(752), *203*
Hedges, R. M., 279(1022), *321*
Heertjes, P. M., 178(999), *209*
Heiligman–Rim, R., 52(251,252), *191*

Hein, D. E., 259(760,761), 260(763,763a), 265(760,761,763), 271(760,761), *313*
Heindel, N. D., 101(481), 110(481), *196*
Heine, H. G., 124(601), 160(601), *200*
Heins, C. F., 165(852,854), *206*
Heisel, F., 273(854), *316*
Heiss, J., 56(281), 71(281), *191,* 230(276), *298*
Heitkamper, P., 133(661), *201*
Helene, C., 162(816), *205*
Heller, A., 73(395), *194,* 268, 269(790), *314*
Helman, W. P., 252(627), 273(627), 284, *308*
Hempel, J. C., 328, *338*
Hems, M. A., 188(572), 155(759), *199, 204*
Henderson, G. L., 219, 291
Henderson, K. O., 116(559), *198*
Henderson, W. A., Jr., 62(326,327), 150 (733), *192, 203*
Henneberg, D., 279(1007), *320*
Hennessy, J., 278(969), *319*
Henry, B. R., 249, 251(618,626), 268, 269 (788), *308, 314*
Henson, K., 239(433), *303*
Henton, D. E., 93(455), *196*
Hentz, R. R., 278(978), *319*
Herbert, D., 162(811), *205*
Herbert, M. A., 235(399,407), *302*
Hercules, D. M., 231(313), 257(719a), *299, 311,* 325(5), 328, *338*
Herlens, D., 10(17), *185*
Hermann, H., 157(169), *204*
Herndon, W. C., 160(799), *205*
Herre, W., 275(908), *317*
Herron, D. K., 69(369), *193*
Hess, D., 160(800), *205*
Hess, L. D., 79(410), *194*
Hess, R. H., 5
Hesse, A., 71(381), *194*
Hesse, R. H., 179(1016), *210*
Hesselmann, I. A. M., 245(553,554), *306*
Hevesi, J., 232(326), 261(326), 276(934), 277(964), 289(964), *300, 318, 319*
Hibaum, G., 244(535,536), *306*
Hida, M., 217, 218, 228(227), *291, 297*
Hikino, H., 134(666), *201*
Hill, J., 104(488), *197*
Hill, M. L., 28(110), *187*
Hill, R. R., 226(192), *296*

Hillers, S., 229(246), *297*
Hinchliffe, A., 228(226), *297*
Hinohara, T., 48(219), *190,* 256(705), *311*
Hinshaw, J. C., 168(885), *207*
Hintz, P. J., 118(571), *199*
Hirai, H., 164(831a), 165(851), *205, 206*
Hirakawa, A. Y., 221(46), *292*
Hirano, H., 130(630), *201*
Hiraoka, H., 42(170), *189*
Hirashima, T., 171(907), *207*
Hirayama, F., 252(622), 270(795a), 273 (795a,849,853), 284(795a,849), *308, 314, 316*
Hiroma, S., 239(442), *303*
Hirooka, T., 275(909), *317*
Hirose, Y., 37(143), *188*
Hirota, Y., 162(807), *205*
Hitch, M. J., *292*
Hiyama, H., 171(907), 207
Ho, S. K., 277(996), *320*
Hoa, G. H. B., 241(481), 261(481), *304*
Hochstetler, A. R., 21(82,84), *187*
Hochstrasser, R. M., 218, 243(528), 244 (541), 246(541,573), 250(9), 270(803), 277(956), 282(803), 287(956), *291, 306, 307, 314, 319*
Hocking, M. B., 55(274a), 130(632), *191, 201*
Hodgkins, J. E., 253(635), 271(635), *309*
Hoefert, M., 278(986), *320*
Hoefnagels, J., 52(254), 131(641), *191, 201*
Hofert, M., 96(462a), *196*
Hoffman, M. Z., 333(44a), 334(44b,49,51), 335(53), *338, 339*
Hoffman, R. W., 121(587), *199*
Hoffman, S., 231(292), *299*
Hohlneicher, G., 223(106), *294*
Hojo, K., 40(155,160), *188*
Holliday, R. E., 177(973), *209*
Hollis, M. L., 231(302), *299*
Holloway, H. E., 223(103), 257(727), 263 (727), *293, 312*
Holmes, E. O., 223(117), *294*
Holmes, H. L., 218, 226(173), *293, 296*
Holovka, J. M., 17(67), 28(110), 42(67), *186, 187*
Holroyd, R. A., 7(1), *185,* 233(359), 273 (359), *301*
Holzbecher, Z., 255(681), *310*
Holzman, P., 273(856), *316*

Honda, M., 220(29), *291*
Horak, V., 219(342), 232(342), *300*
Hornfeld, Y., 162(814), *205*
Horoupian, S., 131(636), 142(636), *201*
Horrocks, D. L., 53(260,261), *191,* 261 (781), 274(863,864), 275(912), 285 (912), *314, 316, 317*
Horsfield, A., 105(494), *197,* 246(559), *307*
Horspool, W. M., 133(660), *201*
Horwell, D. C., 121(588), 165(588), *199*
Hoshino, O., 11(20,21), 172(20), *185*
Hosokawa, T., 170(902a), 176(902a), *207*
Hostettler, H. U., 81(416a), *195*
Houdard-Pereyre, J., 183(1050), *211*
Houghton, J. T., 269(792), *314*
Houlden, S. A., 222(95), *293*
Hourdin, D., 27(108), *187*
House, D. A., 335(53), *339*
Howarth, G., 178(995), *209*
Hoytink, G. J., 240(471), *304*
Hrdlovic, P., 4
Hromatka, O., 180(1037a), *210*
Hruban, L., 223(121), *294*
Hsieh, H. H., 26(101), *187*
Hsu, Y. F. L., 19(76), *187*
Huang, C.-S., 240(463), *304*
Huang, C. W., 51(246), 62(246), *191*
Huang, K.-T., 244(539), 246, *306*
Huang, P. C., 168(880), *206*
Hubbard, R., 49(232), 50(232), *190*
Huber, R., 8(5), *185*
Hubert, J., 173(927,928), *208*
Hudec, J., 51(242), 52(258), *191*
Hudson, J. A., 279(1022,1023), *321*
Huffman, K. R., 150(733), *203*
Hughes, R. G., 333(44a), 334(44b), 335, *338, 339*
Huisgen, R., 122(596), *199*
Hummer, B. E., 47(198), *189*
Hunt, J. W., 235(399,407), *302*
Hunter, T. F., 239(445), 277(952,956), 287, 280(1038), *303, 319, 321*
Hunziker, H. E., 279(1009,1010), *320*
Hurley, R. B., 60(320), *192*
Husain, D., 185(1070), *211,* 274(892), 279 (1002), *317, 320*
Huybrechts, G., 34(130), *188*
Huyffer, P. S., 146(706), *202*
Huysmans, W. G. B., 157(768), *204*

Hydson, A., 157(770), *204*
Hyndman, H. L., 29(117), *188*

Ichibori, K., 176(965), *209*
Ievins, A., 231(308-310), *299*
Iffland, D. C., 221(43), *292*
Igeta, H., 181(1043), *210*
Ignat'eva, S. N., 230(279), *298*
Ihaya, Y., 220(24), *291*
Iida, Y., 224(145,146), 226(184), 240 (145,146), *295, 296*
Ikari, K., 172(917), *207,* 222(96), *293*
Ikeler, T. J., 172(918), *207*
Iles, O. H., 48(215), *190*
Il'iukevic, L. A., *339*
Imai, H., 238(431a), *303*
Imhof, R., 86(431), *195*
Inaba, T., 11(22), *185*
Ingraham, L. L., 229(255), 230(273), *298*
Innes, K. K., 220(35), 222(73), 231(301), *291, 293, 299*
Inokuchi, H., 227(208), 259(761a), 265 (761a), 275(903,909), 285(903), *296, 313, 317*
Inone, E., 52(256), *191*
Inoue, E., *198*
Inukai, Y., 130(635), *201*
Inwood, R. N., 170(900,901), *207*
Iogansen, A. V., 180(1032), *210*
Ipaktschi, J., 84(423), 90(441), *195*
Iriarte, J., 126(616), *200*
Irick, G., Jr., 171(904,905,906), *207*
Irmscher, K., 179(1018), *210*
Irving, C. S., 48(208), *190,* 237, *302*
Isayan, G. A., 180(1030), *210*
Ishikawa, M., 181(1041,1045), 182(1048), *210, 211*
Isomura, K., 172(913), *207*
Ito, M., 231(316,317), 233(354), 235(354), 251(602), 260(768), 265(768), 273(354), *300, 301, 308, 313*
Ito, H., 257(729), *312*
Itoh, M., 128(623), *200,* 240(466), *304*
Itoh, T., 113(544), *198*
Ivanov, V. L., 63(339), *193*
Ivanov, V. M., 229(238,244), *297*
Ivanov, V. V., 238(420), *303*
Iwashima, S., 227(208), *296*
Iwata, M., 9(9), *185*
Iwata, S., 246(567), 276(924), 285(924),

307, 318
Izawa, H., 164(846), 165(846), *206*
Izawa, Y., 46(187), 79(412), 104(412), 111 (412), 113(544), *189, 195, 198*
Izvekov, V. P., 230(280), *298*
Izzo, P. T., 144(696,697), *202*

Jablonski, A., 248(590), 250(590), 264 (590), *307*
Jaffe, H. H., 218(25), 220(25), 239(435), *291, 303*
Jaiswal, R. M. P., 225(160), *295*
Jakob, W., 333(38), *338*
James, S. P., 107(47), *186*
Jankowski, W. A., 254(662), *310*
Janssen, J. F., 168(882), *207*
Janzen, E. G., 178(997), *209*
Jardon, P., 256(706), 278(989), *311, 320*
Jarnagin, R. C., 273(856), *316*
Jaseja, T. S., 255(686), *310*
Jasper, S. W., 77(402), *194*
Jauquet, M., 58(311,312), *192*
Jayswal, M. G., 255(695), *310, 311*
Jeger, O., 68(368), 77(398), 78(368), 91 (398), *193, 194*
Jelliner, H. H. G., 122(594), *199*
Jenevein, R. M., 96(469), *196*
Jenny, W., 65(352), *193*
Jewel, J. S., 71(382), *194*
Jirkovski, I., 143(695), *202*
Job, V. A., 220(35), *291*
Johaitis, H., 232(335,336), *300*
Johansen, H., 229(255), 230(273), *298*
Johansson, N. G., 160(792), *204*
Johns, H. E., 4, 235(399,407), 271(814, 815), *302, 315*
Johnson, A. L., 168(885), *207*
Johnson, G. M., 79(411), *194*
Johnson, P. M., 280(1039), *321*
Johnson, R. C., 165(853), 167(867), *206*
Johnstone, R. A. W., 232(339), *300*
Jonaitis, H., 238(423), *303*
Jonas, E., 34(130), *188*
Jones, G., 148(716), *203*
Jones, G., II, 254(660), *310*
Jones, G. W., 137(674), *202*
Jones, J. K. N., 178(995), *209*
Jones, L. B., 5, 38(149,150), 179(1014), *188, 210*
Jones, M., 175(946,947), 176(946,947), *208*

Jones, P. F., 253(637,638), 271(637,638, 811), 281(811), *309, 315*
Jones, V. K., 5, 38(149,150), *188*
Jorgenson, M. J., 128(624), 129(625), 149 (723), *200, 203*
Jori, G., 53(270), *191*
Jortner, J., 222(76), 227(214), 242(496), 248, 250(576,577,578), 272(828), *293, 297, 305, 307, 315*
Jouet, M. C., 220(38), *292*
Joullie, M. M., 174(936), *208*
Joussot-Dubien, J., 183(1050), *211*, 233 (358), 238(429), *301, 303*
Joyce, T. A., 96(462), 106(462), *196*, 251 (605,615), *308*
Judeikis, H. S., 236, *301*
Judge, D. L., 254(671), *310*
Julia, M., 71(383), *194*
Jullien, R. F., 220(39), *292*
Jurasek, A., 230(285), *299*
Jurd, L., 128(622), *200*

Kabrt, L., 255(681), *310*
Kada, R., 230(285), *299*
Kaemmerer, H., 224(135), *295*
Kagan, J., 72(391), 79(413), 115(413), *194, 195*
Kaizu, Y., 231(317), *300*
Kajiwara, T., 227(208), 259(761a), 265 (761a), 275(903), 285(903), *296, 313, 317*
Kakac, B., 219(342), 232(342), *300*
Kakehi, A., 183(1056), *211*
Kalabina, L. V., 250(587), *307*
Kalantar, A. H., 259(753), 265, 270(753, 798), *312, 314*
Kale, J., 52(255), *191*
Kalibabchuk, V. A., 255(694a), *311*
Kalle, A. G., 108(520), *197*
Kallman, H., 272(832), *315*
Kamano, Y., 150(729,730), *203*
Kameo, T., 171(907), *207*
Kaminski, W., 70(379), *194*
Kaminskii, A. Ya., 224(149), *295*
Kan, R. O., *3*
Kanada, Y., 251(609), *308*
Kanaoka, Y., 11(19), 163(825), *185, 205*
Kanda, Y., 103(485), 257(716), 259(775), 265, 273(839), 277(954), 287, *197, 311, 313, 315, 319*

AUTHOR INDEX

Kane, M., 144(698,699), *202*
Kaneko, C., 181(1041,1045), 182(1048), *210, 211*
Kanematsu, K., 183(1056), *211*
Kano, H., 106(496,507), 108(496,507), 167(862), 168(873), *197, 206*
Kano, K., 178(996,1001), 179(1008), *209, 210*
Kantardzhyan, L. T., 277(959), *319*
Kantrowitz, E. R., 334(51a,51b), 335(51b), *339*
Kapecki, J., 131(643), *201*
Kaplan, B. E., 82(418), *195*
Kaplan, L. R., 171(903), *207*
Kaplin, Yu. A., 251(608), *308*
Karafiath, E., 27(105,106), *187*
Karle, I. L., 11(19), *185*
Karliner, J., 175(938), *208*
Karyakin, A. V., 235(408), *302*
Kasai, N., 239(440), *303*
Kasai, P. H., 168(891), *207*
Kasha, M., 184(1061), 231(305), 257(305), 268, 269(788), *211, 299, 314*
Kashima, C., 126(613,614), 131(638), 139(681), 152(613), *200, 201, 202*
Kataoka, H., 126(613,614,618), 139(681), 152(613), *200, 202*
Kato, H., 44(179), 124(602), 168(893), 169(893), *189, 200, 207*
Kato, M., 93(454), 185(1069), *196, 211*
Kato, N., 49(224a), *190*
Kato, S., 229(250), 235(409), *298, 302*
Katz, B., 222(76), *293*
Kaufman, J. A., 176(967), *209*
Kaupp, G., 113(540), 152(540,746), 153(746), *198, 203*
Kawanisi, M., 13(52), 94(52), 165(851), 185(1069), *186, 206, 211*
Kawaoka, K., 72(390), *194*
Kawashima, K., 173(922,925), 221(46), *292, 207, 208*
Kazachkov, V. G., 280(1037), *321*
Kazanova, N. N., 220(27), *291*
Kearney, P. C., 183(1055), *211*
Kearns, D. R., 72(389,390), 35(428), 140(428), 256(710), 276(927), *194, 195, 311, 318*
Keehn, P. M., 60(322), *192*
Kekudo, M., 239(440), *330*
Keller, O. L., 329(17), *338*

Keller, R. A., 270(799,800), 279(1026), 282, *314, 321*
Kellogg, R. E., 278(982), 281(982), 288, *320*
Kellogg, R. M., 42(168), 183(1051), *189, 211*
Kelly, D. P., 114(553), *198*
Kelly, J. M., 134(665), *201*
Kelso, P. A., 97(477,478), 98(477,478), 99(478), *196*
Kembrovskii, G. S., 241(484), 258(484), *304*
Kemp, D. R., 157(771), 234(390), *204, 302*
Kemp, T. J., 64(349), 274(884), *193, 317*
Kemppainen, A. E., 97(478,479), 98(478), 99(478), 100(479), *196*
Kende, A. S., 132(659), 144(696,697), *201, 202*
Kennedy, F. S., 336(56), *339*
Kern, D. H., 92(445), 96(465,469), 103(445), *195, 196*
Kessler, H., 225(158), *295*
Ketskemety, I., 252(621), *308*
Keyes, T. F., III, 225(157), *295*
Khachaturova, G. T., 256(707), *311*
Khan, A. M., 48(208), *190*
Khan, A. U., 276(927), *318*
Khan, M. A. O., 18(73), 19(74), *186*
Khanna, B. N., 223(104), 227(204), *294, 296*
Kharasch, N., 47(193), *4, 189*
Kharidia, S. P., 131(639), *201*
Kheifits, L. A., 224(131,131a), *294*
Khesina, A. Ya., 258(745), *312*
Khizhnyi, V. A., 49(224), *190*
Khmel'nitskii, L. I., 230(284), *299*
Khomich, M. I., 255(688), 261(783,784), 264(783,784), 266(783,784), *310, 314*
Khromov-Borisov, N. V., 48(221), *190*
Khudyakova, L. P., 233(352), 242(352), 258(352), *301*
Khuong-Huu, J., 10(17), *185*
Kiefer, E. J., 32(125,126), *188*
Kikuchi, K., 280(1032), 290, *321*
Kikuchi, S., 173(926), *208*
Kim, B., 73(394), 74(394), *194*
Kim, J. J., 272(842), 284, *315*
Kim, Y. S., 240(457,460), *304*
Kimoto, K., 13(52), 94(52), *186*
Kimura, H., 179(1025), *210*
Kimura, K., 224(142), *295*

King, T. A., 252(619), *308*
Kinoshita, M., 280(1034), *321*
Kiprianov, A. I., 225(168), 229(240-243), 232(319-321), *295, 297, 300*
Kira, A., 48(213), 184(213), *190*
Kirby, G. H., 244(532), *306*
Kirchenko, L. N., 260(772), 266, *313*
Kirk, A. D., 327, *338*
Kirmse, W., 175(942,943), *208*
Kirsch, G., 40(157), *188*
Kiryukhin, Y. I., 48(214), *190*
Kiryukhin, Yu. I., 274(885), *317*
Kishore, J., 274(865), *316*
Kislyak, G. M., 253(651), 276(918), 285(918), *309, 318*
Kistiakowsky, G. B., 271(807), 282(807), *315*
Kita, S., 30(120), *188*
Kita, T., 257(729), *312*
Kitaura, V., 101(480,482), 109(480), 110(482), *196, 197*
Kitaurd, Y., 108(521), 110(521), *197*
Kittler, L., 163(824), *205*
Klapyta, Z., 225(163), *295*
Klasek, A., 223(121), *294*
Klasic, L., 228(225), *297*
Klasine, L., 234(378), *301*
Klaus, M., 152(748,750), *203*
Klein, H., 176(953a), *208*
Klein, J., 273(854), *316*
Klein, W., 18(71), *186*
Kleinwachter, V., 240(465), *304*
Klemenko, S. L., 220(26), *291*
Klemenkova, Z. S., 237, 240(475), *304*
Kliger, D. S., 233(361), 234(361), 271(361), *301*
Klimova, L. A., 241(478,485), 258(485), 261(478), 264, *304*
Klimusheva, G. V., 223(116), 243(514), *294, 305*
Klingbell, U. I., 54(272), *191*
Klirgebiel, U. I., 183(1055), *211*
Klochkov, V. P., 255(687), *310*
Kloepffer, W., 277(946,947), 287, *319*
Kloosterziel, H., 39(151), *188*
Knoll, J. E., 253(654), *309*
Knollmuller, M., 180(1037a), *210*
Knop, A., 256(700), *311*
Knop, J. V., 256(800), *311*
Knox, R. S., 277(942), 287, *318*

Knudsen, R., 119(578), *199*
Kobayashi, H., 44(175), 278(990), *189, 320*
Kobayashi, M., 116(559d), *199*
Kobyshev, G. I., 280(1031), *321*
Kobzina, J. W., 126(611), *200*
Koch, K. F., *4*
Koch, T. H., 135(670), *201*
Kochevar, I., 97(472), *196*
Kochi, J. K., 116(562), *199*
Kochi, M., 275(909), *317*
Kocki, M., 272(825), 283, *315*
Kocma, K., 226(191), *296*
Koga, G., 172(910), *207*
Koga, N., 172(910), *207*
Kogawa, S., 131(653), *201*
Koizumi, M., 48(213), 145(703,704), 184(213,1067), 185(704), 246(574), 272(817), 276(928), 278(976), 280(1032), 285, 288, 290, *190, 202, 211, 307, 315, 318, 319, 321*
Kojikawa, T., 48(206), *190*
Kojima, K., 68(367), 89(439), 90(367,439), *193, 195*
Kojima, M., 166(858), *206*
Kojima, Y., 217, *291*
Kokado, H., 52(256), 125(605a), *191, 198, 200*
Kokubun, H., 261(782a), 272(817), 280(1032), 290, *314, 315, 321*
Kokubun, K., 145(703,704), 185(704), *202*
Koltzenburg, G., 106(499,506), *197*
Komatsu, M., 150(729,730), *203*
Komori, S., 12(28), 108(28), *186*
Kondo, M., 257(722), *312*
Kondo, S., 176(966), *209*
Kondo, Y., 162(822), *205*
Kopecki, J. A., 107(516), *197*
Kopecky, K. R., 70(376), 168(878), 170(895), *194, 206, 207*
Kopytina, M. V., 229(259), *298*
Koren, J. G., 233(353), 235, *301*
Koritskii, A. T., 116(560), *199*
Kormendy, K., 227(198), *296*
Kornis, G., 134(666), *201*
Korotkov, S. M., 255(687), *310*
Korovina, V. M., 237, 238(416), 241(476), *302, 304*
Korshunov, S. P., 229(246), *297*
Korte, F., 18(70,71,72), 19(72), 51(243), 130(243), 131(637), 150(637), 155(762),

186, 191, 201, 204
Kortum, G., 57(301), *192*
Koser, G. F., 148(711), 181(711), *203*
Koser, G. K., 147(708), *203*
Kosonocky, W. F., 254(663), *310*
Kosower, E. M., 168(880), *206*
Kostko, M. Ya., 253(649), 254(659), 261 (782), 270(649), 279(1012), *309, 314, 320*
Kostyanovsky, R. G., 173(931), *208*
Kostyshin, A. P., 275(905), *317*
Kotaki, A., 261(779a), 270(779a), *314*
Kothe, W., 57(297), 151(743), *192, 203*
Koudelka, J., 240(465), *304*
Kovalev, A. F., 227(201), *296*
Kovaliv, Yu. D., 229(239), *297*
Kovbasyuk, A. S., 62(331), *193*
Kovrizhnykh, N. A., 241(492), *305*
Koyanagi, M., 273(839), 277(954), 287, *315, 319*
Kozachenko, A. I., 244(540), 245(542), 247(542), *306*
Koziol, J., 232(323), *300*
Kozlov, I. N., 255(688), *310*
Kozlov, Yu. I., 275(911), 285(911), *317*
Kozma, L., 252(621), *308*
Kracmar, J., 231(295), *299*
Kracmarova, J., 231(295), *299*
Kramer, H. E. A., 278(998), *320*
Kramling, R. W., 220(33a), *291*
Krasnoshchekova, G. S., 238(419), *303*
Krasovitskii, B. M., 225(165), 230(288), 255(690), 256(165), *295, 299, 310*
Krauss, H. J., 16(62), 20(62), *186*
Krespan, C. G., 23(89), 46(89), 64(89), *187*
Kresze, G., 217, *291*
Krieg, P., 57(301), *192*
Krishna, V. G., 233(348), 235, *301*
Krishnamurty, K. V., 337(65), *339*
Kristian, P., 229(245), *297*
Kristinsson, H., 48(211), *190*
Krohn, A., 223(107), *294*
Krongauz, V. A., 223(115), 255(115), 270 (115), 271(115), *294*
Kropp, J. L., 66(360,361), 234(397), 253 (397,640-646), 270(642-646), 271(640-643,645,646), 281(640,641), *193, 302, 309*
Kropp, P. J., 16(62), 20(62,79), 51(79), *4, 186, 187*

Krow, G. R., 164(835), *206*
Krubsack, A. J., 116(558), *198*
Krueger, K., 272(819), 282(819), *315*
Krueger, S. M., 33(129), 90(440), 131(643), *188, 195, 201*
Kruglik, E. K., 272(818), *315*
Krull, I. S., 117(567), 148(710), *199, 203*
Krusic, P. J., 116(562), *199*
Kryman, J. F., 122(574), *199*
Kubota, T., 68(366), *193*
Kubota, Y., 246(566), 280(1041), *307, 321*
Kuboyama, A., 225(156), 259(758), 266 (758), *295, 313*
Kucharczyk, N., 219(342), 232(342), *300*
Kuder, J. E., 156(763), 184(1062), *204, 211*
Kudrna, J. C., 179(1014), *210*
Kuebler, N. A., 217(64,65), 220(17), 221 (64,65), *291, 292*
Krueger, K., 57(288), *192*
Kuhla, D. E., 43(173), *189*
Kuhnle, W., 44(178), *189*
Kules, I., 279(1018), *321*
Kulik, L. N., 243(511), *305*
Kulis, Y. Y., 63(339), *193*
Kul'nevich, V. G., 232(330), *300*
Kumari, D., 179(1023), *210*
Kumler, P. L., 181(1044,1046), 182(1044), *210*
Kummler, R. H., 276(929b), *318*
Kunec-Vajic, E., 274(883), *317*
Kuntz, E., 276(940), 286(940), *318*
Kurik, M. V., 237, 239(447), 243(524,526), *303, 305, 306*
Kuroda, H., 239(440,442,443), *303*
Kuroi, T., 240(455), *304*
Kuroki, M., 230(266), *298*
Kurtin, W. E., 184(1066), 254(657,658), 256(703), 272(658), 282(658), *211, 309, 311*
Kurtz, D. W., 137(674), *202*
Kusabayashi, S., 239(440), *303*
Kushner, A. S., 164(840), *206*
Kushnikov, Yu. A., 222(80), *293*
Kusmierczyk, W., 224(148), *295*
Kuthan, J., 229(256), *298*
Kutsyna, L. M., 230(290), 245(290), 247, *299*
Kuwahara, M., 49(224a), *190*
Kuwano, H., 257(722), *312*

Kuwata, S., 89(437), 140(437), *195*
Kuz'min, M. G., 63(339), 238(421), 244 (540), 245(542), 247(542), 274(871, 872), *193, 303, 306, 316*
Kuznetsov, V. A., 251(608), *308*
Kuznetsova, K. E., 180(1030), *210*
Kwiatkowski, J. S., 229(262), *298*
Kwiram, A. L., 244, 246(572), *307*
Kyba, E. P., 164(838), 172(838), *206*
Kydd, R. A., 242(503), *305*
Kyoto, K., 160(801a), *205*

Laban, G., 232(340), *300*
Labarre, J. F., 232(343), *300*
L'Abbe, 5, 172(909), *207*
Labhart, H., 164(843), *206*, 218(254), 229 (254), *298*
Lablance-Combier, A., 42(169), *189*
Labrence, I., 231(308), *299*
Lacko, R., 19(76), *187*
Lagercrantz, C., 178(980,983), *209*
Laipanov, R. Z., 258(748), *312*
Laird, T., 10(13), 104(489), *185, 197*
Lal, B. B., 255(693), *311*
Lalande, R., 15(58), *186*
LaLonde, R. T., 131(640), *201*
Lambeth, P. F., 106(498,500,509), 108 (498,500,522), 111(498,509,522), *197*, 274(891), *317*
Lami, H., 273(854), *316*
Lamola, A. A., 5, 162(810,812), *205*, 279, (1021), *321*
Lamotte, M., 238(429), *303*
Lance, D. G., 178(995), *209*
Land, E. J., 158(781), *204*, 234(394), *302*
Land, H. B., 166(861), *206*
Landesberg, J. M., 16(15), 84(5), *186*
Lane, A. G., 12(26), *185*
Lang, H., 238(417), 241(482), *303, 304*
Lange, G., 134(666), *201*
Langer, H., 170(899), *207*
Langford, C. H., 330, 332(30), 337, *338*
Langmuir, M. E., 178(1006), *209*
Lantzke, I., 334(50), 335(50), *339*
Lapointe, J.-P., 176(959), *208*
Lapouyade, R., 64(345-347), *193*
Lappin, G. R., 47(196), 49(196), 110(524), 111(524), *189, 197*
Larson, C. W., 70(374), *194*
Lashkov, G. I., 231(311), 275(915), 285, *299, 318*
Lassila, J. D., 80(414), 88(436), 135(436), 142(414), 144(698,699), 151(737), 180 (1027), *195, 202, 203, 210*
Latowski, T., 47(201), *189*, 274(878), *316*
Lau, N. K., 148(719a), *203*
Laukin, D. C., 131(639), *201*
Laurenzi, B. J., 232, 233(346), *301*
Laurushin, V. F., 230(289), *299*
Laustriat, G., 273(854), *316*
Lavalette, A., 234(371), *301*
Lavalette, D., 234(377), *301*
Lavrushin, V. F., 230(280), 232(345), *298, 301*
Lawson, C., 270(795a), 273(795a), 284, *314*
Lay, W. P., 168(888), *207*
Leach, S., 242(497), 258(497,736a), *305, 312*
Leaver, I. H., 108(517), 178(986,987,988a), *197, 209*
Lebedev, O. V., 230(284), *299*
Ledwith, A., 13(51), 48(215), *186, 190*
Lee, A. C. H., 178(991), *209*
Lee, A. T. H., 220(36), 222(36), *292*
Lee, E. K. C., 252(625), 269(794), 271 (794,806), 278(992), 279(1011), 282 (806), *308, 314, 315, 320*
Lee, G. A., 14(53), 25(99), 45(99), *186, 187*
Lee, K., 64(341), *193*
Lee, N. E., 269(794), 271(794), *314*
Lee, S. K., 269(793a), 280, *314*
Lee, T. J., 158(789), *204*
Lee, Y. J., 58(309), *192*
Leermakers, P. A., 29(118), 116(559c), *188, 199*, 237, 275(896), 279(1017,1019, 1024), 289, *302, 317, 321*
Lee-Ruff, E., 127(620), *200*
Le Goff, M.-T., 21(85,86), 113(541), *187, 198*
Leibovici, C., 222(83,93,94), *293*
Leistner, S., 230(286), *299*
Leite, M. S. S. C., 275(901), *317*
Lemaire, J., 158(785), *204*
Lemal, D. M., 49(231), *190*
Lenhert, P. G., 137(677), *202*
Leppard, D. G., 149(726), *203*
Le Roux, J. P., 9(8), *185*
Le Roux, Y., 163(830), *205*

Leska, J., 229(257), *298*
Leterrier, F., 237, 240(472a), 246(562), 261(472a), *304, 307*
Letsinger, R. L., 179(1012,1013,1015), *210*
Leubner, I., 232(322), 253(633,635), 256(322), 265, 271(635), *300, 309*
Levashova, L. A., 180(1030,1032), *210*
Levin, P. I., 110(525), *198*
Levina, P. Y., 4
Levina, R. Ya., 117(536), *198*
Levisalles, J., 176(956), *208*
Levshin, L. V., 237, 240(474,475), 274(866), *304, 316*
Levshin, V. L., 277(960,961), *319*
Lewenz, W., 224(135), *295*
Lewin, N., 136(672), *202*
Lewis, E. S., 177(973), *209*
Lewis, F. D., 84(421a), 172(915), *195, 207*, 278(970,997), *319, 320*
Lewis, J. W., 157(770), *204*
Lewis, R. G., 238(428), *303*
Lexa, D., 246(563), *307*
Lhoste, J. M., 234(372), 235(372), 246(563), 256(372), *301, 307*
Li, R., 249, 251(607), *308*
Li, Y. H., 249, 251(607), 256(702), 272(702), 282(702), *308, 311*
Libman, J., 115(556), 138(678), *198, 202*
Liebman, S. A., 177(972), *209*
Lim, E. C., 103(485), *197*, 249, 251(607, 607a), 256(702), 257(716), 270(801, 804), 272(702), 273(843,844), 282, 284, *308, 311, 314, 316*
Lin, L. C., 129(627), *200*
Lin, T.-S., 246(573), *307*
Lindholm, R. D., 332(32), 333(34), 334(34), 336(34), *338*
Link, J. W., 178(1002), *209*
Linke, H. A. B., 225(167), *295*
Lipnitskii, I. V., 329, *338*
Lippert, E., 57(288), *192*, 260(778), 266, 272(819), 282(819), *313, 315*
Lipsett, F. R., 260(765), 271(765), 281(765), *313*
Lipsett, R. F., 280(1046), 290(1046), *322*
Lipsky, S., 252(622), 270(795a), 273(795a, 849,853), 284, *308, 314, 316*
Liptay, W., 238(418), *303*
Lisewski, T., 162(815), *205*

Lissi, E. A., 269(792a), 276(920), 285(920), *314, 318*
Litt, A. D., 96(460,461), 106(460,461), *196*, 253(656), *309*
Litvinenko, V. I., 227(201), *296*
Liu, K., 231(293), *299*
Liu, R. S. H., 22(87), 23(89,90), 45(90), 46(89), 64(89,90), *187*, 278(982,983), 281(982,983), 288, *320*
Lober, G., 163(824), *205*
Lochte, W., 243(530), *306*
Loeber, G., 238(417), 241(482), *303, 304*
Loeliger, H., 337, *339*
Loeschen, R. L., 144(698,699), *202*
Loewenthal, E., 272(841), 284(841), *315*
Logan, L. M., 251(600), 255(674a), 256(699), 260(765), 269(699), 271(765), 281(765), *308, 310, 311, 313*
Logvinenko, P. N., 234(391), *302*
Loho, M. I., 323(332), *300*
Lohse, C., 181(1046), *210*
Lomakin, A. N., 231(314), *300*
Lombardi, J. R., 230(282), 244(538,539), 246, *298, 306*
Long, K. M., 335(55), *339*
Looker, J. J., 151(734), *203*
Loos, K. R., 220(18), *291*
Lopez-Campillo, A., 258(736a), *312*
Lopez-Delgado, R., 258(736a), *312*
Lopp, J. G., 58(313), *192*
Lopresti, R., 62(327), *192*
Loutfy, R. O., 131(650), 132(650), *201*
Loveridge, E. L., 119(578), 125(605), *199, 200*
Low, J. N., 149(726), *203*
Lower, S. K., 237, 239(448), *303*
Lowrance, W. W., 69(369), 134(668), *193, 201*
Loy, L., 150(733), *203*
Lucatu, E., 274(858,867), *316*
Ludi, A., 327, *338*
Ludwig, P. K., 278(978), *319*
Lugt, van der W. T. A. M., 29(113), *187*
Lugtenburg, J., 62(328), *193*
Lukashin, A. V., 218, 220(20), *291*
Lumb, M. D., 253(634), 269(634), 273(851), 275(900), *309, 316, 317*
Lumer, Z., 64(340), *193*
Lumma, W. C., Jr., 92(447), 138(447), *196*
Lungle, M. L., 94(457,458), *196*

Luongo, J. P., 12(31), *186*
Lutskii, A. E., 238(427), 239(450), *303*
Lwowski, W., 172(919), *207*
Lynch, J., 120(586), *199*
Lysenko, G. M., 253(651), 276(918), 285 (918), *309, 318*
Lytle, F. E., 328, *338*
Lyubopytova, N. S., 221(56), *292*
Lyuts, A. E., 222(80), *293*

MacBride, J. A. H., 183(1054), *211*
Mac Gregor, D. J., 138(680), *202*
Mackenzie, K., 168(888), *207*
Mackie, J. C., 158(787), 275(917), 285 (917), *204, 318*
Mackor, A., 177(977), 178(984), *209*
Maclaren, J. A., 276(935), 286(935), *318*
Mader, H., 122(596), *199*
Maeda, M., 166(858), *206*
Maggiora, G., 229(255), *298*
Maheshwari, K. K., 92(448), *196*
Maier, G., 85(424), 118(424,573,574), *195, 199*
Maki, A. H., 245(544,545), 246(568), 276 (927), *306, 307, 318*
Maki, Y., 184(1064), *211*
Makishima, S., 164(831a), *205*
Maldonado, P., 52(249), *191*
Malkes, L. Ya., 223(124), 241(488,490), 258(124,490), *294, 305*
Malley, M. M., 268, 269(791), *314*
Malpass, J. R., 164(835), *206*
Mal'tseva, N. I., 225(165), 256(165), *295*
Malykhin, N. N., 233(352), 242(352), 258 (352), *301*
Mamedov, Kh. I., 258(739,740,748), *312*
Manable, O., 171(907), *207*
Manfrin, M. F., 331(26), *338*
Mani, A., 230(282), *298*
Mani, I., 233(357), *301*
Manikowski, H., 232(328), 261(328), *300*
Manitto, P., 113(543), 120(543), 151(735, 738), *198, 203*
Manmade, A., 48(210), 119(575), *190, 199*
Mansour, S., 277(951), 287, *319*
Mantecon, J., 16(64), *186*
Mantsch, H., 182(1049), 228(221), 256 (221), 257(269), *211, 297, 298*
Mar, T., 252(617), 262, *308*
Marakami, Y., 259(775), 265, *313*

Marcheville, H. C., de, 21(83), 113(541), *187, 198*
Marciani, S., 162(817), *205*
Mare, G. R., de, 34(130), *188*
Margerum, J. D., 115(557), *198*
Maria, H. J., 252(628,629), 259(629), 266, 271(628), *308, 309*
Mariano, P. S., 55(273), *191*
Mark, F., 131(646), *201*
Mark, G., 131(646), *201*
Marky, M., 174(937), *208*
Marples, B. A., 23(92), *187*
Marshall, J. A., 19(78), 20(80,82), *5, 187*
Martelli, G., 53(263), 56(279), *191*
Martelli, J., 260(764), *313*
Martensson, O., 223(98), *293*
Martin, E. A., 256(689), *310*
Martin, G. J., 220(38), *292*
Martin, H. D., 168(884), *207*
Martin, J. E., 330(22), 331(22), *338*
Martin, M. L., 220(38), *292*
Martin, R. H., 57(302,303), 62(303), 64 (302,303,351), 65(303), *192, 193*
Martin, T. E., 259(753), 265, 270(753,798), *312, 314*
Martin, W., 41(164,165), *189*
Martinaud, M., 241(479), *304*
Martinez, A., 259(755), 265, 277(945), 287, 280(1033), 290, *313, 318, 321*
Martz, P., 181(1042), *210*
Marx, J. N., 148(718), *203*
Marzzacco, C., 218, 233(379), 235(379), 243(379), 250(9), *291, 301*
Masamune, S., 35(136), 40(155,160), 116 (561), *188, 199*
Masamune, T., 179(1019-1021,1026), 180 (1026), *210*
Mashenkov, V. A., 232(324), 252(324), 272 (823), 282(823), *300, 315*
Maskill, H., 56(283), *191*
Mason, M. M., 69(372), 77(372), 105(372), *194*
Mason, S. F., 237(410), 256(216), 257 (714), *297, 302, 311*
Massot, F., 175(944), 176(944), *208*
Masuzaki, S., 259(758), 266(758), *313*
Maszew, J., 224(148), *295*
Mataga, N., 62(332), 226(194), 254(194), 274(881), 275(906), 277(953), *193, 296, 317, 319*

Mathis, P., 278(999), *320*
Matsen, F. A., 328, *338*
Matsubara, A., 166(857), *206*
Matsuda, H., 242(494), *305*
Matsui, H., 274(881), *317*
Matsui, K., 256(705), *311*
Matsumoto, H., 106(496,507), 108(496, 507), 167(862), 168(873), 272(822), 274(881), 282(822), *197, 206, 315, 317*
Matsumoto, M., 124(602), *200*
Matsumoto, T., 131(652,653), *201*
Matsumura, Y., 124(602), *200*
Matsunaga, Y., 224(145,146), 240(145, 146), *295*
Matsuo, T., 230(281), *298*
Matsusaki, H., 243(531), *306*
Matsuura, T., 47(204), 50(235), 101(480, 482), 108(521), 109(480), 110(482,521), 147(709), *190, 196, 197, 203*
Matsuyama, M., 53(268), *191*
Matthias, G., 106(506), *197*
Mattingly, T. W., 59(316), *192*
Matzat, N., 131(637), 150(637), *201*
Maulding, D. R., 66(355), 227(205), 255 (205), *193, 296*
Maumy, M., 71(383), *194*
Mauser, H., 180(1036), *210*
Maycock, J. N., 243(530), *306*
Mayer, C. F., 78(406), *194*
Mayer, R., 232(337), *300*
Mayer, W., 152(750), *203*
Mayo, G. O., 14(54), 25(99), 45(99), *186, 187*
Mayo, P., de, 131(649-651), 132(649-651), 134, 164(846), 165(846), *201, 206*
Mazur, Y., 115(556), 138(678), *198, 202*
Mazzocchi, P. H., 33(128), *188*
Mazzucato, U., 57(299), *192*
McAlpine, R. D., 222(90a), 255(90a), 277 (956), 287, *293, 319*
McAneny, M. P., 221(43), *292*
McBeath, M. E., 271(808), *315*
McCain, J. H., 179(1015), *210*
McCall, M. T., 57(286,290), 58(309,310), 245(548), 256(698), 270(698), 271(698), *192, 306, 311*
McCapra, F. A., 70(378), *194*
McCarthy, W. J., 254(661), 256(709), *310, 311*
McCarville, M., 280(1034), *321*

McCarville, M. E., 280(1040), *321*
McClain, W. M., 280(1039), *321*
McClure, D. S., 259(752), *312*
McConaghy, J. S., 35(133), *188*
McCulloch, A. W., 152(747), *203*
McCullough, J. J., 51(246,247), 62(246), 126(607), 134(663,665), 245(547), *191, 200, 201, 306*
McDaniel, D. M., 80(414a), *195*
McDiarmid, R., 218, 222(75,79), *293*
McDonald, D. P., 154(757), 336(60), *204, 339*
McDonald, J. R., 223(103a), 253(655), 255 (103a), *294, 309*
McDonald, R. J., 62(323), *192*
McDonald, S., 31(122), *188*
McEwen, W. E., 39(152), *188*
McFarlane, P. H., 178(1007), *210*
McGhie, J. F., 179(1023,1024), *210*
McGlynn, S. P., 252(628,629), 253(655), 259(629), 266, 271(628), 280(1034, 1040), *308, 309, 321*
McGreer, D. E., 126(609), *200*
McHugh, A. J., 244(534), *306*
McInnes, A. G., 152(747), *203*
McIntosh, C. L., 117(566), *199*
McIntosh, J. M., 173(923), *207*
McKellar, J. F., 106(500), 108(500), 157 (775), *197, 204*
McKendry, L. H., 150(731), *203*
McKenna, J., 172(914), *207*
McKenna, J. M., 172(914), *207*
McKinney, T. M., 221(48), *292*
McKinney, W. J., 240(454), *304*
McKnight, W. B., 269(792), *314*
McLean, A., 164(834), *205*
McLean, S., 173(932), *208*
McSwinney, H. D., Jr., 231(301), *299*
Meehan, G. V., 26(104), 37(104), 89(438), 93(453), 142(453), *187, 195, 196*
Megarity, E. D., 57(292), 278(971), *192, 319*
Megrabyan, R. L., 180(1030,1032), *210*
Mehlhorn, A., 220(30), 228(228), *291, 297*
Mehrhof, W., 179(1018), *210*
Mehta, P. C., 231(300), *299*
Meijere, A., de, 168(887), 176(958), *207, 208*
Meinwald, J., 63(336), 69(370a), 126(611, 612), *193, 194, 200*

Melhuish, W. H., 234(393), *302*
Mellugh, A. J., 227(209), *296*
Mende, U., 85(424), 118(424,573,574), *195, 199*
Mendez, J., 232(332), *300*
Men'shova, I. I., 239(450), *303*
Menter, J. M., 103(486), 279(1015), 289, *197, 321*
Merceille, J. B., 234(372), 235(372), 256(372), *301*
Mercier, C., 178(1004), 180(1038), *209, 210*
Merer, A. J., 12(37), 216, 218, 221(62), 222(74,77), *186, 292, 293*
Merkelo, H., 252(617), 261, *308*
Merrifield, R. E., 280(1049,1050), 290, *322*
Meshkova, G. N., 240(464), *304*
Messer, W., 166(859,860), *206*
Messer, W. R., *5*
Mestechkin, M. M., 222(82), *293*
Metelitsa, D. I., 46(188), *189*
Meth – Cohn, O., 120(586), *199*
Metts, L., 29(116), *188*
Metzger, J., 52(249), *191*
Metzger, J. L., 249, 251(610), *308*
Metzler, D. E., 217, *291*
Metzner, W., 51(244), 124(601), 160(601), *191, 200*
Meunier, J., 246(562), *307*
Mews, R., 12(29), *186*
Meyer, B., 249, 251(610), *308*
Meyer, G., 64(350), *193*
Meyer, J. M., 226(176), *296*
Meyer, Y., 234(370), *301*
Meyer, Y. H., 234(383), *302*
Michl, J., 227(211), *297*
Midwinter, J. E., 236(411), *302*
Migita, T., 176(963,964), *208, 209*
Mika, N., 223(106), *294*
Mikawa, H., 239(440), *303*
Mikhailenka, V. I., 258(735), 264, *312*
Mikhailova, I. F., 228(232), 256(232), *297*
Miki, F., 87(433), *195*
Miki, T., 139(682,683), *202*
Milazzo, G., 216, 221(66), 232(344), *292, 300*
Miles, D. M., 9(10), 182(10), *185*
Mileshina, L. A., 256(707), *311*
Mill, T., 168(881), *206*

Miller, A. C., 131(651), 132(651), 278(984), *201, 320*
Miller, D. L., 116(559b), *198*
Miller, E. D., 153(752), *203*
Miller, K., 244(532), *306*
Miller, L. L., 54(271), *191*
Milne, G. S., 168(883), *207*
Minami, K., 49(229a), *190*
Minato, H., 116, *199*
Minn, F. L., 106(510), *197*
Misner, R. E., 185(1071), *211*
Misra, T. N., 280(1036), 290, *321*
Mitchell, G. H., 226(192), *296*
Mitra, R. P., 332(27), *338*
Miura, T., 240(462), *304*
Miura, M., 256(566), 280(1041), *307, 321*
Miwa, T., 93(454), *196*
Miwa, Ta Kuji, 279(1000), *320*
Miwa, Tomoko, 279(1000), *320*
Miyadera, T., 55(277), *191*
Miyamoto, R., 230(281), *298*
Miyashita, M., 34(131), *188*
Miyata, T., 242(494), *305*
Mizuno, M., 243(515), *305*
Mobius, D., 277(943a), 287, *318*
Mobius, K., 70(379), *194*
Modica, A. P., 220(34), *291*
Moebius, C., 329, *338*
Moggi, L., 330(24), 331(26), 336(58), *338, 339*
Mohan, H., 332(27), *338*
Moisya, E. G., 242(500), *305*
Mollicone, B. M., 334(49), *339*
Momicchioli, F., 228(224), *297*
Money, T., 150(732), *203*
Monroe, B. M., 29(117), 161(804), *188, 205*
Montandon, E., 15(58), *186*
Monties, B., 112(537), *198*
Montillier, J. P., 29(118), 279(1019), 289, *188, 321*
Mookherji, T., 243(516), *305*
Moomaw, R., 260(766,767), 265, *313*
Moomaw, W. R., 253(653), 260(769a), 262, *309, 313*
Moore, J. A., 81(416), *195*
Moore, N. A., 180(1028a), *210*
Moore, W. M., 168(872), *206*
Mori, T., 13(52), 94(52), 95(458b), 120(581,585), *186, 196, 199*

AUTHOR INDEX

Mori, Y., 113(545), 120(580), *198, 199*
Moriarty, R. M., 167(866), *206*
Moriconi, E. J., 185(1071), *211*
Morita, M., 235(409), *302*
Morita, T., 113(545), *198*
Moritani, I., 28(111), 53(268), 87(433), 170(902a), 176(902a), *187, 191, 195, 207*
Morlot, G., 243(513), *305*
Morren, G., 57(302), 64(302), *192*
Morris, G. C., 227(214), *297*
Morris, M. R., 149(727), *203*
Morris, R. V., 223(119a), *294*
Morrison, H., 44(177), *189*
Morrison, H. A., 274(893), 285, *317*
Morrison, W. H., 8(6), *185*
Morton, D. R., 127(620), *200*
Morton, W. D., 173(929), *208*
Moser, J.-F., 92(449-451), *196*
Mosier, A. R., 54(271), *191*
Moskaleva, L. A., 231(314), *300*
Moszew, J., 225(163), *295*
Motuz, A. A., 239(447), *303*
Moule, D. C., 220(37), *292*
Mousseron, M., *4*
Mousseron-Canet, M., 38(147), 126(608, 610), 141(688), *188, 200, 202*
Movsesyan, M. E., 279(1001), *320*
Moyer, E. S., 254(661), *310*
Mozumder, A., 278(978), *319*
Muel, B., 251(614), *308*
Mueller, A., 227(198), 233(362), 234(362), *296, 301*
Muhlstadt, M., *4*
Mujamoto, T., 113(546), 114(546), *198*
Muk, A. A., 221(47), *292*
Mukai, T., 56(278), 164(849), 165(849), 166(857), *191, 206*
Mukherjee, R., 167(866), *206*
Mukherjee, R. K., 243(519), *305*
Mukherjee, T. K., 240(458), *304*
Mulder, B. J., 272(830), *315*
Mullen, P. A., 223(108), 230(277), *294, 298*
Muller, A., 66(359), *193*
Muller, E., 56(280,281), 71(281), *191*
Muller, L. L., *4*
Muller, R., 163(831), *205*
Mulliken, R. S., 12(37), 216, 218, 221 (62), 222(77), 237, 238(432), *186, 292, 293, 303*
Mullis, D. P., 151(742), *203*
Mumford, C., 70(376), *194*
Munakata, K., 49(224a), *190*
Munchausen, L., 121(590), *199*
Munekata, S., 173(926), *208*
Munro, I. H., 234(396), 255(396), 260 (765), 271(765), 281(765), *302, 313*
Murakami, Y., 251(609), *308*
Muralidharan, V. P., 96(468), 108(519), 111(519), 274(887), 285, *196, 197, 317*
Murata, Y., 275(906), *317*
Murov, S. L., 252(631), 271(631), *309*
Murray, R. K., 23(93,94), 45(94), 90(442, 442a), *187, 195*
Musa, W. E., 164(837), *206*
Musajo, L., 162(818), *205*
Musgrave, W. K. R., 183(1054), *211*
Mushkalo, I. L., 229(242), *297*
Mustafa, A., 224(147), 229(236), *295, 297*
Muszkat, K. A., 57(291,296), *192*
Myers, L. S., Jr., 231(302), *299*

Naboikin, Yu. V., 250(580), *307*
Nafissi-V., M., 120(582), *199*
Nagai, Y., 230(287), *299*
Nagakura, S., 233(354), 235(354), 251 (601,602), 272(354), 276(924), 285 (924), *301, 308, 318*
Nagao, Y., 51(237), *190*
Nagy, O. B., 239(441), *303*
Nair, K. P. R., 255(692,693), *311*
Naito, T., 120(581,585), *199*
Nakabayashi, N., 225(157), *295*
Nakadaira, Y., 139(684,685), 162(807), *202, 205*
Nakagura, S., 246(567), *307*
Nakahara, A., 277(954), 287, *319*
Nakai, F., 113(547), 114(547), *198*
Nakai, H., 11(19), *185*
Nakaido, S., 176(964-966), *209*
Nakamura, K., 246(574), *307*
Nakamura, T., 277(955), *319*
Nakamaru, K., 184(1067), *211*
Nakanishi, A., 255(680), *310*
Nakanishi, H., 58(306), 146(306), *192*
Nakanishi, K., 139(684,685), 162(807), *202, 205*
Nakano, Y., 253(652), 260(768,769), 262, 265, *309, 313*

Nakatsukasa, K., 229(250), *298*
Nakayama, K., 176(965), *209*
Nakhimovskaya, L. A., 233(355), 241(355, 483,486,487), 257(486,487,732), 258 (483), 259(486,487,732), 264, *301, 304, 312*
Naoi, M., 261(779a), 270(779a), *314*
Naqvi, K. R., 248, 250(581), 273(581), 275(901), 284, *307, 317*
Naruto, S., 119(578a), 178(578a), *199*
Nasibou, I. K., 258(739,740), *312*
Nasielski, J., 51(240), 58(311,312), *191, 192*
Nathan, E. C., 131(642), *201*
Nauman, R. V., 223(103), 257(727), 263 (727), *293, 312*
Naumova, T. M., 233(368), 234(382), 236, 241(382,485), 257(382), 258(382,485), 264, 280(1037), *301, 302, 304, 321*
Navatte, J. C., 258(738,750), *312*
Navon, G., 276(937), 286(937), *318*
Nazarenko, A. I., 223(124), 225(165), 241 (488-490), 255(690), 256(165), 258 (124,490), 294, *295, 305, 310*
Neadle, D. J., 178(1005), *209*
Nebe, W. J., 36(139), *188*
Neckers, D. C., 3, 49(230), 53(264,266), *190, 191*
Nelander, B., 239(253,452), *304*
Nelsen, S. F., 118(571), 153(751,753), 154(756), *199, 203, 204*
Nelson, A. J., 53(267,269), *191*
Nelson, D. H., 106(511), 108(511), *197*
Nelson, P. H., 126(616), *200*
Nelson, P. J., 88(436), 135(436), *195*
Nemodruk, A. A., *338*
Nemoto, M., 145(703,704), 185(704), *202, 272(817), 315*
Nersesova, G. N., 241(478,485), 261(478), 264, *304*
Neuberger, K. R., 12(46), 77(46), 96(46), 131(651), 132(651), *186, 201,* 278(984, 993), 288, *320*
Neugebauer, F. A., 167(864), *206*
Neumeyer, J. L., 47(202), *189*
Neustadt, B. R., 47(202), *189*
Ng Lim, L. S., 15(61), *186*
Nicholson, A. A., 131(649), 132(649), *201*
Niclause, M., 158(785), *204*
Nicoderm, D. C., 96(468), *196*

Nieman, G. C., 251(612), 253(636), 257 (726), 259(754), 263(726), 271(754), *308, 309, 312, 313*
Nieman, Z., 231(306), *299*
Niemczk, M., 102(483), *197*
Nieswandt, K., 279(1007), *320*
Niizuma, S., 246(574), *307*
Nikoloff, P., 41(163), *189*
Ninomiya, I., 120(581,585), *199*
Nishida, S., 131(653), *201*
Nishikawa, M., 279(1013), *321*
Nishimoto, K., 229(250), *298*
Nitta, M., 56(278), *191*
Niyakov, E. N., 258(740), *312*
Nizamov, N., 274(866), *316*
Nnadi, J. C., 123(599), 162(599,823), *200, 205*
Noda, H., 182(1047), *211*
Noe, L. J., 244(541), 246(541), *306*
Nofre, C., 163(830), *205*
Noltes, J. G., 221(52), *292*
Nomi, T., 52(248), *191*
Nonnenmacher, R., 141(690), *202*
Nordin, S., 278(1004), 288, *320*
Norfolk, D. H., 239(445), *303*
Norton, G. D., 106(511), 108(511), *197*
Noskova, M. P., 220(27), *291*
Nouchi, G., 234(376,380,389), 241(479), *301, 302, 304*
Nouguier, R., 180(1028), *210*
Novak, J. R., 233(364), 234(364), *301*
Novikova, T. S., 230(284), *299*
Noyes, W. A., Jr., 3
Noyori, R., 44(179), 177(974), 185(1069), *189, 209, 211*
Nozaki, H., 13(52), 44(179), 52(248), 94 (52), 95(458b), 165(851), 177(974), 185 (1069), *186, 189, 191, 196, 206, 209, 211*
Nugent, L. J., 329(17), *338*
Nuhn, P., 230(286), *299*
Nurmukametov, R. N., 57(294), *192,* 226 (189), 227(206), 228(231), 229(247), 241(231), 248, 250(578a), 255(696), 256 (247,707), 258(189,231), *296, 297, 307, 311*
Nuzuma, S., 184(1067), *211*
Nyburg, S. C., 162(813,815a), *205*
Nyman, C. J., 336, *339*

AUTHOR INDEX

Obara, H., 130(628,629,630), 179(1025), *200, 201, 210*
Obata, N., 165(855), 170(902a), 176 (902a), *206, 207*
Obashi, H., 277(953), *319*
Obolentsev, R. D., 221(56), *292*
Obyknovennaya, I. E., 275(902), 285(902), *317*
Ochiai, E., 335(55), *339*
Ochiai, H., 163(826), *205*
O'Connell, E. J., 108(518), *197,* 259(759), *313*
Odaira, Y., 113(546-548), 114(546,547), *198*
Odiot, S., 220(38), *292*
O'Donnell, C. M., 96(468), *196*
Odum, R. A., 168(874), 172(916), *206, 207*
O'Dwyer, M. F., 277(949,950), 287, *319*
Oettle, W. F., 117(570), *199*
Offen, H. W., 259(760,761), 260(762,763, 763a), 265, 271(760,761), 272(842), 284, *313, 315*
Ogata, M., 106(496), 108(496), 167(862), 168(873), *197, 206*
Ogata, N., 106(507), 108(507), *197*
Ogata, Y., 46(187), 48(205), 79(412), 104(412), 111(412), 113(544), *189, 190, 195, 198*
Ogawa, M., 222(68), 254(68,671), *292, 310*
Ogi, Y., 116(559d), *199*
Ogihara, Y., 157(777), *204*
Ogloblina, A. I., 241(485), 258(485), 264, *304*
Ogura, K., 147(709), *203*
Ogurtsova, L. A., 250(580), *307*
Oh, D. Y., 113(545a), *198*
Oh, J. H., 240(457,460), *304*
Oh, K. H., 47(202), *189*
O'Hare, D. O., 222(69), *293*
Ohashi, T., 12(28), 108(28), *186*
Ohishi, N., 261(779a), 270(779a), *314*
Ohnishi, Y., 123(600,600b), 124(600a), *200*
Ohki, E., 122(597), *200*
Ohloff, G., 26(103), *187*
Ohnesorge, W. E., 328, *338*
Ohno, A., 123(600,600b), 124(600a), *200*
Ohoronyk, H., 126(607), *200,* 245(547), *306*

Ohta, M., 168(893), 169(893), *207*
Ohta, Y., 37(143), *188*
Oida, S., 122(597), *200*
Oine, T., 164(849), 165(849), 166(857), *206*
Oka, K., 256(705), *311*
Oka, S., 21(85,86), *187*
Okada, M., 172(913), *207*
Okada, T., 165(851), *206,* 274(881), 277 (953), *317, 319*
Okahara, M., 12(28), 108(28), *186*
Okamura, S., 15(59), *186*
Okano, T., 240(462,467), 272(822), 274 (882), 282(822), *304, 315, 317*
Okumura, K., 50(234), *190*
Okutsu, E., 120(580), *199*
Olander, C. R., 49(229), *190*
Olive, J.-L., 126(608,610), *200*
Olmsted, J., 66(354), *193*
Olszowski, A., 241(493), 258(493), 259 (493), 264, *305*
Omura, K., 47(204), 50(235), *190*
Onari, S., 222(81), *293*
O'Neal, H. E., 70(374), *194*
Ono, H., 179(1021), *210*
Ono, K., 160(801a), *205*
Onodera, J., 130(628,629), *200*
Oohari, H., 274(881), *317*
Oosterhoff, L. J., 29(113), *187,* 222(91), *293*
Orfanos, V., 41(166), *189*
Orger, B. H., 46(191), *189*
Orlando, C. M., 77(401), 82(401), 95(401), 108(401), *194,* 276(932), *318*
Orloff, M. K., 223(108), 230(277), 233 (366), 234(366), 235(366), 244, 245 (552a), 252(623), 262, 271(623), 282, *298, 301, 306, 308*
Orlov, V. D., 232(345), *301*
Osawa, T., 19(75), *187*
Ostapenko, N. I., 242(499), *305*
Osterroht, C., 125(604), *200*
Ostertag, R., 259(759a), *313*
Ostrem, D., 88(436), 135(436), *195*
Osugi, J., 160(801a), *205*
Oth, J. F. M., 40(157), 41(163,165), *188, 189*
Ottolenghi, M., 49(223), 164(836), *190, 206,* 257(717), 276(923), 279(1030), 285(923), *311, 318, 321*

Oudman, D., 160(802), *205*
Ourisson, G., 117(564), *199*
Owens, D. V., 270(802), 282(802), *314*
Ozerova, G. A., 50(233), *190,* 234(388), 255(696), *302, 311*

Pac, C., 48(218,220), 64(348), 70(220), 114(552), *190, 193, 198,* 274(876,877), *316*
Pacific, J. G., 81(417), 171(904-906), *195, 207*
Padhye, M. R., 256(697), *311*
Padwa, A., 4, 92(452), 96(470), 100(470), 102(483,484), 164(832), *196, 197, 205,* 276(919), 285(919), *318*
Page, B. D., 126(609), *200*
Pagni, R. M., 23(95), *187*
Paice, J. C., 160(795), 161(795,803), *205*
Paioni, R., 65(352), *193*
Pai Verneker, V. R., 243(530), *306*
Pak, M. A., 50(233), *190,* 234(388), *302*
Pakula, B., 257(721,723), *311, 312*
Pal'chevskii, V. V., 238(425), *303*
Palmowski, J., 46(189a), *189*
Pancir, J., 223(107), *294*
Pande, C. D., 114(551), *198*
Pandey, B. R., 231(296), *299*
Pant, D. D., 240(473), 261(473), *304*
Pant, H. C., 240(473), 261(473), *304*
Pant, K. C., 240(473), 261(473), *304*
Paoloni, L., 228(220), *297*
Papaconstantinou, E., 334(49)
Pape, M., 4, 106(506), *197*
Pappas, B. C., 156(765,766), *204*
Pappas, S. P., 156(765,766), 160(793), *204*
Paquette, L. A., 26(104), 35(137,138), 37 (104), 43(173), 89(438), 93(453), 94 (456a), 142(453), 164(835), *187-189, 195, 196, 206*
Park, C. H., 228(234), *297*
Park, J. D., 15(60), *186*
Parkanyi, C., 223(105), *294*
Parkash, V., 255(686), *310*
Parker, C. A., 5, 96(462), 106(462), *196,* 248, 251(604,605,613,615), 278(975), 288, *308, 319,* 333(41), 334(41), *338*
Parkin, J. E., 217, *291*
Parmenter, C. S., 255(674), 269(674), 271 (805), 276(926), 281, *310, 315, 318*
Pashayan, D., 96(470), 100(470), 164 (832), *196, 205,* 276(919), 285(919), *318*
Pashkevich, Yu. M., 226(183), *296*
Paspaleev, E., 239(438), *303*
Pasternack, G., 221(54,55), *292*
Pasto, D. J., 4
Pastohr, H., 71(381), *194*
Pasztor-Bartoszewicz, B., 245(541a), *306*
Pathak, A. N., 224(127), *294*
Patsko, A. I., 272(818), *315*
Patterson, J. M., 42(167), 94(176), *189*
Paul, J. C., 131(643), *201*
Paulson, D. R., 14(55), *186*
Pavlopoulos, T. G., 249, 251(606), *308*
Payo, E., 16(64), *186*
Paxton, J., 105(494), *197,* 246(559), *307*
Pearlstein, R. M., 277(944a), *318*
Pearson, E. F., 222(73), *293*
Pecchold, E., 126(617), 128(617), 168 (876), *200, 206*
Pedersen, L., 57(290), *192*
Pedersen, L. G., 245(548), *306*
Pedler, A. E., 47(192), *189*
Pellois, A., 258(738,742,743,750), *312*
Pereira, L. C., 253(634), 269(634), *309*
Perkampus, H. H., 256(700), *311*
Perkins, M. J., 4, 173(935), *208*
Perkins, P. G., 223(100), *293*
Permogorov, V. I., 217, 218, 220(20), 250 (8), *291*
Perold, G. W., 117(564), *199*
Perona, M. J., 85(429), *195*
Perrin, C. L., 230(274), *298*
Perrin, M. H., 218, 230(274), *298*
Perrins, N. C., 46(186), *189*
Personov, R. I., 258(733,734,746), 264, *312*
Perumareddi, J. R., 325, *338*
Pesteil, L., 233(350), 235(350), 273(350), 283(350), *301*
Pesteil, P., 233(350), 235(350), 273(350), 283(350), *301*
Pete, J. P., 88(435), *195*
Peters, A. T., 219(143), 224(143), *295*
Peterson, F. C., 231(302), 233(359), 273 (359), *299, 301*
Peterson, O. G., 272(816), *315*
Petrenko, G. P., 62(331), *193*
Petrov, I., 238(431), *303*
Petrowski, G. E., 37(142), *188*

AUTHOR INDEX

Petrusis, C. T., 115(557), *198*
Petterson, R. C., 48(208), *190*
Peuker, K., 260(771), 265, *313*
Peyerimhoff, S. D., 217(102), 223(102), *293*
Pfau, M., 101(481), 110(481), *196*
Pfefferkorn, R., 275(898), *317*
Pflughaupt, K. H. W., 163(831), *205*
Pfundt, G., 57(297), *192*
Phillips, D. T., 272(842), 284, *315*
Phillips, G. O., 157(775), *204*
Phillips, J. C., 35(137), *188*
Phillips, R. W., 12(25), *185*
Philpott, M. R., 237, 242(509), 243(523), *305*
Piek, H. J., 157(772), *204*
Pierce, R. A., 235(403), *302*
Pietra, F., 5, 47(203), *190*
Pietra, S., 51(239), *191*
Pikulik, L. G., 253(649), 254(659), 261 (782), 270(649), 274(869), 279(1012), *309, 314, 316, 320*
Pinhey, J. T., 47(199,200), 114(553), *189, 198*
Pirkle, W. H., 147(708), 150(731), *203*
Piskunov, A. K., 255(696), *311*
Pispisa, B., 230(278), *298*
Pistic, S., 221(47), *292*
Piterskaya, I. V., 260(776), 266, *313*
Pitt, C. G., 221(51), *292*
Pitts, J. N., 117(569), *199*
Pitts, J. N., Jr., 3, 4, 79(410), *194,* 223 (105), *294,* 334(48), *339*
Plank, D. A., 114(550), 131(645), 146 (707), *198, 201, 202*
Platanova, N. V., 227(202), *296*
Pleiss, M. G., 163(826), *205*
Plepys, R. A., 19(77), *187*
Plesch, P. H., 223(119,122), *294*
Pliev, T. N., 224(130-131a), *294*
Plimmer, J. R., 47(198), 54(272), 183 (1055), *189, 191, 211*
Plunkett, A. O., 154(755), *204*
Pobedimskii, D. G., 12(35), *186*
Poc, C., 47(197), *189*
Podgornyi, A. P., 250(580), *307*
Poduzhailo, V. F., 228(232), 256(232), *297*
Pohl, L., 179(1018), *210*
Pokhodenko, V. D., 49(224), *190*

Pokrovskaya, F. S., 250(580), *307*
Poland, H. M., 271(805), 281, *315*
Polansky, O. E., 131(646), *201,* 223(101), *293*
Poletova, V. N., 224(131,131a), *294*
Pollitt, R. L., 178(1005), *209*
Pomerantz, M., 52(259), *191*
Pond, D. M., 77(403), 84(421a), 131(651), 132(651), *194, 195, 201,* 278(984), *320*
Ponochovnyi, V. I., 253(651), 276(918), 285(918), *309, 318*
Pool, G. L., 278(985), *320*
Poole, J. A., 254(666,667), 263, *310*
Poon, L., 143(695), *202*
Pope, M., 242(504), 272(832), *305, 315*
Pope, M. E., 280(1051,1052), 290, *322*
Popov, A. I., 240(454), *304*
Popov, A. K., 217, 250(15), *291*
Popov, K. R., 224(150), *295*
Popova, K. R., 227(202), *296*
Popova, T. Ya., 217, 250(15), *291*
Porter, G., 157(771), *204,* 233(363), 234 (363,390), 257(715,718), *301, 302, 311*
Porter, G. B., 330, 335, *338*
Porter, G. G., 219, 226(174), *296*
Portnoy, N. A., 156(765), *204*
Postashnik, R., 164(836), *206*
Postawka, A., 226(191), *296*
Potashnik, R., 276(923), 279(1030), 285 (923), *318, 321*
Poznyak, A. L., 337(68), *339*
Praefcke, K., 57(284), *192*
Prajer-Janczewska, L., 226(191), *296*
Pramer, D., 225(167), *295*
Prell, G., 182(1049), 184(1068), *211*
Preuss, J., 243(525), *306*
Priimachek, V. R., 275(904), 280(1035), *317, 321*
Prikhot'ko, A. F., 243(520,527), 258(744), *305, 306, 312*
Prinzbach, H., 113(540), 149(725), 152 (540,725,744-746,748), 153(746,749), 168(884), *198, 203, 207*
Pritchard, G. O., 85(429), *195*
Prochazkova, J., 229(256), *298*
Prochorow, J., 240(461), 254(461), *304*
Proskuryakova, N. S., 227(206), 257(732), 259(732), 264(732), *296, 312*
Przybytek, J. T., 72(391), *194*
Ptak, M., 246(563), *307*

Pullman, B., 229(263), 231(306), *298, 299*
Pushkina, L. N., 228(231), 241(231), 258(231), *297*
Put, J., 155(760), *204*

Quadrifoglio, F., 230(278), *298*
Queguiner, G., 151(741), *203*
Quinkert, G., 46(189a), *189*

Rabalais, J. W., 252(628,629), 259(629), 266, 271(628), 280(1034), *308, 309, 1034*
Rabinowitch, E., 232(327), 256(327), 261(327), *300*
Raciszewski, Z., 164(842), *206,* 260(779), 266(779), 270(799), *313*
Radvilavicius, C., 242(507), *305*
Rahn, R. O., 162(819), *205*
Rajagopal, S., 232(331), *300*
Rama Rao, C. G., 224(140), *295*
Ramiah, K. V., 239(451), *304*
Ramme, G., 142(693), *202,* 234(374), 236(374), 278(374), *301*
Ramsay, C. C. R., 160(795), 161(795), *205*
Ramsay, D. A., 227(209), *296*
Ramsay, G. C., 178(986,987,988a), *209*
Ramsay, I. A., 234(396), 255(396), *302*
Rao, C. N. R., 221(41,42), *292*
Rasmussen, R. W. W., 134(663,665), *201*
Rassat, A., 178(985), *209*
Rathi, S. S., 274(865), *316*
Rau, J. D., 276(926), *318*
Rauh, R. D., 29(118), *188,* 279(1017,1019, 1024), 289(1017,1019), *321*
Rautian, S. G., 217, 250(15), *291*
Raymonda, J. W., 221(61), *292*
Razumova, V. P., 225(169,170), *295*
Read, I. A., 221(44), *292*
Rees, C. W., 121(588), 165(588), 175(941), *199, 208*
Rees, L. W., 4
Reetz, M., 176(953), *208*
Regitz, M., 176(955), *208*
Rehm, D., 274(888), *317*
Reid, S. T., 178(994), *209*
Reilhac, L. De., 216, 222(70), *293*
Reilly, C., 272(821), 281(821), 282(821), *315*
Reineke, C. E., 19(77), *187*
Reinisch, R. F., 12(32), *186*

Reisch, V. J., 134(662a), *201*
Remizov, A. B., 230(279), *298*
Renkes, G. D., 70(375), 73(392a), *194*
Rentzepis, P. M., 254(665), 263(665), 268, 269(791), 270(665), 280, 281, *310, 314*
Reske, G., 66(358), *193,* 274(862), *316*
Rettig, K. R., 175(947), 176(947), *208*
Reutov, G. A., 238(418a), *303*
Reznikova, I. I., 257(728), 263(728), *312*
Rhee, J. J.-Y., 218(248), 229(248), *297*
Rhodes, W., 268, 269(787,788), *314*
Rice, S., 272(826), 283(826), *315*
Rice, S. A., 57(289), *192*
Richards, F. F., 176(951), *208*
Richborn, B., 85(429), *195*
Richtol, H. H., 111(529), 142(693), *198, 202,* 234(373,374), 236, 278(374), *301*
Rieckhoff, K. E., 256(704), *311*
Rieke, P. D., 180(1028a), *210*
Rieker, A., 225(158), *295*
Rigaudy, J., 8(7), *185*
Rigby, R. D. G., 47(199,200), 114(553), *189, 198*
Riley, R. F., 244, 245(552), *306*
Rings, M., 42(169a), *189*
Ripoche, J., 258(738,742,743,750), *312*
Ritchie, R. K., 222(69), *293*
Riter, J. R., 233(347), *301*
Rivas, C., 16(64), *186*
Riveros, J. M., 222(71), *293*
Roazhkova, N. G., 229(259), *298*
Roberts, B. G., 66(355), *193,* 227(205), 252(623), 255(205), 262(623), 271(623), 282, *296, 308*
Roberts, D. B., 53(265), *191*
Roberts, D. R., 279(1016), 289, *321*
Roberts, G. L., 223(118), *294*
Roberts, J. P., 64(349), *193,* 274(884), *317*
Roberts, T. D., 121(589,590), *199*
Robin, M. B., 217(64,65), 220(17), 221(64,65), *291, 292*
Robinson, G. E., 47(194), *189*
Robinson, G. W., 233(351), 235(351), *301*
Rockley, M. G., 73(392,392a), *194*
Rodighiero, G., 162(817), *205*
Roe, D. K., 231(313), *299*
Roebber, J. L., 221(59), 254(59), *292*
Rogers, T. R., 145(701), *202*
Rohatgi, K. K., 276(931), 286, *318*
Rohatigi, K. K., 277(1006), *320*

AUTHOR INDEX

Rokos, K., 261(780), *314*
Romand, J., 216, 222(70), *293*
Romanyuk, N. A., 243(511), *305*
Romine, H. E., 15(60), *186*
Rondeau, R., 52(257), 57(304), *191, 192*
Roselius, E., 106(499,506), 197
Rosen, J. D., 18(73), 19(74), *186*
Rosenberg, H. M., 52(257), 57(304), 164 (845), 165, *191, 192, 206,* 228(219), 233 (219), *297*
Rosenthal, I., 163(827,828), 184(827,828, 1063), *205, 211*
Rosenthal, J., 244, 245(552), *306*
Roshchupkin, D. I., 224(133), 234(398), 254(133), *294, 302*
Rosiello, L. A., 329, *338*
Ross, I. G., 227(209), 244(534), 251(600), 255(674a), 256(699), 269(699), *296, 306, 308, 310, 311*
Rossiter, B. W., 106(504), 108(504), 116 (504), *197*
Roswell, D. F., 70(377), *194,* 278(972), 279(1016), 289, *319, 321*
Roth, H. J., 178(1000), *209*
Roth, W. R., 170(898), *207*
Rothman, L. J., 25(97,98), 45(97), *187*
Rotkiewicz, K., 257(719), *311*
Rottele, H., 41(163,164,165), *189*
Rousset, Y., 273(838), 283(838), *315*
Roy, S. K., 149(721), *203*
Rozum, Yu. S., 230(280), *298*
Rubin, M. B., 5, 35(134,135), 36(134), 127(621), 158(778), 161(805), *188, 200, 204, 205*
Rubinstein, H., 71(380), *194*
Rudik, K. I., 251(596-598), 254(597), *308*
Rudolph, K., 226(191), *296*
Rudowska, D., 333(38), *338*
Ruf, H. H., 272(868), 283(868), *316*
Ruff, F., 227(198), *296*
Rumin, R., 55(276), *191*
Rumyantsev, B. M., 280(1045), *321*
Rundel, W., 49(225-227), *190,* 225(158), *295*
Russell, D. W., 178(1007), *210*
Russo, G., 113(543), 120(543), 151(735, 1738), *198, 203*
Ruszkay, R. J., 156(764), *204*
Rutter, J., 176(955), *208*
Ruziewicz, Z., 241(493), 258(493), 259 (493,749), 264, *305, 312*
Rygalov, L. N., 235(408), *302*

Saboz, J. A., 141(691), *202*
Saburi, Y., 49(229a), *190*
Sadlej, A. J., 220(16), 240(16), 273(845), 284, *291, 316*
Sadovskii, N. A., 274(872), *316*
Sakai, K., 68(367), 89(439), 90(367,439), *193, 195*
Sakai, T., 37(143), 177(974), *188, 209*
Sakaino, Y., 230(287), *299*
Sakamoto, K., 220(24), *291*
Saklej, A. J., 257(719), *311*
Sakurai, H., 12(34), 47(197), 48(218,220), 50(234), 51(237), 64(348), 67(365a), 68 (366), 70(220), 114(552), 226(179), 274 (876,877), *186, 189, 190, 193, 198, 296, 316*
Salamon, Z., 232(328), 261(328), *300*
Salem, L., 141(689), *202*
Salna, V., 232(334-336), 238(423), *300, 303*
Salovey, R., 12(31), *186*
Saltiel, J., 12(46), 15(61), 29(114,116), 30 (119), 57(292), 77(46), 96(46), 131(651), 132(651), 160(796), 161(796), 270 (804a), 278(971,984,993), 288, *186, 187, 188, 192, 201, 205, 314, 319, 320*
Saltzman, M. D., 106(514), 108(514), *197*
Salzman, W. R., 233(348), 235, *301*
Samuelson, G. E., 36(141), *188*
Sanders, D. C., 63(337), *193*
Sandorfy, C., 64(342), 222(72), 227(207), *193, 293, 296*
Sandros, K., 60(317), 106(317), 158(779), *192, 204*
Sanford, E. C., 32(123), *188*
Sanjiki, T., 168(893), 169(893), *207*
Sankawa, U., 157(777), *204*
Santavy, F., 223(121), *294*
Santry, D. P., 126(607), 245(547), *200, 306*
Santucci, L., 226(180), *296*
Sanyal, N. K., 224(136), *295*
Sargent, M. V., 42(171), *189*
Sarholz, W., 239(433), *303*
Sarkar, I. M., 48(208), *190*
Sarver, E. W., 101(481), 110(481), *196*
Sarzhevskii, A. M., 255(688), 261(783,784), 264(783,784), 266(783,784), *310, 314*

Sasaki, T., 164(844), 183(1056), *206, 211*
Sasse, W. H. F., 53(265), 56(282), 60(282), 65(353), *191, 193*
Sastri, V. S., 330, 337, *338*
Sato, E., 163(825), *205*
Sato, H., 48(206), *190*
Sato, N., 179(1019,1020,1026), 180(1026), *210*
Sato, S., 12(44,45), 44(175), 278(990,991), 279(1008), *186, 189, 320*
Sato, T., 28(111), 45(180), 47(195), 51(238), 53(268), *187, 189, 190, 191*
Sato, Y., 126(618), 131(638), 139(681), *200, 201, 202*
Satoh, S., 279(1008), *320*
Sauer, M. C., Jr., 279(1013), *321*
Sauerbrier, M., 56(280,281), 71(281), *191*
Sauers, R. R., 69(371,372), 77(371,372), 78(371,372), 103(371,372), 105(372), *194*
Saunders, W. H., Jr., 278(997), *320*
Sauvageau, P., 222(72), *293*
Savin, F. A., 218, 250(11,12), *291*
Sawaki, S., 11(20,21), 172(20), *185*
Saxena, J. P., 231(300), *299*
Scala, A. A., 113(544a), *198*
Scandola, F., 330(24), 333(42), 337(62, 63), *338, 339*
Schaffner, K., 68(368), 77(398), 78(368), 85(428), 86(431), 89(437), 91(398), 96(471), 105(471), 106(471), 140(428,437, 686), 141(691), 148(717), *4, 193, 194, 195, 196, 202, 203*
Schaffner, T. J., 137(677), *202*
Schaffner-Sabba, K., 148(713), *203*
Scharf, D. H., *5*
Scharf, H. D., 12(38), 51(243), 130(243), 155(762), *186, 191, 204*
Schauer, A., 175(940), *208*
Schechter, H., 63(337), 121(589), 177(971), *193, 199, 209*
Scheffer, J. R., 94(457,458), *196*
Scheibe, G., 230(276), 232(322), 256(322), *298, 300*
Scheinbaum, M. L., 180(1037), *210*
Scheiner, P., 167(871), 173(930), 175(871), 180(1027), *206, 208, 210*
Scheithauer, S., 232(337), *300*
Schenck, G. H., 239(449), *303*
Schenck, G. O., 38(146), 106(499,506), 157(769), *188, 197, 204*
Scheppers, G., 175(948), *208*
Scherr, V. M., 253(655), *309*
Schiemenz, G. P., *302*
Schier, W., 122(596), *199*
Schinski, W., 69(372), 77(372), 78(372), 103(372), 105(372), *194*
Schlaefer, H. L., 325, 327, 329, 332, 333(35), *338*
Schlag, E. W., 253(648), 268, 269(632,648, 786), *309, 314*
Schlessinger, R. H., 62(334,335), *193*
Schmall, B., 168(874), *206*
Schmid, H., 174(937), *208*
Schmidt, J., 246(569,570), *307*
Schmidt, M. W., 279(1011), *320*
Schmidt, P. P., 272(824), 283, *315*
Schmidt, U., 125(604), *200*
Schneider, A. S., 229(265), *298*
Schneider, B., 230(267,268), 256(267), *298*
Schneider, F., 260(778), 266, *313*
Schneider, S., 223(106), *294*
Schollkopf, U., 176(953), *208*
Scholz, K.-H., 105(495), 160(495), *197*
Scholz, M., *4*
Schomburg, G., 279(1007), *320*
Schonberg, A., 57(284), *3, 192*
Schonleber, D., 175(945), 176(945), *208*
Schoonveld, L., 222(74), *293*
Schott, H. N., 97(473,474), *196*
Schrader, L., 175(948-950), *208*
Schreiber, K., 10(18), *185*
Schroder, G., 40(157), 41(163-165), *188, 189*
Schroeter, S. H., 71(387), 77(399,401), 82(401), 95(401), 108(399,401), *194*
Schryver, F. C., de, 164(833), *205*
Schulte-Elte, K. H., 26(103), *187*
Schulte-Frohlinde, D., 275(898), *317*
Schultz, A. G., 62(334,335), *193*
Schurter, J. J., 57(302,303), 62(303), 64(302,303), 65(303), *192*
Schuster, D. I., 146(705), 148(710,719a), *202, 203*
Schuster, G. B., 126(615), *200*
Schuster, H., 276(922), 285(922), *318*
Schuster, P., 223(101), *293*
Schutte, L., 78(407,408), 88(408), *194*
Schwartz, J., 40(156), 41(156), *188*
Schwoerer, M., 60(321), 246(571), 263

AUTHOR INDEX

(571), *192, 307*
Scott, A. R., 158(780), 234(381), *204, 302*
Scott, J. D., 226(195), 255(195), *296*
Scribe, P., 27(108), *187*
Sears, A. B., 132(658), *201*
Seebach, D., 12(41), 29(41), *186*
Seeley, D. A., 38(148), *188*
Seeley, G. G., 178(998), *209*
Seelye, R. N., 178(990), *209*
Seguchi, K., 128(623), *200*
Segupta, S. K., 229(237), *297*
Seibold, K., 218(254), 229(254), *298*
Seidl, H., 221(49,50), 223(99), *292, 293*
Seidner, R. T., 35(136), *188*
Seiffert, W., 182(1049), 230(267-270), 256 (267), 257(269), *298*
Seite, J., 260(764), *313*
Sekanove, S., 239(440), *303*
Selinger, B., 60(318), 66(356), *192, 193*
Selinger, B. K., 59(315), 62(323), 64(343), 273(847), 284, *192, 193, 316*
Seliskar, C. J., 260(777), 266, 271(777), *313*
Selvarajan, R., 172(912), 173(934,934a), *207, 208*
Semeluk, G. P., 158(786), 255(675), 271 (808), 279(675), *204, 310, 315*
Semina, G. N., 180(1032), *210*
Senkowski, T., 333(38,39), *338*
Seo, S., 157(777), *204*
Serdyukova, L. A., 217, 250(8), *291*
Serebryakova, I. K., 225(155), *295*
Serve, P., 52(257), 57(304), *191, 192*
Sethuraman, V., 220(35), *291*
Seto, S., 151(736), *203*
Sevchenko, A. N., 232(325), 241(325,484), 258(484), *300, 304*
Sevcik, J., 225(164), *295*
Sevost'yanova, V. V., 230(284), *299*
Seybold, P. G., 248, 250(583), 254(583, 664), *307, 310*
Seyfarth, H. E., 71(381), *194*
Sgamelloti, A., 222(87), *293*
Shablya, A. V., 231(311), 275(915), 285 (915), *299, 318*
Shaffer, G. W., 69(373), 137(675), *194, 202*
Shagisultanova, G. A., *339*
Shah, S., 176(961), *208*
Shain, A. L., 279(1020), *321*

Shakhverdov, T. A., 235(404), 278(967), *302, 319*
Shani, A., 129(627), *200*
Shannon, P. U. R., 88(434), *195*
Sharma, B. K., 332(27), *338*
Sharma, D., 223(123), 224(136), *294, 295*
Sharma, R. K., *4*
Sharnoff, M., 245(555,558,560), 246(558, 560), *306, 307*
Sharpe, R. R., 157(775), *204*
Shchegloa, N. A., 255(694), *311*
Sheehan, J. C., 120(582), *199*
Shefter, E., 92(452), *196*
Sheglova, N. A., 275(911), 285(911), *317*
Sheheglova, N. A., 227(196,197), 255(197), *296*
Sheiner, P., 175(939), *208*
Sheka, E. F., 272(827), *315*
Shekk, Yu. B., 277(994,995), *320*
Shen, K., 39(152), *188*
Shepel, A. V., 230(290), 245(290), 247, *299*
Sheridan, J. B., 16(63), 152(63), *186*
Sherwin, M. A., 23(88,91), *187*
Shetter, M. D., 75(396), 78(396), *194*
Shevchenko, E. A., 230(288), *299*
Shibata, S., 157(777), *204*
Shibata, T., 46(185), 61(185), *189*
Shieh, T.-C., 252(631), 271(631), *309*
Shigemitsu, Y., 113(546,547), 114(546, 547), *198*
Shigeoka, T., 273(839), *315*
Shigorin, D. N., 50(233), 227(196,197), 229(247), 234(588), 248, 250(578a), 250(578a), 255(197,694,696), 256(247, 707), 275(911), 285(911), *190, 296, 297, 302, 307, 311, 317*
Shim, S. C., 62(325), 254(668), 263, *192, 310*
Shimada, R., 251(609), *308*
Shimada, S., 51(238), *190*
Shimizu, I., 52(256), *191*
Shimizu, K., 180(1031), *210*
Shimka, K., 51(237), 67(365a), 68(366), *190, 193*
Shindo, N., 49(224a), *190*
Shine, H. J., 49(222a), *190*
Shinomiya, M., 80(1031), *210*
Shirkov, V. I., 257(728), 263(728), *312*
Shirotani, I., 227(208), 259(761a), 265

(761a), 275(903), 285(903), *296, 313, 317*
Shizuka, H., 113(545), 119(576,577,577a), 120(580), *198, 199*
Shkirmn, S. F., 232(324), 252(324), *300*
Shogorin, D. N., 226(189), 258(189), *296*
Short, G. D., 276(921), 285(921), *318*
Shortridge, R. G., Jr., 279(1011), *320*
Shosenji, H., 230(281), *298*
Shpak, M. T., 242(499), *305*
Shubert, T. A., 225(171), *295*
Shukla, P. R., 275(914), 285(914), *317*
Shusheriva, N. P., 117(536), *4, 198*
Siano, D. B., 217, *291*
Sichkar, O. N., 234(392), 241(392,405), 258(392,405), 264, *302*
Siebrand, W., 249, 251(618,626), 270(797), *308, 314*
Sieczkowski, J., 16(65), 84(65), *186*
Siegel, S., 236, 253(637), 271(637,811), 277(996), 281(811), *301, 309, 315, 320*
Siegoczynski, R., 240(461), 254(461), *304*
Signore, R., 279(1003), *320*
Sigwalt, C., 181(1040), 183(1040), *210*
Sigwart, C., 336, *339*
Silvie, C., 234(389), *302*
Sim, G. A., 49(229), *190*
Simmonds, D., 231(301), *299*
Simmons, J. D., 221(67), *292*
Simonaitis, R., 117(565,569), *199*
Simonetta, M., 223(114a), *294*
Simons, J. P., 46(186), *189*
Simpson, W. T., 242(510), *305*
Simson, J. M., 172(919), *207*
Singer, L. A., 108(519), 111(519,534), 177(969), 274(886,887), 285, *197, 198, 209, 317*
Singh, A., 158(780), 234(381), *204, 302*
Singh, A. N., 254(669,670), *310*
Singh, B., 167(865), *206*
Singh, I. S., 254(669,670), *310*
Singh, K., 224(128), *294*
Singh, O. N., 255(678), *310*
Singh, P., 149(721-723), *203*
Singh, P. D., 224(127), *294*
Singh, R. N., 226(182), *296*
Singh, R. S., 226(193), 227(200), 255(200, 678,695), *296, 310, 311*
Singh, S., 64(342), 227(207), *193, 296*
Singh, S. N., 226(193), 227(200), 255(200, 692,695), *296, 311*
Singh, S. P., 72(391), 79(413), 115(413), *194, 195*
Singh, V. B., 224(128), *294*
Singhal, G. S., 232(326,327), 252(617), 256(327), 261(326,327), 262, 276(931), 277(1006), 286, *300, 308, 318, 320*
Sircar, J. C., 36(140), *188*
Sixl, H., 60(321), 246(571), 263, *192, 307*
Sjoberg, B., 160(792), *204*
Skorobogat'Ko, A. F., 243(520,527), 258(744), *305, 306,* 312
Skripko, L. A., 224(144), *295*
Skrowaczewska, Z., 231(298,299), *299*
Skulski, L., 226(188), *296*
Skvortsov, B. V., 235(408), *302*
Slamnik, M., 219, 230(283), *298*
Slifkin, M. A., 239(446), *303*
Slovetskii, V. I., 230(284), *299*
Small, G. J., 227(203), 252(203), 262, *296*
Smelyakova, V. B., 225(165), 256(165), *295*
Smets, G., 52(254), 131(641), *191, 201*
Smirnov, E. A., 260(772), 266, *313*
Smirnov, L. V., 227(202), *296*
Smirnov, V. A., 277(948), 288, *319*
Smith, A. B., 132(659), *201*
Smith, B. E., 48(222), 249, 251(610), 257(714), *190, 308, 311*
Smith, B. M., 239(446), *303*
Smith, H. E., 137(677), *202*
Smith, J. H., 29(114), *187*
Smith, K. C., 162(820), 163(829), *205*
Smith, N. H. P., 225(152), *295*
Smith, R. H., 179(1009), *210*
Smith, R. H., Jr., 179(1010), *210*
Smith, R. L., 48(210), 119(575), 168(889), *190, 199, 207*
Smith, S. G., 147(708), *203*
Smith, W. F., 106(504,505), 108(504), 116(504,505), *197*
Snatzke, G., 170(899), 176(953a), *207, 208*
Snavely, B. B., 272(816), *315*
Snieckus, V., 177(975), 183(1058), *209, 211*
Sodtke, U., 57(284), *192*
Sokolovskii, R. I., 217, 250(15), *291*
Solodonov, V. V., 258(746), *312*
Solodunov, R. I., 258(733), *312*
Solomon, B. S., 60(319), 168(883), 275

AUTHOR INDEX

(894), *192, 207, 217*
Solov'ev, K. N., 232(324), 252(324), 272(823), 282(823), *300, 315*
Sommer, B.-S., 242(496), 272(828), *305, 315*
Sommer, U., 66(359), 233(362), 234(362, 378), *193, 301*
Song, P.-S., 184(1066), 230(275), 254(657, 658), 256(703), 272(658), 282(658), *211, 298, 309, 311*
Sonntag, C., von, 12(36), *186*
Sonoda, A., 87(433), *195*
Soos, Z. G., 273(836), 283, *315*
Sopchyshyn, F., 158(780), 234(381), *204, 302*
Soptrajanov, B., 238(431), *303*
Sosa, R., 228(220), *297*
Sousa, J. A., 178(993), *209*
Sovocool, G. W., 278(981), 288, *320*
Spagnolo, P., 53(263), 56(279), *191*
Spector, R. H., 174(936), *208*
Speed, R., 66(356), 273(847), 284, *193, 316*
Spence, J., 336, *339*
Spencer, H. E., 333(40), 335(40), *338*
Spencer, R. D., 252(616), 261, *308*
Sperati, C. R., 335(55), *339*
Sperling, J., 71(384-386), *194*
Speyer, J. L., 248(593), 251(593), *308*
Spoerke, R. W., 82(419), 83(419), *195*
Sprecher, M., 115(556), 138(678), *198, 202*
Squire, R. H., 180(1035), *210*
Srinivasacharya, K. G., 223(126), 225(161), *294, 295*
Srinivasan, B. N., 280(1034), *321*
Srinivasan, R., 42(170), 69(370), 85(428a), *4, 189, 193, 195*
Stamm, R. F., 252(623), 262, 271(623), 282, *308*
Stanislaus, J., 103(485), 257(716), *197, 311*
Stankorb, J. W., 94(458a), *196*
Stanley, W. H., 106(511), 108(511), *197*
Starkey, R. H., 143(694), *202*
Starnes, W. H., 131(645), 146(707), *201, 202*
Starvos, J. V., 49(231), *190*
Starzak, M. E., 253(632a), *309*
Staude, H., 224(134), *294*

Staudenmayer, R., 121(590), *199*
Stauff, J., 276(929), *318*
Steel, C., 60(319), 96(460,461), 106(460, 461), 168(883), 253(656), 275(894), *192, 196, 207, 309, 317*
Stegemeyer, H., 62(329), 223(110), *193, 294*
Stein, N., 106(513), 108(513), *197*
Steinmaus, H., 163(827), 184(827,1063), *205, 211*
Steinmetz, R., *4*
Steller, K. E., 179(1012), *210*
Stenberg, V. I., 164(837), *206*
Stengl, J., 229(258), *298*
Stepanov, B. I., 226(178), *296*
Stephen, J. F., 164(842), 260(779), 266(779), 270(779), *206, 313*
Stephenson, L. M., 67(364), 275(895), 283(941), *193, 317, 318*
Stermitz, F. R., 96(468), 184(1065), *4, 196, 211*
Stern, R. L., 56(283), *191*
Sterns, M., 59(315), *192*
Stevens, B., 272(820), 279(1028), 282(820), 290, *315, 321*
Stevens, R. D. S., 255(675), 279(675), *310*
Stevenson, C. D., 274(892), *317*
Stevenson, K. L., *338*
Steward, O. W., 221(53), *292*
Sticzay, T., 230(285), *299*
Still, I. W. J., 149(721), *203*
Stocker, J. H., 92(445), 96(465,469), 103(445), *195, 196*
Stolyarova, L. A., 232(329), *300*
Stolzle, K. G., 277(957), *319*
Story, P. R., 8(6), *185*
Stowell, J. C., 35(138), *188*
Strating, J., 161(806), *205*
Strausz, O. P., 115(555), *198*
Streith, J., 181(1040,1042), 183(1040, 1057), *210, 211*
Streitweiser, A., Jr., 223(109), *294*
Strel'tsova, A. A., 180(1030,1032), *210*
Strickler, S. J., 252(630), *309*
Strimer, P., 243(513), *305*
Stringham, R. S., 168(881), *206*
Strong, R. L., 111(529), 142(693), 234(373,374), 236, 278(374,1004), 288, *198, 202, 301, 320*
Strow, C. B., 38(110), *187*

Stryer, L., 277(963), *319*
Stubbs, J. K., 154(754), *204*
Studenov, V. I., 260(776), 266, *313*
Stuermer, D. H., 154(757), 336(60), *204, 339*
Sturis, A., 231(307,308), *299*
Stuzka, V., 228(235), 256(235), *297*
Suciu, P., 274(858), *316*
Sugie, M., 12(28), 108(28), *186*
Sugimori, A., 114(549), 125(549), 183 (1053), *198, 211*
Sugimura, T., 19(75), *187*
Suginome, H., 179(1017,1019-1022,1026), 180(1026), *210*
Sugiyama, H., 151(736), *203*
Sugiyama, N., 126(613,614,618), 131(638), 139(681), 152(613), *200, 201, 202*
Sugowdz, G., 65(353), *193*
Suguyama, N., 9(9), *185*
Sullivan, J. F., 49(222a), *190*
Sullivan, P. J., 275(914), 285(914), *317*
Sundberg, R. J., 179(1009,1010), *210*
Sunder-Plassman, P., 126(616), *200*
Sung, M., 130(634), *201*
Sunjik, V., 219, 230(283), *298*
Sunkel, J., 224(134), *294*
Suppan, P., 236, 247, *302*
Suprunova, A. I., 232(330), *300*
Surzur, J. M., 10(11,12), 180(1028), *185, 210*
Sussman, D. H., 151(739), *203*
Sutcliffe, L. H., 238(424), *303*
Sutherland, D. J., 18(73), 19(74), *186*
Suzuki, A., 128(623), *200*
Suzuki, E., 178(986), *209*
Suzuki, F., 58(306), 146(306), *192*
Suzuki, H., 220(29), *291*
Suzuki, S., 248, 250(589), *307*
Suzuki, Y., 58(306,307), 146(306), *192*
Sveshnikova, E. B., 271(810), *315*
Swallow, A. J., 234(394), *302*
Swan, G. A., 155(761), *204*
Swenberg, C. E., 280(1043), 290, *321*
Swenson, C. A., 221(40), *292*
Swenton, J., 125(606,606a), *200*
Swenton, J. S., 12(39), 25(100), 29(39), 51 (247), 67(363), 116(558), 172(918), *5, 186, 191, 193, 198, 207*
Sykes, A., 234(384), *302*
Syks, A., 234(406), 278(406), *302*

Symaelloti, A., 220(32), *291*
Szabo, A. G., 104(491), *197*
Szalay, L., 251(599), *308*
Szarek, W. A., 71(382), 178(995), *194, 209*
Szerenyi, P., 245(557), *306*
Szmuszkovicz, J., 45(181), *189*
Szoc, K., 111(531), *198*

Tabata, T., 37(144), 164(846), 165(846), *188, 206*
Tachikawa, R., 55(277), *191*
Tai, J. C., 221(63), *292*
Takada, S., 40(160), *188*
Takahashi, H., 130(628,629,630), *200, 201*
Takahashi, M., 164(844), *206*
Takamuku, S., 50(234), *190*
Takaya, H., 177(974), *209*
Takaya, T., 50(236), 51(236), *190*
Takeda, Y., 19(75), *187*
Takemura, T., 231(318), 254(318), 271 (318), *300*
Takenada, S., 151(736), *203*
Takizawa, T., 165(855), *206*
Talaty, E. R., 120(583,584), 121(583), *199*
Talukdar, P. B., 229(237), *297*
Tam, J. N. S., 178(989,991), *209*
Tamano, T., 93(454), *196*
Tamamura, T., 239(440), *303*
Tambovtsev, V. S., 257(731,732), 259 (731), 264(731,732), *312*
Tamura, M., 243(531), *306*
Tanabe, K., 68(367), 89(439), 90(367,439), *193, 195*
Tanaka, I., 113(545), 120(580), *198, 199*
Tanaka, J., 119(577a), 237, 242(508), 243 (515,517), *199, 305*
Tanaka, M., 237, 242(508), *305*
Tanaka, Y., 150(730), *203*
Tandon, K., 231(300), *299*
Tandon, S. P., 231(300), *299*
Tang, C. S., 96(459b), 123(599a), *196, 200*
Tanida, H., 170(896,897), 176(954), *207, 208*
Tanielian, C., 275(899), 276(899), 278 (974), 289, *317, 319*
Taniguchi, H., 172(913), *207*
Taniguchi, K., 128(623), *200*
Tanikaga, R., 177(979), *209*
Tanna, C. H., 32(125), *188*
Tanner, D. D., 12(33), *186*

AUTHOR INDEX

Tanner, S. P., 329(17), *338*
Tanquary, R. L., 260(762), 265(762), *313*
Tao, T., 248(591), 251(591), *307*
Tarrant, J. R., 329(17), *338*
Tatlow, J. C., 47(192), *189*
Taylor, C. A., 184(1061), 231(305), 257 (305), *211, 299*
Taylor, E. A., 175(940), *208*
Taylor, E. C., 184(1064), *211*
Tazuke, S., 15(59), *186*
Tchir, M. F., 131(649-651), 132(649-651), 278(984), *201, 320*
Teague, M. W., 279(1020), *321*
Tegner, L., 223(98), *293*
Telford, J. R., 168(888), *207*
Temperilli, A., 113(538), *198*
Teplaykov, P. A., 258(735), 264, *312*
Teplitskaya, T. A., 258(734), 264, *312*
Terao, S., 139(682,683), *202*
Terashima, S., 172(920,921), *207*
Teratake, S., 170(896,897), *207*
Terauchi, K., 12(34), 219(179), 226(179), *186, 296*
Ter Borg, A. P., 39(151), *188*
Terenin, A. N., 275(915), 285(915), *318, 321*
Terenzani, A. J., 225(166a), *295*
Terry, G. C., 234(385), 253(385), *302*
Terskoi, Ya. A., 277(944), 287, *318*
Testa, A. C., 250(584), 278(969), *307, 319*
Tezuka, H., 67(362), *193*
Tezuka, T., 67(365), *193*
Thakur, S. N., 224(132), *294*
Theard, L. M., 231(302), 233(359), 273 (359), *299, 301*
Thielecke, W., 12(27), *185*
Thiery, C., 246(562), *307*
Thiry, P., 51(240), *191*
Thomas, J. K., 233(357,360), *301*
Thomas, J. M., 64(340), *193*
Thomas-Magos, M. C., 258(736a), *312*
Thomaz, M. F., 272(820), 282(820), *315*
Thomson, A. J., 257(712), *311*
Thomson, J. B., 11(24), 141(687), *185, 202*
Thomson, R., 222(78), *293*
Throckmorton, J. R., 117(568), *199*
Tiacco, M., 53(263), *191*
Tichy, M., 222(92), 255(92), *293*
Tiecco, M., 56(279), *191*

Tilford, S. G., 221(67), 231(301), *292, 299*
Timmons, C. J., 57(293), *4, 192*
Timmons, R. B., 120(579), *199*
Ting, C.-H., 268, 269(785), *314*
Tinland, B., 218(84,85), 221(60), 222(84, 85,87,90), 228(222), 229(251), *292, 293, 297, 298*
Tint, D. S., 260(769a), *313*
Tinti, D. S., 246(568), 253(653), 262, 270 (802), 282(802), *307, 309, 314*
Tishchenko, V. G., 230(290), 245(290), 247, *299*
Tiwari, S. K., 224(132), *294*
Tobin, M., 46(184), *189*
Toda, S., 217, *291*
Toda, T., 56(278), *191*
Toki, S., 68(366), *193*
Tokuda, M., 128(623), *200*
Tokumaru, K., 130(633), *201*
Tolstoi, N. A., 280(1044), *321*
Tolstorozhev, G. B., 253(639), 271(639), 281, *309*
Tomasik, P., 231(297-299), *299*
Tombacz, E., 251(599), *308*
Tomeoka, H., 79(412), 104(412), 111(412), *195*
Tomimatsu, T., 257(729), *312*
Tominaga, T., 123(598), *200*
Tomioka, H., 46(187), *189*
Tomkiewicz, Y., 260(774), 266(774), 272 (841), 274(860,861), 278(968,1005), 284 (841), 287, *313, 315, 316, 319, 320*
Tonkyn, R. G., 68(368a), *193*
Topp, M. R., 233(363), 234(363), *301*
Torihashi, Y., 62(332), 226(194), 254(194), *193, 296*
Tosa, T., 47(197), 274(877), *189, 316*
Toshima, N., 164(831a), 165(851), *205, 206*
Tosi, C., 225(166), *295*
Toth, M., 34(130), *188*
Toy, M. S., 33(127), *188*
Tozune, S., 176(963,964), *208, 209*
Trachtenber, I., 177(978), *209*
Travecedo, E. F., 164(837), *206*
Traverso, O., 333, 337(62), *338, 339*
Traynard, P., 106(512), 108(512), *197*
Traynham, J. G., 12(26), 26(101), *185, 187*
Trecker, D. J., *5*
Tret'yak, Z. A., 238(427), *303*

Triboulet, C., 226(180), *296*
Triff, D., 239(449), *303*
Trifonov, E. D., 260(771), 265, *313*
Trifunac, A. D., 168(877), 171(877), *206*
Trinajstic, N., 223(113), 228(226), *294, 297*
Trindle, C., 131(644), *201*
Tripathi, B. N., 109(523), 114(551), 255 (683), *197, 198, 310*
Tripathi, G. N. R., 225(151), *295*
Tripathi, L. N., 224(137), *295*
Trishchenko, V. G., 260(773), *313*
Trombetti, A., 238(422), *303*
Trost, B. M., 17(68), 162(806a), 176(968), *186, 205, 209*
Trozzolo, A. M., 223(120), 254(120), *294*
Truscott, T. G., 234(384,406), 278(406), *302*
Tsai, L., 158(780a), *204*
Tsai, Lin, 254(673), 271(673), 281, *310*
Tshekhanskii, R. S., 225(155), *295*
Tsikora, L. I., 243(526,527), 258(744), *306, 312*
Tsuboi, M., 221(46), *292*
Tsubomura, H., 48(206), 224(142), 234 (387), *190, 295, 302*
Tsuchihashi, G., 123(600,600b), 124(600a), *200*
Tsuchiya, T., 181(1043), *210*
Tsuda, M., 130(631), 277(958), *201, 319*
Tsuji, K., 240(467), *304*
Tsuji, T., 87(433), 168(880), *195, 206*
Tsukerman, S. V., 230(280,289), 232(345), *298, 299, 301*
Tsunashima, S., 279(1008), *320*
Tsunoda, T., 172(917), 222(96), *207, 293*
Tsushima, S., 139(682,683), *202*
Tsutsumi, S., 114(552), 123(598), *198, 200*
Tsvirko, M. P., 272(823), 282(823), *315*
Tulp, A., 240(471), *304*
Turner, D. C., 260(777), 266, 271(777), *313*
Turner, D. W., 221(64), *292*
Turner, L. T., 82(418), *195*
Turner, P. H., 106(500), 108(500), *197*
Turoverov, K. K., 276(933), 285, *318*
Turro, N., 244, *306*
Turro, N. J., 12(42), 46(184), 77(400,403, 404), 80(414a,415), 82(400), 84(421a), 127(620), 158(782,783,784,789), 274

(889,890), *3, 4, 5, 186, 189, 194, 195, 200, 204, 317*
Tushishvii, L. Sh., 227(196), *296*
Tyutyulkov, N., 229(252), 233(252), 244 (535,536), *298, 306*

Uda, H., 34(131), *188*
Udagawa, Y., 233(354), 235(354), 273 (354), *301*
Udding, A. C., 161(806), *205*
Ueda, K., 178(1001), *209*
Uehara, Y., 225(156), *295*
Uekama, K., 240(462), 274(882), *304, 317*
Uff, A. J., 47(192), *189*
Uffindell, V. E., 234(385), 253(385), *302*
Ukena, T., 116(559b), *198*
Ukigai, T., 46(187), *189*, 48(187), *190*
Ukita, T., 19(75), *187*
Ullman, E. F., 130(634), 150(733), 178 (988b), *201, 203, 209*
Ulstrup, J., 330(23), 331(23), *338*
Umezawa, B., 11(20,21), 172(20), *185*
Umreiko, D. S., 329, *338*
Unger, I., 158(786), *204*, 255(675,684,685), 271(808), 279(675), *310, 315*
Urba, V., 232(334-336), 238(423), *300, 303*
Uri, N., 334(47), *339*
Usherwood, E. W., 164(834), *205*
Ustyugova, L. N., 241(486), 257(486,732), 259(486,732), 264(486,732), *304, 312*
Usui, Y., 276(928), 285, *318*
Utkina, L. P., 241(491), *305*
Uzhinov, B. M., 244(540), 245(542), 247 (540,542), *306*

Vajda, M., 227(198), *296*
Vala, M., Jr., 243(517), *305*
Valcher, S., 333(42), *338*
Val'dman, M. M., 233(352), 242(352), 258 (352), *301*
Valentine, D. A., Jr., 325(1), *337*
Van, S. P., 62(325), *192*, 254(668), 263, *310*
Van Auken, T. V., 28(110), *187*
Van Beek, H., 178(999), *209*
Van Bergen, T. J., 183(1051), *211*
Van-Catledge, F. A., 223(114), *294*
Van der Deijl, G. M., 271(809), 281, *315*
Vander Donckt, E., 51(240), 58(311,312),

191, 192
Vanderhoek, J. Y., 134(667), *201*
Van der Linde, R., 223(125), *294*
Van der Lugt, W. Th. A. M., 222(91), *293*
Van der Waals, J. H., 245(553,554), 246 (569,570), *306, 307*
Vandewyer, P. H., 52(254), 131(641), *191, 201*
Van Eldik, R., 337, *339*
VanLear, G. E., 175(938), *208*
Van Sinoy, A., 58(311,312), *192*
Van Vliet, A., 179(1011), *210*
Van Zwet, H., 271(804b), *314*
Varkonyi, Z., 277(965), 289(965), *319*
Vartanyan, A. T., 240(464), *304*
Varvoglis, G. A., 178(1003), *209*
Vaughan, W. M., 253(650), *309*
Vay, P. M., 220(21), *291*
Vedernikov, G. S., 258(747), 265, *312*
Vedrilla, D., 223(101), *293*
Veenland, J. U., 177(977), *209,* 223(125), *294*
Venkataramani, B., 114(551), *198*
Venters, K., 229(246), *297*
Verbeek, M., 155(762), *204*
Verbovskaya, T. M., 229(245), *297*
Verdieck, J. F., 254(662), *310, 338*
Vereshchagin, L. I., 229(246), *297*
Verhoeven, J. W., 239(436), 240(456), *303, 304*
Vermes, J. P., 164(848), 165(848), *206*
Vermont, G. B., 80(415a), *195*
Vernon, J. M., 47(194), *189*
Verzilina, M. K., 224(138), *295*
Veschambre, H., 176(959,960), *208*
Veselova, T. V., 257(728), 263(728), *312*
Vidal, M., 175(944), 176(944), *208*
Viblyi, I. V., 243(511), *305*
Vietmeyer, N. D., 137(675), *202*
Viktorova, E. N., 270(795), 271(795), *314*
Villaume, M. L., 88(435), *195*
Vinogradov, I. P., 227(215), 255(215), *297*
Vishnevskii, V. N., 243(511), *305*
Vize, L., 251(599), *308*
Vladzimirskaya, E. V., 226(183), *296*
Vocelle, D., 167(959,960), 173(923), *207, 208*
Voevoda, L. V., 230(290), 245(290), 247, *299*
Vogel, F., 280(1051,1052), 290, *322*

Vogel, P., 149(725), 152(725,745), *203*
Vogler, A., 334(50), 335(50), *339*
Voigt, E. M., 256(704), *311*
Volker, E. J., 81(416), *195*
Volkert, O., 228(225), *297*
Vollner, L., 18(71), *186*
Volman, D. H., 12(25), *185*
Von Buenau, G., 279(1007), *320*
Von Weyessenhoff, H., 253(632), 268, 269 (632,786), *309, 314*
Vorob'eva, R. P., 229(259), *298*
Voronkov, M. G., 239(450), *303*
Voss, A. J. R., 12(43), *186,* 273(852), *316*
Voznyak, V. M., 255(696), *311*
Vukov, R., 116(561), *199*

Wagenaar, A., 161(806), *205*
Wagner, F., 122(595a), *199*
Wagner, G., 230(286), *299*
Wagner, P. J., 82(419), 83(419), 97(472-475,477-479), 98(477), 99(475), 100(475), 106(503), 108(503), 131(648, 656), 132(648,656), 278(987), 288, *4, 195, 196, 197, 201, 320*
Wagner, R., 276(940), 286(940), *318*
Wagniere, G., 218(254), 229(254), *298*
Waiss, A. C., Jr., 28(109), *187*
Wajer, Th. A. J. W., 178(984), *209*
Wakabayashi, M., 45(180), *189*
Wakamatsu, S., 180(1031), *210*
Wall, E. N., 172(914), *207*
Walmsley, R. H., 239(446), *303*
Walsh, A. D., 222(69), *293*
Walsh, T. D., 46(183), *189*
Waltz, W. L., 332(28), 333(34), 334(34), 336(34), *338*
Wamhoff, H., 131(637), 150(637), *201*
Wampfler, G., 96(463), 108(463), 278(988), *196, 320*
Wan, J. K. S., *4*
Wang, S. Y., 123(599), 162(599,823), *200, 205*
Ward, H. R., 27(105,106), *187*
Ward, S. D., 232(339), *300*
Ward, T. M., 231(303), *299*
Wardzinski, H., 253(647), *309*
Ware, W. R., 269(793a), 275(914), 280, 285 (914), *314, 317*
Waring, A. J., 149(727), *203*
Warnet, R. J., 49(229), *190*

Warrener, R. N., 85(427), 152(427), *4, 195*
Warsop, P. A., 222(69,78), *293*
Warwick, D. A., 95(459), *196*
Wasqestian, H. F., 333(35), *338*
Wasserman, A. M., 178(988), *209*
Wasserman, H. H., 60(322), 167(863), *192, 206*
Watanabe, H., 151(736), *203*
Watanabe, M., 170(896,897), *207*
Waters, J. A., 20(81), 145(702), 148(702), *187, 202*
Watkins, D. A. M., 96(459a), *196*
Watkins, W. B., 178(990), *209*
Watmann-Grajcar, L., 222(86), 258(737), *293, 312*
Watson, W. H., 226(195), 255(195), *296*
Watts, D. W., 333(34), 334(34), 336(34), *338*
Watts, R. J., 252(630), *309*
Way, H., 333, 334, *339*
Weber, D. J., 221(43), *292*
Weber, G., 248(592), 251(592), 252(616), 253(650), 261, 277(592), *307, 308, 309*
Weber, J. B., 231(303), *299*
Weber, K., 274(883), *317*
Wegner, G., 225(157), *295*
Wehrli, H., 86(431), *195*
Wei, C. C., 184(1065), *211*
Wei, K. S., 276(925), 285, *318*
Weinblum, D., 162(815a), *205*
Weiner, M. A., 221(54), *292*
Weiner, S. A., 161(804), 168(879), *205, 206*
Weinhardt, K. K., 47(202), *189*
Weininger, S. J., 176(967), *209*
Weinreb, A., 255(682), 260(774), 266, 272 (841), 277(951), 278(968,1005), 284 (841), 287, *310, 313, 315, 319, 320*
Weinryb, I., 276(936), 286(936), *318*
Weinstein, J., 178(993), *209*
Weiss, D. S., 84(421a), *5, 195*
Weiss, F., 15(57), *186*
Weiss, J. J., 325, *338*
Weiss, K., 160(797), 234(395), 276(932), *205, 302, 318*
Weisstuch, A., 278(969), *319*
Weller, A., 60(319), 274(880,888), 275 (894,916), 284, 285(916), *192, 316, 317, 318*
Weller-Feilchenfeld, H., 231(306), *299*

Wells, C. H. J., 95(459), 105(494), 246 (559), *196, 197, 307*
Welzel, P., 179(1023), *210*
Wendisch, D., 51(244), *191*
Wendling, H., 243(513), *305*
Weringa, W. D., 49(228), *190*
Werner, A., 255(682), *310*
Werner, G. K., 329(17), *338*
Werner, T. C., 257(719a), *311*
Westberg, H. H., 85(427), 152(427), *195*
Wettack, F. S., 70(375), 73(392,392a), 254 (672), 269(672), *194, 310*
Wettermark, G., 178(1006), 223(98), *209, 293*
Weyl, D. A., 275(900), *317*
Wharton, J. H., 53(262), 223(103), 257 (727), 263(727), 275(913), 285(913), *191, 293, 312, 317*
Wheeler, D. M. S., 49(229), *190*
Whillans, D. W., 235(399), *302*
Whipple, E. B., 105(493), *197*
Whistler, R. L., 172(911), *207*
White, A. H., 255(674), 269(674), *310*
White, E. H., 56(283), 70(377), 148(718), 279(1016), 289, *191, 194, 203, 321*
White, J. D., 31(121), *188*
Whitesides, G. M., 26(102), *187*
Whiting, D. A., 126(619), 127(619), *200*
Whitman, P. J., 176(968), *209*
Whitten, D. G., 57(286,290), 58(309,310, 313), 245(548), 256(698), 270(698), 271 (698), *192, 306, 311*
Whitten, J. L., 217, 220(19), *291*
Whitten, W. B., 280(1048), 290(1048), *322*
Whittle, J. A., 69(371), 77(371), 78(371), 103(371), *194*
Wiberg, K., 176(958), *208*
Wiberg, K. B., 168(887), 221(65), *207, 292*
Wiecko, J., 70(377), *194*
Wiegand, D. A., 256(709), *311*
Wierdorff, W. W., 46(189a), *189*
Wierzchowski, K. L., 162(815), 257(724), *205, 312*
Wiesmann, J., 27(108), *187*
Wiesner, K., 11(22), 143(695), *185, 202*
Wijnberger, C., 231(291), *299*
Wild, U. P., 220(18), 223(111), 234(111), *291, 294*
Wildes, P. D., 58(313), *192*
Wiles, D. M., 75(397), *194*

AUTHOR INDEX

Wiley, M., 28(109), *187*
Wiliams, J. O., 64(340), *193*
Willets, F. W., 234(385), 253, *302*
Willhalm, B., 149(725), 152(725), *203*
Williams, B. H., 172(918), *207*
Williams, D. F., 260(765), 271(765), 281 (765), *313*
Williams, D. W., 271(814), *315*
Williams, H., 10(13), *185*
Williams, J. C., 135(669), *201*
Williams, J. L. R., 57(287), 58(314), 184 (1060), *192, 211,* 257(720), *311*
Williams, K. L., 19(76), *187*
Williams, N., 104(489), *197*
Williams, W. M., 181(1039,1039a), *210*
Williams, W. P., 232(327), 256(327), 261 (327), *300*
Williams-Smith, D. L., 168(888), *207*
Wilson, D. T., 64(343), *193*
Wilson, R., 48(207), 106(498,500), 107 (207), 108(498,500,522), 111(498,522), *190, 197*
Windsor, M. W., 233(364), 234(364,397), 253(397,640,641), 271(640,641), 281 (640,641), *301, 302, 309*
Winkler, M. H., 248(593), 251(593), *308*
Winnik, M. A., 110(526), *198*
Winter, A. M., 245(556), *306*
Witkop, B., 11(19), 20(81), 145(702), 148 (702), 162(822), *185, 187, 202, 205*
Wittig, G., 42(169a), *189*
Witzke, H., 329(19), *338*
Woerz, O., 230(276), *298*
Wohl, A. J., 218(13), *291*
Wolf, A. P., 39(152), *188*
Wolf, D., 256(711), *311*
Wolf, F. T., 225(172), 255(172), *295*
Wolf, H. C., 245(551), 259(759a), 272 (831), 273(835), 283(831,835), *306, 313, 315*
Wolf, H. P., 160(797), *205,* 234(395), *302*
Wolf, R. E., 78(407,408), 88(408), *194*
Wolfe, G. A., 26(103), *187*
Wood, J. M., 336(56), *339*
Worthington, N. W., 157(775), *204*
Wratten, R. J., 223(113), *294*
Wriede, P. A., 77(400,403), 82(400), *194*
Wright, H. E., 144(698,699), *202*
Wright, J. S., 223(109), *294*
Wright, T. R., 168(892), 169(892), *207,*
275(897), 285(897), *317*
Wrighton, M., 12(46), 29(116), 77(46), 96 (46), *186, 188,* 278(993), 288(993), *320*
Wyatt, J., 223(105), *294*
Wyatt, J. R., 79(410), *194*
Wynberg, H., 42(168), 53(264,266), 160 (802), 183(1051), *189, 191, 205, 211*
Wyncke, B., 243(513), *305*

Yabe, S., 225(156), *295*
Yager, W. A., 12(31), *186*
Yagi, K., 261(779a), 270(779a), *314*
Yagihara, T., 176(963,964), *208, 209*
Yagupol'skii, L. M., 223(116), *294*
Yakovenko, V. A., 252(649), 254(659), 270(649), 279(1012), *309, 320*
Yakovenko, V. N., 223(124), 241(488-490), 258(124,490), *294, 305*
Yamada, K., 9(9), 126(613,614,618), 139 (681), 152(613), *185, 200, 202*
Yamada, S., 172(920,921), 181(1041,1045), 182(1048), *207, 210, 211*
Yamamoto, H., 95(458b), *196*
Yamamoto, M., 131(638), *201*
Yamamoto, N., 48(206), *190,* 234(387), *302*
Yamaoka, T., 172(917), *207,* 222(96), *293*
Yamase, T., 125(605a), *198, 200*
Yamazaki, H., 13(50), *186*
Yamazaki, I., 179(1017), *210,* 231(318), 254(318), 271(318), *300*
Yamazaki, R., 225(156), *295*
Yamazaki, Y., 274(882), *317*
Yang, N.-C., 73(393,394), 74(394), 96 (467), 106(467), 129(627), *194, 196, 200,* 252(631), 255(677), 271(631), *309, 310*
Yang, S. S., 129(627), *200*
Yankelevich, S., 85(425), *195*
Yao, S. J., 268, 269(786), *314*
Yaremko, R. V., 223(116), *294*
Yates, K., 220(26), *291*
Yates, P., 83(420), 138(680), 149(721-723), *195, 202, 203*
Yatsenko, L. P., 243(514), *305*
Yguerabide, J., 277(963), *319*
Yildiz, A., 272(821), 281(821), 282(821), *315*
Yokoe, I., 181(1041,1045), *210*
Yomosa, S., 220(29), *291*

Yonemitsu, O., 11(19), 119(578a), 178 (578a), *185, 199*
Yonezawa, T., 124(602), *200*
Yoshida, M., 240(462), *304*
Yoshihara, K., 37(143), *188*
Yoshikoshi, A., 34(131), *188*
Yoshimoto, T., 49(229a), *190*
Yoshioka, M., 9(9), 126(618), *185, 200*
Youn, C. H., 19(76), *187*
Young, R. N., 227(212), *297*
Younis, F. A., 48(207), 106(498), 107(207), 108(498,522), 111(498,522), *190, 197*
Yu, J. M. H., 251(607a), *308*
Yu, S., 131(653), *201*
Yudovich, E. E., 238(425), *303*
Yunes, R. A., 225(166a), *295*

Zafirious, O. C., 30(119), *188,* 270(804a), *314*
Zahlan, A. B., 279(1027), 290, *321*
Zahradnik, R., 222(92), 223(107), 228(228), 255(92), *293, 294, 297*
Zaidi, Z. H., 223(104), 227(204), *294, 296*
Zalewski, E. F., 259(752), *312*
Zalkow, L. H., 173(933), *208*
Zander, M., 5, 48(216,217), 111(535), *190, 198,* 231(312), 256(312,701), 260(701), 279(312,701,1025), *299, 311, 321*
Zandstra, P. J., 271(809), 281, *315*
Zanker, V., 48(212), 182(1049), 184(1068), *190, 211,* 230(267-270), 235(401), 243(525), 256(267), 257(269), *298, 302, 306*
Zannucci, J. S., 47(196), 49(196), 110(524), 111(524), *189, 197*
Zauli, C., 226(175), 238(422), *296, 303*
Zderic, J. A., 126(616), *200*
Zeeck, E., 217(33), 220(33), *291*
Zelenskaya, L. G., 180(1032), *210*
Zelikman, Z. I., 232(330), *300*
Zelinskii, V. V., 274(869), *316*
Zhigunova, I. A., 243(514), 244(533), *305, 306*
Zhukova, N. A., 226(190), *296*
Zhuravleva, T. S., 233(367), 236, *301*
Zich, W., 7(2), *185*
Ziegler, G. R., 17(66), *186*
Ziegler, S. M., 260(770), *313*
Ziffer, H., 84(422), *195*
Zimin, Y. B., 110(525), *198*
Zimmer, H., 131(639), *201*
Zimmerman, A. A., 12(48), *186,* 276(932), *318*
Zimmerman, H. E., 23(88), 24(96), 36(141), 55(273), 61(96), 137(672,673), 145(700), 146(706), 148(716), *187, 188, 191, 202, 203,* 254(660), *310*
Zinato, E., 332, 333(34), 334(34), 335(52), 336(34), *338, 339*
Zountsas, G., 56(280), *191*
Zubkov, A. V., 116(560), *199*
Zuegel, M., 278(998), *320*
deZwaan, J., 49(230), *190*
Zwanenburg, B., 161(806), *205*
Zwarich, R., 242(502), 243(521), 272(834), 283, (834), *305, 315*
Zweig, A., 59(316), 62(326,327), *192,* 226(187), 255(187), *296*
Zyka, J., 231(295), *299*

QD
601
A1
A55
v.3
1969

NOV 1 1971